Frequent Pattern Mining

Charu C. Aggarwal • Jiawei Han

Editors

Frequent Pattern Mining

 Springer

Editors
Charu C. Aggarwal
IBM T. J. Watson Research Center
Yorktown Heights
New York
USA

Jiawei Han
University of Illinois at Urbana-Champaign
Urbana
Illinois
USA

ISBN 978-3-319-34689-2   ISBN 978-3-319-07821-2 (eBook)
DOI 10.1007/978-3-319-07821-2
Springer Cham Heidelberg New York Dordrecht London

Printed on acid-free paper

Springer is part of Springer Science+Business Media (www.springer.com)

Preface

The field of data mining has four main "super-problems" corresponding to clustering, classification, outlier analysis, and frequent pattern mining. Compared to the other three problems, the frequent pattern mining model for formulated relatively recently. In spite of its shorter history, frequent pattern mining is considered the marquee problem of data mining. The reason for this is that interest in the data mining field increased rapidly soon after the seminal paper on association rule mining by Agrawal, Imielinski, and Swami. The earlier data mining conferences were often dominated by a large number of frequent pattern mining papers. This is one of the reasons that frequent pattern mining has a very special place in the data mining community. At this point, the field of frequent pattern mining is considered a mature one.

While the field has reached a relative level of maturity, very few books cover different aspects of frequent pattern mining. Most of the existing books are either too generic or do not cover frequent pattern mining in an exhaustive way. A need exists for an exhaustive book on the topic that can cover the different nuances in an exhaustive way.

This book provides comprehensive surveys in the field of frequent pattern mining. Each chapter is designed as a survey that covers the key aspects of the field of frequent pattern mining. The chapters are typically of the following types:

- *Algorithms*: In these cases, the key algorithms for frequent pattern mining are explored. These include join-based methods such as *Apriori*, and pattern-growth methods.
- *Variations*: Many variations of frequent pattern mining such as interesting patterns, negative patterns, constrained pattern mining, or compressed patterns are explored in these chapters.
- *Scalability*: The large sizes of data in recent years has led to the need for big data and streaming frameworks for frequent pattern mining. Frequent pattern mining algorithms need to be modified to work with these advanced scenarios.
- *Data Types*: Different data types lead to different challenges for frequent pattern mining algorithms. Frequent pattern mining algorithms need to be able to work with complex data types, such as temporal or graph data.

- *Applications*: In these chapters, different applications of frequent pattern mining are explored. These includes the application of frequent pattern mining methods to problems such as clustering and classification. Other more complex algorithms are also explored.

This book is, therefore, intended to provide an overview of the field of frequent pattern mining, as it currently stands. It is hoped that the book will serve as a useful guide for students, researchers, and practitioners.

Contents

Contributors

Charu C. Aggarwal IBM T. J. Watson Research Center, Yorktown Heights, NY, USA

Gagan Agrawal Ohio State University, Columbus, OH, USA

Luiza Antonie University of Guelph, Guelph, Canada

David C. Anastasiu Department of Computer Science and Engineering, University of Minnesota, Minneapolis, USA

Ira Assent Department of Computer Science, Aarhus University, Aarhus, Denmark

Mansurul A. Bhuiyan Indiana University–Purdue University, Indianapolis, IN, USA

Hong Cheng Department of Systems Engineering and Engineering Management, The Chinese University of Hong Kong, Hong Kong, China

Aris Gkoulalas-Divanis IBM Research-Ireland, Damastown Industrial Estate, Mulhuddart, Dublin, Ireland

Jiawei Han University of Illinois at Urbana-Champaign, Urbana, IL, USA
Department of Computer Science, University of Illinois at Urbana-Champaign, Champaign, USA

Jayant Haritsa Database Systems Lab, Indian Institute of Science (IISc), Bangalore, India

Mohammad Al Hasan Indiana University–Purdue University, Indianapolis, IN, USA

Jeremy Iverson Department of Computer Science and Engineering, University of Minnesota, Minneapolis, USA

Ruoming Jin Kent State University, Kent, OH, USA

Murat Kantarcioglu UTD Data Security and Privacy Lab, University of Texas at Dallas, Texas, USA

Victor E. Lee John Carroll University, University Heights, OH, USA

Matthijs van Leeuwen KU Leuven, Leuven, Belgium

Carson Kai-Sang Leung University of Manitoba,Winnipeg, MB, Canada

Jundong Li University of Alberta, Alberta, Canada

Zhenhui Li Pennsylvania State University, University Park, USA

George Karypis Department of Computer Science and Engineering, University of Minnesota, Minneapolis, USA

Siegfried Nijssen KU Leuven, Leuven, Belgium

Universiteit Leiden, Leiden, The Netherlands

Jian Pei Simon Fraser University, Burnaby, BC, Canada

Shaden Smith Department of Computer Science and Engineering, University of Minnesota, Minneapolis, USA

Wei Shen Tsinghua University, Beijing, China

Nikolaj Tatti HIIT, Department of Information and Computer Science, Aalto University, Helsinki, Finland

Jilles Vreeken Max-Planck Institute for Informatics and Saarland University, Saarbrücken, Germany

Jianyong Wang Tsinghua University, Beijing, China

Xifeng Yan Department of Computer Science, University of California at Santa Barbara, Santa Barbara, USA

Osmar Zaiane University of Alberta, Alberta, Canada

Feida Zhu Singapore Management University, Singapore, Singapore

Albrecht Zimmermann INSA Lyon, Villeurbanne CEDEX, France

Arthur Zimek Ludwig-Maximilians-Universität München, Munich, Germany

Chapter 1
An Introduction to Frequent Pattern Mining

Charu C. Aggarwal

Abstract The problem of frequent pattern mining has been widely studied in the literature because of its numerous applications to a variety of data mining problems such as clustering and classification. In addition, frequent pattern mining also has numerous applications in diverse domains such as spatiotemporal data, software bug detection, and biological data. The algorithmic aspects of frequent pattern mining have been explored very widely. This chapter provides an overview of these methods, as it relates to the organization of this book.

Keywords Frequent pattern mining · Association rules

1 Introduction

The problem of frequent pattern mining is that of finding relationships among the items in a database. The problem can be stated as follows.

Given a database D with transactions $T_1 \ldots T_N$, determine all patterns P that are present in at least a fraction s of the transactions.

The fraction s is referred to as the *minimum support*. The parameter s can be expressed either as an absolute number, or as a fraction of the total number of transactions in the database. Each transaction T_i can be considered a sparse binary vector, or as a set of discrete values representing the identifiers of the binary attributes that are instantiated to the value of 1. The problem was originally proposed in the context of market basket data in order to find frequent groups of items that are bought together [10]. Thus, in this scenario, each attribute corresponds to an item in a superstore, and the binary value represents whether or not it is present in the transaction. Because the problem was originally proposed, it has been applied to numerous other applications in the context of data mining, Web log mining, sequential pattern mining, and software bug analysis.

In the original model of frequent pattern mining [10], the problem of finding *association rules* has also been proposed which is closely related to that of frequent

C. C. Aggarwal (✉)
IBM T. J. Watson Research Center, Yorktown Heights, NY 10598, USA
e-mail: charu@us.ibm.com

C. C. Aggarwal, J. Han (eds.), *Frequent Pattern Mining*,
DOI 10.1007/978-3-319-07821-2_1, © Springer International Publishing Switzerland 2014

patterns. In general association rules can be considered a "second-stage" output, which are derived from frequent patterns. Consider the sets of items U and V. The rule $U \Rightarrow V$ is considered an *association rule* at minimum support s and minimum confidence c, when the following two conditions hold true:

1. The set $U \cup V$ is a frequent pattern.
2. The ratio of the support of $U \cup V$ to that of U is at least c.

The minimum confidence c is always a fraction less than 1 because the support of the set $U \cup V$ is always less than that of U. Because the first step of finding frequent patterns is usually the computationally more challenging one, most of the research in this area is focussed on the former. Nevertheless, some computational and modeling issues also arise during the second step, especially when the frequent pattern mining problem is used in the context of other data mining problems such as classification. Therefore, this book will also discuss various aspects of association rule mining along with that of frequent pattern mining.

A related problem is that of *sequential pattern mining* in which an order is present in the transactions [5]. Temporal order is quite natural in many scenarios such as customer buying behavior, because the items are bought at specific time stamps, and often follow a natural temporal order. In these cases, the problem is redefined to that of sequential pattern mining, in which it is desirable to determine relevant and frequent *sequences* of items.

Some examples of important applications are as follows;

- *Customer Transaction Analysis:* In this case, the transactions represent sets of items that co-occur in customer buying behavior. In this case, it is desirable to determine frequent patterns of buying behavior, because they can be used for making decision about shelf stocking or recommendations.
- *Other Data Mining Problems:* Frequent pattern mining can be used to enable other major data mining problems such as classification, clustering and outlier analysis [11, 52, 73]. This is because the use of frequent patterns is so fundamental in the analytical process for a host of data mining problems.
- *Web Mining:* In this case, the Web logs may be processed in order to determine important patterns in the browsing behavior [24, 63]. This information can be used for Web site design. recommendations, or even outlier analysis.
- *Software Bug Analysis:* Executions of software programs can be represented as graphs with typical patterns. Logical errors in these bugs often show up as specific kinds of patterns that can be mined for further analysis [41, 51].
- *Chemical and Biological Analysis:* Chemical and biological data are often represented as graphs and sequences. A number of methods have been proposed in the literature for using the frequent patterns in such graphs for a wide variety of applications in different scenarios [8, 29, 41, 42, 69–75].

Since the publication of the original article on frequent pattern mining [10], numerous techniques have been proposed both for frequent and sequential pattern mining [5, 4, 13, 33, 62]. Furthermore, many variants of frequent pattern mining, such as sequential pattern mining, constrained pattern mining, and graph mining have been proposed in the literature.

Frequent pattern mining is a rather broad area of research, and it relates to a wide variety of topics at least from an application specific-perspective. Broadly speaking, the research in the area falls in one of four different categories:

- **Technique-centered:** This area relates to the determination of more *efficient* algorithms for frequent pattern mining. A wide variety of algorithms have been proposed in this context that use different enumeration tree exploration strategies, and different data representation methods. In addition, numerous variations such as the determination of compressed patterns of great interest to researchers in data mining.
- **Scalability issues:** The scalability issues in frequent pattern mining are very significant. When the data arrives in the form of a stream, multi-pass methods can no longer be used. When the data is distributed or very large, then parallel or big-data frameworks must be used. These scenarios necessitate different types of algorithms.
- **Advanced data types:** Numerous variations of frequent pattern mining have been proposed for advanced data types. These variations have been utilized in a wide variety of tasks. In addition, different data domains such as graph data, tree structured data, and streaming data often require specialized algorithms for frequent pattern mining. Issues of interestingness of the patterns are also quite relevant in this context [6].
- **Applications:** Frequent pattern mining have numerous applications to other major data mining problems, Web applications, software bug analysis, and chemical and biological applications. A significant amount of research has been devoted to applications because these are particularly important in the context of frequent pattern mining.

This book will cover all these different areas comprehensively, so as to provide a comprehensive overview of this broader area.

This chapter is organized as follows. The next section discusses algorithms for the frequent pattern mining problem, and its basic variations. Section 3 discusses scalability issues for frequent pattern mining. Frequent pattern mining methods are advanced data types are discussed in Sect. 4. Privacy issues of frequent pattern mining are addressed in Sect. 5. The applications are discussed in Sect. 6. Section 7 gives the conclusions and summary.

2 Frequent Pattern Mining Algorithms

Most of the algorithms for frequent pattern mining have been designed with the traditional support-confidence framework, or for specialized frameworks that generate more interesting kinds of patterns. These specialized framework may use different types of interestingness measures, model negative rules, or use constraint-based frameworks to determine more relevant patterns.

2.1 Frequent Pattern Mining with the Traditional Support Framework

The support framework is designed to determine patterns for which the raw frequency is greater than a minimum threshold. Although this is a simplistic way of defining frequent patterns, this model has an algorithmically convenient property, which is referred to as the *level-wise* property. The level-wise property of frequent pattern mining is algorithmically crucial because it enables the design of a bottom-up approach to exploring the space of frequent patterns. In other words, a $(k + 1)$-pattern may not be frequent when any of its subsets is not frequent. This is a crucial observation that is used by virtually all the efficient frequent pattern mining algorithms.

Since the problem of frequent pattern mining was first proposed, numerous algorithms have been proposed in order to make the solutions to the problem more efficient. This area of research is so popular that an annual workshop *FIMI* was devoted to implementations of frequent pattern mining for a few years. This site [77] is now organized as a repository, where many efficient implementations of frequent pattern mining are available. The techniques for frequent pattern mining started with *Apriori*-like join-based methods. In these algorithms, candidate itemsets are generated in increasing order of itemset size. The generation in increasing order of itemset size is referred to as *level-wise exploration*. These itemsets are then tested against the underlying transaction database and the frequent ones satisfying the minimum support constraint are retained for further exploration. Eventually, it was realized that these *Apriori*-like methods could be more systematically explored as *enumeration trees*. This structure will be explained in detail in Chap. 2, and provides a methodology to perform systematic and non-redundant frequent pattern exploration. The enumeration tree provides a more flexible framework for frequent itemset mining because the tree can be explored in a variety of different strategies such as depth-first, breadth-first, or other hybrid strategies [13]. One property of the breadth-first strategy is that level-wise pruning can be used, which is not possible with other strategies. Nevertheless, strategies such as depth-first search have other advantages, especially for maximal pattern mining. This observation for the case of maximal pattern mining was first stated in [12]. This is because long patterns are discovered early, and they can be used for downward closure-based pruning of large parts of the enumeration tree that are already known to be frequent. It should be pointed out, that for the case where *all* frequent patterns are mined, the order of exploration of an enumeration tree does not affect the number of candidates that are explored because the size of the enumeration tree is fixed.

Join-based algorithms are always level-wise, and can be viewed as equivalent to breadth-first enumeration tree exploration. The algorithm proposed in the first frequent pattern mining paper [10] was an enumeration-tree based algorithm, whereas the second algorithm proposed was referred to as *Apriori*, and was a join-based algorithm [4]. Both algorithms are level-wise algorithms. Subsequently, many algorithms have been proposed in order to improve the implementations based on the enumeration tree paradigm with the use of techniques such as lookahead [17], depth-first search [12, 13, 33] and vertical exploration [62]. Some of these methods such as *TreeProjection*, *DepthProject* and *FP-growth* [33] use a projection strategy in which smaller transaction databases are explored at lower levels of the tree.

One of the challenges of frequent pattern mining is that a large number of redundant patterns are often mined. For example, the subset of a frequent pattern is also guaranteed to be frequent and by mining a maximal itemset, one is assured that the other frequent patterns can also be generated from this smaller set. Therefore, one possibility is to mine for only *maximal* itemsets [17]. However, the mining of maximal itemsets loses information about the exact value of support of the subsets of maximal patterns. Therefore, a further refinement would be to find *closed* frequent itemsets [58, 74]. Closed frequent itemsets are defined as frequent patterns, no superset of which have the same frequency as that itemset. By mining closed frequent itemsets, it is possible to significantly reduce the number of patterns found, without losing any information about the support level. Closed patterns can be viewed as the maximal patterns from each group of *equi-support* patterns (i.e., patterns with the same support). All maximal patterns are, therefore, closed.

The depth-first method has been shown to have a number of advantages in maximal pattern mining [12], because of the greater effectiveness of the pruning-based lookaheads in the depth-first strategy. Different techniques for frequent pattern mining will be discussed in Chaps. 2 and 3. The former chapter will generally focus on frequent pattern mining algorithms, whereas the latter chapter will focus on pattern-growth algorithms. An additional chapter with greater detail has been devoted to pattern-growth methods, because of it is considered a state-of-the-art technique in frequent pattern mining. The efficiency in frequent pattern mining algorithms can be gained in several ways:

1. Reducing the size of the candidate search space, with the use of pruning methods, such as maximality pruning. The notion of *closure* can also be used to prune large parts of the search space. However, these methods often do not exhaustively return the full set of frequent patterns. Many of these methods returned condensed representations such as maximal patterns or closed patterns.
2. Improving the efficiency of *counting*, with the use of database projection. Methods such as *TreeProjection* speed up the rate at which each pattern is counted, by reducing the size of the database with respect to which patterns are compared.
3. Using more efficient data structures, such as vertical lists, or an FP-Tree for more compressed database representation. In frequent pattern mining, both memory and computational speeds can be improved by judicious choice of data structures.

A particular scenario of interest is one in which the patterns to be mined are very long. In such cases, the number of subsets of frequent patterns can be extremely large. Therefore, a number of techniques need to be designed in order to mine very long patterns. In such cases, a variety of methods are used to explore the long patterns early, so that their subsets can be pruned effectively. The scenario of long pattern generation is discussed in detail in Chap. 4, though it is also discussed to some extent in the earlier Chaps. 2 and 3.

2.2   Interesting and Negative Frequent Patterns

A major challenge in frequent pattern mining is that the rules found may often not be very interesting, when quantifications such as support and confidence are used. This is because such quantifications do not normalize for the original frequency of the underlying items. For example, an item that occurs very rarely in the underlying database would naturally also occur in itemsets with lower frequency. Therefore, the *absolute* frequency often does not tell us much about the likelihood of items to *co-occur* together, because of the biases associated with the frequencies of the individual items. Therefore, numerous methods have been proposed in the literature for finding interesting frequent patterns that normalize for the underlying item frequencies [6, 26]. Methods for finding interesting frequent patterns are discussed in Chap. 5. The issue of interestingness is also related to compressed representations of patterns such as closed or maximal itemsets. These issues are also discussed in the chapter.

In negative associative rule mining, we attempt to determine rules such as *Bread* $\Rightarrow$ $\neg Butter$, where the symbol $\neg$ indicates negation. Therefore, in this case $\neg Butter$ becomes a pseudo-item denoting a "negative item." One possibility is to add negative items to the data, and perform the mining in the same way as one would determine rules in the support-confidence framework. However, this is not a feasible solution. This is because traditional support frameworks are not designed for cases where an item is presented in the data 98 % of the time. This is the case for "negative items." For example, most transactions may not contain the item *Butter*, and therefore even positively correlated items may appear as negative rules. For example, the rule *Bread* $\Rightarrow$ $\neg Butter$ may have confidence greater than 50 %, even though *Bread* is clearly correlated in a positive way with *Butter*. This is because, the item $\neg Butter$ may have an even higher support of 98 %.

The issue of finding negative patterns is closely related to that of finding interesting patterns in the data [6] because one is looking for patterns that satisfy the support requirement in an interesting way. This relationship between the two problems tends to be under-emphasized in the literature, and the problem of negative pattern mining is often treated independently from interesting pattern mining. Some frameworks, such as collective strength, are designed to address both issues simultaneously. Methods for negative pattern mining are addressed in Chap. 6. The relationship between interesting pattern mining and negative pattern mining will be discussed in the same chapter.

2.3 Constrained Frequent Pattern Mining

Off-the-shelf frequent pattern mining algorithms discover a large number of patterns which are not useful when it is desired to determine patterns on the basis of more refined criteria. Frequent pattern mining methods are often particularly useful in the context of constrained applications, in which rules satisfying particular criteria are discovered. For example, one may desire specific items to be present in the rule. One solution is to first mine all the itemsets, and then enable online mining from this set of base patterns [3]. However, pushing constraints directly into the mining process has several advantages. This is because when constraints are pushed directly into the mining process, the mining can be performed at much lower support levels than can be performed by using a two-phase approach. This is especially the case when a large number of intermediate candidates can be pruned by the constraint-based pattern mining algorithm.

A variety of arbitrary constraints may also be present in the patterns. The major problem with such methods is that the constraints may result in the violation of the downward closure property. Because most frequent pattern mining algorithms depend crucially on this property, its violation is a serious issue. Nevertheless, many constraints have specialized properties because of which specialized algorithms can be developed. Methods for constrained frequent pattern mining method have been discussed in [55, 57, 60]. Constrained methods have also been developed for the sequential pattern mining problem [31, 61]. In real applications, the output of the vanilla frequent pattern mining problem may be too large, and it is only by pushing constraints into the pattern mining process, that useful application-specific patterns can be found. Constrained frequent pattern mining methods are closely related to the problem of pattern-based classification, because the latter problem requires us to discover discriminative patterns from the underlying data. Methods for constrained frequent pattern mining will be discussed in Chap. 2.

2.4 Compressed Representations of Frequent Patterns

A major problem in frequent pattern mining algorithms is that the volume of the mining patterns is often extremely large. This scenario creates numerous challenges for using these patterns in a meaningful way. Furthermore, different kinds of redundancy are present in the mined patterns. For example, maximal patterns imply the presence of all their subsets in the data. There is some information loss in terms of the exact support values of these subsets. Therefore, if it is not needed to preserve the values of the support across the patterns, then the determination of concise representations can be very useful.

A particularly interesting form of concise representation is that of *closed patterns* [56]. An itemset X is set to be closed if none of its supersets have the same support as X. Therefore, by determining all the closed frequent patterns, one can derive not only the exhaustive set of frequent itemsets, but also their supports. Note that support values are lost by maximal pattern mining. In other words, the set of maximal patterns cannot be used to derive the support values of missing subsets. However, the support values of closed frequent itemsets can be used to derive the support values of missing subsets. Many interesting methods [58, 67, 74] have been designed for identifying frequent closed patterns. The general principle of determining frequent closed patterns has been generalized to that of determining δ-freesets [18]. This issue is closely related to that of mining all *non-derivable frequent itemsets* [20]. A survey on this topic may be found in [21]. These different forms of compression are discussed in Chaps. 2 and 5.

Finally, a formal way of viewing compression is from the perspective of information-theoretic models. Information-theoretic models are designed for compressing different kinds of data, and can therefore be used to compress itemsets as well. This basic principle has been used for methods such as *Krimp* [66]. The problem of determining compressed representations of frequent itemsets is discussed in Chap. 8. This chapter focusses mostly on the information-theoretic issues of frequent itemset compression.

3 Scalability Issues in Frequent Pattern Mining

In the modern era, the ability to collect large amounts of data has increased significantly because of advances in hardware and software platforms. The amount of data is often so large that specialized methods are required for the mining process. The streaming and big-data architectures are slightly different and pose different challenges for the mining process. The following discussion will address each of these challenges.

3.1 Frequent Pattern Mining in Data Streams

In recent years, data stream have become very popular because of the advances in hardware and software technology that can collect and transmit data continuously over time. In such cases, the major constraint on data mining algorithms is to execute the algorithms in a single pass. This can be significantly challenging because frequent and sequential pattern mining methods are generally designed as level-wise methods. There are two variants of frequent pattern mining for data streams:

- *Frequent Items or Heavy Hitters:* In this case, frequent 1-itemsets need to be determined from a data stream in a single pass. Such an approach is generally needed when the total number of distinct items is too large to be held in main memory. Typically, sketch-based methods are used in order to create a compress data structure in order to maintain *approximate* counts of the items [23, 27].
- *Frequent itemsets:* In this case, it is not assumed that the number of distinct items are too large. Therefore, the main challenge in this case is computational, because

the typical frequent pattern mining methods are multi-pass methods. Multiple passes are clearly not possible in the context of data streams [22, 39].

The streaming scenario also presents numerous challenges in the context of data of advanced types. For example, graph streams are often encountered in the context of network data. In such cases, methods need to be designed for determining dense groups of nodes in real time [16]. Methods for mining frequent items and itemsets in data streams are discussed in Chap. 9.

3.2   Frequent Pattern Mining with Big Data

The big data scenario poses numerous challenges for the problem of frequent pattern mining. A major problem arises when the data is large enough to be stored in a distributed way. Therefore, significant costs are incurred in shuffling around data or intermediate results of the mining process across the distributed nodes. These costs are also referred to as data transfer costs. When data sets are very large, then the algorithms need to designed to take into account both the disk access constraint and the data transfer costs. In addition, many distributed frameworks such as *MapReduce* [28] require specialized algorithms for frequent pattern mining. The focus of big-data framework is somewhat different from streams, in that it is closely related to the issue of shuffling large amounts of data around for the mining process. Interestingly, it is sometimes easier to process the algorithms in a single pass in streaming fashion, than when they have already been stored in distributed frameworks where access costs become a major issue. Algorithms for frequent pattern mining with big data are discussed in detail in Chap. 10. This chapter discusses both the parallel algorithms and the big-data algorithms that are based on the *MapReduce* framework.

4   Frequent Pattern Mining with Advanced Data Types

although the frequent pattern mining problem is naturally defined on sets, it can be extended to various advanced data types. The most natural extension of frequent pattern mining algorithms is to the case of temporal data. This was one of the earliest proposed extensions and is referred to as *sequential pattern mining*. Subsequently, the problem has been generalized to other advanced data types, such as spatiotem-poral data, graphs, and uncertain data. Many of the developed algorithms are basic variations of the frequent pattern mining problem. In general, the basic frequent pattern mining algorithms need to be modified carefully to address the variations required by the advanced data types.

4.1 Sequential Pattern Mining

The problem of sequential pattern mining is closely related to that of frequent pattern mining. The major difference in this case is that record contain *baskets of items* arranged sequential. For example, each record R_i may be of the following form:

$$R_i = \langle \{Bread\}, \{Butter, Cake\}, \{Chicken, Yogurt\} \rangle$$

In this case, each entity within {} is a basket of items that are bought together and, therefore, do not have a temporal ordering. This basket of items is collectively referred to as an *event*. The length of a pattern is equal to the sum of the lengths of the complex items in it. For example, R_i is a 5-pattern, even though it has 3 events. The different complex entities (or events) do have a temporal ordering. In the aforementioned example, it is clear that $\{Bread\}$ has been bought earlier than $\{Butter, Cake\}$. The problem of sequential pattern mining is that of finding sequences of events that are present in at least a fraction s of the underlying records [5]. For example, the sequence $\langle \{Bread\}, \{Butter\}, \{Chicken\} \rangle$ is present in the afore-mentioned record, but not the sequence $\langle \{Bread\}, \{Cake\}, \{Butter\} \rangle$. The pattern may also contain complex events. For example, the pattern $\langle \{Bread\}, \{Chicken, Yogurt\} \rangle$ is present in R_i. The problem of sequential pattern mining is closely related to that of frequent pattern mining except that it is somewhat more complex to account for both the presence of complex baskets of items in the database, and the temporal ordering of the individual baskets. An extension of a sequential pattern may either be a set-wise extension of a complex item, or a temporal extension with an entirely new event. This affects the nature of the extensions of items in the transactions. Numerous modifications of known frequent pattern mining methods such as *Apriori* and its variants, *TreeProjection* and its variants [32], and the *FP-growth* method and its variants, can be used in order to solve the sequential pattern mining problem [5, 35, 36]. The enumeration tree concept can also be generalized to sequential pattern mining [32]. Therefore, in principle, all enumeration tree algorithms can be generalized to sequential pattern mining. This is a powerful ability because, as we will see in Chap. 2 all frequent pattern mining algorithms are, implicitly or explicitly, enumeration-tree algorithms. Sequential pattern mining methods will be discussed in detail in Chap. 11.

4.2 Spatiotemporal Pattern Mining

The advent of GPS-enabled mobile phones and wearable sensors has enabled the collection of large amounts of spatiotemporal data. Such data may include trajectory data, location-tagged images, or other content. In some cases, the spatiotemporal data exists in the form of RFID data [37]. The mining of patterns from such spatiotemporal data provides numerous insights in a wide variety of applications, such as traffic control and social sensing [2]. Frequent patterns are also used for trajectory clustering classification and outlier analysis [38, 45–48]. Many trajectory analysis problems can be approximately transformed to sequential pattern mining with the use of appropriate transformations. Algorithms for spatiotemporal pattern mining are discussed in Chap. 12.

4.3 Frequent Patterns in Graphs and Structured Data

Many kinds of chemical and biological data, XML data, software program traces, and Web browsing behaviors can be represented as structured graphs. In these cases, frequent pattern mining is very useful for making inferences in such data. This is because frequent structural patterns provide important insights about the graphs. For example, specific chemical structures result in particular properties, specific program structures result in software bugs, and so on. Such patterns can even be used for clustering and classification of graphs![14, 73].

A variety of methods for structural frequent pattern mining are discussed in [41, 69–71, 72]. A major problem in the context of graphs is the problem of *isomorphism*, because of which there are multiple ways to match two graphs. An *Apriori*-like algorithm can be developed for graph pattern mining. However, because of the complexity of graphs and and also because of issues related to isomorphism, the algorithms are more complex. For example, in an *Apriori*-like algorithm, pairs of graphs can be joined in multiple ways. Pairs of graphs can be joined when they have $(k-1)$ nodes in common, or they have $(k-1)$ edges in common. Furthermore, either kind of join between a pair of graphs can have multiple results. The counting process is also more challenging because of isomorphism. Pattern mining in graphs becomes especially challenging when the graphs are large, and the isomorphism problem becomes significant. Another particularly difficult case is the streaming scenario [16] where one has to determine dense patterns in the graphs stream. Typically, these problems cannot be solved exactly, and approximations are required.

Frequent pattern mining in graphs has numerous applications. In some cases, these methods can be used in order to perform classification and clustering of structured data [14, 73]. Graph patterns are used for chemical and biological data analysis, and software bug detection in computer programs. Methods for finding frequent patterns in graphs are discussed in Chap. 13. The applications of graph pattern mining are discussed in Chap. 18.

4.4 Frequent Pattern Mining with Uncertain Data

Uncertain or probabilistic data has become increasingly common over the last few years, as methods have been designed in order to collect data with very low quality. The attribute values in such data sets are *probabilistic*, which implies that the values are represented as probability distributions. Numerous algorithms have been proposed in the literature for uncertain frequent pattern mining [15], and a computational evaluation of the different techniques is provided in [64]. Many algorithms such as *FP-growth* are harder to generalize to uncertain data [15] because of the difficulty in storing probability information with the FP-Tree. Nevertheless, as the work in [15] shows, other related methods such as *H-mine* [59] can be generalized easily to the case of uncertain data. Uncertain frequent pattern mining methods have also been extended to the case of graph data [76]. A variant of uncertain graph pattern mining discovers highly *reliable* subgraphs [40]. Highly reliable subgraphs are subgraphs that are hard to disconnect in spite of the uncertainty associated with the edges. A discussion of the different methods for frequent pattern mining with uncertain data is provided in Chap. 14.

5 Privacy Issues

Privacy has increasingly become a topic of concern in recent years because of the wide availability of personal data about individuals [7]. This has often led to reluctance to share data, share it in a constrained way, or share downgraded versions of the data. The additional constraints and downgrading translate to challenges in discovering frequent patterns. In the context of frequent pattern and association rule mining, the primary challenges are as follows:

1. When privacy-preservation methods such as randomization are used, it becomes a challenge to discover associations from the underlying data. This is because a significant amount of noise has been added to the data, and it is often difficult to discover the association rules in the presence of this noise. Therefore, one class of association rule mining methods [30] proposes effective methods to perturb the data, so that meaningful patterns may be discovered while retaining privacy of the perturbed data.
2. In some cases, the output of a privacy-preserving data mining algorithm can lead to violation of privacy. This is because association rules can reveal sensitive information about individuals when they relate sensitive attributes to other kinds of attributes. Therefore, one class of methods focusses on the problem of *association rule hiding* [65].
3. In many cases, the data to be mined is stored in a distributed way by competitors who may wish to determine global insights without, at the same time, revealing their local insights. This problem is referred to as that of distributed privacy preservation [25]. The data may be either horizontally partitioned across rows (different records) or vertically partitioned (across attributes). Each of these forms of partitioning require different methods for distributed mining.

Methods for privacy-preserving association rule mining are addressed in Chap. 15.

6 Applications of Frequent Pattern Mining

Frequent pattern mining has applications of two types. The first type of application is to other major data mining problems such as clustering, outlier detection, and classification. Frequent patterns are often used to determine relevant clusters from the underlying data. In addition, rule-based classifiers are often constructed with the use of frequent pattern mining methods. Frequent pattern mining is also used in generic applications, such as Web log analytics, software bug analysis, chemical, and biological data.

6.1 Applications to Major Data Mining Problems

Frequent pattern mining methods can also be applied to other major data mining problems such as clustering [9, 19], classification and outlier analysis. For example, frequent pattern mining methods are often used for subspace clustering [11], by discretizing the quantitative attributes, and then finding patterns from these discrete values. Each such pattern, therefore, corresponds to a rectangular region in a subspace of the data. These rectangular regions can then be integrated together in order to create a more comprehensive subspace representation.

Frequent pattern mining is also applied to problems such as classification, in which rules are generated by using patterns on the left hand side of the rule, and the class variable on the right hand side of the rule [52]. The main goal here is to find *discriminative* patterns for the purpose of classification, rather than simply patterns that satisfy the support requirements. Such methods have also been extended to structured XML data [73] by finding discriminative graph-structured patterns. In addition, sequential pattern mining methods can be applied to other temporal mining methods such as event detection [43, 44, 53, 54] and sequence classification [68]. Frequent pattern mining has also been applied to the problem of outlier analysis [1], by determining deviations from the expected patterns in the underlying data. Methods for clustering based on frequent pattern mining are discussed in Chap. 16, while rule-based classification are discussed in Chap. 17. It should be pointed out that constrained frequent pattern mining is closely related to the problem of classification with frequent patterns, and therefore both are discussed in the same chapter.

6.2 Generic Applications

Frequent pattern mining has applications to a variety of problems such as clustering, classification and event detection. In addition, specific application areas such as Web mining and software bug detection can also benefit from frequent pattern mining methods. In the context of Web mining, numerous methods have been proposed for finding useful patterns from Web logs in order to make recommendations [63]. Such techniques can also be used to determine outliers from Web log sequences [1]. Frequent patterns are also used for trajectory classification and outlier analysis [49–48]. Frequent pattern mining methods can also be used in order to determine relevant rules and patterns in spatial data, as they related to spatial and non-spatial properties of objects. For example, an association rule could be created from the relationships of land temperatures of "nearby" geographical locations. In the context of spatiotemporal data, the relationships between the motions of different objects could be used to create spatiotemporal frequent patterns. Frequent pattern mining methods have been used for finding patterns in biological and chemical data [42, 29, 75]. In addition, because software programs can be represented as graphs, frequent pattern mining methods can be used in order to find logical bugs from program execution traces [51]. Numerous applications of frequent pattern mining are discussed in Chap. 18.

7 Conclusions and Summary

Frequent pattern mining is one of four major problems in the data mining domain. This chapter provides an overview of the major topics in frequent pattern mining. The earliest work in this area was focussed on determining the efficient algorithms for frequent pattern mining, and variants such as long pattern mining, interesting pattern mining, constraint-based pattern mining, and compression. In recent years scalability has become an issue because of the massive amounts of data that continue to be created in various applications. In addition, because of advances in data collection technology, advanced data types such as temporal data, spatiotemporal data, graph data, and uncertain data have become more common. Such data types have numerous applications to other data mining problems such as clustering and classification. In addition, such data types are used quite often in various temporal applications, such as the Web log analytics.

References

1. C. Aggarwal. Outlier Analysis, *Springer*, 2013.
2. C. Aggarwal. Social Sensing, *Managing and Mining Sensor Data*, Springer, 2013.
3. C. C. Aggarwal, and P. S. Yu. Online generation of Association Rules, *ICDE Conference*, 1998.
4. R. Agrawal, and R. Srikant. Fast Algorithms for Mining Association Rules in Large Databases, *VLDB Conference*, pp. 487–499, 1994.
5. R. Agrawal, and R. Srikant. Mining Sequential Patterns, *ICDE Conference*, 1995.
6. C. C. Aggarwal, and P. S. Yu. A New Framework for Itemset Generation, *ACM PODS Conference*, 1998.
7. C. Aggarwal and P. Yu. Privacy-preserving data mining: Models and Algorithms, *Springer*, 2008.
8. C. C. Aggarwal, and H. Wang. Managing and Mining Graph Data, *Springer*, 2010.
9. C. C. Aggarwal, and C. K. Reddy. Data Clustering: Algorithms and Applications, *CRC Press*, 2013.

10. R. Agrawal, T. Imielinski, and A. Swami. Database Mining: A Performance Perspective. *IEEE Transactions on Knowledge and Data Engineering*, 5(6), pp. 914–925, 1993.
11. R. Agrawal, J. Gehrke, D. Gunopulos, P. Raghavan. Automatic Subspace Clustering of High Dimensional Data for Data Mining Applications, *ACM SIGMOD Conference*, 1998.
12. R. Agarwal, C. C. Aggarwal, and V. V. V. Prasad. Depth-first Generation of Long Patterns, *ACM KDD Conference*, 2000: Also appears as IBM Research Report, RC, 21538, 1999.
13. R. Agarwal, C. C. Aggarwal, and V. V. V. Prasad. A Tree Projection Algorithm for Generation of Frequent Itemsets, *Journal of Parallel and Distributed Computing*, 61(3), pp. 350–371, 2001. Also appears as IBM Research Report, RC 21341, 1999.
14. C. C. Aggarwal, N. Ta, J. Wang, J. Feng, M. Zaki. Xproj: A framework for projected structural clustering of XML documents, *ACM KDD Conference*, 2007.
15. C. C. Aggarwal, Y. Li, J. Wang, J. Feng. Frequent Pattern Mining with Uncertain Data, *ACM KDD Conference*, 2009.
16. C. Aggarwal, Y. Li, P. Yu, and R. Jin. On dense pattern mining in graph streams, *VLDB Conference*, 2010.
17. R. J. Bayardo Jr. Efficiently mining long patterns from databases. *ACM SIGMOD Conference*, 1998.
18. J.-F. Boulicaut, A. Bykowski, and C. Rigotti. Free-sets: A Condensed Representation of Boolean data for the Approximation of Frequency Queries. *Data Mining and Knowledge Discovery*, 7(1), pp. 5–22, 2003.
19. G. Buehrer, and K. Chellapilla. A Scalable Pattern Mining Approach to Web Graph Compression with Communities. *WSDM Conference*, 2009.
20. T. Calders, and B. Goethals. Mining all non-derivable frequent itemsets, *Principles of Knowledge Discovery and Data Mining*, 2006.
21. T. Calders, C. Rigotti, and J. F. Boulicaut. A survey on condensed representations for frequent sets. In *Constraint-based mining and inductive databases*, pp. 64–80, Springer, 2006.
22. J. H. Chang, W. S. Lee. Finding Recent Frequent Itemsets Adaptively over Online Data Streams. *ACM KDD Conference*, 2003.
23. M. Charikar, K. Chen, and M. Farach-Colton. Finding Frequent Items in Data Streams, *Automata, Languages and Programming*, pp. 693–703, 2002.
24. M. S. Chen, J. S. Park, and P. S. Yu. Efficient data mining for path traversal patterns, *IEEE Transactions on Knowledge and Data Engineering*, 10(2), pp. 209–221, 1998.
25. C. Clifton, M. Kantarcioglu, J. Vaidya, X. Lin, and M. Zhu. Tools for privacy preserving distributed data mining. *ACM SIGKDD Explorations Newsletter*, 4(2), pp. 28–34, 2002.
26. E. Cohen. M. Datar, S. Fujiwara, A. Gionis, P. Indyk, R. Motwani, J. Ullman, and C. Yang. Finding Interesting Associations without Support Pruning, *IEEE TKDE*, 13(1), pp. 64–78, 2001.
27. G. Cormode, S. Muthukrishnan. What's hot and what's not: tracking most frequent items dynamically, *ACM TODS*, 30(1), pp. 249–278, 2005.
28. J. Dean and S. Ghemawat. *MapReduce*: Simplified Data Processing on Large Clusters. *OSDI*, pp. 137–150, 2004.
29. M. Deshpande, M. Kuramochi, N. Wale, and G. Karypis. Frequent substructure-based approaches for classifying chemical compounds. *IEEE TKDE.*, 17(8), pp. 1036–1050, 2005.
30. A. Evfimievski, R. Srikant, R. Agrawal, and J. Gehrke. Privacy preserving mining of association rules. *Information Systems*, 29(4), pp. 343–364, 2004.
31. M. Garofalakis, R. Rastogi, and K. Shim.: Sequential Pattern Mining with Regular Expression Constraints, *VLDB Conference*, 1999.
32. V. Guralnik, and G. Karypis. Parallel tree-projection-based sequence mining algorithms. *Parallel Computing*, 30(4): pp. 443–472, April 2004.
33. J. Han, J. Pei, and Y. Yin. Mining Frequent Patterns without Candidate Generation, *ACM SIGMOD Conference*, 2000.
34. J. Han, H. Cheng, D. Xin, and X. Yan. Frequent Pattern Mining: Current Status and Future Directions, *Data Mining and Knowledge Discovery*, 15(1), pp. 55–86, 2007.

35. J. Han, J. Pei, B. Mortazavi-Asl, Q. Chen, U. Dayal, and M. C. Hsu. FreeSpan: frequent pattern-projected sequential pattern mining. *ACM KDD Conference*, 2000.
36. J. Han, J. Pei, H. Pinto, B. Mortazavi-Asl, Q. Chen, U. Dayal, and M. C. Hsu. PrefixSpan: Mining sequential patterns efficiently by prefix-projected pattern growth. *ICDE Conference*, 2001.
37. J. Han, J.-G. Lee, H. Gonzalez, X. Li. Mining Massive RFID, Trajectory, and Traffic Data Sets (Tutorial). *ACM KDD Conference*, 2008. Video of Tutoral Lecture at: http://videolectures.net/kdd08_han_mmrfid/
38. H. Jeung, M. L. Yiu, X. Zhou, C. Jensen, H. Shen, Discovery of Convoys in Trajectory Databases, *VLDB Conference*, 2008.
39. R. Jin, G. Agrawal. Frequent Pattern Mining in Data Streams, *Data Streams: Models and Algorithms*, pp. 61–84, Springer, 2007.
40. R. Jin, L. Liu, and C. Aggarwal. Discovering highly reliable subgraphs in uncertain graphs. *ACM KDD Conference*, 2011.
41. G. Kuramuchi and G. Karypis. Frequent Subgraph Discovery, *ICDM Conference*, 2001.
42. A. R. Leach and V. J. Gillet. *An Introduction to Chemoinformatics*. Springer, 2003.
43. W. Lee, S. Stolfo, and P. Chan. Learning Patterns from Unix Execution Traces for Intrusion Detection, *AAAI workshop on AI methods in Fraud and Risk Management*, 1997.
44. W. Lee, S. Stolfo, and K. Mok. A Data Mining Framework for Building Intrusion Detection Models, *IEEE Symposium on Security and Privacy*, 1999.
45. J.-G. Lee, J. Han, K.-Y. Whang, Trajectory Clustering: A Partition-and-Group Framework, *ACM SIGMOD Conference*, 2007.
46. J.-G. Lee, J. Han, X. Li. Trajectory Outlier Detection: A Partition-and-Detect Framework, *ICDE Conference*, 2008.
47. J.-G. Lee, J. Han, X. Li, H. Gonzalez. TraClass: trajectory classification using hierarchical region-based and trajectory-based clustering. *PVLDB*, 1(1): pp. 1081–1094, 2008.
48. X. Li, J. Han, and S. Kim. Motion-alert: Automatic Anomaly Detection in Massive Moving Objects, *IEEE Conference in Intelligence and Security Informatics*, 2006.
49. X. Li, J. Han, S. Kim and H. Gonzalez. ROAM: Rule- and Motif-based Anomaly Detection in Massive Moving Object Data Sets, *SDM Conference*, 2007.
50. Z. Li, B. Ding, J. Han, R. Kays. Swarm: Mining Relaxed Temporal Object Moving Clusters, *VLDB Conference*, 2010.
51. C. Liu, X. Yan, H. Lu, J. Han, and P. S. Yu. Mining Behavior Graphs for "backtrace" of non-crashing bugs, *SDM Conference*, 2005.
52. B. Liu, W. Hsu, Y. Ma. Integrating Classification and Association Rule Mining, *ACM KDD Conference*, 1998.
53. S. Ma, and J. Hellerstein. Mining Partially Periodic Event Patterns with Unknown Periods, *IEEE International Conference on Data Engineering*, 2001.
54. H. Mannila, H. Toivonen, and A. I. Verkamo. Discovering Frequent Episodes in Sequences, *ACM KDD Conference*, 1995.
55. R. Ng, L. V. S. Lakshmanan, J. Han, and A. Pang. Exploratory mining and pruning optimizations of constrained associations rules. *ACM SIGMOD Conference*, 1998.
56. N. Pasquier, Y. Bastide, R. Taouil, and L. Lakhal. Discovering frequent closed itemsets for association rules. *International Conference on Database Theory*, pp. 398–416, 1999.
57. J. Pei, and J. Han. Can we push more constraints into frequent pattern mining? *ACM KDD Conference*, 2000.
58. J. Pei, J. Han, R. Mao. CLOSET: An Efficient Algorithms for Mining Frequent Closed Itemsets, *DMKD Workshop*, 2000.
59. J. Pei, J. Han, H. Lu, S. Nishio, S. Tang, and D. Yang. H-mine: Hyper-structure mining of frequent patterns in large databases. In Data Mining, *ICDM Conference*, 2001.
60. J. Pei, J. Han, and L. V. S. Lakshmanan. Mining Frequent Patterns with Convertible Constraints in Large Databases, *ICDE Conference*, 2001.

61. J. Pei, J. Han, and W. Wang. Constraint-based Sequential Pattern Mining: The Pattern-Growth Methods, *Journal of Intelligent Information Systems*, 28(2), pp. 133–160, 2007.
62. P. Shenoy, J. Haritsa, S. Sudarshan, G. Bhalotia, M. Bawa, D. Shah. Turbo-charging Vertical Mining of Large Databases. *ACM SIGMOD Conference*, pp. 22–33, 2000.
63. J. Srivastava, R. Cooley, M. Deshpande, and P. N. Tan. Web usage mining: Discovery and applications of usage patterns from Web data. *ACM SIGKDD Explorations Newsletter*, 1(2), pp. 12–23, 2000.
64. Y. Tong, L. Chen, Y. Cheng, P. Yu. Mining Frequent Itemsets over Uncertain Databases. *PVLDB*, 5(11), pp. 1650–1661, 2012.
65. V. S. Verykios, A. K. Elmagarmid, E. Bertino, Y. Saygin, and E. Dasseni. Association rule hiding. *IEEE Transactions on Knowledge and Data Engineering*, pp. 434–447, 16(4), pp. 434–447, 2004.
66. J. Vreeken, M. van Leeuwen, and A. Siebes. Krimp: Mining itemsets that compress. *Data Mining and Knowledge Discovery*, 23(1), pp. 169–214, 2011.
67. J. Wang, J. Han, and J. Pei. CLOSET+: Searching for the Best strategies for mining frequent closed itemsets. *ACM KDD Conference*, 2003.
68. Z. Xing, J. Pei, and E. Keogh. A Brief Survey on Sequence Classification, *ACM SIGKDD Explorations*, 12(1), 2010.
69. X. Yan, P. S. Yu, and J. Han, Graph indexing: A frequent structure-based approach. *ACM SIGMOD Conference*, 2004.
70. X. Yan, P. S. Yu, and J. Han. Substructure similarity search in graph databases. *ACM SIGMOD Conference*, 2005.
71. X. Yan, F. Zhu, J. Han, and P. S. Yu. Searching substructures with superimposed distance, *ICDE Conference*, 2006.
72. M. Zaki. Efficiently mining frequent trees in a forest: Algorithms and applications. *IEEE Transactions on Knowledge and Data Engineering*, 17(8), pp. 1021–1035, 2005.
73. M. Zaki, C. Aggarwal. XRules: An Effective Classifier for XML Data, *ACM KDD Conference*, 2003.
74. M. Zaki, C. J. Hsiao. CHARM: An Efficient Algorithm for Closed Frequent Itemset Mining, *SDM Conference*, 2002.
75. S. Zhang, T. Wang. Discovering Frequent Agreement Subtrees from Phylogenetic Data. *IEEE Transactions on Knowledge and Data Engineering*, 20(1), pp. 68–82, 2008.
76. Z. Zou, J. Li, H. Gao, and S. Zhang. Mining Frequent Subgraph Patterns from Uncertain Graph Data, *IEEE Transactions on Knowledge and Data Engineering*, 22(9), pp. 1203–1218, 2010.
77. http://fimi.ua.ac.be/

Chapter 2
Frequent Pattern Mining Algorithms: A Survey

Charu C. Aggarwal, Mansurul A. Bhuiyan and Mohammad Al Hasan

Abstract This chapter will provide a detailed survey of frequent pattern mining algorithms. A wide variety of algorithms will be covered starting from *Apriori*. Many algorithms such as *Eclat, TreeProjection*, and *FP-growth* will be discussed. In addition a discussion of several maximal and closed frequent pattern mining algorithms will be provided. Thus, this chapter will provide one of most detailed surveys of frequent pattern mining algorithms available in the literature.

Keywords Frequent pattern mining algorithms · *Apriori* · *TreeProjection* · *FP-growth*

1 Introduction

In data mining, frequent pattern mining (FPM) is one of the most intensively investigated problems in terms of computational and algorithmic development. Over the last two decades, numerous algorithms have been proposed to solve frequent pattern mining or some of its variants, and the interest in this problem still persists [45, 75]. Different frameworks have been defined for frequent pattern mining. The most common one is the support-based framework, in which itemsets with frequency above a given threshold are found. However, such itemsets may sometimes not represent interesting positive *correlations* between items because they do not normalize for the absolute frequencies of the items. Consequently, alternative measures for interestingness have been defined in the literature [7, 11, 16, 63]. This chapter will focus on the support-based framework because the algorithms based on the interestingness

C. C. Aggarwal (✉)
IBM T. J. Watson Research Center, Yorktown Heights, NY 10598, USA
e-mail: charu@us.ibm.com

M. A. Bhuiyan · M. A. Hasan
Indiana University–Purdue University, Indianapolis, IN, USA
e-mail: mbhuiyan@cs.iupui.edu

M. A. Hasan
e-mail: alhasan@cs.iupui.edu

C. C. Aggarwal, J. Han (eds.), *Frequent Pattern Mining*,
DOI 10.1007/978-3-319-07821-2_2, © Springer International Publishing Switzerland 2014

Algorithm *Baseline Mining*(Database: $\mathcal{T}$, Minimum Support: s)
  begin
    $\mathcal{FP} = \{\}$;
    Insert length-one frequent pattern in $\mathcal{FP}$
    until all frequent patterns in $\mathcal{FP}$ are explored **do**
    begin
      Generate a candidate pattern P from one (or more) frequent
          pattern(s) in $\mathcal{FP}$
      if support$(P, \mathcal{T}) \geq s$
        Add P to frequent pattern set $\mathcal{FP}$;
    end
  end

Fig. 2.1 A generic frequent pattern mining algorithm

framework are provided in a different chapter. Surveys on frequent pattern mining may be found in [26, 33].

One of the main reasons for the high level of interest in frequent pattern mining algorithms is due to the computational challenge of the task. Even for a moderate sized dataset, the search space of FPM is enormous, which is exponential to the length of the transactions in the dataset. This naturally creates challenges for itemset generation, when the support levels are low. In fact, in most practical scenarios, the support levels at which one can mine the corresponding itemsets are limited (bounded below) by the memory and computational constraints. Therefore, it is critical to be able to perform the analysis in a space- and time-efficient way. During the first few years of research in this area, the primary focus of work was to find FPM algorithms with better computational efficiency.

Several classes of algorithms have been developed for frequent pattern mining, many of which are closely related to one another. In fact, the execution tree of all the algorithms is mostly different in terms of the order in which the patterns are explored, and whether the counting work done for different candidates is independent of one another. To explain this point, we introduce a primitive "baseline" algorithm that forms the heart of most frequent pattern mining algorithms.

Figure 2.1 presents the pseudocode for a very simple "baseline" frequent pattern mining algorithm. The algorithm takes the transaction database $\mathcal{T}$ and a user-defined support value s as input. It first populates all length-one frequent patterns in a frequent pattern data-store, $\mathcal{FP}$. Then it generates a candidate pattern and computes its support in the database. If the support of the candidate pattern is equal or higher than the minimum support threshold the pattern is stored in $\mathcal{FP}$. The process continues until all the frequent patterns from the database are found.

In the aforementioned algorithm, candidate patterns are generated from the previously generated frequent patterns. Then, the transaction database is used to determine which of the candidates are truly frequent patterns. The key issues of computational efficiency arise in terms of generating the candidate patterns in an orderly and carefully designed fashion, pruning irrelevant and duplicate candidates, and using well chosen tricks to minimize the work in counting the candidates. Clearly, the effectiveness of these different strategies depend on each other. For example, the effectiveness of a pruning strategy may be dependent on the order of exploration of the candidates (level-wise vs. depth first), and the effectiveness of counting is also dependent on the order of exploration because the work done for counting at the higher levels (shorter itemsets) can be reused at the lower levels (longer itemsets) with certain strategies, such as those explored in *TreeProjection* and *FP-growth*. Surprising as it might seem, virtually all frequent pattern mining algorithms can be considered complex variations of this simple baseline pseudocode. The major challenge of all of these methods is that the number of frequent patterns and candidate patterns can sometimes be large. This is a fundamental problem of frequent pattern mining although it is possible to speed up the counting of the different candidate patterns with the use of various tricks such as database projections. An analysis on the number of candidate patterns may be found in [25].

The candidate generation process of the earliest algorithms used joins. The original *Apriori* algorithm belongs to this category [1]. Although *Apriori* is presented as a join-based algorithm, it can be shown that the algorithm is a breadth first exploration of a structured arrangement of the itemsets, known as a *lexicographic tree* or *enumeration tree*. Therefore, later classes of algorithms explicitly discuss tree-based enumeration [4, 5]. The algorithms assume a lexicographic tree (or enumeration tree) of candidate patterns and explore the tree using breadth-first or depth-first strategies. The use of the enumeration tree forms the basis for understanding search space decomposition, as in the case of the *TreeProjection* algorithm [5]. The enumeration tree concept is very useful because it provides an understanding of how the search space of candidate patterns may be explored in a systematic and non-redundant way. Frequent pattern mining algorithms typically need to evaluate the support of frequent portions of the enumeration tree, and also rule out an additional layer of infrequent extensions of the frequent nodes in the enumeration tree. This makes the candidate space of all frequent pattern mining algorithms virtually invariant unless one is interested in particular types of patterns such as maximal patterns.

The enumeration tree is defined on the prefixes of frequent itemsets, and will be introduced later in this chapter. Later algorithms such as *FP-growth* perform suffix-based recursive exploration of the search space. In other words, the frequent patterns with a particular pattern as a suffix are explored at one time. This is because *FP-growth* uses the opposite item ordering convention as most enumeration tree algorithms though the recursive exploration order of *FP-growth* is similar to an enumeration tree.

Note that all classes of algorithms, implicitly or explicitly, explore the search space of patterns defined by an enumeration tree of frequent patterns with different strategies such as joins, prefix-based depth-first exploration, or suffix-based depth-first exploration. However, there are significant differences in terms of the order in which the search space is explored, the pruning methods used, and how the counting is performed. In particular, certain projection-based methods help in reusing the counting work for k-itemsets for $(k + 1)$-itemsets with the use of the notion of projected databases. Many algorithms such as *TreeProjection* and *FP-growth* are able to achieve this goal.

Table 2.1 Toy transaction
database and frequent items
of each transaction for a
minimum support of 3

tid	Items	Sorted frequent items
2	a,b,c,d,f,h	a,b,c,d,f
3	a,f,g	a,f
4	b,e,f,g	b,f,e
5	a,b,c,d,e,h	a,b,c,d,e

This chapter is organized as follows. The remainder of this chapter discusses notations and definitions relevant to frequent pattern mining. Section 2 discusses join-based algorithms. Section 3 discusses tree-based algorithms. All the algorithms discussed in Sects. 2 and 3 extend prefixes of itemsets to generated frequent patterns. A number of methods that extend suffixes of frequent patterns are discussed in Sect. 4. Variants of frequent pattern mining, such as closed and maximal frequent pattern mining, are discussed in Sect. 5. Other optimized variations of frequent pattern mining algorithms are discussed in Sect. 6. Methods for reducing the number of passes, with the use of sampling and aggregation are proposed in Sect. 7. Finally, Sect. 8 concludes chapter with an overall summary.

1.1 Definitions

In this section, we define several key concepts of frequent pattern mining (FPM) that we will use in the remaining part of the chapter.

Let, $\mathcal{T} = \{T_1, T_2, \ldots, T_n\}$ be a transaction database, where each $T_i \in \mathcal{T}, \forall i = \{1 \ldots n\}$ consists of a set of items, say $T_i = \{x_1, x_2, x_3, \ldots x_l\}$. A set $P \subseteq T_i$ is called an itemset. The size of an itemset is defined by the number of items it contains. We will refer an itemset as l-$itemset$ (or l-$pattern$), if its size is l. The number of transactions containing P is referred to as the *support* of P. A pattern P is defined to be frequent if its support is at least equal to the the minimum threshold.

Table 2.1 depicts a toy database with 5 transactions (T_1, T_2 T_3, T_4 and T_5). The second column shows the items in each transaction. In the third column, we show the set of items that are frequent in the corresponding transaction for a minimum support value of 3. For example, the item h in transaction with *tid* value of 2 is an infrequent item with a support value of 2. Therefore, it is not listed in the third column of the corresponding row. Similarly, the pattern $\{a, b\}$ (or, ab in abbreviated form) is frequent because it has a support value of 3.

The frequent patterns are often used to generate *association rules*. Consider the rule $X \Rightarrow Y$, where X and Y are sets of items. The confidence of the rule $X \Rightarrow Y$ is the equal to the ratio of the support of $X \cup Y$ to that of the support of X. In other words, it can be viewed as the conditional probability that Y occurs, given that X has occurred. The support of the rule is equal to the support of $X \cup Y$. Association rule-generation is a two-phase process. The first phase determines all the frequent patterns at a given minimum support level. The second phase extracts all the rules from these patterns. The second phase is fairly trivial and with limited sophistication. Therefore, most of the algorithmic work in frequent pattern mining focusses on the

Fig. 2.2 The lattice of
itemsets

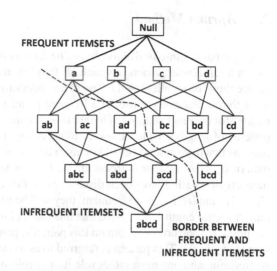

first phase. This chapter will also focus on the first phase of frequent pattern mining,
which is generally considered more important and non-trivial.

Frequent patterns satisfy a *downward closure property*, according to which every
subset of a frequent pattern is also frequent. This is because if a pattern P is a
subset of a transaction, then every pattern $P' \subseteq P$ will also be a subset of T.
Therefore, the support of P' can be no less than that of P. The space of exploration
of frequent patterns can be arranged as a lattice, in which every node is one of the 2^d
possible itemsets, and an edge represents an immediate subset relationship between
these itemsets. An example of a lattice of possible itemsets for a universe of items
corresponding to $\{a, b, c, d\}$ is illustrated in Fig. 2.2. The lattice represents the search
of frequent patterns, and all frequent pattern mining algorithms must, in one way or
another, traverse this lattice to identify the frequent nodes of this lattice. The lattice is
separated into a frequent and an infrequent part with the use of a *border*. An example
of a border is illustrated in Fig. 2.2. This border must satisfy the downward closure
property.

The lattice can be traversed with a variety of strategies such as breadth-first or
depth-first methods. Furthermore, *candidate nodes* of the lattice may be generated
in many ways, such as using joins, or using lexicographic tree-based extensions.
Many of these methods are conceptually equivalent to one another. The following
discussion will provide an overview of the different strategies that are commonly
used.

2   Join-Based Algorithms

Join-based algorithms generate $(k + 1)$-candidates from frequent k-patterns with the
use of joins. These candidates are then validated against the transaction database.
The *Apriori* method uses joins to create candidates from frequent patterns, and is
one of the earliest algorithms for frequent pattern mining.

2.1 Apriori Method

The most basic join-based algorithm is the *Apriori* method [1]. The *Apriori* approach uses a *level-wise* approach in which all frequent itemsets of length k are generated before those of length $(k + 1)$. The main observation which is used for the *Apriori* algorithm is that every subset of a frequent pattern is also frequent. Therefore, *candidates* for frequent patterns of length $(k + 1)$ can be generated from *known* frequent patterns of length k with the use of joins. A join is defined by pairs of frequent k-patterns that have at least $(k - 1)$ items in common. Specifically, consider a frequent pattern $\{i_1, i_2, i_3, i_4\}$ that is frequent, but has not yet been discovered because only itemsets of length 3 have been discovered so far. In this case, because the patterns $\{i_1, i_2, i_3\}$ and $\{i_1, i_2, i_4\}$ are frequent, they will be present in the set $\mathcal{F}_3$ of all frequent patterns with length $k = 3$. Note that this particular pair also has $k - 1 = 2$ items in common. By performing a join on this pair, it is possible to create the *candidate* pattern $\{i_1, i_2, i_3, i_4\}$. This pattern is referred to as a *candidate* because it might *possibly* be frequent, and one most either rule it in or rule it out by support counting. Therefore, this candidate is then *validated* against the transaction database by counting its support. Clearly, the design of an efficient support counting method plays a critical role in the overall efficiency of the process. Furthermore, it is important to note that the same candidate can be produced by joining multiple frequent patterns. For example, one might join $\{i_1, i_2, i_3\}$ and $\{i_2, i_3, i_4\}$ to achieve the same result. Therefore, in order to avoid duplication in candidate generation, two itemsets are joined only whether first $(k - 1)$ items are the same, based on a lexicographic ordering imposed on the items. This provides all the $(k + 1)$-candidates in a non-redundant way.

It should be pointed out that some candidates can be pruned out in an efficient way, without validating them against the transaction database. For any $(k + 1)$-candidates, it is checked whether *all* its k subsets are frequent. Although it is already known that two of its subsets contributing to the join are frequent, it is not known whether its remaining subsets are frequent. If all its subsets are not frequent, then the candidate can be pruned from consideration because of the downward closure property. This is known as the *Apriori* pruning trick. For example, in the previous case, if the itemset $\{i_1, i_3, i_4\}$ does not exist in the set of frequent 3-itemsets which have already been found, then the candidate itemset $\{i_1, i_2, i_3, i_4\}$ can be pruned from consideration with no further computational effort. This greatly speeds up the overall algorithm. The generation of 1-itemsets and 2-itemsets is usually performed in a specialized way with more efficient techniques.

Therefore, the basic *Apriori* algorithm can be described recursively in level-wise fashion. the overall algorithm comprises of three steps that are repeated over and over again, for different values of k, where k is the length of the pattern generated in the current iteration. The four steps are those of (i) generation of candidate patterns $\mathcal{C}_{k+1}$ by using joins on the patterns in $\mathcal{F}_k$, (ii) the pruning of candidates from $\mathcal{C}_{k+1}$, for which all subsets to not lie in $\mathcal{F}_k$, and (iii) the validation of the patterns in $\mathcal{C}_{k+1}$ against the transaction database $\mathcal{T}$, to determine the subset of $\mathcal{C}_{k+1}$ which is truly frequent. The algorithm is terminated, when the set of frequent k-patterns $\mathcal{F}_k$ in a given iteration is empty. The pseudo-code of the overall procedure is presented in Fig. 2.3.

Fig. 2.3 The *Apriori*
algorithm

Algorithm *Apriori*(Database: $\mathcal{T}$, Support: s)
begin
    Generate frequent 1-patterns and 2-patterns
    using specialized counting methods and
    denote by $\mathcal{F}_1$ and $\mathcal{F}_2$;
    $k := 2$;
    while $\mathcal{F}_k$ is not empty **do**
    begin
        Generate $\mathcal{C}_{k+1}$ by using joins on $\mathcal{F}_k$;
        Prune $\mathcal{C}_{k+1}$ with *Apriori* subset pruning trick;
        Generate $\mathcal{F}_{k+1}$ by counting candidates in
          $\mathcal{C}_{k+1}$ with respect to $\mathcal{T}$ at support s;
        $k := k + 1$;
    end
    return $\cup_{i=1}^{k}\mathcal{F}_i$;
end

The computationally intensive procedure in this case is the counting of the candidates in $\mathcal{C}_{k+1}$ with respect to the transaction database $\mathcal{T}$. Therefore, a number of optimizations and data structures have been proposed in [1] (and also the subsequent literature) to speed up the counting process. The data structure proposed in [1] is that of constructing a *hash-tree* to maintain the candidate patterns. A leaf node of the hash-tree contains a list of itemsets, whereas an interior node contains a hash-table. An itemset is mapped to a leaf node of the tree by defining a path from the root to the leaf node with the use of the hash function. At a node of level i, a hash function is applied to the ith item to decide which branch to follow. The itemsets in the leaf node are stored in sorted order. The tree is constructed recursively in top–down fashion, and a minimum threshold is imposed on the number of candidates in the leaf node.

To perform the counting, all possible k-itemsets which are subsets of a transaction are discovered in a *single* exploration of the hash-tree. To achieve this goal *all possible* paths in the hash tree that could correspond to subsets of the transaction, are followed in recursive fashion, to determine which leaf nodes are relevant to that transaction. After the leaf nodes have been discovered, the itemsets at these leaf nodes that are subsets of that transaction are isolated and their count is incremented. The actual selection of the relevant leaf nodes is performed by recursive traversal as follows. At the root node, all branches are followed such that *any* of the items in the transaction hash to one of branches. At a given interior node, if the ith item of the transaction was last hashed, then all items *following it* in the transaction are hashed to determine the possible children to follow. Thus, by following all these paths, the relevant leaf nodes in the tree are determined. The candidates in the leaf node are stored in sorted order, and can be compared efficiently to the hashed sequence of items in the transaction to determine whether they are relevant. This provides a count of the itemsets relevant to the transaction. This process is repeated for each transaction to determine the final support count for each itemset. It should be pointed out that the reason for using a hash function at the intermediate nodes is to reduce the branching factor of the hash tree. However, if desired, a trie can be used explicitly, in which the degree of a

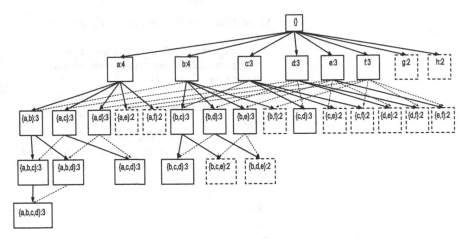

Fig. 2.4 Execution tree of *Apriori* algorithm

node is potentially of the order of the total number of items. An example of such an implementation is provided in [12], and it seems to work quite well. An algorithm that shares some similarities to the *Apriori* method, was independently proposed in [44], and subsequently a combined work was published in [3].

Figure 2.4 illustrates the execution tree of the join-based *Apriori* algorithm over the toy transaction database mentioned in Table 2.1 for minimum support value 3. As mentioned in the pseudocode of *Apriori*, a candidate k-patterns are generated by joining two frequent itemset of size $(k - 1)$. For example, at level 3, the pattern $\{a, b, c\}$ is generated by joining $\{a, b\}$ and $\{a, c\}$. After generating the candidate patterns, the support of the patterns is computed by scanning every transaction in the database and determining the frequent ones. In Fig. 2.4, a candidate patterns is shown in a box along with its support value. A frequent candidate is shown in a solid box, and an infrequent candidate is shown in a dotted box. An edge represents the join relationship between a candidate pattern of size k and a frequent pattern of size $(k - 1)$ such that the latter is used to generate the earlier. The figure also illustrates the fact that a pair of frequent patterns are used to generate a candidate pattern, whereas no candidates are generated from an infrequent pattern.

2.1.1 Apriori Optimizations

Numerous optimizations were proposed for the *Apriori* algorithm [1] that are referred to as *AprioriTid* and *AprioriHybrid* respectively. In the *AprioriTid* algorithm, each transaction is replaced by a shorter transaction or null transaction) during the kth phase. Let the set of $k + 1$-candidates in $\mathcal{C}_{k+1}$ that are contained in transaction T be denoted by $\mathcal{R}(T, \mathcal{C}_{k+1})$. This set $\mathcal{R}(T, \mathcal{C}_{k+1})$ is added to a newly created transaction database $\mathcal{T}_k'$. If the set $\mathcal{R}(T, \mathcal{C}_{k+1})$ is null, then clearly, a number of different tradeoffs exist with the use of such an approach.

- Because each newly created transaction in $\mathcal{T}_k'$ is much shorter, this makes subsequent support counting more efficient.
- In some cases, no candidate may be a subset of the transaction. Such a transaction can be dropped from the database because it does not contribute to the counting of support values.
- In other cases, more than one candidate may be a subset of the transaction, which will actually increase the overhead of the algorithm. Clearly, this is not a desirable scenario.

Thus, the first two factors improve the efficiency of the new representation, whereas the last factor worsens it. Typically, the impact of the last factor is greater in the early iterations, whereas the impact of the first two factors is greater in the later iterations. Therefore, to maximize the overall efficiency, a natural approach would be to *not* use this optimization in the early iterations, and apply it only in the later iterations. This variation is referred to as the *AprioriHybrid* algorithm [1]. Another optimization proposed in [9] is that the support of many patterns can be inferred from those of key patterns in the data. This is used to significantly enhance the efficiency of the approach.

Numerous other techniques have been proposed that use different techniques to optimize the original implementation of the *Apriori* algorithm. As an example, the method in [1] and [44] share a number of similarities but are somewhat different at the implementation level. A work that combines the ideas from these different pieces of work is presented in [3].

2.2 DHP Algorithm

The DHP algorithm, also known as the *Direct Hashing and Pruning* method [50], was proposed soon after the *Apriori* method. It proposes two main optimizations to speed up the algorithm. The first optimization is to prune the candidate itemsets in each iteration, and the second optimization is to trim the transactions to make the support-counting process more efficient.

To prune the itemsets, the algorithm tracks partial information about candidate $(k+1)$-itemsets, while explicitly counting the support of candidate k-itemsets. During the counting of candidate k-itemsets, all $(k+1)$ subsets of the transaction are found and hashed into a table that maintains the counts of the number of subsets hashed into each entry. During the phase of counting $(k+1)$-itemsets, the counts in the hash table are retrieved for each itemset. Clearly, these counts are overestimates because of possible collisions in the hash table. Those itemsets for which the counts are below the user-specified support level are then pruned from consideration.

A second optimization proposed in DHP is that of transaction trimming. A key observation here is that if an item does not appear in at least k frequent itemsets in $\mathcal{F}_k$, then no frequent itemset in $\mathcal{F}_{k+1}$ will contain that item. This follows from the fact that there should be at least k (immediate) subsets of each frequent pattern in $\mathcal{F}_{k+1}$

containing a particular item that also occur in $\mathcal{F}_k$ and also contain that item. This implies that if an item does not appear in at least k frequent itemsets in $\mathcal{F}_k$, then that item is no longer relevant to further support counting for finding frequent patterns. Therefore, that item can be trimmed from the transaction. This reduces the width of the transaction, and increases the efficiency of processing. The overhead from the data structures is significant, and most of the advantages are obtained for patterns of smaller length such as 2-itemsets. It was pointed out in later work [46, 47, 60] that the use of triangular arrays for support counting of 2-itemsets in the context of the *Apriori* method is even more efficient than such an approach.

2.3 Special Tricks for 2-Itemset Counting

A number of special tricks can be used to improve the effectiveness of 2-itemset counting. The case of 2-itemset counting is special and is often similar for the case of join-based and tree-based algorithms. As mentioned above, one approach is to use a triangular array that maintains the counts of the k-patterns explicitly. For each transaction, a nested loop can be used to explore all pairs of items in the transaction and increment the corresponding counts in the triangular array. A number of caching tricks can be used [5] to improve data locality access during the counting process. However, if the number of possible items are very large, this will still be a very significant overhead because it is needed to maintain an entry for each pair of items. This is also very wasteful, if many of the 1-items are not frequent, or some of the 2-item counts are zero. Therefore, a possible approach would be to first prune out all the 1-items which are not frequent. It is simply not necessary to count the support of a 2-itemset unless both of its constituent items are frequent. A hash table can then be used to maintain the frequency counts of the corresponding 2-itemsets. As before, the transactions are explored in a double nested loops, and all pairs of items are hashed into the table, with the caveat, that each of the individual items must be frequent. The set of itemsets which satisfy the support requirements are reported.

2.4 Pruning by Support Lower Bounding

Most of the pruning tricks discussed earlier prune itemsets when they are guaranteed *not* meet the required support threshold. It is also possible to skip the counting process for an itemset if the itemset is guaranteed to meet the support threshold. Of course, the caveat here is that the exact support of that itemset will not be available, beyond the knowledge that it meets the minimum threshold. This is sufficient in the case of many applications.

Consider two k-itemsets A and B that have $k - 1$ items $A \cap B$ in common. Then, the union of the items in A and B, denoted by $A \cup B$ will have exactly $k + 1$ items. Then, if $sup(\cdot)$ represent the support of an itemset, then the support of $A \cup B$ can be lower bounded as follows:

$$sup(A \cup B) \geq sup(A) + sup(B) - sup(A \cap B) \tag{2.1}$$

This condition follows directly from set-theoretic considerations. Thus, the support of $(k+1)$-candidates can be lower bounded in terms of the (already computed) support values of itemsets of length k or less. If the computed value on the right-hand side is greater than the required minimum support, then the counting of the candidate does not need to be performed explicitly, and therefore considerable savings can be achieved. An example of a method which uses this kind of pruning is the *Apriori_LB* method [10].

Another interesting rule is that if the support of an itemset X is the same as that of $X \cup Y$, then for any superset $X' \supseteq X$, it is the case that the support of the itemset X' is the same as that of $X' \cup Y$. This rule can be shown directly as a corollary of the equation above. This is very useful in a variety of frequent pattern mining algorithms. For example, once the support of $X \cup \{i\}$ has been shown to be the same as that of X, then, for any superset X' of X, it is no longer necessary to explicitly compute the support of $X' \cup \{i\}$, after the support of X' has already been computed. Such optimizations have been shown to be quite effective in the context of many frequent pattern mining algorithms [13, 51, 17]. As discussed later, this trick is not exclusive to join-based algorithms, and is often used effectively in tree-based algorithms such as *MaxMiner*, and *MAFIA*.

2.5   Hypercube Decomposition

One feasible way to reduce the computation cost of support counting is to find support of multiple frequent patterns at one time. LCM [66] devise a technique referred to as hypercube decomposition in this purpose. The multiple itemsets obtained at one time, comprise a hypercube in the itemset lattice. Suppose that P is a frequent pattern, $tidset(P)$ contains the transactions that P is part of, and $tail(P)$ denotes the latest item extension to the itemset P. $H(P)$ is the set of items e satisfying $e > tail(P)$ and $tidset(P) = tidset(P \cup e)$. The set $H(P)$ is referred to as the hypercube set. Then, for any $P' \subseteq H(P)$, $tidset(P \cup P') = tidset(P)$ is true, and $P \cup P'$ is frequent. The work in [66] uses this property in the candidate generation phase. For two itemsets P and $P \cup P'$, we say that P'' is between P and $P \cup P'$ if $P \subseteq P'' \subseteq P \cup P'$. In the phase with respect to P, we output all P'' between P and $P \cup H(P)$. This technique saves significant time in counting.

3   Tree-Based Algorithms

The tree-based algorithm is based on set-enumeration concepts. The candidates can be explored with the use of a subgraph of the lattice of itemsets (see Fig. 2.2), which is also referred to as the lexicographic tree or enumeration tree [5]. These terms will,

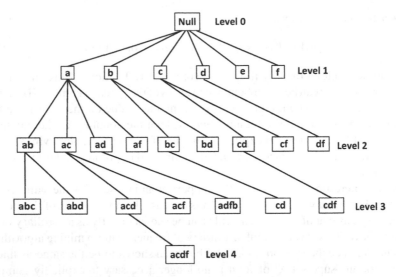

Fig. 2.5 The lexicographic tree (also known as enumeration tree)

therefore, be used interchangeably. Thus, the problem of frequent itemset generation is equivalent to that of constructing the enumeration tree. The tree can be grown in a wide variety of ways such as breadth-first or depth-first order. Because most of the discussion in this section will use this structure as a base for algorithmic development, this concept will be discussed in detail here. The main characteristic of tree-based algorithms is that the enumeration tree (or lexicographic tree) provides a certain order of exploration that can be extremely useful in many scenarios.

It is assumed that a lexicographic ordering exists among the items in the database. This lexicographic ordering is essential for efficient set enumeration without repetition. To indicate that an item i occurs lexicographically earlier than j, we will use the notation $i \leq_L j$. The lexicographic tree is an abstract representation of the large itemsets with respect to this ordering. The lexicographic tree is defined in the following way:

- A node exists in the tree corresponding to each large itemset. The root of the tree corresponds to the *null* itemset.
- Let $I = \{i_1, \ldots i_k\}$ be a large itemset, where $i_1, i_2 \ldots i_k$ are listed in lexicographic order. The parent of the node I is the itemset $\{i_1, \ldots i_{k-1}\}$.

This definition of ancestral relationship naturally defines a tree structure on the nodes that is rooted at the *null* node. A frequent 1-extension of an itemset such that the last item is the contributor to the extension will be called a *frequent lexicographic tree extension*, or simply a tree extension. Thus, each edge in the lexicographic tree corresponds to an item which is the frequent lexicographic tree extension to a node. The frequent lexicographic extensions of node P are denoted by $E(P)$. An example of the lexicographic tree is illustrated in Fig. 2.5. In this example, the frequent lexicographic extensions of node a are b, c, d, and f.

Let Q be the immediate ancestor of the itemset P in the lexicographic tree. The set of *prospective branches* of a node P is defined to be those items in $E(Q)$ which occur lexicographically after the node P. These are the *possible* frequent lexicographic extensions of P. We denote this set by $F(P)$. Thus, we have the following relationship: $E(P) \subseteq F(P) \subset E(Q)$. The value of $E(P)$ in Fig. 2.5, when $P = ab$ is $\{c, d\}$. The value of $F(P)$ for $P = ab$ is $\{c, d, f\}$, and for $P = af$, $F(P)$ is empty.

It is important to point out that virtually all non-maximal and maximal algorithms, starting from *Apriori*, can be considered enumeration-tree methods. In fact, there are few frequent pattern mining algorithms which do not use the enumeration tree, or a subset thereof (in maximal pattern mining) for frequent itemset generation. However, the *order of exploration* of the different algorithms of the lexicographic tree is quite different. For example, *Apriori* uses a breadth-first strategy, whereas other algorithms discussed later in this chapter use a depth-first strategy. Some methods are explicit about the relationship about the candidate generation process with the enumeration tree, whereas others, such as *Apriori*, are not. For example, by examining Fig. 2.4, it is evident that *Apriori* candidates can be generated by joining two frequent siblings of a lexicographic tree. In fact, all candidates can be generated in an exhaustive and non-redundant way by joining frequent siblings. For example, the two itemsets $acdfh$ and $acdfg$ are siblings, because they are children of the node $acdf$. By joining them, one obtains the candidate pattern $acdfgh$. Thus, while the *Apriori* algorithm is a join-based algorithm, it can also be explained in terms of the enumeration tree.

Parts of the enumeration tree may be removed by some of the algorithms by pruning methods. For example, the *Apriori* algorithm uses a levelwise pruning trick. For *maximal* pattern mining the advantages gained from pruning tricks can be very significant. Therefore, the number of candidates in the execution tree of different algorithms is different only because of pruning optimization tricks. However, some methods are able to achieve better *counting* strategies by using the structure of the enumeration tree to avoid re-doing the counting work already done for k-candidates to $(k+1)$-candidates. Therefore, explicitly introducing the enumeration tree is helpful because it allows a more flexible way to visualize candidate exploration strategies than join-based methods. The explicit introduction of the enumeration tree also helps in understanding whether the gains in different algorithms arise as a result of fewer number of candidates, or whether they arise as a result of better counting strategies.

3.1   AIS Algorithm

The original *AIS* algorithm [2] is a simple version of the lexicographic-tree algorithm, though it is not directly presented as such. In this approach, the tree is constructed in levelwise fashion and the corresponding itemsets at a given level are counted with the use of the transaction database. The algorithm does not use any specific optimizations to improve the efficiency of the counting process. As will be discussed later, a variety of methods can be used to further improve the efficiency of tree-based algorithms. Thus, this is a primitive approach that explores the entire search space with no optimization.

3.2   *TreeProjection Algorithms*

Two variants of an algorithm which use recursive projections of the transactions
down the lexicographic tree structure are proposed in [5] and [4], respectively. The
goal of using these recursive projections is to reuse the counting work down at a given
level for lower levels of the tree. This reduces the counting work at the lower levels
by orders of magnitude, as long as it is possible to successfully manage the memory
requirements of the projected transactions. The main difference between the different
versions of *TreeProjection* is the exploration strategy used. *TreeProjection* can be
viewed as a generic framework that advocates the notion of database projection, in
the context of several different strategies for constructing the enumeration tree, such
as a breadth-first, depth-first, or a combination of the two. The depth-first version,
described in detail in [4], also incorporates maximal pruning, though the disabling of
the pruning options can also materialize all the patterns. The breadth-first and depth-
first algorithms have different advantages. The former allows level-wise pruning
which is not possible in depth-first methods though it is often not used in projection-
based methods. The depth-first version allows better memory management. The
depth-first approach works best when the itemsets are very long, and it is desirable
to quickly discover maximal patterns, so that portions of the lexicographic tree can
be pruned off quickly during exploration and it can also be used for discovering
all patterns including non-maximal ones. When all patterns are required, including
non-maximal ones, the primary difference between different strategies is not one
of the size of the candidate space, but that of effective memory management of the
projected transactions. This is because the size of the candidate space is defined by
the size of the enumeration tree, which is fixed, and is agnostic to the strategy used for
tree exploration. On the other hand, memory management of projected transactions
is easier with the depth-first strategy because one only needs to maintain a small
number of projected transaction sets along the depth of the tree. The notion of
database projection is common to *TreeProjection* and *FP-growth*, and helps reduce
the counting work by restricting the size of the database used for support counting.
TreeProjection was developed independently from *FP-growth*. While the *FP-growth*
paper provides a brief discussion of *TreeProjection*, this chapter will provide a more
detailed discussion of the similarities and differences between the two methods. One
major difference between the two methods is that the internal representation of the
corresponding projected databases is different in the two cases.

The basic database projection approach is very similar in both cases of *TreeProjection* and *FP-growth*. An important observation is that if a transaction is not relevant
for counting at a given node in the enumeration tree, then it will not be relevant
for counting in any descendent of that node. Therefore, only those transactions are
retained that contain all items in P for counting at the node P in the projected transactions. Note that this set strictly reduces as we move to lower levels of the tree, and
the set of relevant transactions at the lower level of the enumeration tree is a subset of
the set at a higher level. Furthermore, only the presence of items corresponding to the
candidate extensions of a node are relevant for counting at any of the subtrees rooted

Fig. 2.6 Enumeration tree exploration

Algorithm *ExplorePrefix*(Database: $\mathcal{T}$, Minimum Support: s,
Current Pattern Prefix: P)
begin
Count support of 1-items in $\mathcal{T}$;
Remove infrequent items from $\mathcal{T}$;
  for each frequent item i in $\mathcal{T}$ **do**
  begin
    Append i to end of P and add to
      set of frequent patterns;
    Construct conditional database $\mathcal{T}_i$
      with all transactions in $\mathcal{T}$ containing item i;
    Remove items lexicographically $\leq i$ from $\mathcal{T}_i$;
    ExplorePrefix($\mathcal{T}_i$, s, $P \cup \{i\}$);
  end
end

at that node. Therefore, the database is also projected in terms of attributes, in which only items which are candidate extensions at a node are retained. The candidate set $F(P)$ of item extensions of node P is a very small subset of the universe of items at lower levels of the enumeration tree. In fact, even the items in the node P need not be retained explicitly in the transaction, because they are known to always be present in all the selected transactions based on the first condition. This projection process is performed recursively in top–down fashion down the enumeration tree for counting purposes, where lower level nodes inherit the projections from higher level nodes and add one additional item to the projection at each level. The idea of this inheritance-based approach is that the projected database remembers the counting work done at higher levels of the enumeration tree by (successively) removing irrelevant transactions and irrelevant items at each level of the projection. Such an approach works efficiently because it never repeats the counting work which has already been done at the higher levels. Thus, the primary savings in the strategy arise from avoiding repetitive and wasteful counting.

A bare-bones depth-first version of *TreeProjection*, that is similar to *DepthProject*, but without maximal pruning, is described in Fig. 2.6. A more detailed description with maximal pruning and other optimizations is provided later in this chapter. Because the algorithm is described recursively, the current prefix P (node of the lexicographic tree) being extended is one of the arguments to the algorithm. In the initial call, the value of P is *null* because one intends to determine all frequent descendants at the root of the lexicographic tree. This algorithm recursively extends frequent prefixes and maintains only the transaction database relevant to the prefix. The frequent prefixes are extended by determining the items i that are frequent in $\mathcal{T}$. Then the itemset $P \cup \{i\}$ is reported. The extension of the frequent prefix can be viewed as a recursive call at a node of the enumeration tree. Thus, at a given enumeration tree node, one now has a completely independent problem of extending the prefix with the projected database that is relevant to all descendants of that node. The conditional database $\mathcal{T}_i$ refers to the subset of the original transaction database $\mathcal{T}$ corresponding to transactions containing item i. Furthermore, the item i and any item occurring lexicographically earlier to it is not retained in the database because these items are not relevant to counting the extensions of $P \cup \{i\}$. This independent problem is similar in structure to the original problem, and can be solved recursively. Although it is natural to use recursion for the depth-first versions of *TreeProjection*, the breadth-first versions are not defined recursively. Nevertheless, the breadth-first versions explore a pattern space of the same size as the depth-first versions, and are no different either in terms of the tree size or the counting work done over the entire algorithm. The major challenge in the breadth-first version is in maintaining the projected transactions along the breadth of the tree, which is storage-intensive. It is shown in [5], how many of these issues can be resolved with the use of a combination of exploration strategies for tree growth and counting. Furthermore, it is also shown in [5] how breadth-first and depth-first methods may be combined.

Note that this concept of database projection is common between *TreeProjection* and *FP-growth* although there are some differences in the internal representation of the projected databases. The aforementioned description is designed for discovering all patterns, and does not incorporate maximal pattern pruning. When generating *all* the itemsets, the main advantage of the depth-first strategy over the breadth-first strategy is that it is less memory intensive. This is because one does not have to *simultaneously* handle the large number of candidates along the breadth of the enumeration tree at any point in the course of algorithm execution when combined with counting data structures. The overall size of the candidate space is fixed, and defined by the size of the enumeration tree. Therefore, over the entire execution of the algorithm, there is no difference between the two strategies in terms of search space size, beyond memory optimization.

Projection-based algorithms, such as *TreeProjection*, can be implemented either recursively or non-recursively. Depth-first variations of projection strategies, such as *DepthProject* and *FP-growth*, are generally implemented recursively in which a particular prefix (or suffix) of frequent items is grown recursively (see Fig. 2.6). For recursive variations, the structure and size of the recursion tree is the same as the enumeration tree. Non-recursive variations of *TreeProjection* methods directly present the projection-based algorithms in terms of the enumeration tree by storing projected transactions at the nodes in the enumeration tree. Describing projection strategies directly in terms of the enumeration tree is helpful, because one can use the enumeration tree explicitly to optimize the projection. For example, one does not need to project at every node of the enumeration tree, but project only when the size of the database reduces by a particular factor with respect to the nearest ancestor node where the last projection was stored. Such optimizations can reduce the space-overhead of repeated elements in the projected databases at different levels of the enumeration (recursion) tree. It has been shown how to use this optimization in different variations of *TreeProjection*. Furthermore, breadth-first variations of the strategy are naturally defined non-recursively in terms of the enumeration tree. The recursive depth-first versions may be viewed either as divide-and-conquer strategies (because they recursively solve a set of smaller subproblems), or as projection-based counting reuse strategies. The notion of projection-based counting reuse clearly describes how computational savings are achieved in both versions of the algorithm.

When generating *maximal* patterns, the depth-first strategy has clear advantages in terms of pruning as well. We refer the reader to a detailed description of the *DepthProject* algorithm, described later in this chapter. This description describes how several specialized pruning techniques are enabled by the depth-first strategy for maximal pattern mining. The *TreeProjection* algorithm has also been generalized to sequential pattern mining [31]. There are many different types of data structures that may be used in projection-style algorithms. The choice of data structure is sensitive to the data set. Two common choices that are used with *TreeProjection* family of algorithms are as follows:

1. *Arrays:* In this case, the projected database is maintained as 2-dimensional array. One of the dimensions of the array is equal to the number of relevant transactions and the other dimension is equal to the number of relevant items in the projected database. Both dimensions of the projected database reduce from top level to lower levels of the enumeration tree with successive projection.

2. *BitStrings:* In this case, the projected database is maintained as a 0–1 bit string whose width is fixed to the total number of frequent 1-items, but the number of projected transactions reduces with successive projection. Such an approach loses the power of item-wise projection, but this is balanced by the fact that the bit-strings can be used more efficiently for counting operations.

   Assume that each transaction T contains n bits, and can therefore be expressed in the form of $\lceil n/8 \rceil$ bytes. Each byte of the transaction contains the information about the presence or absence of eight items, and the integer value of the corresponding bitstring can take on any value from 0 to $2^8 - 1 = 255$. Correspondingly, for each byte of the (projected) transaction at a node, 256 counters are maintained and a value of 1 is added to the counter corresponding to the integer value of that transaction byte. This process is repeated for each transaction in the projected database at node P. Therefore, at the end of this process, one has $256 * \lceil d/8 \rceil$ counts for the d different items. At this point, a postprocessing phase is initiated in which the support of an item is determined by adding the counts of the $256/2 = 128$ counters which take on the value of 1 for that bit. Thus, the second phase requires $128 * d$ operations only, and is independent of database size. The first phase, (which is the bottleneck) is the improvement over the naive counting method because it performs only one operation for each *byte* in the transaction, which contains eight items. Thus, the method would be a factor of eight faster than the naive counting technique, which would need to scan the entire bitstring. Projection is also very efficient in the bitstring representation with simple AND operations.

The major problem with fixed width bitstrings is that they are not efficient representations at lower levels of the enumeration tree at which only a small number of items are relevant, and therefore most entries in these bitstrings are 0. One approach to speed this up is to perform the item-wise projection only at selected nodes in the tree, when the reduction in the number of items from the last ancestor at which the item-wise projection was performed is at particular multiplicative factor. At this point, a shorter bit string is used for representation for the descendants at that node,

	Item	tidlist
Table 2.2 Vertical representation of transactions. Note that the support of itemset *ab* can be computed as the length of the intersection of the *tidlists* of *a* and *b*	a	1, 2, 3, 5
	b	1, 2, 4, 5
	c	1, 2, 5
	d	1, 2, 5
	e	1, 4, 5
	f	2, 3, 4
	g	3, 4
	h	2, 5

until the width of the bitstring is reduced even further by the same multiplicative factor. This ensures that the bit strings representations are not sparse and wasteful.

The key issue here is that different representations provide different tradeoffs in terms of memory management and efficiency. Later in this chapter, an approach called *FP-growth* will be discussed which uses the trie data structure to achieve compression of projected transactions for better memory management.

3.3 Vertical Mining Algorithms

The vertical pattern mining algorithms use a vertical representation of the transaction database to enable more efficient counting. The basic idea of the vertical representation is that one can express the transaction database as an inverted list. In other words, for each transaction identifiers, one can have a list of items that are contained in it. This is referred to as a *tidset* or *tidlist*. An example of a vertical representation of the transactions in Table 2.1 is illustrated in Table 2.2.

The key idea in vertical pattern mining algorithms is that the support of k-patterns can be computed by intersection of the underlying *tidlists*. There are two different ways in which this can be done.

- The support of a k-itemset can be computed as a k-way set intersection of the lists of the individual items.
- The support of a k-itemset can be computed as an intersection of the *tidlists* two $(k-1)$-itemsets that join to that k-itemset.

The latter approach is more efficient. The credit for both the notion of vertical tidlists and the advantages of recursive intersection of tidlists is shared by the *Monet* [56] and the *Partition* algorithms [57]. Not all vertical pattern mining algorithms use an enumeration tree concept to describe the algorithm. Many of the algorithms directly use joins to generate a $(k+1)$-candidate pattern from a frequent k-pattern, though even a join-based algorithm, such as *Apriori*, can be explained in terms of an enumeration tree. Many of the later variations of vertical methods use an enumeration tree concept to explore the lattice of itemsets more carefully and realize the full power of the vertical approach. The indvidual ensemble component of Savasere et al.'s [57] *Partition* algorithm is the progenitor of all vertical pattern mining algorithms today, and the original *Eclat* algorithm is a memory-optimized and candidate partitioned version of this Apriori-like algorithm.

3.3.1   Eclat

Eclat uses a breadth-first approach like Savasere et al.'s algorithm [57] on lattice partitions, after partitioning the candidate set into disjoint groups, using a candidate partitioning approach similar to earlier parallel versions of the *Apriori* algorithm. The *Eclat* [71] algorithm is best described with the concept of an enumeration tree because of the wide variation in the different strategies used by the algorithm. An important contribution of *Eclat* [71] is to recognize the earlier pioneering work of the *Monet* and *Partition* algorithms [56, 57] on recursive intersection of tid lists, and propose many efficient variants of this paradigm.

Different variations of *Eclat* explore the candidates in different strategies. The earliest description of *Eclat* may be found in [74]. A journal paper exploring different aspects of *Eclat* may be found in [71]. In the earliest versions of the work [74], a breadth-first strategy is used. The journal version in [71] also presents experimental results for only the breadth-first strategy, although the possibility of a depth-first strategy is mentioned in the paper. Therefore, the original *Eclat* algorithm should be considered a breadth-first algorithm. More recent depth-first versions of *Eclat*, such as *dEclat*, use recursive *tidlist* intersection with differencing [72], and realize the full benefit of the depth-first approach. The *Eclat* algorithm, as presented in [74], uses a levelwise strategy in which all $(k + 1)$-candidates within a lattice partition are generated from frequent k-patterns in level-wise fashion, as in *Apriori*. The *tidlists* are used to perform support counting. The frequent patterns are determined from these *tidlists*. At this point, a new levelwise phase is initiated for frequent patterns of size $(k + 1)$.

Other variations and depth-first exploration strategies of *Eclat*, along with experimental results, are presented in later work such as *dEclat* [72]. The *dEclat* work in [72] presents some additional enhancements such as *diffsets* to improve counting. In this chapter, we present a simplified pseudo-code of this version of *Eclat*. The algorithm is presented in Fig. 2.8. The algorithm is structured as a recursive algorithm. A pattern set $\mathcal{FP}$ is part of the input, and is set to the set of all frequent 1-items at the top level call. Therefore, it may be assumed that, at the top level, the set of frequent 1-items and *tidlists* have already been computed, though this computation is not shown in the pseudocode. In each recursive call of *Eclat*, a new set of candidates $\mathcal{FP}_i$ is generated for every pattern (itemset) P_i, which extends the itemset by one unit. The support of a candidate is determined with the use of *tidlist* intersection. Finally, if P_i is frequent, it is added to a pattern set $\mathcal{FP}_i$ for the next level.

Figure 2.7 illustrates the itemset generation tree with support computation by *tidlist* intersection for the sample database from Table 2.1. The corresponding *tidlists* in the tree are also illustrated. All infrequent itemsets in each level are denoted by dotted, and bordered rectangles. For example, an itemset ab is generated by joining b to a. The *tidlist* of (a) is $\{1, 2, 3, 5\}$, and the *tidlist* of b is $\{1, 2, 4, 5\}$. We can determine the support of ab by intersecting the two *tidlists* to obtain the *tidlist* $\{1, 2, 5\}$ of these candidates. Therefore, the support of ab is given by the length of this *tidlist*, which is 3.

Further gains may be obtained with the use of the notion of *diffsets* [72]. This approach realizes the true power of vertical pattern mining. The basic idea, in *diffsets* is to maintain only the portion of the *tidlists* at a node, that correspond to the change in the inverted list from the parent node. Thus, the *tidlists* at a node can be reconstructed by examining the *tidlists* at the ancestors of a node in the tree. The major advantage

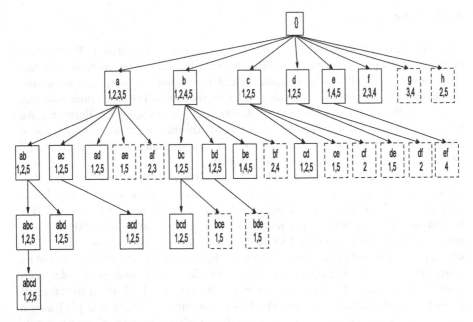

Fig. 2.7 Execution of *Eclat*

Fig. 2.8 The *Eclat* algorithm

Algorithm *Eclat*($\mathcal{FP}$, Support: s)
begin
  for each $P_i \in \mathcal{FP}$ **do**
  begin
    $\mathcal{FP}_i = \{\}$
    for each $P_j \in \mathcal{FP}$, such that $j > i$ **do**
    begin
      $P_{ij} = P_i \cup P_j$
      tidset(P_{ij}) = tidset(P_i) $\cap$ tidset(P_j)
      support(P_{ij}) = |tidset(P_{ij})|
      if (support(P_{ij})$\geq$ s)
        Add P_{ij} to $\mathcal{FP}_i$;
    end
    Eclat($\mathcal{FP}_i, s$)
  end
end

of *diffsets* is that they save significant storage in requirements in terms of the size of the data structure required (Fig. 2.8).

Fig. 2.9 Suffix-based pattern
exploration

Algorithm *SuffixGrowth*(Database of Frequent Items: $\mathcal{T}$,
                  Minimum Support: s, Current Pattern Suffix: P)
begin
  for each item i in $\mathcal{T}$ **do**
  begin
    Append i to beginning of P and add to
      set of frequent patterns;
    Let $\mathcal{T}_i$ be transactions of $\mathcal{T}$ containing i;
    Remove any item lexicographically $\geq i$ from $\mathcal{T}_i$;
    Remove infrequent items in $\mathcal{T}_i$;
    if $\mathcal{T}_i \neq \phi$ *SuffixGrowth*($\mathcal{T}_i$, s, $\{i\} \cup P$);
  end
end

3.3.2   VIPER

The *VIPER* algorithm [58] uses a vertical approach to mining frequent patterns.
The basic idea in the *VIPER* algorithm is ro represent the vertical database in the
form of compressed bit vectors that are also referred to as *snakes*. These snakes are
then used for efficient counting of the frequent patterns. The different compressed
representation of the *tidlists* provide a number of optimization advantages that are
leveraged by the algorithm. Intrinsically, *VIPER* is not very different from *Eclat*
in terms of the basic counting approach. The major difference is in terms of the
choice of the compressed bit vector representation, and the efficient handling of this
representation. Details may be found in [58].

4   Recursive Suffix-Based Growth

In these algorithms recursive suffix-based exploration of the patterns is performed.
Note that in most frequent pattern mining algorithms, the enumeration tree (execution
tree) of patterns explores the patterns in the form of a lexicographic tree of itemsets
built on the prefixes. Suffix-based methods use a different convention in which the
suffixes of frequent patterns are extended. As in all projection-based methods, one
only needs to use the transaction database containing itemset P in order to count
itemsets that have the suffix P. Itemsets are extended from the suffix backwards. In
each iteration, the conditional transaction database (or projected database) of transactions containing the current suffix P being explored is an input to the algorithm.
Furthermore, it is assumed that the conditional database contains only frequent extensions of P. For the top-level call, the value of P is null, and the frequent items are
determined using a single preprocessing pass that is not shown in the pseudo-code.
Because each item is already known to be frequent, the frequent patterns $\{i\} \cup P$
can be immediately generated for each item $i \in \mathcal{T}$. The database is projected further
to include only transactions containing i, and a recursive call is initiated with the
pattern $\{i\} \cup P$. The projected database $\mathcal{T}_i$ corresponding to transactions containing
$\{i\} \cup P$ is determined. Infrequent items are removed from $\mathcal{T}_i$. Thus, the transactions
are recursively projected to reflect the addition of an item in the suffix. Thus, this is a

smaller subproblem that can be solved recursively. The *FP-growth* approach uses the suffix-based pattern exploration, as illustrated in Fig. 2.9. In addition, the *FP-growth* approach uses an efficient data structure, known as the FP-Tree to represent the conditional transaction database $\mathcal{T}_i$ with the use of compressed *prefixes*. The *FP-Tree* will be discussed in more detail in a later section. The suffix in the top level call to the algorithm is the null itemset.

Recursive suffix-based exploration of the pattern space is, in principle, no different from prefix-based exploration of the enumeration tree space with the ordering of the items reversed. In other words, by using a reverse ordering of items, suffix-based recursive pattern space exploration can be simulated with prefix-based enumeration tree exploration. Indeed, as discussed in the last section, prefix-based enumeration tree methods order items from the least frequent to the most frequent, whereas the suffix-based methods of this section order items from the most frequent to the least frequent, to account for this difference. Thus, suffix-based recursive growth has an execution tree that is identical in structure to a prefix-based enumeration tree. This is a difference only of convention, but it does not affect the pattern space that is explored.

It is instructive to compare the suffix-based exploration with the pseudocode of the prefix-based *TreeProjection* algorithm in Fig. 2.6. The two pseudocodes are structured differently because the initial pre-processing pass of removing frequent items is not assumed in the *TreeProjection* algorithm. Therefore, in each recursive call of the prefix-based *TreeProjection*, frequent itemsets must be counted before they are reported. In suffix-based exploration, this step is done as a preprocessing step (for the top-level call) and just before the recursive call for deeper calls. Therefore, each recursive call always starts with a database of frequent items. This is, of course, a difference in terms of how the recursive calls are structured but is not different in terms of the basic search strategy, or the amount of overall computational work required, because infrequent items need to be removed in either case. A few other key differences are evident:

- *TreeProjection* uses database projections on top of a prefix-based enumeration tree. Suffix-based recursive methods have a recursion tree whose structure is similar to an enumeration tree on the frequent suffixes instead of the prefixes. The removal of infrequent items from $\mathcal{T}_i$ in *FP-growth* is similar to determining which branches of the enumeration tree to extend further.
- The use of suffix-based exploration is a difference only of convention from prefix-based exploration. For example, after reversing the item order, one might implement *FP-growth* by growing patterns on the prefixes, but constructing a compressed FP-Tree on[1] the suffixes. The resulting exploration order and execution in the two different implementations of *FP-growth* will be identical, but the latter can be more easily related to traditional enumeration tree methods.

---

[1] The resulting FP-Tree will be a suffix-based trie.

- Various database projection methods are different in terms of the specific data structures used for the projected database. The different variations of *TreeProjection* use arrays and bit strings to represent the projected database. The *FP-growth* method uses an FP-Tree. The FP-Tree will be discussed in the next section. Later variations of FP-Tree also use combinations of arrays and pointers to represent the projected database. Some variations, such as *OpportuneProject* [38], combine different data structures in an optimized way to obtain the best result.

- Suffix-based recursive growth is inherently defined as a depth-first strategy. On the other hand, as is evident from the discussion in [5], the specific choice of exploration strategy on the enumeration tree is orthogonal to the process of database projection. The overall size of the enumeration tree is the same, no matter how it is explored, unless maximal pattern pruning is used. Thus, *TreeProjection* explores a variety of strategies such as breadth-first and depth-first strategies, with no difference to the (overall) work required for counting. The major challenge with the breadth-first strategy is the simultaneous maintenance of projected transaction sets along the breadth of the tree. The issue of effective memory management of breadth-first strategies is discussed in [5], which shows how certain optimizations such as cache-blocking can improve the effectiveness in this case. Breadth-first strategies also allow certain kinds of pruning such as level-wise pruning.

- The major advantages of depth-first strategies arise in the context of maximal pattern mining. This is because a depth-first strategy discovers the maximal patterns very early, which can be used to prune the smaller non-maximal patterns. In this case, the size of the search space explored truly reduces because of a depth-first strategy. This issue is discussed in the section on maximal pattern mining. The advantages for maximal pattern mining were first proposed in the context of the *DepthProject* algorithm [4].

Next, we will describe the FP-Tree data structure that uses compressed representations of the transaction database for more efficient counting.

4.1   The FP-Growth Approach

The *FP-growth* approach combines suffix-based pattern exploration with a compressed representation of the projected database for more efficient counting. The prefix-based FP-Tree is a compressed representation of the database which is built by considering a fixed order among the items in an itemset [32]. This tree is used to represent the conditional transaction sets $\mathcal{T}$ and $\mathcal{T}_i$ of Fig. 2.9. An FP-Tree may be viewed as a prefix-based trie data structure of the transaction database of frequent items. Just as each node in a trie is labeled with a symbol, a node in the FP-Tree is labeled with an item. In addition, the node holds the support of the itemset defined by the items of the nodes that are on the path from the root to u. By consolidating the prefixes, one obtains compression. This is useful for effective memory management. On the other hand, the maintenance of counts and pointers with the prefixes is an

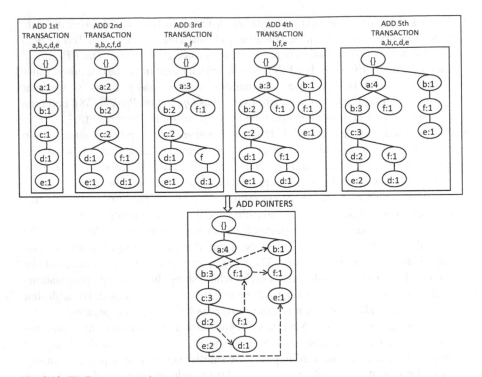

Fig. 2.10 FP-Tree construction

additional overhead. This results in a different set of trade-offs as compared to the array representation.

The initial FP-Tree is constructed as follows. We start with the empty FP-Tree FPT. Before constructing the FP-Tree, the database is scanned and infrequent items are removed. The frequent items are sorted in decreasing order of support. The initial construction of FP-Tree is straightforward, and similar to how one might insert a string in a trie. For every insertion, the counts of the relevant nodes that are affected by the insertion are incremented by 1. If there has been any sharing of prefix between the current transaction t being inserted, and a previously inserted transaction then t will be in the same path until the common prefix. Beyond this common prefix, new nodes are inserted in the tree for the remaining items in t, with support count initialized to 1. The above procedure ends when all transactions have been inserted.

To store the items in the final FP-Tree, a list structure called header table is maintained. A chain of pointers threads through the occurrence of the item in the FP-Tree. Thus, this chain of pointers need to be constructed in addition to the trie data structure. Each entry in this table stores the item label and pointers to the node representing the leftmost occurrence of the item in the FP-Tree (first item in the pointer chain). The reason for maintaining these pointers is that it is possible to determine the conditional FP-Tree for an item by chasing the pointers for that item. An example of the initial construction of the FP-Tree data structure from a

Fig. 2.11 The *FP-growth* algorithm

```
Algorithm FP-growth(FP-Tree on Frequent Items: FPT,
            Minimum Support; s, Current Itemset Suffix: P)
begin
    if FPT is a single path or empty
        for each combination C of nodes in path do
        report all patterns C ∪ P;
    else
    for each item i in FPT do
    begin
        Generate pattern Pᵢ = {i} ∪ P;
        report pattern Pᵢ as frequent;
        Use pointer-chasing to extract conditional
            prefix paths for item i;
        Construct conditional FP-Tree FPTᵢ from conditional
            prefix paths after removing infrequent items;
        if (FPTᵢ ≠ φ) FP-growth(FPTᵢ, Pᵢ, s)
    end
end
```

database of five transactions is illustrated in Fig. 2.10. The ordering of the items is a, b, c, d, e, f. It is clear that a trie data structure is created, and the node counts are updated by the insertion of each transaction in the FP-Tree. Figure 2.10 also shows all the pointers between the different items. The sum of the counts on the items on this pointer path is the support of the item. This support is always larger than the minimum support because a full constructed FP-Tree (with pointers) contains only frequent items. The actual counting of the support of item-extensions and the removal of infrequent items must be done during conditional transaction database (and the relevant FP-Tree) creation. The pointer paths are not available during the FP-Tree creation process. For example, the item e has two nodes on this pointer path, corresponding to $e : 2$ and $e : 1$. By summing up these counts, a total count of three for the item e is obtained. It is not difficult to verify that three transactions contain the item e.

With this new compressed representation of the conditional transaction database of frequent items, one can directly extract the frequent patterns. The pseudo-code of the *FP-growth* algorithm is presented in Fig. 2.11. Although this pseudo-code looks much more complex to understand than the earlier pseudocode of Fig. 2.9, the main difference is that more details of the data structure (FP-Tree), used to represent the conditional transaction sets, have been added.

The algorithm accepts a FP-Tree FPT, current itemset suffix P and user defined minimum support s as input. The additional suffix P has been added to the parameter set P to facilitate the recursive description. At the top level call made by the user, the value of P is ϕ. Furthermore, the conditional FP-Tree is constructed on a database of frequent items rather than all the items. This property is maintained across different recursive calls.

For an FP-Tree FPT, the conditional FP-Trees are built for each item i in FPT (which is already known to be frequent). The conditional FP-Trees are constructed by chasing pointers for each item in the FP-Tree. This yields all the conditional prefix

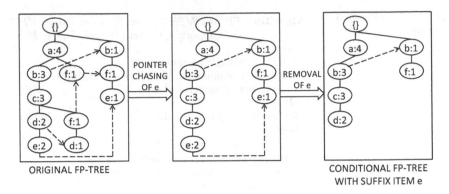

ORIGINAL FP-TREE   CONDITIONAL FP-TREE
                                                                   WITH SUFFIX ITEM e

Fig. 2.12 Generating a conditional FP-Tree by pointer chasing

paths for the item i. The infrequent nodes from these paths are removed, and they are put together to create a conditional FP-Tree FPT_i. Because the infrequent items have already been removed from FPT_i the new conditional FP-Tree also contains only frequent items. Therefore, in the next level recursive call, any item from FPT_i can be appended to P_i to generate another pattern. The supports of those patterns can also be reconstructed via pointer chasing during the process of reporting the patterns. Thus, the current pattern suffix P is extended with the frequent item i appended to the front of P. This extended suffix is denoted by P_i. The pattern P_i also needs to be reported as frequent. The resulting conditional FP-Tree FPT_i is the compressed database representation of T_i of Fig. 2.9 in the previous section. Thus, FPT_i is a smaller conditional tree that contains information relevant only to the extraction of various prefix paths relevant to different items that will extend the suffix P_i further in the backwards direction. Note that infrequent items are removed from FPT_i during this step, which requires the support counting of all items in FPT_i. Because the pointers have not yet been constructed for FPT_i, the support of each item-extension of $\{i\} \cup P$ corresponding to the items in FPT_i must be explicitly determined by locating each instance of an item in FPT_i. This is the primary computational bottleneck step. The removal of infrequent items from FPT_i may result in a different structure of the FP-Tree in the next step.

Finally, if the conditional FP-Tree FPT_i is not empty, the *FP-growth* method is called recursively with parameters corresponding to the conditional FP-Tree FPT_i, extended suffix P_i, and minimum support s. Note that successive projected trans-action databases (and corresponding conditional FP-Trees) in the recursion will be smaller because of the recursive projection. The base case of the recursion occurs when the entire FP-Tree is a single path. This is likely to occur when the projected transaction database becomes small enough. In that case, *FP-growth* determines all combinations of nodes on this path, appends the suffix P to them, and reports them.

An example of how the conditional FP-Tree is created for a minimum support of 1 unit, is illustrated in Fig. 2.12. Note that if the minimum support were 2, then the right branch (nodes b and f) would not be included in the conditional FP-Tree. In this case, the pointers for item e are chased in the FP-Tree to create the conditional prefix paths of the relevant conditional transaction database. This represents all transactions containing e. The counts on the prefix paths are re-adjusted because many branches are pruned. The removal of infrequent items and that of the item e might lead to a conditional FP-Tree that looks very different from the conditional prefix-paths. These kinds of conditional FP-trees need to be generated for each conditional frequent item, although only a single item has been shown for the sake of simplicity. Note that, in general, the pointers may need to be recreated every time a conditional FP-Tree is created.

4.2   Variations

As the database grows larger, the construction of the FP-Tree become challenging both from runtime and space complexity. There have been many works [8, 24, 27, 29, 30, 36, 39, 55, 59, 61, 62] to tackle these challenges. These variations of *FP-growth* method can be classified into two categories. Methods belonging to the first category design memory-based mining process using a memory-resident data structure that holds partitioned database. Methods belonging to the second category improve the efficiency of the FP-Tree representation. In this subsection, we will present these approaches briefly.

4.2.1   Memory-Resident Variations

In the following, a number of different memory-resident variations of the basic *FP-growth* idea will be described.

CT-PRO Algorithm   In this work [62], the authors introduced a new FP-Tree like data structure called Compact FP-Tree (CFP-Tree) that holds the same information as FP-Tree but with 50 % less storage. They also designed a mining algorithm called CT-PRO which follows a non-recursive procedure unlike *FP-growth*. As discussed earlier, during the mining process, *FP-growth* constructs many conditional FP-Trees, which becomes an overhead as the patterns get longer or the support gets lower. To overcome this problem, the *CT-PRO* algorithm divides the database into several disjoint projections where each projection is represented as a CFP-Tree. Then a non-recursive mining process is executed over each projection independently. Significant modifications were made to the header Table 4.1 data structure. In the original FP-Tree, the nodes store the support and item label. However, in the CFP-Tree, item labels are mapped to an increasing sequence of integers that is actually the index of the header table. The header table of CFP-Tree stores the support of each item. To compress the original FP-Tree, all identical subtrees are removed by accumulating them and storing the relevant information in the leftmost branch. The header table contains a pointer to each node on the leftmost branch of the CFP-Tree, as these nodes are roots of subtrees starting with different items.

The mining process starts from the pointers of the least frequent items in the header table. This prunes a large number of nodes at an early stage and shrinks the tree structure. By following the pointers to the same item, a projection of all transactions ending with the corresponding item is built. This projection is also represented as a CFP-Tree called local CFP-Tree. The local CFP-Tree is then traversed to extract the frequent patterns in the projection.

H-Mine Algorithm The authors in [54] proposed an efficient algorithm called *H-Mine*. It uses a memory efficient hyper-structure called H-Struct. The fundamental strategy of *H-Mine* is to partition the database and mine each partition in the memory. Finally, the results from different partitions are consolidated into global frequent patterns. An intelligent module of *H-Mine* is that it can identify whether the database is dense or sparse, and it is able to make dynamic choices between different data structures based on this identification. More details may be found in Chap. 3 on pattern-growth methods.

4.2.2 Improved Data Structure Variations

In this section, several variations of the basic algorithm by improving the underlying data structure will be described.

Using Arrays A significant part of the mining time in *FP-growth* is spent on travers-ing the tree. To reduce this time, the authors in [29] designed an array based implementation of *FP-growth*, named *FP-growth** which drastically reduces the traversal time of the mining algorithm. It uses the FP-Tree data structure in com-bination with an array-like data structure and it incorporates various optimization schemes. It should be pointed out that the *TreeProjection* family of algorithms also uses arrays, though the optimizations used are quite different.

When the input database is sparse, the array based technique performs well be-cause the array saves the traversal time for all the items; moreover the initialization of the next level of FP-Trees is easier using an array. But in case of dense database, the tree base representation is more compact. To deal with the situation, *FP-growth** devises a mechanism to identify whether the database is sparse or not. To do so, *FP-growth** counts the number of nodes in each level of the tree. Based on experiments, they found that if the upper quarter of the tree contains less than 15% of the total number of nodes, then the database is most likely dense. Otherwise, it is sparse. If the database turns out to be sparse, *FP-growth** allocates an array for each FP-Tree in the next level of mining.

The nonordfp Approach This work [55] presented an improved implementation of the well known *FP-growth* algorithm using an efficient FP-Tree like data structure that allows faster allocation, traversal and optional projection. The tree nodes do not store their labels (item identifiers). There is no concept of header table. The data structure stores less administrative information in the tree node which allow the recursive step of mining without rebuilding the tree.

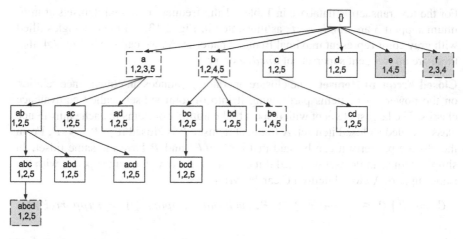

Fig. 2.13 Frequent, maximal and closed itemsets

5   Maximal and Closed Frequent Itemsets

One of the major challenges of frequent itemset mining is that, most of the itemsets mined are subset of the set of single length frequent items. Therefore, a significant amount of time is spent on counting redundant itemsets. One solution to this problem is to discover condensed representations of the frequent itemsets. It will be such representations that synopsizes the property of the set of itemsets completely or partially. The compact representation not only save computational and memory resource but also paved a much easier way towards knowledge discovery stage after mining. Another interesting observation by [53] was that, instead of mining the complete set of frequent itemsets and their associations, association mining only needs to find frequent closed itemsets and their corresponding rules. So, mining frequent closed itemset can fulfill the objectives of mining all frequent itemsets but with less redundancy and better efficiency and effectiveness in mining. In this section, we will discuss two types of condensed representation of itemset: maximal and closed frequent itemset.

5.1   Definitions

Maximal Frequent Itemset   Suppose, $\mathcal{T}$ is the transaction database, $\mathcal{I}$ is the set of all items in the database and $\mathcal{F}$ is the set of all frequent itemsets. A frequent itemset $P \in \mathcal{F}$ is called maximal if it has no frequent superset. let $\mathcal{M}$ be the set of all frequent maximal itemsets, which is denoted by

$$\mathcal{M} = \{P \mid P \in \mathcal{F} \text{ and } \nexists Q \supset P, \text{ such that } Q \in \mathcal{F}\}$$

For the toy transaction database in Table 2.1 the frequent maximal itemsets at minimum support 3 are $abcd$, e, f, as illustrated in Fig. 2.13. All the rectangles filled with grey color represent maximal frequent patterns. As we can see in Fig. 2.4, that there are no frequent supersets of $abcd$, e or f.

Closed Frequent Itemset The closure operator γ induces an equivalence relation on the power set of items partitioning it into disjoint subsets called equivalence classes. The largest element with respect to the number of items in each equivalence class is called a closed itemset. A frequent itemset P is closed if $\gamma(P) = P$. From the closure property it can be said that both $\gamma(P)$ and P have the same tidset. In simpler terms, an itemset is closed if it does not have any frequent superset with the same support. A closed itemset C can be written as:

$$C = \{P \mid P \in \mathcal{F} \text{ and } \nexists Q \supset P, \text{ such that } support(Q) = support(P)\}$$

Because maximal itemsets have no frequent superset, they are vacuously closed frequent itemsets. Thus, all maximal patterns are closed. However, there is a key difference between mining maximal itemsets and closed itemsets. Mining maximal itemsets loses information about the support of the underlying itemsets. On the other hand, mining closed itemsets does not lose any information about the support. The support of the missing subsets can be derived from the closed frequent pattern database. One way of viewing closed frequent patterns is as the maximal patterns from each *equi-support* group of frequent patterns. Closed frequent itemsets are a condensed representation of frequent itemsets that is lossless.

For the toy transaction database of Table 2.1 the frequent closed patterns are a, b, $abcd$, be for minimum support value of 3, as illustrated in Fig. 2.13. All the rectangles with dotted border represent closed frequent patterns. The remaining nodes in the tree (not filled and dotted border) represent frequent itemsets.

5.2 Frequent Maximal Itemset Mining Algorithms

In this subsection, we will discuss some of maximal frequent itemset mining algorithms.

5.2.1 MaxMiner Algorithm

The *MaxMiner* algorithm was the first algorithm that used a variety of optimizations to improve the effectiveness of tree explorations [10]. This algorithm is generally focussed on determining maximal patterns rather than all patterns. The author of [10] observed that it is usually sufficient to only report maximal patterns, when frequent patterns are long. This is because of the combinatorial explosion in examining all subsets of patterns. Although the exploration of the tree is still done in breadth-first fashion, a number of optimizations are used to improve the efficiency of exploration:

- The concept of *lookaheads* is defined. Let $F(P)$ be the set of candidate items that might extend node P. Before counting, it is checked whether $F \cup F(P)$ is a subset of any of the frequent patterns found so far. If such is indeed the case, then it is known that the entire subtree rooted at P is frequent, and can be pruned from consideration (for maximal pattern mining). During counting the support of individual item extensions of P, the support of $P \cup F(P)$ is also determined. If the set $P \cup F(P)$ is frequent, then it is known that all itemsets in the entire subset rooted at that node are frequent. Therefore, the tree does not need to be explored further, and can be pruned.
- The support lower bounding trick discussed earlier can be used to quickly determine patterns which are frequent without explicit counting. The counts of extensions of nodes can be determined without counting in many cases, where the count does not change by extending an item.

It has been shown in [10], that these simple optimizations can improve over the original *Apriori* algorithm by orders of magnitude.

5.2.2 DepthProject Algorithm

The *DepthProject* algorithm is based on the notion of the lexicographic tree, defined in [5]. Unlike *TreeProjection*, the approach aggressively explores the candidates in a depth-first strategy both to ensure better pruning and faster counting. As in *TreeProjection*, the database is recursively projected down the lexicographic tree to ensure more efficient counting. This kind of projection ensures that the counting information for k-candidates is reused for $(k + 1)$-candidates, as in the case of *FP-growth*.

For the case of the *DepthProject* method [4], the lexicographic tree is explored in depth-first order to maximize the advantage of lookaheads in which entire subtrees can be pruned because it is known that all patterns in them are frequent. The overall pseudocode for the depth-first strategy is illustrated in Fig. 2.14. The pseudocodes for candidate generation and counting are not provided because they are similar to the previously discussed algorithms. However, one important distinction in counting is that projected databases are used for counting. This is similar to the *FP-growth* class of algorithms. Note that the recursive transaction projection is particularly effective with a depth-first strategy because a smaller number of projected databases need to be stored along a path in the tree, as compared to the breadth of the tree.

To reduce the overhead of counting long patterns, the notion of lookaheads are used. At any node P of the tree, let $F(P)$ be its possible (candidate) item extensions. Then, it is checked whether $P \cup F(P)$ is frequent in two ways:

1. Before counting the support of the individual extensions of P (i.e., $\{P \cup \{i\} : \forall i \in F(P)\}$), it is checked whether $P \cup F(P)$ occurs as subset of a frequent itemset that has already been discovered earlier during depth-first exploration. If such is the case, then the entire subtree rooted at P is pruned because it is known

to be frequent and it is not a maximal pattern. This type of pruning is particularly effective with a depth-first strategy.

2. During support counting of the item extensions, the support of $P \cup F(P)$ is also determined. If after support counting, $P \cup F(P)$ turns out to be frequent, then the entire subtree rooted at node P can be pruned. Note that the projected database at node P (as in *TreeProjection*) is used.

Although lookaheads are also used in the *MaxMiner* algorithm, it should be pointed out that the effectiveness of lookaheads is maximized with a depth-first strategy. This is true of the first of the two aforementioned strategies, in which it is checked whether $P \cup F(P)$ is a subset of an already existing frequent pattern. This is a because a depth-first strategy tends to explore the itemsets in dictionary order. In dictionary order, maximal itemsets are usually explored much earlier than *most* of their subsets. For example, for a 10-itemset $abcdefghij$, only 9 of the 1024 subsets of the itemsets will be explored before exploring the itemset $abscdefghij$. These 9 itemsets are the immediate prefixes of the itemset. When, the longer itemsets are explored early they become available to prune shorter itemsets.

The following information is stored at each node during the process of construction of the lexicographic tree:

1. The itemset P at that node.
2. The set of lexicographic tree extensions at that node which are $E(P)$.
3. A pointer to the projected transaction set $T(Q)$, where Q is some ancestor of P (including itself). The root of the tree points to the entire transaction database.
4. A bitvector containing the information about which transactions contain the itemset for node P as a subset. The length of this bitvector is equal to the total number of transactions in $T(Q)$. The value of a bit for a transaction is equal to one, if the itemset P is a subset of the transaction. Otherwise it is equal to zero. Thus, the number of 1 bits is equal to the number of transactions in $T(Q)$ which project to P. The bitvectors are used to make the process of support counting more efficient.

After all the projected transactions at a given node have been identified, then finding the subtree rooted at that node is a completely independent itemset generation problem with a *substantially reduced* transaction set. The number of transactions at a node is proportional to the support at that node.

The description in Fig. 2.14 shows how the depth first creation of the lexicographic tree is performed. The algorithm is described recursively, so that the call from each node is a completely independent itemset generation problem that finds all frequent itemsets that are descendants of a node. There are three parameters to the algorithm, a pointer to the database T, the itemset node N, and the bitvector B. The bitvector B contains one bit for each transaction in $T \in T$, and indicates whether or not the transaction T should be used in finding the frequent extensions of N. A bit for a transaction T is one, if the itemset at that node is a subset of the corresponding transaction. The first call to the algorithm is from the *null* node, the parameter T is the entire transaction database. Because each transaction in the database is relevant to perform the counting, the bitvector B consists of all "one" values. One property

Fig. 2.14 The depth first strategy

Algorithm *DepthFirst(Itemset Node: N,*
　　　　PointerToDatabase: $\mathcal{T}$, Bitvector : B)
　begin
　$C = GenerateCandidates(N)$;
　$E = Count(N, \mathcal{T}, B, C)$;
　{ Let $E = \{i_1, \ldots, i_{|E|}\}$, in lexicographic order }
　Store frequent itemsets $N \cup \{i_r\}$ for $r \in \{1, \ldots, |E|\}$;
　$B' = CreateBitvector(N, B, \mathcal{T})$;
　if *(ProjectionCondition)* **then**
　　begin
　　$\mathcal{T}' = Project(\mathcal{T}, E, N, B')$;
　　Modify B' to be a set of $|\mathcal{T}'|$ ones;
　　end;
　　　　else $\mathcal{T}' = \mathcal{T}$;
　for $r := 1$ to $|E|$ **do** $DepthFirst(N \cup \{i_r\}, \mathcal{T}', B')$;
　end

Subroutine *Project(Database: $\mathcal{T}$,*
　　　　FrequentExtensions : E, Bitvector: B)
　begin
　$\mathcal{T}' =$ Empty set of transactions;
　for each transaction $T \in \mathcal{T}$ **do**
　　begin
　　if corresponding bit in B is 1 **then** add $T \cap E$ to $\mathcal{T}'$;
　　end
　return$(\mathcal{T}')$;
　end

Subroutine *CreateBitvector(N, B, $\mathcal{T}$)*
　begin
　Initialize $B' = B$;
　Let n be the lexicographically largest item in N;
　for each transaction $T \in \mathcal{T}$ **do**
　　if $n \notin T$ **then** set the corresponding bit in B' to 0;
　return(B');
　end

of the *DepthProject* algorithm is that the projection is performed only when the transaction database reduces by a certain size. This is the *ProjectionCondition* in Fig. 2.14.

Most of the nodes in the lexicographic tree correspond to the lower levels. Thus, the counting times at these levels account for most of the CPU times of the algorithm. For these levels, a strategy called bucketing can substantially improve the counting times. The idea is to change the counting technique at a node in the lexicographic tree, if $|E(P)|$ is less than a certain value. In this case, an upper bound on the number of distinct *projected* transactions is $2^{|E(P)|}$. Thus, for example, when $|E(P)|$ is nine, then there are only 512 distinct projected transactions at the node P. Clearly, this is because the projected database contains several repetitions of the same (projected)

Fig. 2.15 Aggregating bucket
counts

Algorithm AggregateCounts(*Counts:bucket*[...])
begin
   { We assume that there are $2^{|E(P)|}$ buckets, one
      corresponding to each bitstring of length $|E(P)|$ }
   $k = |E(P)|$;
   for $i := 1$ **to** k **do**
      begin
      for $j := 1$ **to** 2^k **do**
      if the ith bit of bitstring representation
         of j is 0 **then**
         begin
         $bucket[j] = bucket[j] + bucket[j + 2^{i-1}]$;
         end
      end
end

transaction. The fact that the number of *distinct* transactions in the projected database is small can be exploited to yield substantially more efficient counting algorithms. The aim is to count the support for the entire subtree rooted at P with a quick pass through the data, and an additional postprocessing phase which is independent of database size. The process of performing bucket counting consists of two phases:

1. In the first phase, the counts of each distinct transaction present in the projected database are determined. This can be accomplished easily by maintaining $2^{|E(P)|}$ buckets or counters, scanning the transactions one by one, and adding counts to the buckets. The time for performing this set of operations is linear in the number of (projected) database transactions.
2. In the second phase, the counts of the $2^{|E(P)|}$ transaction are used to determine the aggregate support counts for each itemset. In general, the support count of an itemset may be obtained by adding the counts of all the supersets of that itemset to it. A skillful algorithm (from the efficiency perspective) for performing these operations is illustrated in Fig. 2.15.

Consider a string composed of 0, 1, and ∗ that refers to an itemset in which the positions with 0 and 1 are fixed to those values (corresponding to presence or absence of items), while a position with a ∗ is a "don't care". Thus, all itemsets can be expressed in terms of 1 and ∗ because itemsets are traditionally defined with respect to presence of items. Consider for example, the case when $|E(P)| = 4$, and there are four items, numbered $\{1, 2, 3, 4\}$. An itemset containing items 2 and 4 is denoted by ∗1∗1. We start off with the information on $2^4 = 16$ bitstrings which are composed of 0 and 1. These represent all possible distinct transactions. The algorithm aggregates the counts in $|E(P)|$ iterations. The count for a string with a "∗" in a particular position may be obtained by adding the counts for the strings with a 0 and 1 in those positions. For example, the count for the string ∗1∗1 may be expressed as the sum of the counts of the strings 01∗1 and 11∗1.

Fig. 2.16 Performing the second phase of bucketing

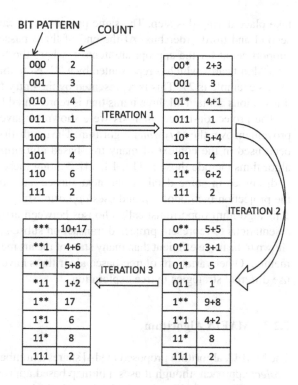

The procedure in Fig. 2.15 works by starting with the counts of the 0–1 strings, and then converts them to strings with 1 and *. The algorithm requires $|E(P)|$ iterations. In the ith iteration, it increases the counts of all those buckets with a 0 in the ith bit, so that the count now corresponds to a case when that bucket contains a * in that position. This can be achieved by adding the counts of the buckets with a 0 in the ith position to that of the bucket with a 1 in that position, with all other bits having the same value. For example, the count of the string 0*1* is obtained by adding the counts of the buckets 001* and 011*. In Fig. 2.15, the process of adding the count of the bucket j to that of the bucket $j + 2^{i-1}$ achieves this.

The second phase of the bucketing operation requires $|E(P)|$ iterations, and each iteration requires $2^{|E(P)|}$ operations. Therefore, the total time required by the method is proportional to $2^{|E(P)|} \cdot |E(P)|$. When $|E(P)|$ is sufficiently small, the time required by the second phase of postprocessing is small compared to the first phase, whereas the first phase is essentially proportional to reading the database for the current projection.

We have illustrated the second phase of bucketing by an example in which $|E(P)| = 3$. The process illustrated in Fig. 2.16 illustrates how the second phase of bucketing is efficiently performed. The exact strings and the corresponding counts in each of the $|E(P)| = 3$ iterations are illustrated. In the first iteration, all those bits with 0 in the lowest order position have their counts added with the count of the bitstring with a 1 in that position. Thus, $2^{|E(P)|-1}$ pairwise addition operations take place during this step. The same process is repeated two more times with the second and third order bits. At the end of three passes, each bucket contains the support count for the appropriate itemset, where the '0' for the itemset is replaced by a "don't care" which is represented by a '*'. Note that the number of transactions in this example is 27. This is represented by the entry for the bucket ***. Only two transactions contain all three items that is represented by the bucket 111.

The projection-based methods were shown to have an order of magnitude improvement over the *MaxMiner* algorithm. The depth-first approach has subsequently been used in the context of many tree-based algorithms. Other examples of such algorithms include those in [17, 18, 14]. Among these, the MAFIA algorithm [14] is discussed in some detail in the next subsection. An approach which varies on the projection methodology, and uses *opportunistic projection* is discussed in [38]. This algorithm opportunistically chooses between array-based and tree-based representations to represent projected transaction subsets. Such an approach has been shown to be more efficient than many state of the art methods such as the FP-Growth method. Other variations of tree-based algorithms have also been proposed [70] that use different strategies in tree exploration.

5.2.3 MAFIA Algorithm

The MAFIA algorithm proposed in [14] shares a number of similarities to the *Depth-Project* approach, though it uses a bitmap based approach for counting, rather than the use of a projected transaction database. In the bitmap-based approach, a sequence of bits is maintained for each itemset that corresponds to whether or not that transaction contains that particular item. Sparse representations (such as a list of transaction identifiers) may also be used, when the fraction of transactions containing the itemset is small. Note that such an approach may be considered a special case of database projection [5], in which vertical projection is used but horizontal projection is not. This has the advantage of requiring less memory, but it reuses a smaller fraction of the counting information from higher level nodes. A number of other pruning optimizations have also been proposed in this work that further improve the effectiveness of the algorithm. In particular, it has been pointed out that when the support of the extension of a node is the same as that of its parent, then that subtree can be pruned away, because of the counts of all the itemsets in the subtree can be derived from those of other itemsets in the data. This is the same as the support lower bounding trick discussed in Sect. 2.4, and also used in *MaxMiner* for pruning. Thus, the approach in [14] uses many of the same strategies used in *MaxMiner* and *TreeProjection*, but with in a different combination, and with some variations on specific implementation details.

5.2.4 GenMax

Like MAFIA, *GenMax* is a uses the vertical representation to speed up counting. Specifically the *tidlists* are used by *GenMax* to speed up the counting approach. In particular the more recent notion of *diffsets* [72] was used, and a depth-first exploration strategy was used. An approach known as successive focussing was used to further improve the efficiency of the algorithm. The details of the *GenMax* approach may be found in [28].

5.3 Frequent Closed Itemset Mining Algorithms

The are several frequent closed itemset mining algorithms [41, 42, 51–53, 64, 66–69, 73] exist to date. Most of the maximal and closed pattern mining algorithms are based on different variations of the non-maximal pattern mining algorithms. Typically pruning strategies are incorporated within the non-maximal pattern mining algorithms to yield more efficient algorithms.

5.3.1 Close

In this algorithm [52] authors apply *Apriori* based patten generation over the closed itemset search space. The usages of closed itemset lattice (search space) significantly reduces the overall search space of the algorithm. *Close* operates in iterative manner. Each iteration consists of three phases, . First, the closure function is applied for obtaining the candidate closed itemsets and their support. Next, the obtained set of candidate closed itemsets are tested against the minimum support constraint. If succeed, the candidates are marked as frequent closed itemset. Finally the same procedure is initiated to generate the next level of candidate closed itemsets. This process continues until all frequent closed itemsets have been generated.

5.3.2 CHARM

CHARM [73] is a frequent closed itemset mining algorithm, that takes advantage of the vertical representation of database as in the case of *Eclat* [71] for efficient closure checking operation. For punning the search space *CHARM* uses the following three properties. Suppose for itemset P and Q, if tidset(P) = tidset(Q), then it replaces every occurrence of P by $P \cup Q$ and prune the whole branch under Q. On the other hand if tidset$(P) \subset$ tidset(Q), it replaces every occurrence of P by $P \cup Q$, but does not prune the branch under Q. Finally if, tidset(P)<>tidset(Q), none of the aforementioned prunings can be applied. The initial call of *CHARM* accepts a set(I) of single length frequent item and minimum support as input. As a first step, it sorts I by the increasing the order of support of the items. For each item P,

CHARM tries to extend it by another item Q from the same set and applies three conditions for pruning. If the newly create itemset by extension is frequent, *CHARM* performs closure-checking to identify whether the itemset is closed. *CHARM* also updates the set I accordingly. In other words, it replaces P with $P \cup Q$, if the corresponding pruning condition is met. If the set I is the not empty, then *CHARM* is called recursively.

5.3.3   CLOSET and CLOSET+

CLOSET [53] and *CLOSET*+ [69] frequent closed itemset mining algorithms are inspired by the *FP-growth* method. The *CLOSET* algorithm makes use of the principles of the FP-Tree data structure to avoid the candidate generation step during the process of mining frequent closed itemsets. This work introduces a technique, referred to as single prefix path compression, that quickly assists the mining process. *CLOSET* also applies partition-based projection mechanisms for better scalability. The mining procedure of *CLOSET* follows the *FP-growth* algorithm. However, the algorithm is able to extract only the closed patterns by careful book-keeping. *CLOSET* treats items appearing in every transaction of the conditional database specially. For example, if Q is the set of items that appear in every transaction of the P conditional database then $P \cup Q$ creates a frequent closed itemset if it is not a proper subset of any frequent closed itemset with the equal support. *CLOSET* also prunes the search space. For example, if P and Q are frequent itemset with the equal support where Q is also a closed itemset and $P \subset Q$, then it does not mine the conditional database of P because the latter will not produce any frequent closed itemsets.

   CLOSET+ is a follow-up work after *CLOSET* by the same group of authors. *CLOSET*+ attempts to design the most optimized frequent closed itemset mining algorithm by finding the best trade-off between depth-first search versus breadth-first search, vertical formats versus horizontal formats, tree structure versus other data structures, top–down versus bottom–up traversal, and pseudo projection versus physical projection of the conditional database. *CLOSET*+ keeps track of the unpromising prefix itemsets for generating potential closed frequent itemsets and prunes the search space by deleting them. *CLOSET*+ also applies "item merging," and "sub-itemset" based pruning. To save the memory of the closure checking operation, *CLOSET*+ uses the combination of the 2-level hash-indexed tree based method and the pseudo-projection based upward checking method. Interested readers are encouraged to refer to [69] for more details.

5.3.4   DCI_CLOSED

DCI_CLOSED [41, 42] uses a bitwise vertical representation of the input database. *DCI_CLOSED* can be executed independently on each partition of the database in any order and, thus, also in parallel. *DCI_CLOSED* is designed to improve memory-efficiency by avoiding the storage of duplicate closed itemsets. *DCI_CLOSED* designs a novel strategy for searching the lattice that can detect and discard duplicate closed patterns on the fly. Using the concept of order-preserving generators

of frequent closed itemsets, a new visitation scheme of the search space is introduced. Such a visitation scheme results a disjoint sub division of the search space.
This also facilitates parallelism.*DCI_CLOSED* applies several optimization tricks
to improve execution time, such as the bitwise intersection of tidsets to compute
support and closure. Where possible, it reuses previously computed intersections to
avoid redundant computations.

6   Other Optimizations and Variations

In this section, a number of other optimizations and variations of frequent pattern
mining algorithms will be discussed. Many of these methods are discussed in detail
in other chapters of this book, and therefore they will be discussed only briefly here.

6.1   *Row Enumeration Methods*

Not all frequent pattern mining algorithms follow the fundamental steps of baseline
algorithm, there exists a number of special cases, for which specialized frequent
pattern mining algorithms have been designed. An interesting case is that of microarray data sets, in which the columns are very long but the number of rows are not
very large. In such cases, a method called *row-enumeration* is used [22, 23, 40, 48,
49] instead of the usual column enumeration, in which combinations of rows are
examined during the search process. There are two categories of row enumeration
algorithm. One category algorithm perform bottom-up [22, 23, 48] search over
the row enumeration tree whereas other category algorithms perform top-down[40]
search strategy.

Row enumeration algorithms perform mining over the transpose of the transaction
database. In transpose database, each transaction id become item and each item corresponds a transaction. Mining over the transposed database is basically the bottom
up search for frequent patterns by enumeration of row sets. However, the bottom-up
search strategy cannot take advantage of user-specified minimum support threshold
to effectively prune the search space, and therefore leads to longer running time
and large memory overhead. As a solution [40] introduce a top-down approach of
mining using a novel row enumeration tree. Their approach can take full advantage
of user-defined minimum support value and prune the search space efficiently hence
lower down the execution time.

Note that, both of the search strategies are applied over the transposed transaction
database. Most of developed algorithm using row enumeration technique concentrate
on mining frequent closed itemset (explained in Sect. 5). The reason behind this
motivation is that due to the nature of micro-array data there exists a large number
of redundancy among the frequent patterns for a minimum support threshold and
closed patterns are capable of summarizing the whole database. These strategies will
be discussed in detail in Chap. 4, and therefore only a brief discussion is provided
here.

6.2 Other Exploration Strategies

The advantage of tree-enumeration strategies is that they facilitate the exploration of candidates in the tree in an arbitrary order. A method known as *Pincer-Search* is proposed in [37] that combines top-down and bottom-up exploration in "pincer" fashion to avail of the advantages of both subset and superset pruning. Two primary observations are used in pincer search:

1. Any subset of a frequent itemset is frequent.
2. Any superset of an infrequent itemset is infrequent.

In pincer-search, top–down and bottom–up exploration are combined and irrelevant itemsets are pruned using both observations. More details of this approach are discussed in [37]. Note that, for sparse transaction data, superset pruning is likely to be inefficient. Other recent methods have been proposed for long pattern mining with methods such as "leap search." These methods are discussed in the chapter on long pattern mining in this book.

7 Reducing the Number of Passes

A major challenge in frequent pattern mining is when the data is disk resident. In such cases, it is desirable to use level-wise methods to ensure that random accesses to disk are minimized. This is the reason that most of the available algorithms use level-wise methods, which ensure that the number of passes over the database are bounded by the size of the longest pattern. Even so, this can be significant, when many long patterns are present in the database. Therefore, a number of methods have been proposed in the literature to reduce the number of passes over the data. These methods could be used in the context of join-based algorithms, tree-based algorithms, or even other classes of frequent pattern mining methods. These correspond to combining the level-wise database passes, using sampling, and using a preprocess-once-query-many paradigm.

7.1 Combining Passes

The earliest work on combining passes was proposed in the original *Apriori* algorithm [1]. The key idea in combing passes is that it is possible to use joins to create candidates of higher order than $(k + 1)$ in a single pass. For example, $(k + 2)$-candidates can be created from $(k + 1)$-candidates before actual validation of the $(k + 1)$-candidates over the data. Then, the candidates of size $(k + 1)$ and $(k + 2)$ can be validated together in a single pass over the data. Although such an approach reduces the number of passes over the data, it has the downside that the number of spurious $(k + 2)$ candidates will be far larger because the $(k + 1)$ candidates were not confirmed to be frequent before they were joined. Therefore, the saving of database passes comes at an increased computational cost. Therefore, it was proposed in [1] that the approach should be used for later passes, when the number of candidates has already reduced significantly. This reduces the likelihood that the number of candidates blows up too much with this approach.

7.2 Sampling Tricks

A number of sampling tricks can be used to greatly improve the efficiency of the frequent pattern mining process. Most sampling methods require two passes over the data, the first of which is used for sampling. An interesting approach that uses two passes with the use of sampling is discussed in [65]. This method generates the approximately frequent patterns over the data, using a sample. False negatives can be reduced by lowering the minimum support level appropriately, so that bounds can be defined on the likelihood of false negatives. False positives can be removed with the use of a second pass over the data. The major downside of the approach is that the reduction in the minimum support level to reduce the number of false negatives can be significant. This also reduces the computational efficiency of the approach. The method however requires only two passes over the data, where the first pass is used to create the sample, and the second pass is used to remove the false positives.

An interesting approach proposed in [57] divides the disk resident database into smaller memory-resident partitions. For each partition, more efficiency algorithms can be used, because of the memory-resident nature of the partition. It should be pointed out that each frequent pattern over the entire database will appear as a frequent pattern in at least one transaction. Therefore, the union of the itemsets over the different transactions provides a superset of the true frequent patterns. A post-processing phase is then used to filter out the spurious itemsets, by counting this candidate set against the transaction database. As long as the partitions are reasonably large, the superset found approximates the true frequent patterns very well, and therefore the additional time spent in counting irrelevant candidates is relatively small. The main advantage of this approach is it requires only two passes over the database. Therefore, such an approach is particularly effective when the data is resident on disk.

The *Dynamic Itemset Counting (DIC)* algorithm [15] divides the database into intervals, and generates longer candidates when it is known that the subsets of these candidates are already frequent. These are then validated over the database. Such an approach can reduce the number of passes over the data, because it implicitly combines the process of candidate generation and counting.

7.3 Online Association Rule Mining

In many applications, a user may wish to query the transaction data to find the association rules or the frequent patterns. In such cases, even at high support levels, it is often impossible to create the frequent patterns in online time because of the multiple passes required over a potentially large database. One of the earliest algorithms for online association rule mining was proposed in [6]. In this approach, an augmented lexicographic tree is stored either on disk or in main-memory. The lexicographic tree is augmented with all the edges represented the subset relationships between itemsets, and is also referred to as the itemset *lattice*. For any given query, the itemset lattice may be traversed to determine the association rules. It has been shown in [6], that such an approach can also be used to determine the non-redundant association rules in the underlying data. A second method [40] uses a condensed frequent pattern tree (instead of a lattice) to pre-process and store the itemsets. This structure can be queried to provide online responses.

A very different approach for online association rule mining has been proposed in [34], in which the transaction database is processed in real time. In this case, an incremental approach is used to mine the transaction database. This is a *Continuous Association Rule Mining Algorithm*, which is referred to as *CARMA*. In this case, transactions are processed as they arrive, and candidate itemsets are generated on the fly, by examining the subsets of that transaction. Clearly, the downside is that such an approach is that it will create a lot more candidates than any of the offline algorithms which use levelwise methods to generate the candidates. This general characteristic is of course true of any algorithm which tries to reduce the number of passes with approximate candidate generation. One interesting characteristic of the CARMA algorithm is that it allows the user to change the minimum support level during execution. In that case, the algorithm is guaranteed to have generated the supersets of the true itemsets in the data. If desired, a second pass over the data can be used to remove the spurious frequent itemsets.

Many streaming methods have also been proposed that use only one pass over the transaction data [19–21, 35, 43]. It should be pointed out that it is often difficult to find even 1-itemsets exactly over a data stream because of the one-pass constraint [21], when the number of distinct items is larger than the main memory availability. This is often true of k-itemsets as well, especially at low support levels. Furthermore, if the patterns in the stream change over time, then the frequent k-itemsets will change significantly as well. These methods therefore have the challenge of finding the frequent itemsets efficiently, maintaining them, and handling issues involving evolution of the data stream. Given the numerous challenges of pattern mining in this scenario, most of these methods find the frequent items approximately. These issues will be discussed in detail in Chap. 9 on streaming pattern mining algorithms.

8 Conclusions and Summary

This chapter provides a survey of different frequent pattern mining algorithms. most frequent pattern algorithms, implicitly or explicitly, explore the enumeration tree of itemsets. Algorithms such as *Apriori* explore the enumeration tree in breadth-first fashion with join-based candidate generation. Although the notion of an enumeration tree is not explicitly mentioned by the *Apriori* algorithm, the execution tree explores the candidates according to an enumeration tree constructed on the prefixes. Other algorithms such as *TreeProjection* and *FP-growth* use the hierarchical relationships between the projected databases for patterns of different lengths, and avoid re-doing the counting work done for the shorter patterns. Maximal and closed versions of frequent pattern mining algorithms are also able to achieve much better pruning performance. A number of efficiency-based optimizations of frequent pattern mining algorithms were also discussed in this chapter.

References

1. R. Agrawal, and R. Srikant. Fast Algorithms for Mining Association Rules in Large Databases, *VLDB Conference*, pp. 487–499, 1994.
2. R. Agrawal, T. Imielinski, and A. Swami. Mining association rules between sets of items in large databases. *ACM SIGMOD Conference*, 1993.
3. R. Agrawal, H. Mannila, R. Srikant, H. Toivonen, and A.I. Verkamo. Fast discovery of association rules, *Advances in Knowledge Discovery and Data Mining*, pp. 307–328, 1996.
4. R. Agarwal, C. C. Aggarwal, and V. V. V. Prasad. Depth-first Generation of Long Patterns, *ACM KDD Conference*, 2000. Also available as IBM Research Report, RC21538, July 1999.
5. R. Agarwal, C. C. Aggarwal, and V. V. V. Prasad. A Tree Projection Algorithm for Generation of Frequent Itemsets, *Journal of Parallel and Distributed Computing*, 61(3), pp. 350–371, 2001. Also available as IBM Research Report, RC21341, 1999.
6. C. C. Aggarwal, P. S. Yu. Online Generation of Association Rules, *ICDE Conference*, 1998.
7. C. C. Aggarwal, P. S. Yu. A New Framework for Itemset Generation, *ACM PODS Conference*, 1998.
8. E. Azkural and C. Aykanat. A Space Optimization for FP-Growth, *FIMI workshop*, 2004.
9. Y. Bastide, R. Taouil, N. Pasquier, G. Stumme, and L. Lakhal. Mining Frequent Patterns with Counting Inference. *ACM SIGKDD Explorations Newsletter*, 2(2), pp. 66–75, 2000.
10. R. J. Bayardo Jr. Efficiently mining long patterns from databases, *ACM SIGMOD Conference*, 1998.
11. J. Blanchard, F. Guillet, R. Gras, and H. Briand. Using Information-theoretic Measures to Assess Association Rule Interestingness. *ICDM Conference*, 2005.
12. C. Borgelt, R. Kruse. Induction of Association Rules: Apriori Implementation, *Conference on Computational Statistics*, 2002. http://fuzzy.cs.uni-magdeburg.de/ borgelt/software. html.
13. J.-F. Boulicaut, A. Bykowski, and C. Rigotti. Free-sets: A Condensed Representation of Boolean data for the Approximation of Frequency Queries. *Data Mining and Knowledge Discovery*, 7(1), pp. 5–22, 2003.
14. D. Burdick, M. Calimlim, and J. Gehrke. MAFIA: A Maximal Frequent Itemset Algorithm for Transactional Databases, *ICDE Conference*, 2000. Implementation URL: http://himalaya-tools.sourceforge.net/Mafia/.
15. S. Brin, R. Motwani, J.D. Ullman, and S. Tsur. Dynamic itemset counting and implication rules for market basket data. *ACM SIGMOD Conference*, 1997.

16. S. Brin, R. Motwani, and C. Silverstein. Beyond Market Baskets: Generalizing Association Rules to Correlations. *ACM SIGMOD Conference*, 1997.
17. T. Calders, and B. Goethals. Mining all non-derivable frequent itemsets *Principles of Data Mining and Knowledge Discovery*, pp. 1–42, 2002.
18. T. Calders, and B. Goethals. Depth-first Non-derivable Itemset Mining, *SDM Conference*, 2005.
19. T. Calders, N. Dexters, J. Gillis, and B. Goethals. Mining Frequent Itemsets in a Stream, *Informations Systems*, to appear, 2013.
20. J. H. Chang, and W. S. Lee. Finding Recent Frequent Itemsets Adaptively over Online Data Streams, *ACM KDD Conference*, 2003.
21. M. Charikar, K. Chen, and M. Farach-Colton. Finding Frequent Items in Data Streams. *Automata, Languages and Programming*, pp. 693–703, 2002.
22. G. Cong, A. K. H. Tung, X. Xu, F. Pan, and J. Yang. FARMER: Finding interesting rule groups in microarray datasets. *ACM SIGMOD Conference*, 2004.
23. G. Cong, K.-L. Tan, A. K. H. Tung, X. Xu. Mining Top-*k* covering Rule Groups for Gene Expression Data. *ACM SIGMOD Conference*, 2005.
24. M. El-Hajj and O. Zaiane. COFI-tree Mining: A New Approach to Pattern Growth with Reduced Candidacy Generation. *FIMI Workshop*, 2003.
25. F. Geerts, B. Goethals, J. Bussche. A Tight Upper Bound on the Number of Candidate Patterns, *ICDM Conference*, 2001.
26. B. Goethals. Survey on frequent pattern mining, *Technical report, University of Helsinki*, 2003.
27. R. P. Gopalan and Y. G. Sucahyo. High Performance Frequent Pattern Extraction using Compressed FP-Trees, *Proceedings of SIAM International Workshop on High Performance and Distributed Mining*, 2004.
28. K. Gouda, and M. Zaki. Genmax: An efficient algorithm for mining maximal frequent itemsets. *Data Mining and Knowledge Discovery*, 11(3), pp. 223–242, 2005.
29. G. Grahne, and J. Zhu. Efficiently Using Prefix-trees in Mining Frequent Itemsets, *IEEE ICDM Workshop on Frequent Itemset Mining*, 2004.
30. G. Grahne, and J. Zhu. Fast Algorithms for Frequent Itemset Mining Using FP-Trees. *IEEE Transactions on Knowledge and Data Engineering*. 17(10), pp. 1347–1362, 2005, vol. 17, no. 10, pp. 1347–1362, October, 2005.
31. V. Guralnik, and G. Karypis. Parallel tree-projection-based sequence mining algorithms. *Parallel Computing*, 30(4): pp. 443–472, April 2004.
32. J. Han, J. Pei, and Y. Yin. Mining Frequent Patterns without Candidate Generation, *ACM SIGMOD Conference*, 2000.
33. J. Han, H. Cheng, D. Xin, and X. Yan. Frequent Pattern Mining: Current Status and Future Directions, *Data Mining and Knowledge Discovery*, 15(1), pp. 55–86, 2007.
34. C. Hidber. Online Association Rule Mining, *ACM SIGMOD Conference*, 1999.
35. R. Jin, and G. Agrawal. An Algorithm for in-core Frequent Itemset Mining on Streaming Data, *ICDM Conference*, 2005.
36. Q. Lan, D. Zhang, and B. Wu. A New Algorithm For Frequent Itemsets Mining Based On Apriori And FP-Tree, *IEEE International Conference on Global Congress on Intelligent Systems*, pp. 360–364, 2009.
37. D.-I. Lin, and Z. Kedem. Pincer-search: A New Algorithm for Discovering the Maximum Frequent Set, *EDBT Conference*, 1998.
38. J. Liu, Y. Pan, K. Wang. Mining Frequent Item Sets by Opportunistic Projection, *ACM KDD Conference*, 2002.
39. G. Liu, H. Lu and J. X. Yu. AFOPT:An Efficient Implementation of Pattern Growth Approach, *FIMI Workshop*, 2003.
40. H. Liu, J. Han, D. Xin, and Z. Shao. Mining frequent patterns on very high dimensional data: a top- down row enumeration approach. *SDM Conference*, 2006.
41. C. Lucchesse, S. Orlando, and R. Perego. DCI-Closed: A fast and memory efficient algorithm to mine frequent closed itemsets. *FIMI Workshop*, 2004.

42. C. Lucchese, S. Orlando, and R. Perego. Fast and memory efficient mining of frequent closed itemsets. *IEEE TKDE Journal*, 18(1), pp. 21–36, January 2006.
43. G. Manku, R. Motwani. Approximate Frequency Counts over Data Streams. *VLDB Conference*, 2002.
44. H. Mannila, H. Toivonen, and A.I. Verkamo. Efficient algorithms for discovering association rules. *Proceedings of the AAAI Workshop on Knowledge Discovery in Databases*, pp. 181–192, 1994.
45. B. Negrevergne, T. Guns, A. Dries, and S. Nijssen. Dominance Programming for Itemset Mining. *IEEE ICDM Conference*, 2013.
46. S. Orlando, P. Palmerini, R. Perego. Enhancing the a-priori algorithm for frequent set counting, *Third International Conference on Data Warehousing and Knowledge Discovery*, 2001.
47. S. Orlando, P. Palmerini, R. Perego, and F. Silvestri. Adaptive and resource-aware mining of frequent sets. *ICDM Conference*, 2002.
48. F. Pan, G. Cong, A. K. H. Tung, J. Yang, and M. J. Zaki. Finding closed patterns in long biological datasets. *ACM KDD Conference*, 2003.
49. F Pan, A. K. H. Tung, G. Cong, X. Xu. COBBLER: Combining column and Row Enumeration for Closed Pattern Discovery. *SSDBM*, 2004.
50. J.-S. Park, M. S. Chen, and P. S. Yu. An Effective Hash-based Algorithm for Mining Association Rules, *ACM SIGMOD Conference*, 1995.
51. N. Pasquier, Y. Bastide, R. Taouil, and L. Lakhal. Discovering frequent closed itemsets for association rules. *ICDT Conference*, 1999.
52. N. Pasquier, Y. Bastide, R. Taouil, and L. Lakhal. Efficient mining of association rules using closed itemset lattices. *Journal of Information Systems*, 24(1), pp. 25–46, 1999.
53. J. Pei, J. Han, and R. Mao. CLOSET: An Efficient Algorithm for Mining Frequent Closed Itemsets, *DMKD Workshop*, 2000.
54. J. Pei, J. Han, H. Lu, S. Nishio, S. Tang, D. Yang. H-mine: Hyper-structure mining of frequent patterns in large databases, *ICDM Conference*, 2001.
55. B. Racz. nonordfp: An FP-Growth Variation without Rebuilding the FP-Tree, *FIMI Workshop*, 2004.
56. M. Holsheimer, M. Kersten, H. Mannila, and H. Toivonen. A Perspective on Databases and Data Mining, *ACM KDD Conference*, 1995.
57. A. Savasere, E. Omiecinski, and S. Navathe. An efficient algorithm for mining association rules in large databases. *VLDB Conference*, 1995.
58. P. Shenoy, J. Haritsa, S. Sudarshan, G. Bhalotia, M. Bawa, D. Shah. Turbo-charging Vertical Mining of Large Databases. *ACM SIGMOD Conference*, pp. 22–33, 2000.
59. Z. Shi, and Q. He. Efficiently Mining Frequent Itemsets with Compact FP-Tree, IFIP International Federation for Information Processing, V-163, pp. 397–406, 2005.
60. R. Srikant. Fast algorithms for mining association rules and sequential patterns. *PhD thesis, University of Wisconsin, Madison*, 1996.
61. Y. G. Sucahyo and R. P. Gopalan. CT-ITL: Efficient Frequent Item Set Mining Using a Compressed Prefix Tree with PatternGrowth, *Proceedings of the 14th Australasian Database Conference*, 2003.
62. Y. G. Sucahyo and R. P. Gopalan. CT-PRO: A Bottom Up Non Recursive Frequent Itemset Mining Algorithm Using Compressed FP-Tree Data Structures. *FIMI Workshop*, 2004.
63. P.-N. Tan, V. Kumar, amd J. Srivastava. Selecting the Right Interestingness Measure for Association Patterns. *ACM KDD Conference*, 2002.
64. I. Taouil, N. Pasquier, Y. Bastide, and L. Lakhal. Mining Basis for Association Rules using Closed Sets, *ICDE Conference*, 2000.
65. H. Toivonen. Sampling large databases for association rules. *VLDB Conference*, 1996.
66. T. Uno, M. Kiyomi and H. Arimura. Efficient Mining Algorithms for Frequent/Closed/Maximal Itemsets, *FIMI Workshop*, 2004.
67. J. Wang, J. Han. BIDE: Efficient Mining of Frequent Closed Sequences. *ICDE Conference*, 2004.

68. J. Wang, J. Han, Y. Lu, and P. Tzvetkov. TFP: An efficient algorithm for mining top-k frequent closed itemsets. *IEEE Transactions on Knowledge and Data Engineering*, 17, pp. 652–664, 2002.
69. J. Wang, J. Han, and J. Pei. CLOSET+: Searching for the Best strategies for mining frequent closed itemsets. *ACM KDD Conference*, 2003.
70. G. I. Webb. Efficient Search for Association Rules, *ACM KDD Conference*, 2000.
71. M. J. Zaki. Scalable algorithms for association mining, *IEEE Transactions on Knowledge and Data Engineering*, 12(3), pp. 372–390, 2000.
72. M. Zaki, and K. Gouda. Fast vertical mining using diffsets. *ACM KDD Conference*, 2003.
73. M. J. Zaki and C. Hsiao. CHARM: An efficient algorithm for closed association rule mining. *SDM Conference*, 2002.
74. M. Zaki, S. Parthasarathy, M. Ogihara, and W. Li. New Algorithms for Fast Discovery of Association Rules. *KDD Conference*, pp. 283–286, 1997.
75. C. Zeng, J. F. Naughton, and JY Cai. On Differentially Private Frequent Itemset Mining. In Proceedings of 39th International Conference on Very Large data Bases, 2012.

Chapter 3
Pattern-Growth Methods

Jiawei Han and Jian Pei

Abstract Mining frequent patterns has been a focused topic in data mining research in recent years, with the development of numerous interesting algorithms for mining association, correlation, causality, sequential patterns, partial periodicity, constraint-based frequent pattern mining, associative classification, emerging patterns, etc. Many studies adopt an Apriori-like, candidate generation-and-test approach. However, based on our analysis, candidate generation and test may still be expensive, especially when encountering long and numerous patterns.

A new methodology, called *frequent pattern growth*, which mines frequent patterns without candidate generation, has been developed. The method adopts a divide-and-conquer philosophy to project and partition databases based on the currently discovered frequent patterns and grow such patterns to longer ones in the projected databases. Moreover, efficient data structures have been developed for effective database compression and fast in-memory traversal. Such a methodology may eliminate or substantially reduce the number of candidate sets to be generated and also reduce the size of the database to be iteratively examined, and, therefore, lead to high performance.

In this paper, we provide an overview of this approach and examine its methodology and implications for mining several kinds of frequent patterns, including association, frequent closed itemsets, max-patterns, sequential patterns, and constraint-based mining of frequent patterns. We show that *frequent pattern growth* is efficient at mining large data-bases and its further development may lead to scalable mining of many other kinds of patterns as well.

Keywords Scalable data mining methods and algorithms · Frequent patterns · Associations · Sequential patterns · Constraint-based mining

J. Han (✉)
University of Illinois at Urbana-Champaign, Urbana, IL 61801, USA
e-mail: hanj@cs.uiuc.edu

J. Pei
Simon Fraser University, Burnaby, BC V5A 1S6, Canada
e-mail: jpei@cs.sfu.ca

C. C. Aggarwal, J. Han (eds.), *Frequent Pattern Mining,*
DOI 10.1007/978-3-319-07821-2_3, © Springer International Publishing Switzerland 2014

1   Introduction

Since the introduction of association mining in [3], there have been many studies on efficient and scalable frequent pattern mining algorithms. A milestone in these studies is the development of an Apriori-based, level-wise mining method for associations [1, 23], which has sparked the development of various kinds of Apriori-like association mining algorithms, as well as its extensions to mining correlation [8], causality [34], sequential patterns [2], episodes [24], max-patterns [5], constraint-based mining [15, 20, 25, 36], associative classification [22], cyclic association rules [26], ratio rules [19], iceberg queries and iceberg cubes [7, 13], partial periodicity [16], emerging patterns [12], and many other patterns.

There is an important, common ground among all these methods developed: the use of an *anti-monotone* Apriori *property* of frequent patterns [1]: *if any length-k pattern is not frequent in the database, none of its length-(k + 1) super-patterns can be frequent.* This property leads to the powerful pruning of the set of itemsets to be examined in the search for longer frequent patterns based on the existing ones.

Besides applying the Apriori property, most of the developed methods adopt a level-wise, candidate generation-and-test approach, which scans the database multiple times (although there have been many techniques developed for reducing the number of database scans). The first scan finds all of the length-1 frequent patterns. The kth (for $k > 1$) scan starts with a *seed set* of length-$(k-1)$ frequent patterns found in the previous pass and generates new potential length-k patterns, called *candidate patterns*. The kth scan of the database finds the *support* of every length k candidate pattern. The candidates which pass the minimum support threshold are identified as frequent patterns and become the seed set for the next pass. The computation terminates when there is no frequent pattern found or there is no candidate pattern that can be generated in any pass.

The candidate generation approach achieves good performance by reducing the number of candidates to be generated. However, when the minimum support threshold is low or the length of the patterns to be generated is long, the candidate generation-based algorithm may still bear the following non-trivial costs, independent of detailed implementation techniques.

1. The number of candidates to be generated may still be huge, especially when the length of the patterns to be generated is long. For example, to generate one frequent pattern of length 100, such as $\{a_1, a_2, \ldots, a_{100}\}$, the number of candidates that has to be generated will be at least $\Sigma_{i=1}^{100} \binom{100}{i} = 2^{100} - 1 \approx 10^{30}$.

2. Each scan of the database examines the entire database against the whole set of current candidates, which is quite costly when the database is large and the number of candidates to be examined is numerous.

To overcome this difficulty, a new approach, called *frequent pattern growth*, has been developed, in a series of studies, such as [17, 18, 27–30], which adopts a divide-and-conquer methodology and mines frequent patterns without candidate generation. The approach has several distinct features:

1. Instead of generating a large number of candidates, the method preserves (in some compressed forms) the essential groupings of the original data elements for mining. Then the analysis is focused on counting the frequency of the relevant data sets instead of candidate sets.
2. Instead of scanning the *entire* database to match against the *whole* corresponding set of candidates in each pass, the method partitions the data set to be examined as well as the set of patterns to be examined by database projection. Such a divide-and-conquer methodology substantially reduces the search space and leads to high performance.
3. With the growing capacity of main memory and the substantial reduction of database size by database projection as well as the space needed for manipulating large sets of candidates, a substantial portion of data can be put into main memory for mining. New data structures and methods, such as FP-tree and pseudo-projection (for mining sequential patterns), have been developed for data compression and pointer-based traversal. The performance studies have shown the effectiveness of such techniques.

A few pieces of work have contributed to the development of the frequent pattern-growth methodology, as illustrated below.

The TreeProjection method [4] proposes a database projection technique which explores the projected databases associated with different frequent itemsets. The FP-growth algorithm [18] performs database projection when the database size is huge and then constructs a compressed data structure, FP-tree, when the compressed tree can fit in main memory. The remaining mining will be focused on the recursively generated, projected FP-trees. Besides mining frequent itemsets, the FP-tree structure can be used for mining frequent closed itemsets, which is presented in the CLOSET algorithm [28].

The frequent pattern-growth methodology influences constraint-based mining of frequent itemsets as well. The constraint-pushing techniques developed for Apriori-based mining [25] can be applied to pattern growth mining. In addition, some complex kinds of constraints, such as convertible constraints, which cannot be pushed deep into the mining process by Apriori, can be done so with frequent pattern growth [29], due to the facts that (1) pattern growth only needs to examine part of the database (the projected one), and (2) data can be organized in a structured way to facilitate the controlled growth of frequent patterns.

Similar divide-and-conquer ideas but different projection techniques have been developed for mining sequential patterns, which are presented in two algorithms, FreeSpan [17] and PrefixSpan [30]. The performance study shows that both methods outperform the classical Apriori-based sequential pattern mining algorithm GSP [35], and PrefixSpan has considerably better performance than FreeSpan.

Table 3.1 The transaction
database TDB

tid	Itemset	(Ordered) frequent items
100	f, a, c, d, g, i, m, p	f, c, a, m, p
200	a, b, c, f, l, m, o	f, c, a, b, m
300	b, f, h, j, o	f, b
400	b, c, k, s, p	c, b, p
500	a, f, c, e, l, p, m, n	f, c, a, m, p

In this chapter, we provide an overview of several recently developed frequent pattern growth mining methods and discuss their implications. The remaining of the paper is organized as follows. In Sect. 2, we examine the FP-growth method for mining frequent itemsets and also mention the CLOSET method for mining frequent closed itemsets. In Sect. 3, we look at the impact of FP-growth to constraint-based mining of frequent patterns and the handling of convertible constraints. In Sect. 4, we introduce two pattern-growth-based methods for mining sequential patterns: FreeSpan [17] and PrefixSpan [30]. In Sect. 6, we discuss the potential extensions of pattern-growth methods and conclude our study.

2 FP-Growth: Pattern Growth for Mining Frequent Itemsets

As shown by many researchers [1, 3], mining frequent itemsets represents the core of mining association rules, correlations, and many other patterns.

Let a *transaction database* TDB consist of a set of transactions in the form of $T = (tid, X)$ where tid is a *transaction-id* and X an itemset (i.e., a set of items). A transaction T is said to *contain* itemset Y if and only if $Y \subseteq X$. The support of an itemset W in TDB, denoted as $sup(W)$, is the number of transactions in TDB containing W. Given a user-specified minimum support threshold, min_sup, W is frequent if and only if $sup(W) \geq min\_sup$. The problem of mining frequent itemsets is to *find the complete set of frequent itemsets in a transaction database TDB w.r.t. given support threshold min_sup*.

Here we examine how one can develop a pattern growth method, FP-growth [18], for efficient mining of frequent itemsets in large databases. FP-growth first performs a frequent item-based database projection when the database is large and then switches to main-memory-based mining by constructing a compact data structure, called FP-tree, and transforming mining database into mining this compact tree. We first show how FP-tree be constructed from a database.

Example 1 (FP-tree)Let the transaction database, DB, be (the first two columns of) Table 3.1 and the minimum support threshold be 3.

First, a scan of DB derives a *list* of frequent items, $\langle (f : 4), (c : 4), (a : 3), (b : 3), (m : 3), (p : 3) \rangle$, (the number after ":" indicates the support), and with items ordered in frequency descending order. This ordering is important since each path of a tree will follow this order. For convenience of later discussions, the frequent items in each transaction are listed in this ordering in the rightmost column of Table 3.1.

Fig. 3.1 FP-tree for
transaction database in
Table 3.1

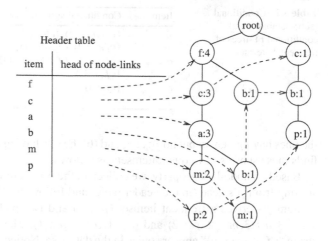

Second, the root of a tree, labeled with "*null*" is created. Scan the DB the second time. The scan of the first transaction leads to the construction of the first branch of the tree: $\langle (f : 1), (c : 1), (a : 1), (m : 1), (p : 1)\rangle$. Notice that the branch is not ordered in $\langle f, a, c, m, p\rangle$ as in the transaction but is ordered according to the order in the *list* of frequent items. For the second transaction, since its (ordered) frequent item list $\langle f, c, a, b, m\rangle$ shares a common prefix $\langle f, c, a\rangle$ with the existing path $\langle f, c, a, m, p\rangle$, the count of each node along the prefix is incremented by 1, and one new node $(b : 1)$ is created and linked as a child of $(a : 2)$ and another new node $(m : 1)$ is created and linked as the child of $(b : 1)$. Remaining transactions can be inserted similarly.

To facilitate tree traversal, an item header table is built, in which each item points, via a head of node-link, to its first occurrence in the tree. Nodes with the same item-name are linked in sequence via node-links. After scanning all transactions in DB, the tree with the associated node-links is shown in Fig. 3.1.

The FP-tree built in Example 2 has some nice properties as follows: (1) *FP-tree contains complete information of TDB w.r.t. frequent itemset mining*: every transaction in TDB is mapped onto one path in the FP-tree, and the frequent itemset information is completely stored in the tree; (2) *FP-tree is a highly compact structure*: since there are often a lot of sharing of frequent items among transactions, the size of the tree is usually much smaller than its original database; and (3) *there is a node-link property*: for every frequent item x, all transactions containing x can be obtained by following x's node-links starting from x's head in the FP-tree header table.

Based on this compact structure, FP-growth mines the complete set of frequent itemsets as follows.

Example 2 (FP-growth) Let us examine the mining process based on the constructed FP-tree (Fig. 3.1).

According to the list of frequent items, the complete set of frequent itemsets can be divided into six subsets without overlap: (1) frequent itemsets having item p; (2)

Table 3.2 Conditional
(sub)-databases and
conditional FP-trees of
frequent 1-itemsets

Item	Conditional sub-database	Conditional FP-tree
p	$\{(fcam:2),(cb:1)\}$	$\{(c:3)\}|p$
m	$\{(fca:2),(fcab:1)\}$	$\{(fca:3)\}|m$
b	$\{(fca:1),(f:1),(c:1)\}$	$\emptyset$
a	$\{(fc:3)\}$	$\{(fc:3)\}|a$
c	$\{(f:3)\}$	$\{(f:3)\}|c$
f	$\emptyset$	$\emptyset$

the ones having item m but no p; ... ; and (6) the one having only item f. FP-growth finds these subsets of frequent itemsets as follows.

Based on node-link property, we collect all the transactions that p participates by starting from p's head (in the header table) and following p's node-links.

Item p derives a frequent itemset $(p:3)$ and two paths in the FP-tree: $\langle f : 4, c : 3, a : 3, m : 2, p : 2\rangle$ and $\langle c : 1, b : 1, p : 1\rangle$. The first path indicates that string "(f, c, a, m, p)" appears twice in the database. Notice although string $\langle f, c, a\rangle$ appears three times and $\langle f\rangle$ itself appears even four times, they only appear twice *together* with p. Thus to study which strings appear together with p, only p's prefix path $\langle fcam : 2\rangle$ counts. Similarly, the second path indicates string "(c, b, p)" appears once in the set of transactions in TDB, or p's prefix path is $\langle cb : 1\rangle$. These two prefix paths of p, "$\{(fcam : 2), (cb : 1)\}$", form p's sub-database, which is called p's *conditional database* (i.e., the sub-database under the condition of p's existence). Construction of an FP-tree on this conditional sub-database (which is called p's conditional FP-tree) leads to only one branch $(c : 3)$. Hence only one frequent itemset $(cp : 3)$ is derived. The search for frequent itemsets having p terminates.

For item m, it derives a frequent itemset $(m : 3)$ and two paths $\langle f : 4, c : 3, a : 3, m : 2\rangle$ and $\langle f : 4, c : 3, a : 3, b : 1, m : 1\rangle$. Notice p appears together with m as well, however, there is no need to include p here in the analysis since any frequent itemsets involving p has been analyzed in the previous examination of p. Similar to the above analysis, m's conditional sub-database is, $\{(fca : 2), (fcab : 1)\}$. Constructing an FP-tree on it, we derive m's conditional FP-tree, $\langle fca : 3\rangle$, a single frequent itemset path.

Since m's conditional FP-tree, $\langle fca : 3\rangle$, has a single branch, instead of recursively constructing its conditional FP-trees, one can simply enumerate all the combinations of its components, i.e., $\{(a : 3), (c : 3), (f : 3), (ca : 3), (fa : 3), (fca : 3), (fc : 3)\}$. Such simple pattern enumeration for single-path FP-trees has been proven truly useful at reducing mining efforts.

Similarly, the remaining frequent itemsets can be mined by constructing corresponding conditional sub-databases and perform mining on them, respectively. The conditional sub-databases and the conditional FP-trees generated are summarized in Table 3.2.

When the database is too big to make its FP-tree fit in memory, the database can be projected into its conditional sub-databases (without constructing disk-based FP-trees). Two methods can be used for the projection of a database into its conditional sub-databases: *parallel projection* and *partition projection*. The former projects each transaction into all of its projected databases in one scan, whereas the latter projects each transaction only to its first projected database (according to the ordering of items). The former facilitates parallel processing but requires large disk space to store all of the projected databases, whereas the latter ensures that the additional disk space required is no more than the original database but it needs additional projections of its (projected) transactions to subsequent projected databases in the later processing.

After one or a few rounds of projections, the corresponding conditional FP-trees should be able to fit in memory. Then a memory-based FP-tree can be constructed for fast mining.

The FP-growth algorithm is presented in [18]. Its performance analysis shows that the FP-growth mining of both long and short frequent itemsets is efficient and scalable. It is about an order of magnitude faster than Apriori [1] and other candidate generation-based algorithms, and is also faster than TreeProjection, a projection-based algorithm proposed in [4].

In comparison with the candidate generation-based algorithms, FP-growth has the following advantages: (1) FP-tree is highly compact, usually substantially smaller than the original database, and thus saves the costly database scans in the subsequent mining process. (2) It avoids costly candidate sets generation and test by successively concatenating frequent 1-itemsets found in the (conditional) FP-trees: It never generates any combinations of new candidate sets which are not in the database because the itemset in any transaction is always encoded in the corresponding path of the FP-trees. In this context, the mining methodology is not Apriori-like (*restricted*) *generation-and-test* but *frequent pattern (fragment) growth only*. The major operations of mining are count accumulation and prefix path count adjustment, which are usually much less costly than candidate generation and itemset matching operations performed in most Apriori-like algorithms. (3) It applies a partitioning-based divide-and-conquer method which dramatically reduces the size of the subsequent conditional sub-databases and conditional FP-trees. Several other optimization techniques, including ordering of frequent items, and employing the least frequent events as suffix, also contribute to the efficiency of the method.

Besides mining frequent itemsets, an extension of the FP-growth method, called CLOSET [28], can be used to mine frequent closed itemsets and max-patterns, where a *frequent closed itemset* is a frequent itemset, c, where there is no proper superset of c sharing the same support count with c, and a *max-pattern* is a frequent pattern, p, such that any proper superpattern of p is not frequent. Max-patterns and frequent closed itemset can be used to reduce the number of frequent itemsets and association rules generated at association mining.

By frequent pattern growth, one can also mine closed frequent itemsets and max-patterns, using the FP-tree structure. Moreover, a single prefix-path compression technique can be developed for compressing FP-trees or conditional FP-trees that contain single prefix paths. This will further enhance the performance and reduce the efforts of redundancy checking at mining closed frequent itemsets and max-patterns.

Table 3.3 The transaction
database $\mathcal{T}$ in Example 3

Transaction ID	Items in transaction
10	a, b, c, d, f
20	b, c, d, f, g, h
30	a, c, d, e, f
40	c, e, f, g

Table 3.4 The profit of each
item in Example 3

Item	a	b	c	d	e	f	g	h
Value	40	0	−20	10	−30	30	20	−10

3 Pushing More Constraints in Pattern-Growth Mining

Frequent pattern mining often generates a large number of frequent itemsets and rules, which reduces not only the efficiency but also the effectiveness of mining since users have to sift through a large number of mined rules to find useful ones.

Recent work has highlighted the importance of the paradigm of constraint-based mining: the user is allowed to express his focus in mining, by means of a rich class of constraints that capture application semantics. Besides allowing user exploration and control, the paradigm allows many of these constraints to be pushed deep inside mining, thus pruning the search space and achieving high performance.

Previous studies [6, 15, 20, 25] have identified three classes of constraints, *anti-monotone, monotone,* and *succinct,* which can be pushed deep in frequent itemset mining. While these cover a large class of useful constraints, many other useful and natural constraints remain. For example, consider the constraints $avg(S)\ \theta\ v$, and $sum(S)\ \theta\ v\ (\theta \in \{\geq, \leq\})$. The first is neither anti-monotone, nor monotone, nor succinct. The second is anti-monotone when θ is $\leq$ and *all items have non-negative values*. But if S can contain items of arbitrary values, the constraint is rather like the first one. This means these constraints are hard to optimize.

With the development of frequent pattern growth method, databases can be projected and partitioned in an organized way, as well as the patterns to be searched for. Thus some constraints which are hard to optimize under the Apriori mining framework can be optimized with the frequent pattern growth method. Let's examine one example.

Example 3 Let Table 3.3 be our transaction database $\mathcal{T}$, with a set of items $I = \{a, b, c, d, e, f, g, h\}$. Let the support threshold be $min\_support = 2$. Also, let each item have an attribute *value* (such as *profit*), with the concrete value shown in Table 3.4.

The constraint $C_{avg} \equiv avg(S) \geq 25$ is not anti-monotone (nor monotone, nor succinct). For example, $avg(df) = (10 + 30)/2 < 25$, violates the constraint. However, upon adding one more item a, $avg(adf) = (40 + 10 + 30)/3 \geq 25$, adf satisfies C_{avg}.

This example shows that a constraint like $avg(S) \geq v$ cannot be pushed deep into the Apriori mining algorithm because the subsets (supersets) of a valid itemset could

Fig. 3.2 Mining frequent
itemsets satisfying constraint
$avg(S) \geq 25$

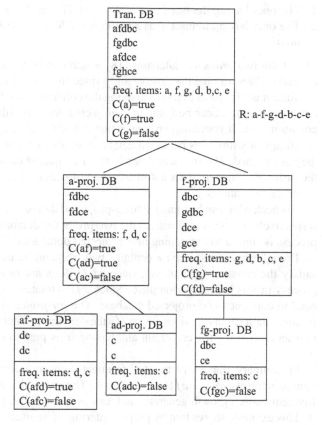

well be invalid and vice versa. Let us examine how to push such a constraint deep
into the mining process in the frequent pattern growth mining.

Example 4 With the same minimum support threshold over transaction database $\mathcal{T}$
in Table 3.3, one can list items in value descending order $\mathcal{R}$: $\langle a(40),\ f(30),\ g(20),$
$d(10),\ b(0),\ h(-10),\ c(-20),\ e(-30)\rangle$.

A database can be then partitioned according to the ordered frequent items. With
the frequent pattern growth mining, the constraint C can be pushed deep into the
mining process, as shown in Fig. 3.2.

By scanning $\mathcal{T}$ once, we find support count for every item. Since h appears in
only one transaction, it is an infrequent item and is thus dropped without further
consideration. The set of frequent 1-itemsets is $\langle a, f, g, d, b, c, e \rangle$, listed in order
$\mathcal{R}$. Among them, only a and f satisfy the constraint. The fact that itemset g does
not satisfy the constraint implies that none of any 1-itemsets after g in order $\mathcal{R}$ can
satisfy the constraint avg. Similarly, itemsets having g, d, b, c or e as prefix cannot
satisfy the constraint. Thus, the set of frequent itemsets satisfying the constraint can
be partitioned into two subsets:

1. The ones having itemset a as a prefix w.r.t. $\mathcal{R}$, i.e., those containing item a; and
2. The ones having itemset f as a prefix w.r.t. $\mathcal{R}$, i.e., those containing item f but no a.

They form two projected databases [18] which can be mined with the constraint C pushed in. We examine the first one only since the second is similar.

Since a is a frequent itemset satisfying the constraint, the frequent itemsets having a as a proper prefix can be found in a-*projected database* (the subset of transactions containing a). It contains two transactions: $bcdf$ and $cdef$. Since items b and e is infrequent within this projected database, neither ab nor ae can be frequent. So, they are pruned. The frequent items in the a-projected database is f, d, c, listed in the order $\mathcal{R}$. Since ac does not satisfy the constraint, there is no need to create an ac-projected database.

To check what can be mined in the a-projected database with af and ad, as prefix, respectively, we need to construct the two projected databases and mine them. This process is similar to the mining of a-projected databases.

The af-*projected database* contains two frequent items d and c, and only afd satisfy the constraint. Moreover, since $afdc$ does not satisfies the constraint, the process in this branch is complete. Since afc violates the constraint, there is no need to construct afc-projected database. The ad-*projected database* contains one frequent item c, but adc does not satisfy the constraint. Therefore, the set of frequent itemsets satisfying the constraint and having a as prefix contains a, af, afd, and ad.

In summary, the complete set of frequent itemsets satisfying the constraint contains 6 itemsets: a, f, af, ad, afd, fg. The method with ordered itemsets and frequent pattern growth generates and tests only a small set of itemsets.

This example shows that by proper ordering of itemsets, frequent pattern growth method may push some tough constraints (called *convertible constraints*) deeper than the Apriori methods. A systematic classification of such constraints and a study of how to push them into the mining process are in [27, 29].

4   PrefixSpan: Mining Sequential Patterns by Pattern Growth

Sequential pattern mining, which discovers frequent subsequences as patterns in a sequence database, is an important data mining problem with broad applications, including the analyses of customer purchase behavior, Web access patterns, scientific experiments, disease treatments, natural disasters, DNA sequences, and so on.

A *sequence database* S is a set of tuples $\langle sid, s \rangle$, where sid is a *sequence_id* and s is a sequence (i.e., an ordered list of itemsets). A tuple $\langle sid, s \rangle$ is said to *contain* a sequence α, if α is a subsequence of s, i.e., $\alpha \sqsubseteq s$. The support of a sequence α in a sequence database S is the number of tuples in the database containing α. Given a positive integer ξ as the *support threshold*, a sequence α is called a *sequential pattern* in sequence database S if the sequence is contained by at least ξ tuples in

Table 3.5 A sequence database

Sequence_id	Sequence
10	$\langle a(abc)(ac)d(cf)\rangle$
20	$\langle (ad)c(bc)(ae)\rangle$
30	$\langle (ef)(ab)(df)cb\rangle$
40	$\langle eg(af)cbc\rangle$

the database, i.e., $support_S(\alpha) \geq \xi$. A sequential pattern with length l is called an l-pattern.

Given a sequence database and a *min_support* threshold, the problem of *sequential pattern mining* is to find the complete set of sequential patterns in the database.

Sequential pattern mining is more challenging than mining frequent itemsets. Sequences allow multiple occurrences of items and combination of items into itemsets, which may lead to a combination explosion. For example, using items a and b, there are only three possible itemsets: a, b and ab. However, even the length of sequences is limited to 3, there are 12 possible sequences: $\langle aaa\rangle$, $\langle aab\rangle$, ..., $\langle bbb\rangle$, $\langle (ab)a\rangle$, ..., $\langle b(ab)\rangle$.

Sequential patterns also have the *Apriori property*: *every non-empty sub-sequence of a sequential pattern is a sequential pattern*. A typical sequential pattern mining algorithm, GSP [35], is based on this *Apriori* property to reduce search space. However, the method bears similar non-trivial, inherent costs as to Apriori in mining frequent itemsets.

Following the similar philosophy of frequent pattern growth, two algorithms, FreeSpan [17] and PrefixSpan [30], are developed for pattern growth-based sequential pattern mining. FreeSpan mines sequential patterns by projecting the sequence database based on any frequent subsequences and growing subsequences in any position; whereas PrefixSpan does it by projecting the database based on only the frequent prefix subsequences and adding postfixes in the growth. Both methods find the complete set of sequential patterns but the latter is more efficient since it involves less database projections and less subsequence combinations to be examined. This analysis has also been verified by the performance results, and thus we examine only PrefixSpan using an example.

Example 5 (PrefixSpan) Suppose we want to mine sequential patterns in a sequence database S, shown in Table 3.5, with the support threshold set to 2. PrefixSpan works as follows.

First, we find length − 1 sequential patterns by scanning S once. This derives the set of frequent items in sequences, i.e., the set of length-1 sequential patterns: $\{(\langle a\rangle : 4), (\langle b\rangle : 4), (\langle c\rangle : 4), (\langle d\rangle : 3), (\langle e\rangle : 3), \text{ and } (\langle f\rangle : 3)\}$.

Then, the search space can be partitioned into the following six subsets: (1) the ones with prefix $\langle a\rangle$; ... ; and (6) the ones with prefix $\langle f\rangle$. The subsets of sequential patterns can be mined by constructing corresponding *projected databases* and mine each recursively. The projected databases as well as sequential patterns found in them are listed in Table 3.6, and the mining process is explained as follows.

Table 3.6 Projected databases and sequential patterns

Prefix	Projected (postfix) database	Sequential patterns
$\langle a \rangle$	$\langle (abc)(ac)d(cf) \rangle$, $\langle (\_d)c(bc)(ae) \rangle$, $\langle (\_b)(df)cb \rangle$, $\langle (\_f)cbc \rangle$	$\langle a \rangle$, $\langle aa \rangle$, $\langle ab \rangle$, $\langle a(bc) \rangle$, $\langle a(bc)a \rangle$, $\langle aba \rangle$, $\langle abc \rangle$, $\langle (ab) \rangle$, $\langle (ab)c \rangle$, $\langle (ab)d \rangle$, $\langle (ab)f \rangle$, $\langle (ab)dc \rangle$, $\langle ac \rangle$, $\langle aca \rangle$, $\langle acb \rangle$, $\langle acc \rangle$, $\langle ad \rangle$, $\langle adc \rangle$, $\langle af \rangle$
$\langle b \rangle$	$\langle (\_c)(ac)d(cf) \rangle$, $\langle (\_c)(ae) \rangle$, $\langle (df)cb \rangle$, $\langle c \rangle$	$\langle b \rangle$, $\langle ba \rangle$, $\langle bc \rangle$, $\langle (bc) \rangle$, $\langle (bc)a \rangle$, $\langle bd \rangle$, $\langle bdc \rangle$, $\langle bf \rangle$
$\langle c \rangle$	$\langle (ac)d(cf) \rangle$, $\langle (bc)(ae) \rangle$, $\langle b \rangle$, $\langle bc \rangle$	$\langle c \rangle$, $\langle ca \rangle$, $\langle cb \rangle$, $\langle cc \rangle$
$\langle d \rangle$	$\langle (cf) \rangle$, $\langle c(bc)(ae) \rangle$, $\langle (\_f)cb \rangle$	$\langle d \rangle$, $\langle db \rangle$, $\langle dc \rangle$, $\langle dcb \rangle$
$\langle e \rangle$	$\langle (\_f)(ab)(df)cb \rangle$, $\langle (af)cbc \rangle$	$\langle e \rangle$, $\langle ea \rangle$, $\langle eab \rangle$, $\langle eac \rangle$, $\langle eacb \rangle$, $\langle eb \rangle$, $\langle ebc \rangle$, $\langle ec \rangle$, $\langle ecb \rangle$, $\langle ef \rangle$, $\langle efb \rangle$, $\langle efc \rangle$, $\langle efcb \rangle$.
$\langle f \rangle$	$\langle (ab)(df)cb \rangle$, $\langle cbc \rangle$	$\langle f \rangle$, $\langle fb \rangle$, $\langle fbc \rangle$, $\langle fc \rangle$, $\langle fcb \rangle$

The sequential patterns with prefix $\langle a \rangle$ are mined in the (prefix) $\langle a \rangle$-projected database. It is the collection that contains only those subsequences prefixed with the first occurrence of $\langle a \rangle$. For example, in sequence $\langle (ef)(ab)(df)cb \rangle$, only the subsequence $\langle (\_b)(df)cb \rangle$ should count. Notice that $(\_b)$ means that the last element in the prefix, which is a, together with b, form one element (i.e., occurring together). Thus the $\langle a \rangle$-*projected database* consists of four postfix sequences: $\langle (abc)(ac)d(cf) \rangle$, $\langle (\_d)c(bc)(ae) \rangle$, $\langle (\_b)(df)cb \rangle$ and $\langle (\_f)cbc \rangle$. By scanning $\langle a \rangle$-projected database once, all the length-2 sequential patterns prefixed with $\langle a \rangle$ can be found. They are: $(\langle aa \rangle : 2)$, $(\langle ab \rangle : 4)$, $(\langle (ab) \rangle : 2)$, $(\langle ac \rangle : 4)$, $(\langle ad \rangle : 2)$, and $(\langle af \rangle : 2)$.

Recursively, all sequential patterns with prefix $\langle a \rangle$ can be partitioned into six subsets: (1) that prefixed with $\langle aa \rangle$, (2) that with $\langle ab \rangle$, ... , and finally, (6) that with $\langle af \rangle$. These subsets can be mined by constructing respective projected databases and mining each recursively.

For example, the $\langle aa \rangle$-projected database consists of only one non-empty (postfix) subsequences prefixed with $\langle aa \rangle$: $\langle (\_bc)(ac)d(cf) \rangle$. Since there is no hope to generate any frequent subsequence from a single sequence, the processing of $\langle aa \rangle$-projected database terminates. Similarly, the $\langle ab \rangle$-projected database consists of three postfix sequences: $\langle (\_c)(ac)d(cf) \rangle$, $\langle (\_c)a \rangle$, and $\langle c \rangle$. Recursively mining it returns four sequential patterns: $\langle (\_c) \rangle$, $\langle (\_c)a \rangle$, $\langle a \rangle$, and $\langle c \rangle$ (i.e., $\langle a(bc) \rangle$, $\langle a(bc)a \rangle$, $\langle aba \rangle$, and $\langle abc \rangle$.)

Using the same method, sequential patterns with prefix $\langle b \rangle$, $\langle c \rangle$, $\langle d \rangle$, $\langle e \rangle$ and $\langle f \rangle$, can be mined from the corresponding projected databases respectively. The projected databases as well as the sequential patterns found are shown in Table 3.6.

The example shows that PrefixSpan examines only the prefix subsequences and projects only their corresponding postfix subsequences into projected databases, and in each projected database, sequential patterns are grown by exploring only local frequent patterns.

To further improve mining efficiency, two kinds of optimizations are explored [30]: (1) *pseudo-projection*, and (2) *bi-level projection*. Pseudo-projection is based on the following idea: When the database can be held in main memory, instead of constructing a *physical* projection by collecting all the postfixes, one can use pointers referring to the sequences in the database as a *pseudo-projection*. Every projection consists of two pieces of information: *pointer* to the sequence in database and *offset* of the postfix in the sequence. This avoids physically copying postfixes. Thus, it is efficient in terms of both running time and space. However, it is not efficient if the pseudo-projection is used for disk-based accessing since random access of disk space is very costly. Therefore, when the sequence database cannot be held in main memory, a *bi-level projection* method is explored, which projects databases not at every level but at every two levels. In comparison with *level-by-level projection*, bi-level projection reduces the cost of database projection and leads to improved performance when the database is huge and the support threshold is low.

A systematic performance study in [30] shows that PrefixSpan with these two optimizations is efficient and scalable. It mines the complete set of patterns and runs considerably faster than both Apriori-based GSP algorithm [35] and FreeSpan [17].

5   Further Development of Pattern Growth-Based Pattern Mining Methodology

Since the proposal of the pattern-growth approach for mining frequent patterns [18], a lot of research has been conducted that extends this methodology in multiple frontiers, including further enhance pattern-growth efficiency at mining frequent patterns (such as [14, 31]), mining structured patterns by pattern growth [37, 38], mining colossal patterns [39], mining approximate patterns [9], mining multi-dimensional patterns [33], pattern-based clustering [32], and pattern-based classification [10, 11].

First, the pattern growth approach has been extended to mine frequent substructures such as frequent subgraph patterns. Graph is a general data structure at modeling sophisticated interconnected data objects, with broad applications including chemical informatics, bioinformatics, computer vision, video indexing, text retrieval, and Web analysis. Among the various kinds of graph patterns, frequent substructures are the very basic patterns that can be discovered in a collection of graphs. The pattern-growth mining algorithms, represented by gSpan [37] for mining frequent subgraph patterns, extends a frequent graph by adding a new edge, in every possible position. A potential problem with the edge extension is that the same graph can be discovered many times. The gSpan algorithm solves this problem by introducing a right-most extension technique, where the extensions will only take place on the right-most path. A right-most path is the straight path from the starting vertex v_0 to the last vertex v_n, according to a depth-first search on the graph. With such an extension, the pattern-growth approach can be extended to mining frequent subgraph patterns with high efficiency. Further, it is also more desirable to mine closed subgraph patterns directly than first mine frequent graph patterns and then filter out those subgraphs that share the same frequency support as their super-graphs. CloseGraph [38] is one such pattern graph algorithm that checks the size of the project graph datasets to prune those paths that cannot generate new closed subgraph patterns.

Second, the pattern growth approach has been extended to pattern-based classification. Frequent patterns have been demonstrated to be useful for classification, where association rules are generated and analyzed for effective classification [21, 22] by discovery of strong associations between frequent patterns and class labels. Moreover, a further study [10] has provided solid reasoning to support the methodology of frequent pattern-based classification. By building a connection between pattern frequency and discriminative measures, such as information gain and Fisher score, it is shown that discriminative frequent patterns are essential for classification, whereas inclusion of infrequent patterns may not improve the classification accuracy due to their limited predictive power. A pattern-growth based methodology, called DDP-Mine [11], is developed that performs a branch-and-bound search for directly mining discriminative patterns without generating the complete pattern set. Instead of selecting best patterns in a batch, a "feature-centered mining approach is introduced that generates discriminative patterns sequentially on a progressively shrinking FP-tree by incrementally eliminating training instances. The instance elimination effectively reduces the problem size iteratively and expedites the mining process. Empirical results show that DDPMine achieves orders of magnitude speedup without downgrading classification accuracy and outperforms the state-of-the-art associative classification methods in terms of both accuracy and efficiency.

6   Conclusions

We have presented a pattern-growth methodology for mining multiple kinds of frequent patterns in large databases. Their associated performance studies show that the algorithms derived from the pattern-growth methodology are more efficient and scalable than many other frequent pattern mining methods.

According to our analysis, the high performance of the pattern-growth methodology is due to the following factors: (1) it adopts a divide-and-conquer strategy to project and partition a large database recursively into a set of progressively smaller ones, and the patterns to be searched for in each corresponding projected database are also reduced substantially; (2) it integrates disk-based database projection algorithms with main memory-based data structures and fast in-memory traversal algorithms, which can be well-tuned to achieve combined high performance by swapping disk-based algorithm into memory-based one when the projected and compressed data set can fit in memory; and (3) it makes good use of the Apriori property implicitly as well as other properties, such as the single tree-path property, but avoids generating a large number of candidates, which ensures each counting and testing is on the real data sets rather than on the potential candidate sets. These several techniques combined lead to high performance mining algorithms.

References

1. R. Agrawal and R. Srikant. Fast algorithms for mining association rules. In *Proc. 1994 Int. Conf. Very Large Data Bases (VLDB'94)*, pages 487–499, Santiago, Chile, Sept. 1994.
2. R. Agrawal and R. Srikant. Mining sequential patterns. In *Proc. 1995 Int. Conf. Data Engineering (ICDE'95)*, pages 3–14, Taipei, Taiwan, Mar. 1995.
3. R. Agrawal, T. Imielinski, and A. Swami. Mining association rules between sets of items in large databases. In *Proc. 1993 ACM-SIGMOD Int. Conf. Management of Data (SIGMOD'93)*, pages 207–216, Washington, DC, May 1993.
4. R. Agarwal, C. C. Aggarwal, and V. V. V. Prasad. Depth-first generation of large itemsets for association rules. In *IBM Technical Report RC21538*, July 1999.
5. R. J. Bayardo. Efficiently mining long patterns from databases. In *Proc. 1998 ACM-SIGMOD Int. Conf. Management of Data (SIGMOD'98)*, pages 85–93, Seattle, WA, June 1998.
6. R. J. Bayardo, R. Agrawal, and D. Gunopulos. Constraint-based rule mining on large, dense data sets. In *Proc. 1999 Int. Conf. Data Engineering (ICDE'99)*, pages 188–197, Sydney, Australia, April 1999.
7. K. Beyer and R. Ramakrishnan. Bottom-up computation of sparse and iceberg cubes. In *Proc. 1999 ACM-SIGMOD Int. Conf. Management of Data (SIGMOD'99)*, pages 359–370, Philadelphia, PA, June 1999.
8. S. Brin, R. Motwani, and C. Silverstein. Beyond market basket: Generalizing association rules to correlations. In *Proc. 1997 ACM-SIGMOD Int. Conf. Management of Data (SIGMOD'97)*, pages 265–276, Tucson, AZ, May 1997.
9. C. Chen, X. Yan, F. Zhu, and J. Han. gApprox: Mining frequent approximate patterns from a massive network. In *Proc. 2007 Int. Conf. Data Mining (ICDM'07)*, Omaha, NE, Oct. 2007.
10. H. Cheng, X. Yan, J. Han, and C.-W. Hsu. Discriminative frequent pattern analysis for effective classification. In *Proc. 2007 Int. Conf. Data Engineering (ICDE'07)*, pages 716–725, Istanbul, Turkey, April 2007.
11. H. Cheng, X. Yan, J. Han, and P. S. Yu. Direct discriminative pattern mining for effective classification. In *Proc. 2008 Int. Conf. Data Engineering (ICDE'08)*, Cancun, Mexico, April 2008.
12. G. Dong and J. Li. Efficient mining of emerging patterns: Discovering trends and differences. In *Proc. 1999 Int. Conf. Knowledge Discovery and Data Mining (KDD'99)*, pages 43–52, San Diego, CA, Aug. 1999.
13. M. Fang, N. Shivakumar, H. Garcia-Molina, R. Motwani, and J. D. Ullman. Computing iceberg queries efficiently. In *Proc. 1998 Int. Conf. Very Large Data Bases (VLDB'98)*, pages 299–310, New York, NY, Aug. 1998.
14. G. Grahne and J. Zhu. Efficiently using prefix-trees in mining frequent itemsets. In *Proc. ICDM'03 Int. Workshop on Frequent Itemset Mining Implementations (FIMI'03)*, Melbourne, FL, Nov. 2003.
15. G. Grahne, L.V.S. Lakshmanan, and X. Wang. Efficient mining of constrained correlated sets. In *Proc. 2000 Int. Conf. Data Engineering (ICDE'00)*, pages 512–521, San Diego, CA, Feb. 2000.
16. J. Han, G. Dong, and Y. Yin. Efficient mining of partial periodic patterns in time series database. In *Proc. 1999 Int. Conf. Data Engineering (ICDE'99)*, pages 106–115, Sydney, Australia, April 1999.
17. J. Han, J. Pei, B. Mortazavi-Asl, Q. Chen, U. Dayal, and M.-C. Hsu. FreeSpan: Frequent pattern-projected sequential pattern mining. In *Proc. 2000 ACM SIGKDD Int. Conf. Knowledge Discovery in Databases (KDD'00)*, pages 355–359, Boston, MA, Aug. 2000.
18. J. Han, J. Pei, and Y. Yin. Mining frequent patterns without candidate generation. In *Proc. 2000 ACM-SIGMOD Int. Conf. Management of Data (SIGMOD'00)*, pages 1–12, Dallas, TX, May 2000.

19. F. Korn, A. Labrinidis, Y. Kotidis, and C. Faloutsos. Ratio rules: A new paradigm for fast, quantifiable data mining. In *Proc. 1998 Int. Conf. Very Large Data Bases (VLDB'98)*, pages 582–593, New York, NY, Aug. 1998.

20. L.V.S. Lakshmanan, R. Ng, J. Han, and A. Pang. Optimization of constrained frequent set queries with 2-variable constraints. In *Proc. 1999 ACM-SIGMOD Int. Conf. Management of Data (SIGMOD'99)*, pages 157–168, Philadelphia, PA, June 1999.

21. W. Li, J. Han, and J. Pei. CMAR: Accurate and efficient classification based on multiple class-association rules. In *Proc. 2001 Int. Conf. Data Mining (ICDM'01)*, pages 369–376, San Jose, CA, Nov. 2001.

22. B. Liu, W. Hsu, and Y. Ma. Integrating classification and association rule mining. In *Proc. 1998 Int. Conf. Knowledge Discovery and Data Mining (KDD'98)*, pages 80–86, New York, NY, Aug. 1998.

23. H. Mannila, H. Toivonen, and A. I. Verkamo. Efficient algorithms for discovering association rules. In *Proc. AAAI'94 Workshop on Knowledge Discovery in Databases (KDD'94)*, pages 181–192, Seattle, WA, July 1994.

24. H. Mannila, H. Toivonen, and A. I. Verkamo. Discovery of frequent episodes in event sequences. *Data Mining and Knowledge Discovery*, 1:259–289, 1997.

25. R. Ng, L.V.S. Lakshmanan, J. Han, and A. Pang. Exploratory mining and pruning optimizations of constrained associations rules. In *Proc. 1998 ACM-SIGMOD Int. Conf. Management of Data (SIGMOD'98)*, pages 13–24, Seattle, WA, June 1998.

26. B. Özden, S. Ramaswamy, and A. Silberschatz. Cyclic association rules. In *Proc. 1998 Int. Conf. Data Engineering (ICDE'98)*, pages 412–421, Orlando, FL, Feb. 1998.

27. J. Pei and J. Han. Can we push more constraints into frequent pattern mining? In *Proc. 2000 ACM SIGKDD Int. Conf. Knowledge Discovery in Databases (KDD'00)*, pages 350–354, Boston, MA, Aug. 2000.

28. J. Pei, J. Han, and R. Mao. CLOSET: An efficient algorithm for mining frequent closed item-sets. In *Proc. 2000 ACM-SIGMOD Int. Workshop on Data Mining and Knowledge Discovery (DMKD'00)*, pages 11–20, Dallas, TX, May 2000.

29. J. Pei, J. Han, and L.V.S. Lakshmanan. Mining frequent itemsets with convertible constraints. In *Proc. 2001 Int. Conf. Data Engineering (ICDE'01)*, pages 433–442, Heidelberg, Germany, April 2001.

30. J. Pei, J. Han, B. Mortazavi-Asl, H. Pinto, Q. Chen, U. Dayal, and M.-C. Hsu. PrefixSpan: Mining sequential patterns efficiently by prefix-projected pattern growth. In *Proc. 2001 Int. Conf. Data Engineering (ICDE'01)*, pages 215–224, Heidelberg, Germany, April 2001.

31. J. Pei, J. Han, H. Lu, S. Nishio, S. Tang, and D. Yang. H-Mine: Hyper-structure mining of frequent patterns in large databases. In *Proc. 2001 Int. Conf. Data Mining (ICDM'01)*, pages 441–448, San Jose, CA, Nov. 2001.

32. J. Pei, X. Zhang, M. Cho, H. Wang, and P. S. Yu. Maple: A fast algorithm for maximal pattern-based clustering. In *Proceedings of the Third IEEE International Conference on Data Mining (ICDM'03)*, Melbourne, Florida, Nov. 2003. IEEE.

33. H. Pinto, J. Han, J. Pei, K. Wang, Q. Chen, and U. Dayal. Multi-dimensional sequential pattern mining. In *Proc. 2001 Int. Conf. Information and Knowledge Management (CIKM'01)*, pages 81–88, Atlanta, GA, Nov. 2001.

34. C. Silverstein, S. Brin, R. Motwani, and J. D. Ullman. Scalable techniques for mining causal structures. In *Proc. 1998 Int. Conf. Very Large Data Bases (VLDB'98)*, pages 594–605, New York, NY, Aug. 1998.

35. R. Srikant and R. Agrawal. Mining sequential patterns: Generalizations and performance improvements. In *Proc. 5th Int. Conf. Extending Database Technology (EDBT'96)*, pages 3-17, Avignon, France, Mar. 1996.

36. R. Srikant, Q. Vu, and R. Agrawal. Mining association rules with item constraints. In *Proc. 1997 Int. Conf. Knowledge Discovery and Data Mining (KDD'97)*, pages 67–73, Newport Beach, CA, Aug. 1997.

37. X. Yan and J. Han. gSpan: Graph-based substructure pattern mining. In *Proc. 2002 Int. Conf. Data Mining (ICDM'02)*, pages 721–724, Maebashi, Japan, Dec. 2002.
38. X. Yan and J. Han. CloseGraph: Mining closed frequent graph patterns. In *Proc. 2003 ACM SIGKDD Int. Conf. Knowledge Discovery and Data Mining (KDD'03)*, pages 286–295, Washington, DC, Aug. 2003.
39. F. Zhu, X. Yan, J. Han, P. S. Yu, and H. Cheng. Mining colossal frequent patterns by core pattern fusion. In *Proc. 2007 Int. Conf. Data Engineering (ICDE'07)*, Istanbul, Turkey, April 2007.

Chapter 4
Mining Long Patterns

Feida Zhu

Abstract The value and importance of long patterns are gaining increasing recognition in a wide range of domains including bioinformatics, social network analysis, software engineering and business intelligence. Yet the task of mining long patterns has remained a challenge due to the prohibitively large number of smaller patterns which often need to be generated first. In this chapter, we first use a pattern lattice model to illustrate and compare various mining paradigms. Then we present recent studies for mining long patterns according to their respective pattern mining paradigms. For each category, we discuss the representative algorithms and the state-of-the-art development.

Keywords Frequent pattern · Long pattern · Colossal pattern · Large pattern

1 Introduction

Pattern mining has been a central theme for data mining since its inception. Among the various constraints often imposed on the patterns to be mined, patterns of large sizes are of growing interest for a number of reasons. Firstly, long patterns are a natural result of ever larger data sets. For example, social network analysis on a network like that of Facebook or Twitter has been shown that functional communities could reach size up to 150, much larger than what most algorithms can typically mine. Similarly, for web structure mining in today's Internet, one should expect the real web structures mined for any domain to be fairly complicated. Secondly, long patterns are more informative in characterizing large data sets and in many cases (e. g., bioinformatics) give more meaningful insights than shorter patterns. For example, in DBLP co-authorship network, small patterns, e.g., several authors collaborate on a paper, are almost ubiquitous. It is shown in [33] that only long patterns would reveal interesting common collaborative patterns, or distinguish distinct patterns, across different research communities. In software engineering, long patterns uncovered

F. Zhu (✉)
Singapore Management University, Singapore, Singapore
e-mail: fdzhu@smu.edu.sg

C. C. Aggarwal, J. Han (eds.), *Frequent Pattern Mining,*
DOI 10.1007/978-3-319-07821-2_4, © Springer International Publishing Switzerland 2014

from program structure data would also reveal software *backbones* which are critical in analyzing large software packages and understanding legacy systems [15, 17].

It should be noted that long patterns are often sought after with another important constraint of frequency, which contributes to much of the mining complexity. In the following, we would denote most of this chapter to large frequent patterns, which are referred to simply as long patterns unless otherwise specified.

Despite its significance in many applications, long patterns are not easy to find. The fact that most existing algorithms discover frequent patterns with increasing sizes means that, before larger patterns can be identified, smaller ones would have to be explored, which typically come in exponentially large numbers. This poses serious challenges to mining algorithms as the swamp of smaller patterns that they have to examine could easily prohibit them from ever reaching the large ones in a reasonable amount of time. While the task is already hard in the case of item sets and sequences, the extra dimension of structural complexity in graph pattern mining exacerbates the situation.

In this chapter, we will first give the preliminaries of large frequent pattern mining, and introduce a pattern lattice model to explain the various algorithms we later present. We then categorise algorithms for mining long patterns by their mining paradigm into three categories: mining by pattern enumeration, mining by pattern merging and mining by pattern traversal with neighborhood adjacency.

2 Preliminaries

Large frequent pattern mining has been primarily studied in three data settings: item sets, sequences and graphs, which we define accordingly as follows.

Item Sets Item sets presents the simplest setting of frequent pattern mining. Let $\mathcal{I}$ be a set of items $\{o_1, o_2, \ldots, o_d\}$. A nonempty subset of $\mathcal{I}$ is called an *itemset*. A transaction dataset D is a collection of itemsets, $D = \{t_1, \ldots, t_n\}$, where $t_i \subseteq \mathcal{I}$. For any itemset α, we denote the set of transactions that contain α as $D_\alpha = \{i | \alpha \subseteq t_i \text{ and } t_i \in D\}$. Define the cardinality of an itemset α as the number of items it contains, i.e., $|\alpha| = |\{o_i | o_i \in \alpha\}|$.

Definition 4.1 (Frequent Itemset) *For a transaction dataset D, an itemset α is frequent in D if $\frac{|D_\alpha|}{|D|} \geq \sigma$, where $\frac{|D_\alpha|}{|D|}$ is called the support of α in D, written $s(\alpha)$, and σ is the* minimum support threshold, $0 \leq \sigma \leq 1$.

Mining long patterns in item set setting is simply to find all frequent item sets with cardinality greater than a user-specified threshold.

Sequences The setting of sequences includes two related yet different cases: frequent substrings and frequent subsequences, the latter being computationally much more challenging than the former. Given a string $S = \langle s_1, \ldots s_n \rangle$ of length n, another string $Z = \langle z_1 \ldots z_m \rangle, m \leq n$ is a *subsequence* of S if there exists a sequence of indices $I = \langle i_1, \ldots, i_m \rangle, i_j < i_{j+1}, 1 \leq j < m$ such that such that $z_j = s_{i_j}$ for all $1 \leq j \leq m$. We call Z a *subsequence* of S, denoted as $S \subseteq Z$. If, in particular, we have $i_j = i_{j+1} - 1, 1 \leq j < m$, then We call Z a *substring* of S. We use $|S|$ to denote the length of a string S.

Definition 4.2 (Frequent Subsequence) *For a string dataset $D = \{S_1, \ldots, S_n\}$ and a string S, let D_S be the subset of D such that $S \subseteq S'$ for all $S' \in D_S$. Then S is frequent in D if $\frac{|D_S|}{|D|} \geq \sigma$, where $\frac{|D_S|}{|D|}$ is called the support of S in D, written $s(S)$, and σ is the* minimum support threshold, $0 \leq \sigma \leq 1$.

Graphs The setting of graphs represent the most complicated case for mining long patterns, which is further divided into two settings: graph transaction setting and single graph setting. As a convention, the *vertex set* of a graph G is denoted by $V(G)$ and the *edge set* by $E(G)$. The size of a graph P is defined by the number of edges of P, written as $|P|$. In frequent graph mining setting, a graph $G = (V(G), E(G))$ is associated with a labeling function $l_G : V(G) \mapsto \Sigma, \Sigma = \{\varsigma_1, \varsigma_2, \ldots, \varsigma_k\}$. Graph isomorphism in our problem setting requires matching of the labels for each mapped pair of vertices. Most methods can also be applied to graphs with edge labels.

Definition 4.3 (Labeled Graph Isomorphism) *Two labeled graphs G and G' are isomorphic if there exists a bijection $f : V(G) \mapsto V(G')$, such that $\forall u \in V(G)$, $l_G(u) = l_{G'}(f(u))$ and $(u, v) \in E(G)$ if and only if $(f(u), f(v)) \in E(G')$.*

We use $G \cong_L G'$ to denote that two labeled graphs G and G' are isomorphic. Given two graphs P and G, a subgraph G' of G is called an *embedding* of P in G if $P \cong_L G'$. For a single graph G and a pattern P, we use e_P to denote a particular embedding of a pattern P, and the set of all embeddings of P is denoted as $E[P]$. We denote as P_{sup} the support set for a pattern P. In single graph setting, $P_{sup} = E[P]$ while in graph transaction setting P_{sup} is the set of graphs of the database each containing at least one embedding of P.

Definition 4.4 (Frequent Graph) *Given D as a graph dataset $D = \{G_1, \ldots, G_n\}$ or a single graph, and a graph G, G is frequent in D if $\frac{|P_{sup}|}{|D|} \geq \sigma$, where σ is the* minimum support threshold, $0 \leq \sigma \leq 1$.

In an effort to reduce the size of the frequent pattern mining result, concepts of *closed* patterns and *maximal* patterns have been proposed which apply to all the different data formats.

Definition 4.5 (Closed Pattern) *A pattern p is a closed pattern in a data set D if p is frequent in D and there exists no proper super-pattern p' such that $p \subset p'$ and p' has the same support as p in D.*

Definition 4.6 (Maximal Pattern) *A pattern p is a maximal pattern in a data set D if p is frequent in D and there exists no super-pattern p' such that $p \subset p'$ and p' is frequent in D.*

It is worth noting that long patterns in a data set are usually the maximal patterns. As such, algorithms mining for maximal patterns would naturally return the long patterns. We therefore give priority to these algorithms in this chapter.

Fig. 4.1 Pattern Lattice
Model

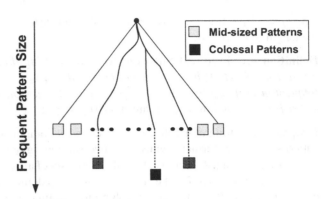

3 A Pattern Lattice Model

To compare different pattern mining algorithms, it is often beneficial to visualize the pattern search space and mining process in a pattern lattice model as illustrated in Fig. 4.1. A pattern lattice is a conceptual structure where each node of the lattice represents a pattern. Nodes at level i represent patterns of size i. A node α at level k is a child of a node β at level $k - 1$ if and only if $\alpha \subset \beta$ and $|\beta| = |\alpha| + 1$.

A mining process can be considered as a traversal along the nodes in the pattern lattice. A node is visited when the algorithm examines the pattern correspondent to the node. As such, different mining algorithms can be visualised as different ways of traversing the pattern lattice. Accordingly, the total mining cost of an algorithm can be evaluated by the average processing time spent for the pattern at each node summed up over all the visits to the nodes during the traversal of the algorithm along the pattern lattice.

The different ways in which mining algorithms traverse the pattern lattice can be classified into the following four main types. (I) Breadth-first enumeration, (II) Depth-first enumeration, (III) Merge-and-Leap, and (IV) Pattern traversal. Both breadth-first and depth-first style mining strategies enumerate all the pattern candidates, and would accordingly have to spend exponential time when the number of closed or maximal mid-sized patterns explodes, even though there are only a few truly long patterns. Section 4 discusses both enumeration approaches. Merge-and-Leap style methods instead aim to reach long patterns as soon as possible by taking leaps in the pattern lattice, visiting only a few nodes along each path toward a long pattern. This is made possible by merging together two or more subgraph patterns already discovered to form a larger one. Section 6 presents representative algorithms in this category. Yet another set of algorithms traverse the pattern lattice by identifying some target patterns and visit adjacent patterns generate the result, examples of which are discussed in Sect. 7.

4 Pattern Enumeration Approach

The most straightforward way of mining the long patterns would be to enumerate all the frequent patterns and report those whose sizes are sufficiently large. Depending on the way they traverse the pattern lattice, they can be categorised into two classes: Breadth-First Approach and Depth-First Approach.

4.1 Breadth-First Approach

The breadth-first approach traverse the pattern lattice level by level, always examining all patterns at one level before going down to the next level. The representative algorithms are the Apriori-based ones. The search for frequent patterns is conducted from patterns of smaller sizes to larger ones by levels of the pattern lattice. At each level, a new frequent pattern is discovered by joining two similar but slightly different frequent patterns discovered at the previous level [1]. To expedite the mining for larger patterns, some typical look-ahead technique has been proposed to identify maximal frequent patterns without visiting every frequent patterns. In the setting of itemset mining, an early work to adopt the look-ahead technique is [30] presenting MaxEclat and MaxClique in which the algorithms look ahead during the initialization stage to identify large frequent itemsets. MaxMiner [4] is an algorithm improved beyond [30] to employ a breadth-first traversal of the pattern lattice for finding maximal itemsets. It uses a look-ahead pruning technique throughout the search to identify long patterns as early as possible, thus reducing database scanning, i.e., if a node with all its extensions can be determined to be frequent, there is no need to further process that node. Besides it also employs item re-ordering heuristic to increase the effectiveness of superset-frequency pruning. As a result, MaxMiner is able to achieve a performance improvement of at least an order of magnitude compared to other look-ahead techniques. In practice, MaxMiner has demonstrated a runtime which is roughly linear in the number of maximal frequent itemsets and the size of the database, irrespective of the size of the largest frequent itemset, which is significantly faster than previous Apriori-based approaches that scale exponentially with the size of the largest pattern.

For sequential pattern mining, long patterns have been studied in a noisy environment such as gene expression analysis in [26], where long patterns are expected yet symbols can be misrepresented to prevent frequent patterns from being correctly discovered. The authors proposed a sampling-based method using the well-known Chernoff bound to estimate the ambiguous patterns whose matches in the sample are very close to the threshold, so that there is no sufficient statistical confidence to tell whether the pattern would be frequent or not in the entire database. In addition, to speed up the pattern frequency verification, a technique called *border collapsing* was proposed based on the observation that the set of all ambiguous patterns occupy a contiguous portion of the pattern lattice by the Apriori property. The border of frequent patterns can then be located efficiently by successively collapsing the gap between the lower and upper borders, minimizing the expected number of scans through the entire database.

In graph mining, AGM [11], the typical Apriori-based frequent graph mining algorithm, generates a new frequent graph pattern candidate of size $k + 1$ by joining two size-k only if they share the same size-$(k - 1)$ subgraph, where the size here is measured by the number of vertices. Similar approach of joining is also adopted by FSG [13] which adopts an edge-based candidate generation scheme.

The advantage of breadth-first approach is the completeness of mining result and minimum number of lattice node visits as a result of the Apriori-style pattern joining in the pattern candidate generation. However, the fact that the breadth-first approaches would exhaust pattern candidates at one level before going down to larger ones at the next level makes reaching long patterns particularly difficult for these algorithms. What is worse, the exponentially large number of potential patterns of medium sizes could make the mining algorithm to get stuck and fail to find any long pattern before draining the system memory. To overcome this issue, algorithms have been proposed to adopt a depth-first approach as discussed below.

4.2 Depth-First Approach

Instead of enumerating all the pattern candidates of size k before exploring larger ones, depth-first approach grows a pattern as much as possible until the frequency threshold cannot be satisfied, adopting a depth-first style of traversal down the pattern lattice. The advantage of such an approach is the following. First, search space pruning could be most effective in this case. This is due to the fact that maximal pattern mining algorithms depend on "look-ahead" technique in which an entire subtree of an enumeration node in the lexicographic tree is pruned if the longest pattern that can possibly be generated from that subtree is a subset of a frequent pattern that has already been found. Clearly, the pruning will be most effective when long or maximal patterns are found earlier in the exploration process. Indeed, a depth-first strategy always explores the itemsets in certain canonical order (after fixing the lexicographic ordering of items). Most of the maximal patterns are always found earlier than their subsets in this order. For a pattern of length l, only $(l - 1)$ of its (immediate prefix) subsets from the 2^l possibilities are explored before discovering the pattern. In comparison, breadth-first algorithms with level-wise exploration are denied the chance of such pruning because they find all patterns of the same size at any given stage of the algorithm. Second, long patterns are more likely to be discovered by avoiding being trapped by the huge number of mid-sized ones. Third, the depth-first exploration strategy better facilitates a memory-efficient reuse of the counting work done at the higher levels of the enumeration tree with the use of projected databases.

As a pioneer, DepthProject [2] finds maximal long itemsets by a depth-first search of a lexicographic tree of itemsets, together with the look-ahead pruning technique with item re-ordering. By post-pruning, it eliminates non-maximal patterns and returns the maximal pattern set. GenMax as proposed in [7] discovers all maximal frequent itemsets by performing a depth-first traversal of the pattern lattice with a backtracking search strategy. In addition, GenMax also adopts a progressive focusing technique to eliminate non-maximal itemsets and uses diffset propagation for fast frequency checking.

Just like in the breadth-first case, algorithm designers adopting a depth-first approach have identified certain enumeration order to guarantee unique pattern generation. gSpan [25] is such an algorithm proposed for frequent graph pattern mining.

gSpan solves the redundant pattern candidate generation problem by proposing a *right-most extension* technique, which imposes an order of pattern growth by only allowing edge extension on the *right-most path* [25]. A right-most path for a given graph is the straight path from the starting vertex v_0 to the last vertex v_n by a depth-first search on the graph. To conduct a depth-first search, it is necessary to identify first a particular node of the graph as the root, which is itself an interesting question given a graph pattern candidate. gSpan proposes a DFS coding technique to order all the different root selections on the same graph. The one with the minimum DFS code is the canonical root of the graph candidate, and only this rooted graph gets extended in the pattern growth. With all these designs, gSpan is able to guarantee the non-redundant generation of all graph pattern candidates without compromising the completeness of the mining result.

Other pattern-growth mining methods proposed along the years include Mafia [6] and FP-growth [9] for itemset mining. In the sequence pattern mining frontier, pioneers along the pattern-growth line include PrefixSpan [21], FreeSpan [8], TreeMiner [28], SPADE [27] and FEQT [3]. For graph mining, we have MoFa [5], FFSM [10], SPIN [22], Gaston [16] and TSMiner [12] for mining large-scale topological structures.

For both breadth-first and depth-first approaches, the inherent computational bottleneck lies in the fact that they still need to examine all the frequent pattern candidates and apply the size constraint to finally find the long patterns. When the total number of frequent patterns is exponentially huge, all these algorithms have difficulties in finish mining.

5   Row Enumeration Approach

There has been another line of work which, instead of enumerating patterns explicitly by enumerating columns, finds frequent patterns by row enumeration. We call this body of work row enumeration approach, which are based on the following observations. While the set of algorithms like Close [20] adopt breadth-first search which is differerent from those adopting depth-first search such as CLOSET+ [24] and CHARM [29], one thing they have in common is that they all perform pattern enumeration by explicitly enumerating the feature sets, which is usually called column

Fig. 4.2 Table transformation

i	$\mathscr{F}(r_i)$
1	a,c,d
2	a,b,d,e
3	b,e
4	b,c,d,e
5	a,b,c,e

f_j	$\mathscr{R}(f_j)$
a	1,2,5
b	2,3,4,5
c	1,4,5
d	1,2,4
e	2,3,4,5

a Original Example Table, T   **b** Transposed Table, TT

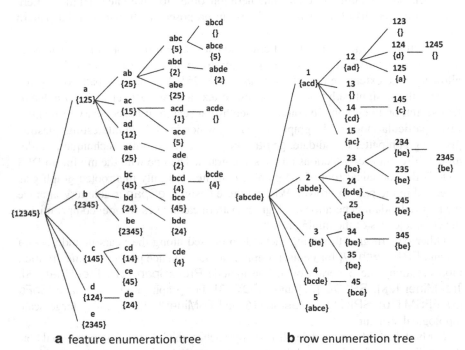

a feature enumeration tree   **b** row enumeration tree

Fig. 4.3 Row enumeration tree

enumeration. Column enumeration works best in data settings with a large number of rows but a small number of columns as their running time increases exponentially with increasing average number of columns. However, the growth of bioinformatics has presented datasets which typically contain a large number of columns and a small number of rows. For example, many gene expression datasets may contain 10,000–100,000 columns but only 100–1000 rows. This has inspired algorithms like CARPENTER [18] to explore row enumeration.

CARPENTER discovers frequent closed patterns by performing depth-first row-wise enumeration instead of the usual column enumeration, and is combined with efficient search pruning techniques. The algorithm starts with transforming the original data table, which corresponds to a column pattern enumeration tree, into a transposed data table where the rows are the columns in the original table, which corresponds to a row enumeration tree. Figure 4.2 illustrates the table transformation and Fig. 4.3 shows the corresponding row enumeration tree.

As each closed pattern corresponds to a unique set of rows as its support, by enumerating all combinations of rows as shown in Fig. 4.3, all closed patterns in the data sets are guaranteed to be enumerated. Note that due to the relatively small number of columns, the row enumeration tree is much smaller compared against the usual column pattern enumeration tree. Yet it is also obvious that a complete traversal of the row enumeration tree is not efficient. CARPENTER proposes three pruning techniques to reduce unnecessary searches.

- **Pruning Step 1.** This pruning is aimed at removing search branches which can never yield closed patterns that satisfy the minimum support threshold. It drops a current search branch if the total number of distinct rows in the current X-conditional transposed table together with those rows already identified as containing X is still less than the minimum support threshold.
- **Pruning Step 2.** This pruning is to expedite the search by identifying the rows that occur in all tuples of the current X-conditional transposed table, from which such rows would be immediately removed.
- **Pruning Step 3.** This pruning is to drop any further search down the branch of a node if it is found that the corresponding column set has been discovered previously in the enumeration tree. It is based on the observation that the set of closed patterns that will be enumerated from the descendants of a node must have been enumerated previously as only closed patterns will be enumerated in the row enumeration search tree.

By exploiting the asymmetry in the row and column sizes, CARPENTER is reported to achieve orders of magnitude faster than her counterpart with column enumeration strategy.

Interestingly, it is natural to raise questions for data sets that have both large number of rows and features since both row enumeration and column enumeration approaches would have difficulty in handling such data. One solution has been proposed in [19] named COBBLER which is designed to dynamically switch between feature enumeration and row enumeration depending on the data characteristic in the process of mining. As such, each portion of the data set can be processed using the most suitable method making the mining more efficient.

COBBLER is based on the notion of a dynamic enumeration tree. There are two motivations for adopting a more dynamic approach.

- First, the characteristics of the conditional tables could be different from the original table. Since the number of rows (or tuples) can be reduced as we move down the enumeration tree, it is possible that a table which has more rows than features initially, could have the characteristic reversed for its conditional tables (i.e. more features than rows). As such, it makes sense to adopt a different enumeration approach as the data characteristic changes.
- Second, for data sets with large number of rows and also large number of features, a combination of row and feature enumeration could help to reduce both the number of rows and features being considered in the conditional tables thus enhancing the efficiency of mining.

The dynamic switching of enumeration method contains two types of operations.

Fig. 4.4 COBBLER main
algorithm

Algorithm
Input: Original table T, transposed table TT, features set F, row set R and support level *minsup*
Output: Complete set of frequent closed patterns, FCP
Method:
1. *Initialization. FCP =* $\emptyset$;
2. *Check switching conditions. SwitchingCondition();*
3. *If mine frequent closed patterns in row enumeration first.* RowMine($TT\vert_\emptyset$,R,FCP);
4. *If mine frequent closed patterns in feature enumeration first.* FeatureMine($T\vert_\emptyset$,F,FCP);

- **Feature to Row Enumeration Switch.** This operation starts with first creating a transposed table for the current feature set, such that we have a tuple for each feature having lower rank than all those in the current feature set; and then performing row enumeration on the transposed table as in CARPENTER.
- **Row to Feature Enumeration Switch.** This operation creates a conditional table such that all feature combinations that is a superset of the feature set represented by the nearest ancestor of the current row enumeration node but a subset of the feature set of the current row enumeration node can be tested systematically based on feature enumeration.

The main algorithm of COBBLER is shown in Fig. 4.4. COBBLER performs a recursive computation of conditional tables and conditional transposed tables for performing a depth-first traversal of the dynamic enumeration tree. Each conditional table represents a feature enumerated node while each conditional transposed table represents a row enumerated node.

One more factor to consider here is the switching condition which are used to decide whether to switch from row enumeration to feature enumeration or vice versa. The main idea adopted in COBBLER is to estimate the enumeration cost for the subtree at a node and select the smaller one between a feature enumeration subtree and a row enumeration subtree.

While COBBLER takes the best part of both the column and row enumeration approach and outperforms previous closed frequent pattern mining algorithms, it has not resolved the inherent complexity barrier for mining large frequent patterns because smaller patterns still need to be generated before larger ones in both enumeration approaches.

6   Pattern Merge Approach

To avoid enumerating all the smaller pattern candidates before reaching the larger ones, a number of algorithms have been proposed which adopt a pattern merge approach. The general idea is to first mine a set of small frequent patterns, and then iteratively merge them into larger ones until no new patterns can be found. From the perspective of the pattern lattice model, pattern merge approach would traverse the pattern lattice in a leaping fashion, jumping from a smaller size pattern to a much larger one directly, and attempt to reach larger patterns as quickly as possible. As they do not attempt to examine the entire pattern candidate space, algorithms of pattern merge approach usually do not aim to return the complete set of long patterns over a certain size constraint. Rather, they either set out to find the top-k largest patterns and guarantee their success by probabilistic arguments, or strive to capture as many long pattern as possible in an best-effort style. Depending on the aggressiveness of the merging, these algorithms can be further put into two types—Piece-wise Merge and Fusion-style Merge.

6.1 Piece-wise Pattern Merge

Piece-wise Pattern Merge algorithms would try to merge two smaller patterns found so far to generate a larger new one. We show two examples here for sequence mining and graph mining respectively.

Long Approximate Sequential Pattern In [32], a piece-wise pattern merge style algorithm is proposed to find the complete set of closed frequent approximate sequential patterns defined with Hamming distance. Hamming distance, defined for two strings of equal length, is the number of substitutions required to change one into the other. Frequent approximate substrings are defined as follows.

Definition 4.7 (Frequent Approximate Substring (FAS)) *Given a string S, a minimum frequency threshold θ and an error tolerance threshold δ, a substring P of S is a frequent approximate substring if and only if there exists a set U of substrings of S and for each $W \in U$, $HammingDist(P, W) \leq |P|\delta$, and $|U| \geq \theta$. U is called the support set of P, denoted as P_{sup}.*

U is represented as a set of indices of S as all substrings in U share the same length as P. Given a input string S, we are interested in finding all frequent approximate substrings of S, i.e., for each such substring, the set of substrings that are considered approximately the same must be sufficiently large. To reduce redundancy in the result, a notion of *closed frequent approximate substring* is also proposed [32].

The design of the algorithm relies on a notion of a *strand*, which is a set of substrings that share one same matching pattern. Consider four substrings S_1, S_2, S_3 and S_4 as shown in Fig. 4.5.

All four substrings are of length 20. If the error tolerance threshold $\delta = 0.1$ and minimum frequency threshold $\theta = 4$, then S_2 is a FAS since the other three substrings are within Hamming distance 2 from S_2. For each substring, the bounding boxes indicate the parts that match exactly with S_2. In this case, S_1 and S_2 have the same matching patterns. S_2, S_3 and S_4 have the same matching patterns. In general, aligning any two substrings W_1 and W_2, one can observe an alternating sequence of maximal matching substrings and gaps of mismatches

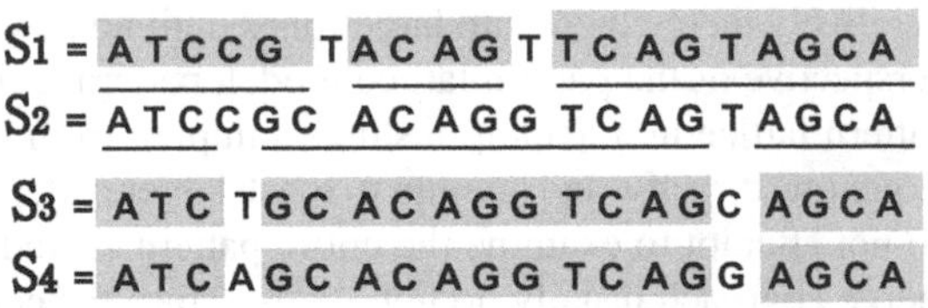

Fig. 4.5 Examples of
biological sequences

S_1 = A T C C G   T A C A G   T T C A G   T A G C A

S_2 = A T C C G C   A C A G G   T C A G   T A G C A

S_3 = A T C   T G C   A C A G G   T C A G C   A G C A

S_4 = A T C   A G C   A C A G G   T C A G G   A G C A

$Pattern(W_1, W_2) = \langle M_1, g_1, M_2, g_2, \ldots, M_k \rangle$, where M_i, $1 \leq i \leq k$ denote the maximal matching substrings shared by W_1 and W_2. g_i, $1 \leq i \leq k$ denote the number of mismatches in the ith gap. In this example of S_1 and S_2, $Pattern(S_1, S_2) = \langle ATCCG, 1, ACAG, 1, TCAGTTGCA \rangle$.

For any two substrings S and P, it is observed that (1) $Pattern(S, P)$ is uniquely defined, and (2) $Pattern(S, P) = Pattern(P, S)$.

One can therefore define the notion of a *strand*, which is a set of substrings that share one same matching pattern.

Definition 4.8 *A set U of substrings $U = \{S_1, \ldots, S_k\}$ is a strand if and only if (1) for any two pairs of substrings $\{S_{i_1}, S_{j_1}\}$ and $\{S_{i_2}, S_{j_2}\}$ of U, $Pattern(S_{i_1}, S_{j_1}) = Pattern(S_{i_2}, S_{j_2})$.*

Based on the idea of a strand, the following approach can be used to decide if a given substring P is a FAS. Find all the closed valid strands of P and let the union of them be X. P is a FAS if and only if the cardinality of X is at least θ. Consider the example of Fig. 4.5 in which the error tolerance is 0.1 and minimum frequency threshold is 4. Both strands $\{S_1, S_2\}$ and $\{S_2, S_3, S_4\}$ are valid. Suppose these two strands are also closed, then combining them one gets a support set of size 4, satisfying the frequency requirement. As such, S_2 is a FAS.

On the highest level, the algorithm works in two steps.

1. **Growing Strand**
   Compute a set of closed valid strands initially. Mine out all closed valid strands by iteratively growing current ones on both ends. Let the result set be X.
2. **Grouping Strand**
   For each distinct substring P in the closed valid strands of X, group all the strands which contain P by taking the union of them. If the result of the union contains at least θ members, report P as a FAS.

The set of initial strands is the set of all maximal exact repeats. More precisely, for each initial strand U, $Pat(U) = \langle M_1 \rangle$, $Miss(U) = 0$ and U is closed. These initial strands are computed by $InitStrand$ using the suffix tree of the input sequence S. Similar approach has been used in REPuter [14] to mine exact repeats. A suffix tree is a data structure that compactly encodes the internal structure of a string. As such, it can be used to solve some complicated string problems in linear time. In particular, it enables us to mine out all frequent maximal exact-matching substrings of S with a running time linear in the length of S. When growing a current strand, the algorithm scans the entire tape and, for each strand encountered, checks on both ends to see if the current strand can be grown by assembling neighboring strands.

Fig. 4.6 Two strands U_1 and U_2 and their substring relation graph

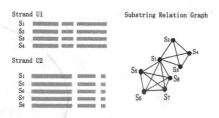

After finding all the closed valid strands in the first step, the algorithm computes the support set for each frequent approximate substring. The idea of grouping the strands is the following. Given the set X of all closed valid strands, we construct a substring relation graph G from X. The vertex set is all the substrings in the strands of X, each vertex representing a distinct substring. There is an edge between two substrings if and only if the Hamming distance between two substrings is within the error tolerance. Since all the substrings in one valid strand share the same distance among each other and the distance is within the error tolerance, all corresponding vertices in G form a clique. After scanning all the strands in X, we would construct a graph G which is a union of cliques. Then a substring is a frequent approximate substring if and only if the degree of the corresponding vertex is greater than or equal to the minimum frequency threshold, as illustrated in Fig. 4.6.

Large Graph Pattern   In [33], a piece-wise pattern-merge style mining algorithm targeted at large graph patterns, called SpiderMine, has been proposed based on the concept of a spider, which is critically important in both identifying and faster reaching the long patterns. Formally, an r-spider is defined as follows.

Definition 4.9   [r-spider] *Given a frequent pattern P in graph G and a vertex $u \in V(P)$, if P is r-bounded from u, we call P an r-spider with head u.*

The SpiderMine algorithm is designed to solve the problem of mining approximate top-K long patterns with bounded diameter in a single large graph. The basic challenges for the problem are two-fold: (1) How to identify the top-K largest patterns with a high probability? and (2) How to quickly reach the long patterns?

As trying all the possible growth paths is unaffordable, one has to identify a small set of highly potential ones which would lead to the long patterns with good chance. The SpiderMine solution is based on the following observation: *long patterns are composed of a large number of small components which would eventually become connected after certain rounds of growth.* The more of such small components of a long pattern we can identify, the higher chance we can recover it. Thus, SpiderMine first mines all such small frequent patterns, which are the *spiders* as defined in Definition 4.9. Compared with small patterns, long patterns contain far more spiders as their subgraphs. It follows that if one picks spiders uniformly at random from the complete spider set, the chance that one would pick some spider within a long pattern is accordingly higher. Moreover, if the algorithm carefully decides on the number of spiders to be randomly picked, the probability that multiple spiders within P would be chosen is higher if P is a larger pattern than a smaller one. Denote the set of all

Fig. 4.7 SpiderMine

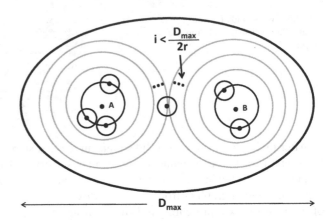

spiders within P which are initially picked in the random draw as H_P. According to the authors' observation, for any two spiders in H_P, there must be a pattern growth path such that along the path their super-patterns will be able to merge. And SpiderMine is going to catch that as follows. Once all the spiders are picked, they will be grown to larger patterns in λ iterations where λ will be determined by D_{max}. In each iteration, each spider will be grown in a procedure called SpiderGrow(), which always expands the current pattern by appending spiders to its boundary such that the pattern's radius is increased by r. Also, in each iteration, two patterns will be merged if some of their embeddings are found to overlap and the resulting merged pattern is frequent enough. Now for any long pattern P, the following Lemma holds,

Lemma 4.10 *For any pattern P with diameter upper-bound D_{max}, let Spider-Grow(Q) be a procedure which grows a pattern Q such that the radius of Q is increased by r, then all patterns growing out of H_P which are sub-patterns of P must have merged into one sub-pattern of P after $\lambda = \frac{D_{max}}{2r}$ iterations of running SpiderGrow(Q).*

This means that as long as one picks more than one spider within a long pattern P in the initial random draw, i. e., $|H_P| > 1$, one can guarantee he will not miss P by retaining all the merged patterns. On the other hand, for smaller patterns, the probability that more than one spider within the pattern get picked in the random draw is much lower than that of long patterns. As such, keeping only the merged patterns at the end of the iterations would highly likely prune away patterns that would grow only toward small patterns. Thus after the pruning, what are left are a small number of candidates each of which, with high probability, is a subgraph of long patterns. We then use SpiderGrow() again to further extend these candidates until no larger patterns can be found.

The SpiderMine algorithm works in the following three stages. An illustration is given in Fig. 4.7.

1. **Stage I: Mining Spiders**

   Mine all r-spiders from the input graph G. By the end of this stage, all the frequent patterns up to a diameter $2r$ with all their embeddings in G are obtained.

2. **Stage II: Large Pattern Identification**

   Randomly pick M spiders from all the spiders obtained in Stage I as the initial set of frequent subgraphs. The next step consists of $\frac{D_{max}}{2r}$ iterations. In each iteration, use SpiderGrow() to grow each of the M subgraphs by extending its boundary with selected spiders such that the radius of the subgraph is increased by r. In each iteration, if it is detected that two frequent subgraphs, whose embeddings are all previously disjoint, begin to overlap on some of their embeddings as a result of growth in this iteration, the algorithm would merge them if the resulting merged subgraph is frequent. Note that one can avoid pair-wise checking for potential merging because all patterns grow with spiders as units and one only has to monitor the same spiders being used by different patterns to detect overlapping. At the end of the $\frac{D_{max}}{2r}$ iterations, keep only those frequent subgraphs which are generated as a result of merging at some iteration. Let the set kept be S. The frequent subgraphs in S are believed to be subgraphs of long patterns with high probability.

3. **Stage III: Large Pattern Recovery**

   With high probability, each one of the top-K long patterns now has some portion of it as a pattern in S. To recover the full patterns, grow each subgraph in S by SpiderGrow() until no more frequent patterns can be found. All the patterns discovered so far are maintained in a list sorted by their size. Return the top-K patterns.

A key question here is that, in Stage II of SpiderMine, how to choose M, the number of initial seed spiders, to achieve the discovery of top-K largest patterns with guaranteed probability. If more than one spider within a pattern P are chosen in the random drawing process, it is said that P is successfully identified. Denote as $P_{success}$ the probability that all the top-K largest patterns are successfully identified. In [33], the authors show the following lemma with detailed proof sketch.

Lemma 4.11 *Given a network G and a user-specified K, we have $P_{success} \geq \left(1 - (M+1)(1 - \frac{V_{min}}{|V(G)|})^M\right)^K$.*

V_{min} is the minimum number of vertices in a long pattern required by users, usually an easy lower bound that a user can specify. Now to compute M, we just need to set $\left(1 - (M+1)(1 - \frac{V_{min}}{|V(G)|})^M\right)^K = 1 - \epsilon$ and solve for M. It follows that, once the user specifies K and ϵ, we could compute M accordingly, and then if we pick M spiders initially in the random drawing process, we are able to return the top-K largest patterns with probability at least $1 - \epsilon$. For example, with $\epsilon = 0.1$, $K = 10$, and $V_{min} = \frac{|V(G)|}{10}$, we get $M = 85$, which means to return top 10 largest patterns(each of size at least $\frac{|V(G)|}{10}$ if any) with probability at least 90 %, we need to randomly draw 85 spiders initially. With the analysis above, it is not hard to prove the following theorem.

Theorem 4.12 *Given a graph G, the error bound ϵ, the diameter upper bound D_{max}, the support threshold σ and K, with probability at least $1 - \epsilon$,* SpiderMine *returns a set S of top-K largest subgraphs of G such that for each $P \in S$, $|P_{sup}| \geq \sigma$ and $diam(P) \leq D_{max}$.*

It has been shown in [33] that the reasons why the spider-based algorithm could recover long patterns efficiently are (1) Spiders reduce combinatorial complexity in recovering long patterns, and (2) Spiders minimize the heavy cost of graph isomorphism checking.

6.2 Fusion-style Pattern Merge

While piece-wise pattern merge could to certain extent expedite the long pattern discovery, some more aggressive solutions have also been proposed. In particular, [31] has proposed a pattern-merge based algorithm for probabilistically finding large frequent itemsets adopting a *fusion*-style approach. The task there is to efficiently find a good approximation to the set of all the large frequent patterns, which are also called *colossal* patterns.

As shown with the pattern lattice model, in previous mining models pattern candidates are examined by implicitly or explicitly traversing a search tree in either a breadth-first or depth-first manner. When the search tree is exponential in size at some level, such exhaustive traversal has to run with an exponential time complexity. A new mining model was therefore developed in [31] to attack the problem. The mining strategy, PatternFusion, distinguishes itself from other methods in that it is able to fuse small frequent patterns in a fusion-style into colossal patterns, which is even more aggressive than piece-wise patter merge approach. It avoids the pitfalls of both breadth-first and depth-first search by applying the following concepts.

1. Pattern-Fusion traverses the tree in a bounded-breadth way. It always pushes down a frontier of a bounded-size candidate pool, i.e., only a fixed number of patterns in the current candidate pool will be used as starting nodes to go downwards in the pattern tree. As such, it avoids the problem of exponential search space.
2. Pattern-Fusion has the capability to identify "shortcuts" whenever possible. The growth of each pattern is not performed with one item addition, but an agglomeration of multiple patterns in the pool. These shortcuts will direct Pattern-Fusion down the search tree much more rapidly toward the colossal patterns.

Figure 4.8 conceptualizes this mining model.

Pattern-Fusion is based on a study on the relationship between the support set of a colossal pattern and those of its subpatterns reveals the notion of *robustness* of colossal patterns. Colossal patterns exhibit *robustness* in the sense that *if a small number of items are removed from the pattern, the resulting pattern would have a similar support set.* The larger the pattern size, the more prominent this robustness

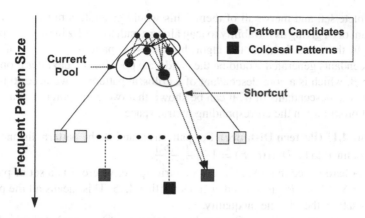

Fig. 4.8 Pattern tree traversal

is observed. Pattern-Fusion captures this relationship between a pattern and its subpattern by the concept of *core pattern*.

Definition 4.13 (Core Pattern) *For a pattern α, an itemset $\beta \subseteq \alpha$ is said to be a τ-core pattern of α if $\frac{|D_\alpha|}{|D_\beta|} \geq \tau, 0 < \tau \leq 1$. τ is called the core ratio.*

For a pattern α, let C_α be the set of all its core patterns, i.e., $C_\alpha = \left\{ \beta | \beta \subseteq \alpha, \frac{|D_\alpha|}{|D_\beta|} \geq \tau \right\}$ for a specified τ. The robustness of a colossal pattern can be further defined as follows.

Definition 4.14 ((d, τ)-Robustness) A pattern α is (d, τ)-*robust* if d is the maximum number of items that can be removed from α for the resulting pattern to remain a τ-core pattern of α, i.e.,

$$d = \max_\beta \{|\alpha| - |\beta| \,| \beta \subseteq \alpha, \text{ and } \beta \text{ is a } \tau - core \text{ pattern of } \alpha\}$$

Due to its robustness, a colossal pattern tends to have a large number of core patterns. Let α be a colossal pattern which is (d, τ)-robust. The following two lemmas show that the number of core patterns of α is at least exponential in d. In particular, it can be shown that for a (d, τ)-robust pattern α, $|C_\alpha| \geq 2^d$.

This core-pattern-based view of the pattern space leads to the following two observations which are essential in Pattern-Fusion design.

Observation 1. Due to the observation that a colossal pattern has far more core patterns than a smaller-sized pattern does, given a small c, a colossal pattern therefore has far more core descendants of size c.

Observation 2. A colossal pattern can be generated by merging a proper set of its core patterns.

These observations on colossal patterns inspires the following mining approach: First generate a complete set of frequent patterns up to a small size, and then randomly pick a pattern, β. By our foregoing analysis β would with high probability be a core-descendant of some colossal pattern α. Identify all α's core-descendants in this complete set, and merge all of them. This would generate a much larger core-descendant of α, giving us the ability to leap along a path toward α in the core-pattern tree T_α. In the same fashion, the algorithm picks K patterns. The set of larger core-descendants generated would be the candidate pool for the next iteration.

Given β, which is a core-descendant of a colossal pattern α, we need to find all the other core-descendants of α. It can be shown that two core patterns of a pattern α exhibit proximity in the corresponding metric space.

Definition 4.15 (Pattern Distance) For patterns α and β, the pattern distance of α and β is defined to be $Dist(\alpha, \beta) = 1 - \frac{|D_\alpha \cap D_\beta|}{|D_\alpha \cup D_\beta|}$.

It is not hard to see that $(S, Dist)$ is a metric space, where S is a set of patterns and $Dist : S \times S \mapsto R^+$ is defined as in Definition 4.15. This means all the pattern distances satisfy the triangle inequality.

For two patterns $\beta_1, \beta_2 \in C_\alpha$, we have $Dist(\beta_1, \beta_2) \leq r(\tau)$, where $r(\tau) = 1 - \frac{1}{2/\tau - 1}$. It follows that all core patterns of a pattern α are bounded in the metric space by a "ball" of diameter $r(\tau)$. This means that given one core pattern $\beta \in C_\alpha$, one can identify all of α's core patterns in the current pool by posing a range query. In Pattern-Fusion, each randomly picked pattern could be a core-descendant of more than one colossal pattern, and as such, when merging the patterns found by the "ball", more than one larger core-descendant could be generated.

The overview of Pattern-Fusion consists of two phases.

1. **Initial Pool:** Pattern-Fusion assumes available an initial pool of small frequent patterns, which is the complete set of frequent patterns up to a small size, e.g., 3. This initial pool can be mined with any existing efficient mining algorithm.

2. **Iterative Pattern Fusion:** Pattern-Fusion takes as input a user-specified parameter, K, which is the maximum number of patterns to be mined. The mining process is conducted iteratively. At each iteration, K seed patterns are randomly picked from the current pool. For each of these K seeds, it find all the patterns within a ball of a size specified by τ as defined in Definition 4.13. All the patterns in each "ball" are then fused together to generate a set of super-patterns. All the super-patterns thus generated are put together as a new pool. If this pool contains more than K patterns, the next iteration begins with this pool for the new round of random drawing. The termination of the iteration process is guaranteed by the fact that the support set of every super-pattern shrinks with each new iteration.

Pattern-Fusion merges all the small subpatterns of a long pattern in one step instead of expanding patterns with additional single items. This gives Pattern-Fusion the advantage to circumvent mid-sized patterns and progress on a path leading to a potential colossal pattern. The idea is illustrated in Fig. 4.9. Each point shown in the metric space represents a core pattern. A larger pattern has far more core patterns close to each other, all of which would be bounded by a ball as shown in dotted line, than a smaller pattern. Since the ball of the larger pattern is much denser, Pattern-Fusion will hit one of its core patterns with a higher probability when performing a random draw from the initial pattern pool.

Fig. 4.9 Pattern metric space

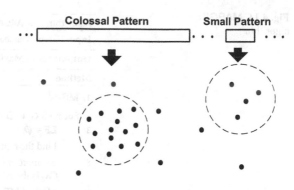

Fig. 4.10 Pattern Space
Explored by MARGIN

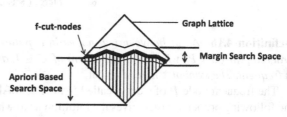

One interesting feature of Pattern-Fusion is that it gives a good approximation to the complete answer set by favoring colossal patterns over smaller-sized ones and catching the outliers.

7   Pattern Traversal Approach

An alternative to finding long patterns one at a time is to identify one long pattern, and try to find all other long patterns by exploring the adjacent ones in the neighborhood. We call this type of long pattern discovery the *Pattern Navigation Approach*. MARGIN as proposed in [23] is a representative of such an approach (Fig. 4.10).

MARGIN is a frequent subgraph mining algorithm to find all maximal frequent subgraphs. As long patterns form a subset of all the maximal frequent subgraphs, MARGIN would return all the largest patterns in a graph data set. In a nutshell, MARGIN is based on the idea that all maximal subgraphs are adjacent in the pattern lattice in the sense that any one of them is reachable from another one by a common child node. The set of all candidate subgraphs which are likely to be maximally frequent are the set of n-edge frequent subgraphs that have a $n + 1$-edge infrequent supergraph. Such a set of nodes in the lattice is referred to as the set of *f-cut-nodes*. Comparing to Apriori based algorithms, MARGIN greatly improves mining efficiency by exloring a much smaller search space by visiting the pattern lattice around the *f-cut-nodes*. A *cut* is defined as follows.

Fig. 4.11 MARGIN
algorithm

Algorithm 1: MARGIN
Input: Graph Database $\mathbb{D} = \{G_1, G_2, \ldots, G_n\}$
Output: Set of Maximal Frequent Graphs $\mathbb{MF}$
Method:
1. $\mathbb{MF} = \emptyset$
2. For each $G_i \in \mathbb{D}$ do
3. $\mathbb{LF} = \phi$
4.   Find the representative R_i of G_i
5.   *ExpandCut* ($\mathbb{LF}$, Cr_i † R_i) where CR_i is the infrequent child of R_i
6.   Merge ($\mathbb{MF}$, $\mathbb{LF}$)

Definition 4.16 *A cut between two nodes in a pattern lattice L is defined as an ordered pair (C † P) where P is the parent of $C \in L$ and C is not frequent while P is frequent. The symbol † is read as* cut.

The frequent node P of a cut is called *f-cut*. The MARGIN algorithm is based on the following property which gives the intuition as to why two maximal patterns can be reached from one to the other.

Upper Diamond Property Any two children C_i, C_j of a node P in a pattern lattice L have a common child node A.

The set of candidate subgraphs that are likely to become maximally frequent are the *f-cut* nodes. This is because they are frequent subgraphs having an infrequent child. MARGIN avoids traversing the lattice bottom up and instead traverses the cuts alone in each lattice of the given graphs. The set of all *f-cut* nodes are further pruned to given the set of all maximal frequent subgraphs. Essentially, MARGIN works in the following two main steps (Fig. 4.11).

Stage I Find the initial *f-cut* nodes by dropping edges one by one from the initial graph G, ensuring that the resulting subgraph is connected until it finds the first frequent subgraph R_i. The frequent subgraph found by such dropping of edges are called the *Representative* R_i of G. Accordingly, the initial cut is thus $(CR_i † R_i)$ where CR_i is the infrequent child of R_i.

Stage II For each cut discovered in G, an algorithm called ExpandCutis used to recursively extends the cut to generate all cuts in G. ExpandCut expands a given cut such that all its neighboring cuts will be explored.

8 Conclusion

With the increasing data size of today's real-life applications, long patterns are gaining increasing recognition in a wide range of domains including bioinformatics, social network analysis, software engineering and business intelligence. Yet the task of mining long patterns has remained a challenge due to the prohibitively large number of smaller patterns which often need to be generated first in traditional mining frameworks. In this chapter, we first use a pattern lattice model to illustrate and compare various mining paradigms. We group existing mining algorithms into three categories based on the way they traverse the pattern lattice, which are pattern enumeration, pattern merging and pattern traversal. We present recent studies for mining long patterns according to their respective pattern mining paradigms. For each category, we discuss the representative algorithms and the state-of-the-art development. These studies provide valuable insight into the problem of long pattern mining and give inspiration for future works.

References

1. R. Agrawal and R. Srikant. Fast algorithms for mining association rules. In *VLDB*, pages 487–499, 1994.
2. R. Agrawal, C. C. Aggarwal, and V. V. V. Prasad. Depth first generation of long patterns. In *Proceedings of the ACM SIGKDD International Conference on Knowledge Discovery and Data Mining (KDD-2000)*, pages 108–118, 2000. Also available as IBM Research Report, RC21538, 1999.
3. T. Asai, K. Abe, S. Kawasoe, H. Arimura, H. Satamoto, and S. Arikawa. Efficient substructure discovery from large semi-structured data. In *SDM*, pages 158–174, 2002.
4. Roberto J. Bayardo. Efficiently mining long patterns from databases, pages 85–93, 1998.
5. C. Borgelt and M. Berthold. Mining molecular fragments: Finding relevant substructures of molecules. In *ICDM*, pages 211–218, 2002.
6. D. Burdick, M. Calimlim, J. Flannick, J. Gehrke, and T. Yiu. Mafia: A maximal frequent itemset algorithm. *IEEE Transactions on Knowledge and Data Engineering*, 17(11):1490–1504, 2005.
7. K. Gouda and M. Zaki. Genmax: An efficient algorithm for mining maximal frequent itemsets. *Data Min. Knowl. Discov.*, 11(3):223–242, November 2005.
8. J. Han, J. Pei, B. Mortazavi-Asl, Q. Chen, U. Dayal, and M.-C. Hsu. Freespan: Frequent pattern-projected sequential pattern mining. In *Proceedings of the Sixth ACM SIGKDD International Conference on Knowledge Discovery and Data Mining*, KDD '00, pages 355–359, New York, NY, USA, 2000. ACM.
9. J. Han, J. Pei, Y. Yin, and R. Mao. Mining frequent patterns without candidate generation: A frequent-pattern tree approach. *Data Min. Knowl. Discov.*, 8(1):53–87, January 2004.
10. J. Huan, W. Wang, and J. Prins. Efficient mining of frequent subgraph in the presence of isomorphism. In *ICDM*, pages 549–552, 2003.
11. A. Inokuchi, T. Washio, and H. Motoda. An apriori-based algorithm for mining frequent substructures from graph data. In *PKDD*, pages 13–23, 2000.
12. R. Jin, C. Wang, D. Polshakov, S. Parthasarathy, and G. Agrawal. Discovering frequent topological structures from graph datasets. In *Proceedings of the Eleventh ACM SIGKDD International Conference on Knowledge Discovery in Data Mining*, KDD '05, pages 606–611, New York, NY, USA, 2005. ACM.
13. M. Kuramochi and G. Karypis. Frequent subgraph discovery. In *ICDM*, pages 313–320, 2001.
14. S. Kurtz, JV. Choudhuri, E. Ohlebusch, C. Schleiermacher, J. Stoye, and R. Giegerich. Reputer: the manifold applications of repeat analysis on a genomic scale. In *Nucleic Acids Research*, number 22, pages 4633–4642, 2001.
15. D. Lo, S-C. Khoo, and C. Liu. Efficient mining of iterative patterns for software specification discovery. In *KDD*, pages 460–469, 2007.

16. S. Nijssen and J. Kok. A quickstart in frequent structure mining can make a difference. In *SIGKDD*, pages 647–652, 2004.
17. C. Olston, S. Chopra, and U. Srivastava. Generating example data for dataflow programs. In *SIGMOD*, pages 245–256, 2009.
18. F. Pan, G. Cong, A. K. H. Tung, J. Yang, and M. J. Zaki. Carpenter: finding closed patterns in long biological datasets. In Lise Getoor, Ted E. Senator, Pedro Domingos, and Christos Faloutsos , editors, *KDD*, pages 637–642. ACM, 2003.
19. F. Pan, G. Cong, X. Xin, and A. K. H. Tung. Cobbler: Combining column and row enumeration for closed pattern discovery. In *In Proc 2004 Int. Conf. on Scientific and Statistical Database Management (SSDBM'04), Santorini Island*, pages 21–30, 2004.
20. J. Pei, J. Han, and R. Mao. CLOSET: An efficient algorithm for mining frequent closed itemsets. In *DMKD*, pages 11–20, 2000.
21. J. Pei, J. Han, B. Mortazavi-Asl, H. Pinto, Q. Chen, U. Dayal, and M. Hsu. PrefixSpan: Mining sequential patterns efficiently by prefix-projected pattern growth. In *ICDE*, pages 215–224, 2001.
22. J. Prins, J. Yang, J. Huan, and W. Wang. Spin: Mining maximal frequent subgraphs from graph databases. In *SIGKDD*, pages 581–586, 2004.
23. L. Thomas, S. Valluri, and K. Karlapalem. Margin: Maximal frequent subgraph mining. In *ICDM*, pages 1097–1101, 2006.
24. J. Wang, J. Han, and J. Pei. Closet+: Searching for the best strategies for mining frequent closed itemsets. In *SIGKDD*, pages 236–245, 2003.
25. X. Yan and J. Han. gSpan: Graph-based substructure pattern mining. In *ICDM*, pages 721–724, 2002.
26. J. Yang, W. Wang, P. S. Yu, and J. Han. Mining long sequential patterns in a noisy environment. In *Proceedings of the 2002 ACM SIGMOD International Conference on Management of Data*, SIGMOD '02, pages 406–417, New York, NY, USA, 2002. ACM.
27. M. J. Zaki. Spade: An efficient algorithm for mining frequent sequences. *Mach. Learn.*, 42(1–2):31–60, January 2001.
28. M. Zaki. Efficiently mining frequent trees in a forest. In *SIGKDD*, pages 71–80, 2002.
29. M. Zaki and C. Hsiao. CHARM: An efficient algorithm for closed itemset mining. In *SDM*, pages 457–473, 2002.
30. M. J. Zaki, S. Parthasarathy, M. Ogihara, and W. Li. New algorithms for fast discovery of association rules. In *In 3rd Intl. Conf. on Knowledge Discovery and Data Mining*, pages 283–286. AAAI Press, 1997.
31. F. Zhu, X. Yan, J. Han, P. Yu, and H. Cheng. Mining colossal frequent patterns by core pattern fusion. In *ICDE*, pages 706–715, 2007.
32. F. Zhu, X. Yan, J. Han, and P. S. Yu. Efficient discovery of frequent approximate sequential patterns. In *ICDM*, pages 751–756, 2007.
33. F. Zhu, Q. Qu, D. Lo, X. Yan, J. Han, and P. S. Yu. Mining top-k large structural patterns in a massive network. *PVLDB*, 4(11):807–818, 2011.

Chapter 5
Interesting Patterns

Jilles Vreeken and Nikolaj Tatti

Abstract Pattern mining is one of the most important aspects of data mining. By far the most popular and well-known approach is frequent pattern mining. That is, to discover patterns that occur in many transactions. This approach has many virtues including monotonicity, which allows efficient discovery of all frequent patterns. Nevertheless, in practice frequent pattern mining rarely gives good results—the number of discovered patterns is typically gargantuan and they are heavily redundant.

Consequently, a lot of research effort has been invested toward improving the quality of the discovered patterns. In this chapter we will give an overview of the interestingness measures and other redundancy reduction techniques that have been proposed to this end.

In particular, we first present classic techniques such as closed and non-derivable itemsets that are used to prune unnecessary itemsets. We then discuss techniques for ranking patterns on how expected their score is under a null hypothesis—considering patterns that deviate from this expectation to be interesting. These models can either be static, as well as dynamic; we can iteratively update this model as we discover new patterns. More generally, we also give a brief overview on pattern set mining techniques, where we measure quality over a set of patterns, instead of individually. This setup gives us freedom to explicitly punish redundancy which leads to a more to-the-point results.

Keywords Pattern mining · Interestingness measures · Statistics · Ranking · Pattern set mining

J. Vreeken (✉)
Max-Planck Institute for Informatics and Saarland University,
Saarbrücken, Germany
e-mail: jilles@mpi-inf.mpg.de

N. Tatti
HIIT, Department of Information and Computer Science,
Aalto University, Helsinki, Finland
e-mail: nikolaj.tatti@aalto.fi

C. C. Aggarwal, J. Han (eds.), *Frequent Pattern Mining*, 105
DOI 10.1007/978-3-319-07821-2_5, © Springer International Publishing Switzerland 2014

1   Introduction

Without a doubt, pattern mining is one of the most important concepts in data mining. In contrast to the traditional task of modeling data—where the goal is to describe all of the data with one model—patterns describe only *part* of the data [27]. Of course, many parts of the data, and hence many patterns, are not interesting at all. The goal of pattern mining is to discover only those that are.

Which brings us to one of the core problem of pattern mining, and the topic of this chapter: interestingness measures. Or, how to determine whether a given pattern is interesting, and how to efficiently mine the interesting patterns from a given dataset. In particular, we find many interesting research challenges in the combination of these two problems.

Before we go into this dual, there is a key problem we have to address first: interestingness is inherently subjective. That is, what is very interesting to one may be nothing but a useless result to another. This goes both between different analysts looking at the same data, but also between different data bases, as well as data mining tasks. As such, we know that our lunch will not be free: there is not a single general measure of interestingness that we can hope to formalize and will satisfy all. Instead, we will have to define task specific interestingness measures.

Foregoing any difficulties in defining a measure that correctly identifies what we find interesting, the second key problem is the exponentially large search space. That is, there are exponentially many *potentially* interesting patterns. Naively evaluating these one by one and only reporting those that meet the criteria is hence infeasible for all but the most trivial of pattern languages [3]. As such, in addition to correctly identifying what is interesting, ideally an interestingness measure also defines a structured, easily traversable search space to find these patterns.

A big breakthrough in this regard was made in 1994 with the discovery by Agrawal and Srikant, and independently by Mannila, Toivonen, and Verkamo [1, 44], that the frequency measure exhibits anti-monotonicity, a property frequently referred to as the A Priori principle. In practice, this property allows to prune very large parts of the search space, making it feasible to mine frequent patterns from very large databases. In subsequent years, many highly efficient algorithms to this end were proposed [78, 76, 26] (See also Chaps. 2 and 3).

Soon after the discovery of the A Priori principle people found that frequency is not a very good measure for interestingness. In particular, people ran into the so-called 'pattern explosion'. While for strict thresholds only patterns expressing common knowledge were discovered, for non-trivial thresholds the exponential space of patterns made that incredibly many patterns were returned as 'interesting'—many of which only variations of the same theme.

In years since, many interestingness measures have been proposed in the literature to tackle these problems; many for specialized tasks, pattern or data types, but we also find highly general frameworks that attempt to approximate the ideal (subjective) interestingness measure. In this chapter we aim to give an overview of the work done in these respects. We will discuss a broad range of interestingness measures, as well as how we can define efficient algorithms for extracting such patterns from data. In order to keep the chapter focused and succinct we will restrict ourselves to measures for unsupervised, or exploratory, pattern mining in binary data—by far the most well studied part of pattern mining. We do note up front, however, that many of the discussed measures and algorithms are highly general and applicable to other settings.

We will discuss the topic in three main parts, loosely following the development of the field over time. That is, in Sect. 2 we discuss relatively simple, absolute measures of interest—of which frequency is a well-known example. As we will see, applying these measures leads to problems in terms of redundancy, difficult parameterization, as well as returning trivial results. In Sect. 3 we discuss, on a relatively high level, the advanced approaches proposed aim to solve these problems. We discuss two of the main proponents in Sects. 4 and 5. In the former we go into detail on approaches that use statistical tests to select or rank patterns based on how significant they are with regard to background knowledge. In the latter we cover the relatively new approach of iterative pattern mining, or, dynamic ranking, where we iteratively update our background knowledge with the most informative patterns so far.

We note that despite our best efforts, we did not find an ideal taxonomy over all methods, as some methods exhibit aspects of more than one of these categories. In such instances we choose to discuss them in the category they fit most naturally, yet will identify alternate ways of looking at these papers. We identify open research challenges and round up with conclusions in Sect. 7.

2   Absolute Measures

In this section we discuss relatively straightforward measures of interestingness. In particular, we focus on what we call *absolute* measures. That is, measures that score patterns using only the data at hand, without contrasting their calculations over the data to any expectation using statistical tests.

More formally, in this section we consider a specific—and perhaps the most well-known—class of pattern mining problems, viz., theory mining [45]. In this setting, a pattern is defined as a description of an interesting subset of the database. Formally, this task has been described by Mannila and Toivonen [43] as follows.

Given a database D, a language $\mathcal{L}$ defining subsets of the data, and a selection predicate q that determines whether an element $\phi \in \mathcal{L}$ describes an interesting subset of D or not, the task is to find

$$T(\mathcal{L}, D, q) = \{\phi \in \mathcal{L} \mid q(D, \phi) \text{ is true}\}$$

That is, the task is to find all interesting subsets.

2.1   *Frequent Itemsets*

The best known instance of theory mining is frequent set mining [3]. The standard example for this is the analysis of shopping baskets in a supermarket. Let I be the set of items the store sells. The database D consists of a set of transactions in which each transaction t is a subset of I. The pattern language $\mathcal{L}$ consists of itemsets, i.e., again sets of items. The support count of an itemset X in D is defined as the number of transactions that contain X, i.e., $supp_D(X) = |\{t \in D \mid X \subseteq t\}|$. We write $fr_D(X)$ to denote the relative support of X in D, i.e., $fr_D(X) = supp_D(X)/|D|$. We do not write D wherever clear from context.

The 'interestingness' predicate is a threshold on the support of the itemsets, the minimal support: $minsup$. In other words, the task in frequent set mining is to compute

$$\{X \in \mathcal{L} \mid supp_D(X) \geq minsup\}$$

The itemsets in the result are called *frequent* itemsets.

Intuition The intuition behind this measure is simple: the more often an itemset occurs in the data, the more interesting it is.

Frequent itemset mining was originally not a goal on itself, but merely a necessary step in order to mine association rules [3]. There, the task is to discover rules $X \to Y$, where X and Y are itemsets with $X \cap Y = \emptyset$, such that when itemset X is a subset of a row $t \in D$, $X \subset t$, with high confidence we will also see itemset $Y \subset t$. Such rules express associations, possibly correlations, and can hence be useful in many applications. A main motivation was supermarket basket analysis, the idea being that by advertising X, people will also buy more of Y.

The basic strategy for mining association rules is to first mine frequent patterns, and then consider all partitionings of each frequent itemset Z into non-overlapping subsets X and Y, to form candidate rules $X \to Y$, while finally keeping only those association rules that satisfy some quality threshold [3]. Though an interesting research topic on itself, interestingness measures for association rules are beyond the scope of this chapter. We refer the interested reader to the recent survey by Tew et al. [69].

A Priori With a search space of $2^{|\mathcal{I}|}$ patterns, the naive approach of evaluating every pattern is infeasible. However, in 1994 it was discovered that support exhibits monotonicity. That is, for two itemsets X and Y, we know

$$X \subset Y \to supp(X) \geq supp(Y) \quad ,$$

which is known as the A Priori property [1, 44], and allows for efficient search for frequent itemsets over the lattice of all itemsets.

The A Priori algorithm was independently discovered by Agrawal and Srikant [1], and by Mannila, Toivonen, and Verkamo [44]. It is a so-called candidate test framework. Given a transaction database D over a set of items $\mathcal{I}$ and a support threshold minsup, it first determines the set of singleton *frequent* itemsets $\mathcal{F}_1 = \{i \in I \mid supp(i) \geq minsup\}$. Then, given a set $\mathcal{F}_k$ of frequent patterns of length k, we can construct the set $\mathcal{C}_{k+1}$ of candidate frequent patterns of length $k + 1$, by considering only itemsets that have all k sub-itemsets of length k included in $\mathcal{F}_k$. We then determine the supports of all candidates in one pass over the data, and obtain $\mathcal{F}_{k+1}$ by keeping only the candidates with $supp(X) \geq minsup$.

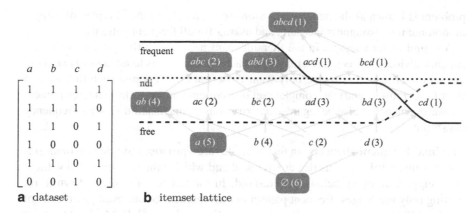

a dataset   b itemset lattice

Fig. 5.1 A dataset of 4 items and 6 transactions and the corresponding lattice. The lattice shows the free, non-derivable, and, for $minsup = 2$, the frequent itemsets. Closed itemsets are *highlighted*

As an example, consider Fig. 5.1, where we depict a toy dataset and the lattice of all itemsets. Say we aim to mine all frequent itemsets with a minimal support of 2, i.e., a minimum frequency of $2/6 = 1/3$. The A Priori algorithm considers the lattice level-wise, and first identifies the frequent singleton itemsets. Here, a, b, c, and d are all frequent. It then constructs the candidate set by taking the Cartesian product with the frequent singletons. In this example this is the full set of itemsets of cardinality 2, i.e., $C_2 = \{ab, ac, bc, ad, bd, cd\}$. We calculate the support of all candidates, and find that all itemsets, except cd, are frequent, i.e., $\mathcal{F}_2 = \{ab, ac, bc, ad, bd\}$. Iterating to the third level, we have $C_3 = \{abc, abd\}$, as all other extensions of $\mathcal{F}_2$ contain cd, of which we know it is not frequent, and hence neither will any larger itemset containing it. We find that the two remaining candidates are frequent, $\mathcal{F}_3 = C_3$. Finally, $C_4 = \emptyset$ as there are no itemsets of size 4 that have all of their sub-itemsets of length 3 in $\mathcal{F}_3$. Hence, the answer to the stated problem, the complete set of frequent itemsets for $minsup = 2$, is hence $\mathcal{F} = \{a, b, c, ab, ac, bc, ad, bd, abc, abd\}$.

The A Priori, or, perhaps more aptly named, level-wise algorithm can be applied for any enumerable pattern language $\mathcal{L}$ and monotonic interestingness measure q. Soon after the discovery of the A Priori property, we see three major focal points for further research. In particular, a lot of attention was given to investigating more efficient algorithms for mining frequent itemsets [24] (see also Chaps. 2 and 3), methods that can mine frequent patterns from data types other than binary (see Chap. 11), and third, on further measures of interestingness (this chapter).

Pattern Explosion Now armed with the ability to mine frequent itemsets in practice, researchers quickly found that frequency is not quite the ideal interestingness measure. That is, we find that for high support thresholds only find patterns representing common knowledge are discovered. However, when we lower the threshold, we are typically quickly flooded with such enormous amounts of results that it becomes impossible to inspect or use them. Moreover, the result set is highly redundant: very many of the returned patterns are simply variations of each other. Combined, this problem is known as the pattern explosion, and stems from the interplay of using a monotonic interestingness measure, and asking for all frequent patterns.

We find many attempts in the literature aiming to solve the pattern explosion, roughly divided between three main approaches. The first is to attempt to *condense* the set of results. That is, we only want those patterns reported such that we can infer (to certain extend) the complete set of frequent patterns. These interestingness measures can hence also be regarded as extra constraints in addition to the frequency constraint.

Maximal Frequent Itemsets In this vein, Bayardo proposed to mine *maximal* frequent itemsets [6]: itemsets that are frequent and which cannot be extended without their support dropping below the threshold. In the lattice, this comes down to reporting only the longest frequent pattern in each branch. In our example, the set of maximal frequent itemsets for $minsup = 2$ is $\mathcal{F}_{max} = \{abc, abd\}$. Maximal frequent itemsets are a lossy representation of the set of frequent itemsets in that all frequent itemsets can be reconstructed, yet the individual frequencies are lost. While maximal itemsets can be useful when we are interested in long patterns, we should be aware that for very low support thresholds complete data records are returned—which beats the purpose of pattern mining. Maximal frequent itemsets can be mined efficiently using, e.g., the Max-Miner [6] and MAFIA [11] algorithms.

Closed Frequent Itemsets In contrast to maximal frequent itemsets, *closed* frequent itemsets [52] provide a lossless representation of the frequent itemsets, as both these itemsets and their frequencies can be reconstructed exactly. The definition of a closed frequent itemset is an itemset X that is frequent, $supp(X) \geq minsup$, and of which there exists no extension for which the support remains the same, i.e., there is no $Y \subsetneq X$ such that $supp(Y) = supp(X)$. Following this definition, in our example, the set of closed frequent itemsets consists of $\mathcal{F}_{closed} = \{a, ab, abc, abd\}$, which is smaller than the complete set of frequent itemsets, yet larger than for maximal itemsets. Efficient algorithms for mining closed frequent itemsets include Charm [77].

Given a set of closed frequent itemsets $\mathcal{F}_{closed}$, we can determine the support of any frequent itemset $X \in \mathcal{F}$ with ease. That is, for a given itemset X, we simply find the smallest superset $Y \in \mathcal{F}_{closed}$, with $X \subseteq Y$, and return the support of Y. If no such superset exists in $\mathcal{F}_{closed}$, X is not frequent. As such, we essentially derive the frequency of X using a very simple rule.

Free Frequent Itemsets Closed itemsets can be seen as the maximal itemsets having among the itemsets having the same support. Closely related are *free* sets [9], which are the *minimal* itemsets among the itemsets having the same support, that is, an itemset X is free if there is no $Y \subsetneq X$ such that $supp(X) = supp(Y)$. Each free itemset X has a unique closure Y, a closed itemset Y such that $X \subseteq Y$. However, a closed itemset may stem from many free itemsets. This means that free itemsets will always be a larger collection than closed itemsets. Free itemsets are handy since they form a monotonically downward closed collection, that is, all sub-itemsets of a free itemset are also free. In our example, frequent free itemsets are $\mathcal{F}_{free} = \{a, b, c, d\}$. Itemset cd is also free but it is not frequent.

Non-Derivable Itemsets Calders and Goethals [12] developed the notion of support derivability a step further, and proposed to mine *non-derivable* frequent itemsets. An itemset is said to be derivable if we can derive its support using inclusion/exclusion rules, i.e., the upper and lower bound of its support are equal. In our example, abc is a derivable itemset: since $supp(bc) = supp(c) = 2$ we know that whenever c appears, b appears as well. Hence, it follows that $supp(abc) = supp(ac) = 2$. In our example the set of non-derivable itemsets for $minsup = 2$ is $\mathcal{F}_{ndi} = \{a, b, c, ab, ac, bc\}$. Like free itemsets, non-derivable itemsets are also monotonically downward closed, which allows us to mine them efficiently.

In practice, for a given database and threshold the number of closed itemsets and non-derivable itemsets is typically comparable; how many exactly depends on the structure of the data. In both cases, for clean data, up to orders of magnitude fewer itemsets are returned than when mining frequent itemsets. However, if the data is noisy, it can be that no reduction can be obtained and we still find millions or billions of itemsets for non-trivial thresholds.

Margin-Closed and Robust Frequent Itemsets Moerchen et al. [50] hence argues to prune more aggressively, and to this end proposes to relax the requirement on maintaining frequencies exactly. That is, to mine *margin-closed* frequent itemsets; essentially reporting only those frequent itemsets for which the support deviates more than a certain amount compared to their subsets. A related, but different approach was recently proposed by Tatti and Moerchen [66], whom acknowledge the data at hand is just a sample; whether a given itemset is frequent, maximal, closed, or non-derivable may just be happenstance. To this end they propose to mine only those itemsets that exhibit a given property *robustly*, i.e., in many random subsamples of the data. For example, the idea is that in the more (sub)samples of the data we find a certain itemset to be closed, the more informative it is to report this particular itemset to the end-user. A happy coincidence of robustness is that the monotonicity of the chosen property propagates. That is, if the property is monotonic, for example, non-derivability or freeness, the robust version is also monotonic, and hence for those measures we can mine robust itemsets efficiently.

Sampling Frequent Itemsets We should stress that A Priori works for any monotonic measure, for example, the Jaccard-distance based measure Cohen et al. [14] propose,

$$supp(X)/|\{t \in D \mid X \cap t \neq \emptyset\}|,$$

the support of the itemset divided by the number of transactions that share at least one element with X. However, while monotonic, in practice A Priori is impractical for this measure: we cannot prune any singletons, and hence have $\mathcal{F}_1 = \mathcal{I}$, by which already at the first step we have to check *all* itemsets of size 2. To circumvent this problem, Cohen et al. first of all consider only itemsets of length 2, and, only calculate the actual score for a sample of the complete candidate set. However, because of the exponential search space, to avoid mining mostly itemsets with very low scores, one will have to be careful what distribution to sample from. Cohen et al. sample according to a hash-based score that estimates the correlation between two items. In theory this approach can be extended to itemsets of arbitrary sizes, but will require a non-trivial extension of this estimate.

A possible solution to this end may have been given by Boley et al. [7], whom proposed a framework that allows to directly sample itemsets proportional to *any* score based on frequency and/or cardinality of a pattern. However, the more different 'components' a score has, the more computationally expensive the pre-processing becomes. In a follow-up paper [8], the same authors refined the procedure and removed the need for this pre-processing by formalizing a coupling-from-the-past MCMC sampler.

Al Hassan and Zaki [4] proposed a different approach that allows for directly sampling the output space of any pattern miner. While the paper discusses patterns in graphs, the same techniques can be applied for mining itemsets. In follow-up work they discuss Origami [5], an approach to sample patterns that are representative, as well as orthogonal to earlier sampled patterns. By sampling patterns not just proportionally to a static distribution, but with regard to earlier sampled results, this process comes rather close to dynamic ranking, which we will discuss in more detail in Sect. 5.

2.2 Tiles

The next class of absolute interestingness measures we consider are not for itemsets, but for *tiles*. In tile mining we are particularly interested in the area a pattern covers in the data. That is, $\mathcal{L}$ consists of tiles $T = (X, Y)$ which are defined by both an *intention*, a subset of all items $X \subseteq \mathcal{I}$, as well as an *extension*, a subset of all rows $Y \subseteq \mathcal{R}$. We then use q to calculate the interestingness over the cells of D identified by $X \times Y$.

Large Tile Mining The most constrained variant of this task is to mine exact tiles, tiles for which in D we find only 1s, that meet a *minarea* threshold. That is, tile for which $area(T) = |X||Y| \geq minarea$. A maximal tile is then a tile T for which we cannot add an element to X or Y, and updating the vice-versa to maintain the all-1s constraint, without the $area(T)$ decreasing. Note that as *area* does not exhibit monotonicity, the level-wise algorithm cannot be applied.

Intuition Large areas of only 1s in D are interesting.

Geerts et al. [21], however, gave a set of constraints that can be used to mine large tiles efficiently in practice; essentially implementing the greedy algorithm for Set Cover [21]. It is interesting to note that every large tile is a closed frequent itemset, an observation Xiang et al. [74] used in their algorithm, first mining closed frequent itemsets and then pruning this set.

Noise and large tile mining do not go together well. That is, given a dataset in which there exists one large tile against an empty background, simply by flipping one 1 to 0 makes it that the complete tile will not be discovered; instead we will find two partitions. Every further flipped value will partition the tile more.

More in particular, it may not be realistic to expect a process to generate a tile full of ones. Instead, we may need to relax our requirement and look for *noisy*, or *dense* tiles instead of *exact* tiles.

Noisy Tile Mining A noisy tile is a tile T associated with a frequency of ones in the data, for which we write, slightly abusing notation,

$$fr(T) = \frac{|\{(i, j) \in (X \times Y) \mid D_{ij} = 1\}|}{|X||Y|} .$$

An exact tile then is a special case, with $fr(T) = 1.0$. When mining noisy tiles we are interested in finding large areas in the data that contain many 1s, or possibly, many 0s.

Intuition The more uniform the values of D over the area identified by T, i.e., the more 1s resp. 0s we find, the more interesting the tile.

We find the problem of mining noisy tiles in many guises and embedded in many problem settings. Examples include dense itemset mining [60], dense tile mining [75], bi-clustering [55], and Boolean Matrix Factorization [49, 40], as well as fault-tolerant itemset mining [54].

Fault tolerant itemset mining for a large part follows the regular frequent itemset mining setting, with, however, the twist that we do not just calculate support over $t \in D$ for which $X \subseteq t$, but also those transactions that *nearly* but not exactly support t. The general approach is that, per itemset X, we are given a budget of ϵ 1s that we may use to maximize the fault-tolerant support of X [54]. Clearly, a fixed budget favors small itemsets, as there per row fewer items can be missing. Poernomo and Gopalkrishnan [56] gave an efficient algorithm for mining fault-tolerant itemsets where the budget is dependent on the cardinality of the itemset.

Seppänen and Mannila [60] generalized the problem of ϵ fault-tolerant itemset mining to *dense* itemset mining. That is, instead of using a fixed budget of flips, the proposed algorithms mine itemsets for which we there exist at least σ rows such that the density of 1s in the data projected over the itemset is at least δ.

In Boolean Matrix Factorization the goal is to find a low-rank approximation of the full data matrix. Optimizing, as well as approximating this problem is NP-hard [49], and hence the standard approach is to iteratively find good rank-1 approximations of the data, i.e., large noisy tiles with high frequency. The Asso algorithm does this by searching for tiles that exhibit high association between the rows and columns, and has been shown to efficient heuristic for finding large noisy tiles [49].

Frequent itemsets can be used to bootstrap the search for dense areas in the data. Xiang et al. [75] gave a fast heuristic for finding dense tiles that first mines closed itemsets, and then iteratively combines them until a density threshold is reached. We find a similar strategy in the PandA algorithm [40].

2.3 Low Entropy Sets

The final absolute measure for interestingness we discuss is *entropy*. Whereas many of the above measures put explicit importance on the associations between 1s in the data, by ignoring or penalizing 0s. This, however, ignores the fact that there may be interesting associations in the data between both the 1s and the 0s. Heikinheimo et al. [30] hence argue to put equal importance on both 0/1, and instead of mining frequent itemsets, propose to mine itemsets for which the counts of the contingency table are highly skewed. That is, for an itemset X we calculate the support of all of its $2^{|X|}$ instances, and calculate the entropy over these counts. The score is minimal (0) when only one instance occurs in the data, e.g., if for itemset $X = abc$ if we find $supp(abc = 110) = |D|$, while the score is maximal ($|X|$) when all instances have the same support.

Intuition An itemset X is interesting if the distribution of the data is highly skewed, i.e., either highly structured or very random.

Using this score, which exhibits monotonicity, we can use the level-wise algorithm to efficiently mine either *low entropy* sets, if one is interested in highly structured parts of the data, or to mine *high entropy* sets if one is interested in identifying the most random parts of the data. Mampaey [41] proposed to speed up the mining by using inclusion-exclusion, making use of the fact that in practice only a fraction of all $2^{|X|}$ possible instances of X occur in the data. The μ-Miner algorithm provides a speed-up of orders of magnitude compared to the level-wise algorithm.

3 Advanced Methods

Though each of the methods described above has nice properties, we find that in practice they do not perform as well as advertised. In general, we find that all absolute measures identify far too many results as interesting, with or without condensation. The key problem is redundancy. Absolute measures have no means of identifying whether the score for a pattern is expected, nor are they able to prune variants of patterns that identify single statistically significant concepts.

We identify three main lines of research aimed at tackling these problems, or in other words, aimed at identifying more interesting patterns. A common theme in these approaches is the reliance on statistical analysis. The main difference between these methods and the methods described in the previous section is that in order to rank patterns we impose a statistical model on our data, and measure how interesting are the patterns given that model.

We can divide the methods into three rough categories:

1. **Static pattern ranking**. Here we assume that we know a simple statistical model, derived from a simple background information. We assume that this model is well-understood, and any pattern that is well-explained by this model should be

discarded. Consequently, we are interested in patterns that the model considers very unlikely.

2. **Iterative pattern ranking**. While static pattern ranking addresses the problem of redundancy with respect to background knowledge, it does not explicitly address the problem of redundancy between patterns. We can approach this problem more directly with dynamic ranking: At the beginning we start with a simple model and find the most surprising pattern(s). Once this pattern is identified, we consider it 'known' and insert the pattern into our model, which updates our expectations— and repeat the process. As a result we get a sequence of patterns that are surprising and non-redundant with regard to the background knowledge and higher ranked patterns.

3. **Pattern set mining**. The methods in the above categories measure interestingness only per individual pattern. The third and last category we consider aims at identifying the best *set* of patterns, and hence propose an interestingness measure over *pattern sets*. As such, these measures directly punish redundancy—a pattern is only as good as its contribution to the set.

4   Static Background Models

In Sect. 2 we discussed absolute interestingness measures, which we can now say are essentially only based on counting. In this section we will cover slightly more advances measures. In particular, we will discuss measures that instead of relying just on absolute measurements, contrast these measurements with the *expected* measurement for that pattern. The basic intuition here is that the more strongly the observation deviates from the expectation, the more interesting the pattern is.

Clearly, there are many different ways to express such expectation. Most often these are calculated using on a probabilistic model of the data. Which model is appropriate depends on the background knowledge we have and/or the assumptions we are willing to make about the data. As such, in this section we will cover a wide range of different models that have been proposed to formalize such expectations.

However, in order to be able to identify whether a pattern is interesting, we need to be able whether the deviation between the observation and the expectation is large enough. That is, whether the deviation, and hence correspondingly the pattern, is significant or not. To this end we will discuss a variety of (statistical) tests that have been proposed to identify interesting patterns.

For clarity, we will start our discussion with the most simple model, the independence model. We will then use this model as an example to discuss a range of significance measures. We will then proceed to discuss more complex models, that can incorporate more background knowledge, for which many of these tests are also applicable. Interleaved we will also discuss interestingness measures specific to particular models and setups.

Before we start, there is one important observation to make. As opposed to the previous section, the measures we will discuss here are typically not used to mine *all* interesting patterns. This is mostly due to that these measures are typically not monotonic—and hence do not easily allow for efficient search—as well as that it is often difficult to express a meaningful threshold (i.e., significance level). Instead, these measures are used to rank a given collection of patterns, e.g., mined all frequent patterns up to a certain support threshold, or, when practical bounds are available, to mine a top-k of the most significant patterns. Many of the authors of work we survey in this section argue that in practice analysts do not want, nor have time, to consider all patterns, and hence a small list of the most interesting patterns is preferable.

4.1 Independence Model

We start with the simplest background model, which is the model where we assume that the individual items are all independent. Under this assumption we expect the frequency of a given itemset $X = x_1 \cdots x_n$ to be equal to

$$ind(X) = \prod_{i=1}^{n} fr(x_i) \quad .$$

The background knowledge we use are simply the frequencies of the individual items, which can be straightforwardly computed from the data. Moreover, it seems reasonable to expect to expect the data analyst (e.g., store manager) to know these margins (e.g., how often each product is sold) and hence be able to make such inferences intuitively. As such, the independence model is expected to correctly identify 'boring' patterns, patterns for which the frequencies follow under the independence model.

Testing Observations against Expectations Now that we have a model, we will use it as an exemplar to discuss a range of widely used methods for comparing the observed measurement with the expectation. After covering these general methods, we will discuss more detailed models, and more specialized measures.

With the above, we can compute both the observed frequency $fr(X)$ and the expectation $ind(X)$ of the independence model. The next step is compare these two quantities. A straightforward way to do this is to consider their ratio, a measure known as *lift* [33], and formally defined as

$$lift(X) = fr(X)/ind(X).$$

Here we consider itemsets that have a high lift to be interesting, that is, itemsets whose observed support is substantially higher than the independence assumption. Hence, a larger ratio implies higher interestingness.

In our example, we have $fr(ab) = 0.66$, while under the independence model we have $ind(ab) = \frac{5 \times 4}{6 \times 6} = 0.55$. As such, we find $lift(ab) = 1.2$. For abc, on the other hand, we have $lift(abc) = 0.33/0.18 = 1.83$. While both patterns have a positive lift score, and are hence potentially interesting, the higher score for abc identifies this pattern as the most interesting of the two.

In this example the outcome follows our intuition, but in practice this is not always the case: lift is a rather ad-hoc score. This is due to it comparing the two absolute values directly, without taking into account how likely these values, or their ratio is, given the background model.

We can, however, also compare the deviation by performing a proper statistical test. In order to do so, note that according the independence model the probability of generating a transaction containing an itemset X is equal to $ind(X)$. Assume that our dataset contains N transactions, and let Z be a random variable stating in how many transactions X occurs. The probability that $Z = M$ is equal to binomial distribution,

$$p(Z = M) = \binom{N}{M} q^M (1 - q)^{N-M}, \quad \text{where} \quad q = ind(X) \quad .$$

Now that we have this probability, we can perform a one-sided statistical test by computing the probability that we observe a support of $fr(X)$ or higher, $p(Z \geq Nfr(X))$. Note that the larger frX, the smaller the p-value is.

Computing the right-hand side amounts to computing a sum of probabilities $p(Z = M)$, which as there are $2^{|Z|}$ possible values for M, may prove to be restrictively slow in practice. However, as exactly in those cases binomial distributions are accurately approximated by a normal distribution, we can perform an alternative test by considering a *normal approximation* of the binomial distribution. In this case, we can obtain the p-value by computing the tail of the normal distribution $N\left(Nq, \sqrt{Nq(1 - q)}\right)$, where $q = ind(X)$. This is estimate is inaccurate if q is very close to 0 or 1 and N is small. One rule of thumb is that if $Nq > 5$ and $N(1 - q)$, then this approximation is fairly accurate.

4.1.1 Beyond Frequency

So far, we only considered comparing the frequency of an itemset against its expected value. Clearly, we do not have to limit ourselves to only this measure (or, better, statistic).

Related to fault-tolerant itemsets we saw in Sect. 2, we can say that an itemset X is a *violation* of a transaction t if t does not contain X, $X \notin t$, yet t does contain some elements from X, $t \cap X \neq \emptyset$. We denote the fraction of transactions being violated by X as $v(X)$. The quantity $1 - v(X)$ is then a fraction of transactions that either contain (X) or do not contain any items from X. If items are highly correlated we expect $1 - v(X)$ to be high and $v(X)$ to be low.

Now, let q be the *expected* value of $v(X)$ based on the independence model. We can now calculate what Aggarwal and Yu call [2] the *collective strength* of a pattern as follows

$$cs(X) = \frac{1 - v(X)}{v(X)} \times \frac{q}{1 - q} \quad .$$

In other words we compare the ratio of $\frac{1-v(X)}{v(X)}$ against the expected value.

4.1.2   Beyond Single Measurements

Instead of comparing just a single statistic, like support or the violation rate, we can consider much richer information. One example is to compare the complete contingency table of an itemset X with an expectation [10].

Assume we are given a distribution p over items in $X = x_1 \cdots x_M$. That is, a distribution over $2^{|X|}$ entries. For convenience, let us write

$$p(X = t) = p(x_1 = t_1, \ldots, x_M = t_M),$$

where t is a binary vector of length M. We now consider two different distributions: the first is the *empirical distribution* computed from the dataset,

$$p_{emp}(X = t) = \frac{|\{u \in D \mid u_X = t\}|}{|D|} \quad,$$

and the second, p_{ind}, is the independence model,

$$p_{ind}(X = t) = \prod_{i=1}^{M} p_{ind}(x_i = t_i) \quad,$$

where the margins (item frequencies) are computed from the input data.

The standard way of comparing these two distributions is by doing a so-called G-test, which essentially is a log-likelihood ratio test,

$$2 \sum_{t \in D} \log p_{emp}(X = t_X) - 2 \sum_{t \in D} \log p_{ind}(X = t_X) \quad.$$

Under the assumption that the items of X are distributed independently (which we here do), this quantity approaches the χ^2 distribution with $2^{|X|} - 1 - |X|$ degrees of freedom. Interestingly, this quantity can also be seen as a (scaled) Kullback-Leibler divergence, $2|D| KL(p_{emp} || p_{ind})$.

Alternatively, we can also compare the two distributions with Pearson's χ^2 test,

$$|D| \sum_{t \in \{0,1\}^{|X|}} \frac{(p_{emp}(X = t) - p_{ind}(X = t)^2)}{p_{emp}(X = t)},$$

which has the same asymptotic behavior as the G-test.

Each of these tests can be used to determine the p-value, or likelihood, of a pattern under an assumed model. In practice these measurements are used to rank the patterns from most surprising (under the model) to least surprising, typically showing the user only the top-k of most surprising patterns.

Next we will now look into more elaborate models, which allow us to make more realistic assumptions about the data than complete independence.

4.2 Beyond Independence

While the independence model has many positive aspects, such as ease of computation, intuitive results, as well as interpretability, it is also fair to say it is overly simplistic: it is naive to assume all item occurrences are independent. In practice, we may want to take known interaction into account as background knowledge.

4.2.1 Partition Models

With the goal of mining interesting associations, Webb [73] discusses 6 principles for identifying itemsets that are *unlikely* to be interesting, and to this end proposes to check whether the frequency of an itemset X can either be closely determined by assuming independence between any of its partitions, or by the frequency of any of the supersets of X.

The so-called *partition* model which is needed to perform these tests is a natural generalization from the independence model. More specifically, if we are given an itemset X, consider a partition $\mathcal{P} = P_1, \ldots, P_M$ of X, with $\bigcup_{i=1}^{M} P_i = X$, and $P_i \cap P_j = \emptyset$ for $i \neq j$. Under this model, we expect the support of an itemset to be equal to the product of the frequencies of its parts, i.e., $\prod_{i=1}^{M} fr(P_i)$. It is easy to see that for the maximal partition, when the partition contains only blocks of size 1, the model becomes equal to the independence model.

We can now compare the expected values and the observations in the same way we compared when were dealing with the independence model. If the partition contains only 2 blocks, $M = 2$, we can use Fisher's exact test [19]. While not monotonic, Hamalainen [25] recently gave a practical bound that allows to prune large parts of the search space.

To use the partition model we need to choose a partition. To do so, we can either construct a global model, i.e., choose a fixed partition of $\mathcal{I}$, or we can construct a local model in which the actual partition depends on the itemset X. As an example of the latter case we can consider find the partition of size 2 that best fits the observed frequency [72].

4.2.2 Bayesian Networks

Another natural extension of the independence model are *Bayesian networks*, where dependencies between items are expressed by a directed acyclic graph. In general, computing an expected support from a global Bayesian network is NP-hard problem, however it is possible to use the network structure to your advantage [15]. Additional speed-ups are possible if we rank itemsets in one batch which allows us to use share some computations [34].

Clearly, the partition model is mostly a practical choice with regard to computability and allowing the Fisher test; it does not allow us to incorporate much more knowledge than the independence model. Bayesian networks are very powerful, on the other hand, but also notoriously hard to infer from data. More importantly, they can be very hard to read. Unless we use very simple networks, it is possible our model can make inferences that are far from the intuition of the analyst, and hence prune itemsets that are potentially interesting. Next we discuss a class of models that can circumvent these problems.

4.3 Maximum Entropy Models

In general, for any knowledge that we may have about the data, there are potentially infinitely many possible distributions that we can choose to test against: any distribution that satisfies our background knowledge goes. Clearly, however, not all of these are an equally good choice. For example, say that all we know about a certain row in the data is that it contains 10 ones out of a possible 100 items. Then, while not incorrect, a distribution that puts all probability mass on exactly one configuration (e.g., the first 10 items), and assigns probability 0 to all other configurations, does make a choice that intuitively seems unwarranted given what we know. This raises the question, how should we choose the distribution to test against?

The answer was given by Jaynes [35] who formulated the Maximum Entropy principle. Loosely speaking, the MaxEnt principle states that given some background knowledge, the best distribution is the one that (1) matches the background knowledge, and (2) is otherwise as random as possible. It is exactly this distribution that makes optimal use of the provided background knowledge, while making no further assumptions.

4.3.1   MaxEnt Models for Transactions

As an example, let us discuss the MaxEnt model for binary data, in which we can incorporate frequencies of itemsets as background knowledge. Formally, let K be the number of items, and let Ω be the space of all possible transactions, that is, $\Omega = \{0, 1\}^K$ is a set of binary vectors of length K. In order to compute the expected support of an itemset, we need to infer a distribution, say p, over Ω.

Our next step is to put some constraints on what type of distributions we consider. More formally, we assume that we are given a set of functions $S_1, \ldots, S_M, S_i : \Omega\mathbb{R}$, accompanied with desired values θ_i. Now let us consider a specific set of distributions, namely

$$Q = \{p \mid E_p[S_i] = \theta_i, \ i = 1, \ldots, M\},$$

where $E_p[S_i]$ is the expected value of S_i w.r.t. p,

$$E_p[S_i] = \sum_{\omega \in \Omega} p(\omega)S_i(\omega) \quad .$$

In other words, Q consists of distributions for which the average value of S_i is equal to θ_i.

Apart from border cases, Q will typically be very large and may even contain infinitely many distributions. We need to have one distribution, and hence our next step is to choose one from Q. We do this by the Maximum Entropy principle. Formally, we identify this distribution by

$$p^* = \arg\max_{p \in Q} - \sum_{\omega \in \Omega} p(\omega) \log p(\omega),$$

where the standard convention $0 \times \log 0 = 0$ is used.

Besides nice information-theoretical properties, the maximum entropy distribution also has many other interesting and practical properties. For instance, it has a very useful regular form [16].

First consider that for some $\omega \in \Omega$, *every* distribution $p \in Q$ has $p(\omega) = 0$. Since $p^* \in \Omega$, this immediately implies that $p^*(\omega) = 0$. Let us define Z be the set of such vectors $Z = \{\omega \in \Omega \mid p(\omega)\text{for all} p \in Q\}$. The probability of the remaining points $\Omega \setminus Z$ can be expressed as follows: there is a set of numbers $r_0, \ldots, r_M$, such that

$$p^*(\omega) = \exp\left(r_0 + \sum_{i=1}^{M} r_i S_i(\omega)\right) \quad \text{for} \quad \omega \in \Omega \setminus Z. \tag{5.1}$$

The coefficient r_0 acts as a normalization constant. This form is the well-known *log-linear* model.

Now that we have established the general form of the maximum entropy model, let us look at some special cases of background information.

Assume that we do not provide any constraints, then the maximum entropy distribution will be the uniform distribution, $p(\omega) = 1/|\Omega|$. Consider now that we limit ourselves by setting K constraints, one for each item, $S_i(\omega) = \omega_i$. Then, $E[S_i]$ is the ith column margin, and we can show by simple manipulation of Eq. 5.1 that p^* corresponds to the independence model.

Consider now the other extreme, where we provide $2^K - 1$ constraints, one for each non-empty itemset, by setting $S_X(\omega)$ to be 1 if ω contains X, and 0 otherwise. We can show using inclusion-exclusion tricks that there is *only one* distribution in Q. If the corresponding targets θ_X were computed from a dataset D, then p^* is equal to the empirical distribution p_{emp} computed from D, $p_{emp}(\omega) = |\{t \in D \mid t = \omega\}|/|D|$.

Consider now that we do not use all itemset constraints. Instead we have a partition P and our itemset constraints consists only of itemsets that are subsets of blocks of P. In this case, p^* will have independent items belonging to different blocks. In other words, p^* is a partition model. Another example of non-trivial itemset constraints is a set consisting of all singleton itemsets and itemsets of size 2 such that these itemsets form a tree when viewed as edges over the items. In such case, p^* corresponds to the Chow-Liu tree model [13], a special case of Bayesian network where items may have only one parent. As an example of constraints not related to itemsets, consider $T_k(t)$, being equal to 1 if and only if t contains k 1 s, and 0 otherwise [65]. In such case, ET_k is the probability that a random transaction has k 1s.

All the cases we describe above have either closed form or can be computed efficiently. In general we can infer the MaxEnt distribution using iterative approaches, such as iterative scaling or a gradient descent [16]. The computational bottleneck in these methods is checking the constraints, namely computing the mean ES, a procedure that may take $O(|\Omega|) = O(2^K)$ time. As such, solving the MaxEnt problem in general for a given set of itemset constraints is computationally infeasible [63]. However, let us recall that in this section we are computing the distribution only to rank itemsets. Hence, we can limit ourselves by considering constraints defined only on items in X, effectively ignoring any item outside X [64, 46, 53]. This effectively brings down the computational complexity down to a much more accessible $O(2^{|X|})$.

4.3.2 MaxEnt and Derivability

Once inferred, we can compare the expectation to the observed supports using the same techniques as we developed above for the independence model. Moreover, there exists an interesting connection between the MaxEnt model and derivability of the support of an itemset. More specifically, for a given itemset X, the MaxEnt model derived using *proper* subsets of X shares a connection with the concept of non-derivable itemsets. An itemset is derivable if and only if its frequency can be deduced from the frequencies of its subsets. This is only possible when any distribution $p \in Q$ produces the same expectation $E_p[S_X]$ as the observed support. This immediately implies that the expected support according to p_{emp} is exactly the same as observed support. In summary, if an itemset is derivable, then a MaxEnt model derived from its subsets will produce the same expectation as the observed support.

4.3.3 MaxEnt Models for Whole Databases

So far we have considered only models on individual transactions. Alternatively, we can consider models on whole datasets, that is, instead of assigning probabilities to individual transactions, we assign probability to whole datasets. The space on which distribution is defined is now $\Omega = \{0, 1\}^{N \times K}$, where K is the number of items and N is the number of transactions, that is, Ω contains all possible binary datasets of size $N \times K$. Note that under this model N is fixed along with K, where as in transaction-based model only K is fixed and we consider dataset to be N i.i.d. samples. That is, while above we considered the data to be a bag of i.i.d. samples, we here assume the *whole* dataset to be one single sample. As such, different from the setting above, here *which* rows support an itemset are also considered to be of interest—and hence, the background knowledge should not just contain patterns, but also their row-sets.

In other words, we can use *tiles* as constraints. Given a tile T, let us write $S_T(D)$ for the number of 1s in entries of D corresponding to T. The mean $E[S_T]$ is then the expected number of 1 s in T. Note that we can easily model column margins using tiles, simply by creating K tiles, ith tile containing ith column and every row. We can similarly create row margins. The maximum entropy model derived from both rows and margins is known as *Rasch model* [57].

Unlike with transaction-based MaxEnt model and itemsets as constraints, discovering the MaxEnt for a set tiles can be done in polynomial time. The reason for this is that tile constraints allow us to factorize the model into a product of individual cells, which in turns allows us to compute the expectations $E[S_T]$ efficiently.

We can use the Rasch model to rank tiles based on the likelihood, that is, the probability that a random dataset will obtain the same values in T as the original dataset. We can show that this is equal to

$$\prod_{(i,j)\in T} p_{emp}(R_{ij} = D_{ij}),$$

where D_{ij} is the (i, j)th entry of the input dataset and R is the variable representing (i, j)th entry in a dataset. The smaller the probability, the more surprising is the tile according to the Rasch model. Unfortunately, this measure is monotonically decreasing. Consequently, the most interesting tile will contain the whole dataset. In order to remedy this problem, Kontonasios and De Bie [37] propose an normalization approach inspired by the Minimum Description Length principle [58], dividing the log-likelihood of the tile by its description length, roughly equal to the size of the transaction set plus the size of the itemset.

4.4   Randomization Approaches

So far, we compute expected frequencies by explicitly inferring the underlying distribution. However, we can avoid this by *sampling* datasets.

More formally, let Ω be the set of all possible binary datasets of size $N \times K$, $\Omega = \{0, 1\}^{N \times K}$. Many of these datasets are not realistic, for example Ω contains a dataset full of 1 s. Hence, we restrict our attention to datasets that have the same characteristics as the input dataset. One particular simple set of statistics is a set containing row and column margins. Assume that we have computed the number of 1s in each row and column. Let us write c_i for the number of 1 s in ith column, and r_j, the number of 1 s in jth row. Now consider, Ω' to be a subset Ω containing only the datasets that have the column margins corresponding to $\{c_i\}$ and row margins corresponding to $\{r_j\}$. Consider a uniform distribution over Ω'. We can now use this distribution to compute the expected value of a pattern. Such a distribution is closely related to the Rasch models explained previously. However, there are some technical differences. Datasets sampled from Ω' are forced to have certain row and column margins exactly while with Rasch models row and column margins are only forced on average. This, however, comes at a cost.

The uniform distribution over Ω' is very complex and, unlike the Rasch model, cannot be used directly. To remedy this problem we will have to sample datasets. Sampling Ω' from scratch is very difficult, hence we will use a MCMC approach [23, 29]. Given a dataset D from Ω', we first sample two columns, i and j, and two rows x and y. If it happens that out of four values D_{ix}, D_{jx}, D_{iy}, and D_{jy}, two are

Fig. 5.2 An example of swap
randomization

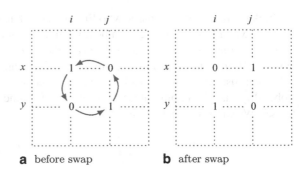

a before swap   **b** after swap

1 s and two are 0 s, and both ones are in opposite corners, see Fig. 5.2, then we swap D_{ix} with D_{iy} and D_{jx} with D_{jy}. It is easy to see that the new dataset will have the same row and column margins. By repeating this process many times, i.e., until the MCMC chain has converged, we can generate random datasets simply by starting the random walk from the input dataset. Note that in practice, however, we do not know when the MCMC chain has converged, and hence have to use a heuristic number of steps to reach a 'random' point in Ω'.

By sampling many random datasets, we can assess how significant a score obtained on the original data is in light of the maintained background knowledge by computing an empirical p-value—essentially the fraction of sampled datasets that produce higher support of an itemset X than the original support. The number of sampled datasets hence determines the resolution of the p-value.

In short, the (swap-)randomization and MaxEnt modelling approaches are very related. The former can be used to sample data that maintains the background knowledge exactly, while in the latter information is only maintained on expectation. The latter has the advantage that exact probabilities can be calculated. By sampling random data, whether by randomization or from a MaxEnt model, we can obtain empirical p-values—also for cases where by the nature of the score we're looking at (e.g., clustering error, classification accuracy, etc), it is impossible, or unknown, how to calculate exact values given a probabilistic model.

5 Dynamic Background Models

So far, we have covered only *static* scores. While within this class method have been proposed that are increasingly good at correctly identifying uninteresting patterns, they can only do so for individual patterns and with regard to a *static* background model. As such, when regarding the top-k result, we may still find patterns that are mostly variants of the same (significant) theme. In this chapter we turn our attention to models that explicitly take relationships between patterns into account.

Note, however, that some of the static models already do this to some extend. For example, the partition model studies subsets of the itemset under investigation. They do not, however, necessarily take all higher ranked itemsets into account.

In general, we can divide the dynamic approach into two categories. The first category is iterative pattern mining, where the goal is to greedily build a sequence of patterns, each pattern being the most surprising given the previous patterns and background knowledge. This is the class we discuss in this chapter. The second category is pattern set mining. There the goal is to produce a *set of patterns* that together optimize a given score over the whole set, as opposed to scoring patterns only individually. We will discuss pattern set mining in the next chapter.

Both categories have many technical similarities and share algorithmic approaches. In fact, some methods are difficult to pigeonhole as they can perform both tasks. The main difference we identify is that in pattern set mining we are looking for a set of patterns, that is, we need to control the number of patterns, whereas in iterative pattern ranking, we are 'simply' ranking patterns.

5.1 The General Idea

The main ingredient needed to perform iterative pattern mining, as opposed to static ranking, is a model that we can update. In particular, we need a model that can incorporate background knowledge in the same shape as what we're mining: patterns.

As such, the general approach here is that in the first iteration we infer the model p_1 according to the basic background knowledge $\mathcal{B}_1$ we may have about the data. We then rank all patterns accordingly, and select the top-k best scoring/most interesting patterns, $\mathcal{X}_1 = \{X_1, \ldots, X_k\}$. We assume the analyst will investigate these in detail, and hence that now onward we may regard these patterns *and* what can be derived from them as 'known'. As such, we update our background knowledge with $\mathcal{X}_1$, and hence for iteration 2 have $\mathcal{B}_2 = \mathcal{B}_1 \cup \mathcal{X}_1$, for which we infer model p_2. We then rank all patterns accordingly, etc, until we're done.

Next we will discuss three methods that allow for dynamic ranking.

5.2 Maximum Entropy Models

Maximum Entropy models, which we've met in the previous section, provide a natural way of constructing a probabilistic model from a given set of itemsets and their frequencies: essentially, each itemset is a constraint. As the technical details of how to infer a model under such constraints are beyond the scope of this chapter, and we refer the interested reader to, for example, Pavlov et al. [53].

Given a model that can incorporate itemsets and frequencies as background knowledge, we need to define a score for ranking candidates. We can use the statistical tests from Sect. 4, but a more intuitive approach is to use the likelihood of the data under the model. That is, the typical goal in dynamic ranking is to find a ranking that of which the top-k is the best explanation of the data in k terms.

To this end we construct a score in a post-hoc fashion. That is, we score a candidate on how much information we would gain *if* we would include it in the background knowledge. We do this as follows.

In general, given a set of itemsets $\mathcal{F}$, and a set of target frequencies θ_X for every $X \in \mathcal{F}$ as background knowledge, we can construct a MaxEnt model p^* such that $E[S_X] = \theta_X$ for every $X \in \mathcal{F}$. In turn, we can use the likelihood $p^*(D \mid \mathcal{F})$ to score the quality of $\mathcal{F}$. In other words, the better p^* can predict all frequencies the more likely is the data according to p^*, the better is the collection is $\mathcal{F}$. To be more precise, we know that $p^*(D \mid \mathcal{F} \cup \{Y\}) \geq p^*(D \mid \mathcal{F})$ and, as a special case, the equality holds whenever the observed frequency of Y is exactly equal to the expectation derived from $\mathcal{F}$. Moreover, the score increases as the observed frequency of Y becomes more distant from the expected value.

Given this likelihood score we can evaluate how informative a pattern Y is about the data in addition to our background knowledge. The question now is, how can find good rankings and pattern sets efficiently?

Wang and Parthasararthy [71] take a pre-mined collection of frequent itemsets as candidates, and consider these in level-wise batches. That is, they first consider the itemsets of size 1, then of size 2, and so on. Per batch they select all itemsets for which the predicted frequencies (L1 distance) deviates more than a given threshold, and add all of these to the background knowledge, after which they update the model and iterate to the next level. In order to make the ranking feasible, i.e., to get around the NP-hardness of inferring frequencies from the MaxEnt model, the authors sample frequencies instead of inferring them exactly.

Alternatively, Mampaey et al. [42] iteratively mine the top-most informative pattern, regardless of its cardinality. To do so efficiently, they propose an efficient convex bound which allows many candidate patterns to be pruned, as well as a method for more efficiently inferring the MaxEnt model using a quick inclusion/exclusion based approach. NP-hardness problems are here circumvented by partitioning the model, either explicitly such that only patterns of up to length k are allowed, or by allowing up to k overlapping itemsets per part.

5.3 Tile-based Techniques

While expressing redundancy with itemsets and generative models can be very complicated, as we have to somehow determine expected frequencies of itemsets given frequencies of other itemsets, tiles provide much more straightforward and natural ways of measuring redundancy. For instance, we can consider the overlap between tiles.

As a basic example of such problem, consider the large tile mining problem we encountered in Sect. 2, where the goal is to find all exact tiles covering at least *minarea* 1s. When we cast this in the dynamic ranking framework, we would want to find the sequence of tiles such that each top-k covers as many 1s as possible using k exact tiles. That is, every tile in the ranking has to cover as many uncovered 1s as possible: any 1s the k-th tile covers already covered by a tile of rank $\leq k$ are simply not counted—and hence redundancy is directly punished. Geerts et al. [21] call this the maximal tiling problem, and identify it as an instance of the set cover problem, which is known to be NP-hard [36].

For noisy tiles, overlap alone does not suffice to identify redundancy, as different tiles may explain areas of the data in more fine or coarse detail. We should therefore consider the *quality* of the tile, for example by punishing the noise, or favoring tiles that are surprising.

To this end we can re-use the Rasch model [18], now using it to discover surprising tile sets. Similar for ranking tiles based on area, we can also rank tilings by computing the likelihood of entries covered by the tile set [37]. Similarly, to rank individual tiles, we need to normalize this probability, as otherwise the best tiling is automatically one tile containing the whole dataset. Kontonasios and De Bie [37] do not explicitly update their model, but instead consider the case where the analyst would investigate every tile exactly. As such, the values (0 s and 1 s) of a processed tile can be assumed known, and hence the likelihoods of already covered cells set to 1. They further show this covering problem is an instance of Weighted Budgeted Maximum Set Cover, which is NP-hard, but for which the greedy approach is known to provide good solutions.

The MaxEnt model composed from tiles can be also used in a similar manner as the MaxEnt model from itemsets. For instance, given a set of tiles and their densities, that is, the fraction of 1 s inside each tile, we can construct the corresponding MaxEnt model and use the likelihood of the data as the goodness of the tile set [67]. Besides for ranking a set of candidate tiles, Tatti and Vreeken show that many (exploratory) data mining results on binary data can be translated into sets of noisy tiles. Hence, through the same machinery, we can dynamically rank results from different algorithms based on their relative informativeness [67].

Whereas the basic MaxEnt allows only frequencies of 1s within tiles as background knowledge, a recent paper by Kontonasios and De Bie [38], demonstrates how more complex information can be incorporated. Examples include frequencies of itemsets—unluckily, however, as we saw for the transaction based MaxEnt model, this does mean that inferring from the model becomes the NP-hard in general.

In the introduction we spoke about the impossibility of formalizing the inherently subjective notion of interestingness in general. Conceptually, however, dynamic ranking comes very close. As long as we can infer the Maximum Entropy model for the background knowledge an arbitrary user has, under the assumption that the user can optimally make all inferences from this knowledge, we know from information theory that our framework will correctly identify the most surprising result. De Bie [17] argues that this setup is one of the most promising for measuring subjective interestingness. The key challenges he identifies are in defining Maximum Entropy models for rich data and pattern types, as well as for efficiently mining those patterns that optimize the score.

5.4 Swap Randomization

As the last model we consider for dynamic ranking, we return to the swap randomization model we first described in Sect. 4. Recall that using this model we can sample random datasets of fixed row and column margins, and that we can use these samples to obtain empirical p-values.

We can extend this model by requiring that the sampled datasets must also have fixed frequencies for a certain given set of itemsets. Unsurprisingly, this makes sampling datasets very difficult, however. In fact, producing a a new dataset satisfying all the constraints is computationally intractable in general, even if we have original dataset at our disposal [29].

Instead of forcing hard constraints, we can relax these conditions and require that the probability of a random dataset R should decrease as the frequencies of given itemsets diverge from the target frequencies. Datasets that satisfy the given itemsets exactly will have the largest probability but other datasets are also possible. This relaxation allows us to use the same, well-understood, MCMC techniques as for standard swap randomization [29].

6 Pattern Sets

Pattern set mining is the fourth and final approach to discovering interesting patterns that we cover, and is also the most recent. It is strongly related to the dynamic modeling approach we met in the previous section, but has a slightly different twist and implications.

The general idea in pattern set mining is simple: instead of measuring the interestingness of each pattern X individually, i.e., through $q(X)$, we now define q over *sets of patterns* $\mathcal{X}$. That is, instead of evaluating a pattern X only locally, e.g., checking whether it describes significant local structure of the data, the goal is now defined globally. As such, we aim to find that set of patterns $\mathcal{X}$ that is optimal with regard to q. For example, we can now say we want to find that set of patterns that together describes the structure of the data best, i.e., that models the full joint distribution of the data best.

By measuring quality over sets instead of individual patterns, we face a combinatorial problem over an exponential space of candidate elements. That is, to find the optimal solution naively we would have to consider every possible subset out of the space of $2^{|\mathcal{I}|}$ possible patterns in the data. Sadly, none of the proposed quality measures for pattern set mining exhibit monotonicity, and hence we have no efficient strategy to obtain the optimal pattern set. Moreover, while for some measures we know that even approximating the optimum within any constant factor is NP-hard [49], for most measures the score landscape is so erratic we so far have no results at all on the complexity of the optimization problem.

In light of the search space and the difficulty of the problem, most pattern set mining methods employ heuristics. In particular, the locally optimal greedy strategy is a popular choice. This means that in practice, pattern set mining methods and dynamic modeling have a lot in common; although in iterative ranking there is no explicit global goal defined, in order to find a ranking, we iteratively greedily find the locally most informative pattern, and update the model. A key difference to pattern set mining is that here we explicitly take the complexity of the pattern set in check; we consider both the gain in quality, as well as the cost in complexity over the full set.

(Though pattern set mining is a combinatorial problem, and intuitively optimizing most instances seems very complex, so far theoretical hardness results have only been found for a handful of cases, including [49, 48, 75].)

6.1 Itemsets

KRIMP is among the most well-known pattern set mining algorithms. Siebes et al. [62] define the best set of itemsets by the Minimum Description Length principle as the set that provides the best lossless compression. Each itemset is assigned a code word, the length of which depends on how frequently the itemset is used when greedily covering the data without overlap. The pattern set is then scored on the number of bits necessary to lossless describe the data. That is, the sum over the number of bits to encode the dictionary, the itemsets and their code words, and number of bits to encode the data using these code words.

The KRIMP algorithm heuristically finds good sets by first mining frequent itemsets up to a given *minsup* threshold, and greedily selecting from these in a fixed order. The resulting pattern sets are typically small (100 s) and have been shown to be useful in many data mining tasks [70] (see also Chap. 8). There exist a number of variants of KRIMP for other data types, including for low-entropy sets [31]. Siebes and Kersten [61] investigated to structure functions, and proposed the GROEI algorithm for approximating the optimal k-pattern set.

Alternative to a descriptive model, we can aim to find a good generative model. The MINI algorithm proposed by Gallo et al. [20] employs a probability distribution based on itemsets and frequencies, and aims finding the set of itemsets that predict the data best.

While intuitively appealing, using likelihood to score pattern sets, however, is not enough since the score increases w.r.t. the inclusion of pattern set, that is, the set containing all itemsets will have the highest likelihood. To control the size of the set we need to exert some control over the output set. For example, we can either ask the user to give a number of patterns k, or automatically control the number of the itemsets by BIC [59] or MDL [58], in which case the improvement of adding a new itemset into a result set must be significant.

The MaxEnt model employed by the MTV algorithm, which we met in the previous section, does not only lend itself for iterative ranking, but can also be straight-forwardly used with either of test two model selection criteria [42]. In practice, (partly due to the partitioning constraints necessary to keep computation feasible) MTV typically finds pattern sets in the order of tens of patterns.

6.2   Tiles

The k maximal tiling problem, as defined by Geerts et al. [21] is perhaps the earliest example of pattern set mining. Xiang et al. [74, 75] expanded on the problem setting and aim to cover as many of the 1s of the data with k noisy tiles. Both problem settings are NP-hard, and are related to Set Cover. As such, the greedy strategy is known to find a solution within $O(\log n)$ of the optimum.

Boolean Matrix Factorization   Strongly related to tiling is the problem of Boolean Matrix Factorization (BMF) [49]. The goal in matrix factorization is to find a low-rank approximation of the data. In case of BMF, each factor can be regarded as a noisy tile, and a factorization hence as a set of tiles. BMF is known to be NP-hard, as well as NP-hard to approximate [49]. The Asso algorithm is a heuristic for finding good k-factorizations, and can be coupled with an MDL strategy to automatically determine the best model order [47]. Lucchese et al. [40] gave the much faster PANDA algorithm, which optimizes a more loose global objective that weighs the number of covered 1s with the number of 1s of the factors.

Geometric Tiles   So far we have only considered unordered binary data in this chapter. When we fix the order of the rows and columns, however, for instance because the data is ordered spatially, it may only make sense to consider tiles that are consecutive under this ordering. This problem setting gives rise to interesting problem settings, as well as algorithmic solutions.

Gionis et al. [22] propose to mine dense *geometric* tiles that stand out from the background distribution, and to do so iteratively in order to construct a tree of tiles. To determine whether a tile is significant, they propose a simple MDL based score. Finding the optimal tile under this score is $O(n^2 m^2)$ and hence infeasible for non-trivial data. To circumvent this problem, they propose a randomized approach. Tatti and Vreeken [68] recently proposed an improved algorithm that can find the optimal sub-tile in only $O(mn\min(m, n))$. The tile trees discovered by these methods typically contain in the order of 10 s to 100 s of tiles.

Gionis et al. [22] also showed that by the same strategy, by first applying spectral ordering, meaningful *combinatorial* tiles can be discovered from unordered binary data.

6.3   Swap Randomization

The final pattern set mining approach we consider is based on swap randomization. Lijffijt et al. [39] aim to find the smallest set of patterns that explains most about the data in terms of a global p-value. By employing the swap randomization framework, and in particular by using empirical p-values they can consider a rich range of patterns, including frequent itemsets as well as clusterings. The goal of finding the smallest set that statistically explains the data is NP-hard in its general form, and there exists no efficient algorithm with finite approximation ratio.

As many random datasets need to be sampled in order to determine significance this approach is not as computationally efficient as some of the methods covered above, however, it should be noted that the framework is highly general. In principle it can be applied to any data and pattern type for which a (swap-)randomization variant has been defined, which includes, among others, real-valued data [51], graphs [28], and sequences [32].

7 Conclusions

With the goal of mining interesting patterns we face two main problems: first, we need to be able to identify whether a pattern is potentially interesting, and second, we do not want the results to be redundant. Both these concepts are subjective, and hence there is no single correct answer to either of the two goals. Consequently, we provide an overview of myriad of different techniques for mining interesting patterns. These techniques range from classic reduction approaches, such as, closed itemsets and non-derivable itemsets to statistical methods, where we either are looking patterns that deviate most from the expectation or looking for a compact set that models the data well.

Unlike when mining frequent itemsets, for more advanced interestingness measures we rarely have monotonicity to our advantage. This means that we cannot prune the search space easily, and mining algorithms are hence significantly more complex. In particular algorithms for discovering pattern sets are often heuristic. Better understanding of these combinatorial problems and their score, for example, by providing theoretical guarantees, is both a promising and necessary line of work toward developing better and faster algorithms.

In this chapter we covered techniques meant only for binary data. In comparison, discovering and defining interesting patterns and pattern sets from other types of data, such as, sequences, graphs, or real-valued datasets is still strongly under-developed. Both in the definition of useful interestingness measures, as well as in the development of efficient algorithms for extracting such patterns directly from data there exist many opportunities for exciting future work.

Regardless of data type, the key idea for future work is developing algorithms for mining *small* and *non-redundant* sets of only the most interesting patterns. Or, to quote Toon Calders at the 2012 ECMLPKDD most-influential paper award: "please, please stop making new algorithms for mining *all* patterns".

Acknowledgments Jilles Vreeken is supported by the Cluster of Excellence "Multimodal Computing and Interaction" within the Excellence Initiative of the German Federal Government. Nikolaj Tatti is supported by Academy of Finland grant 118653 (ALGODAN).

References

1. R. Agrawal and R. Srikant. Fast algorithms for mining association rules. In *VLDB*, pages 487–499, 1994.
2. C. C. Aggarwal and P. S. Yu. A new framework for itemset generation. In *PODS*, pages 18–24. ACM, 1998.
3. R. Agrawal, T. Imielinksi, and A. Swami. Mining association rules between sets of items in large databases. In *SIGMOD*, pages 207–216. ACM, 1993.
4. M. Al Hasan and M. J. Zaki. Output space sampling for graph patterns. *PVLDB*, 2(1):730–741, 2009.
5. M. Al Hasan, V. Chaoji, S. Salem, J. Besson, and M. J. Zaki. Origami: Mining representative orthogonal graph patterns. In *ICDM*, pages 153–162. IEEE, 2007.
6. R. Bayardo. Efficiently mining long patterns from databases. In *SIGMOD*, pages 85–93, 1998.
7. M. Boley, C. Lucchese, D. Paurat, and T. Gärtner. Direct local pattern sampling by efficient two-step random procedures. In *KDD*, pages 582–590. ACM, 2011.
8. M. Boley, S. Moens, and T. Gärtner. Linear space direct pattern sampling using coupling from the past. In *KDD*, pages 69–77. ACM, 2012.
9. J.-F. Boulicaut, A. Bykowski, and C. Rigotti. Free-sets: a condensed representation of boolean data for the approximation of frequency queries. *Data Min. Knowl. Disc.*, 7(1):5–22, 2003.
10. S. Brin, R. Motwani, and C. Silverstein. Beyond market baskets: Generalizing association rules to correlations. In *SIGMOD*, pages 265–276. ACM, 1997.
11. D. Burdick, M. Calimlim, J. Flannick, J. Gehrke, and T. Yiu. MAFIA: A maximal frequent itemset algorithm. *IEEE TKDE*, 17(11):1490–1504, 2005.
12. T. Calders and B. Goethals. Mining all non-derivable frequent itemsets. In *PKDD*, pages 74–85, 2002.
13. C. Chow and C. Liu. Approximating discrete probability distributions with dependence trees. *IEEE TIT*, 14(3):462–467, 1968.
14. E. Cohen, M. Datar, S. Fujiwara, A. Gionis, P. Indyk, R. Motwani, J. D. Ullman, and C. Yang. Finding interesting associations without support pruning. *IEEE TKDE*, 13(1):64–78, 2001.
15. R. G. Cowell, A. P. Dawid, S. L. Lauritzen, and D. J. Spiegelhalter. Probabilistic networks and expert systems. In *Statistics for Engineering and Information Science*. Springer-Verlag, 1999.
16. I. Csiszár. I-divergence geometry of probability distributions and minimization problems. *Annals Prob.*, 3(1):146–158, 1975.
17. T. De Bie. An information theoretic framework for data mining. In *KDD*, pages 564–572. ACM, 2011.
18. T. De Bie. Maximum entropy models and subjective interestingness: an application to tiles in binary databases. *Data Min. Knowl. Disc.*, 23(3):407–446, 2011.
19. R. A. Fisher. On the interpretation of χ^2 from contingency tables, and the calculation of P. *J. R. Statist. Soc.*, 85(1):87–94, 1922.
20. A. Gallo, N. Cristianini, and T. De Bie. MINI: Mining informative non-redundant itemsets. In *ECML PKDD*, pages 438–445. Springer, 2007.
21. F. Geerts, B. Goethals, and T. Mielikäinen. Tiling databases. In *DS*, pages 278–289, 2004.
22. A. Gionis, H. Mannila, and J. K. Seppänen. Geometric and combinatorial tiles in 0-1 data. In *PKDD*, pages 173–184. Springer, 2004.
23. A. Gionis, H. Mannila, T. Mielikäinen, and P. Tsaparas. Assessing data mining results via swap randomization. *TKDD*, 1(3):167–176, 2007.
24. B. Goethals and M. Zaki. Frequent itemset mining dataset repository (FIMI). http://fimi.ua.ac.be/, 2004.

25. W. Hämäläinen. Kingfisher: an efficient algorithm for searching for both positive and negative dependency rules with statistical significance measures. *Knowl. Inf. Sys.*, 32(2):383–414, 2012.
26. J. Han, J. Pei, and Y. Yin. Mining frequent patterns without candidate generation. In *SIGMOD*, pages 1–12. ACM, 2000.
27. D. Hand, N. Adams, and R. Bolton, editors. *Pattern Detection and Discovery*. Springer-Verlag, 2002.
28. S. Hanhijärvi, G. C. Garriga, and K. Puolamäki. Randomization techniques for graphs. In *SDM*, pages 780–791. SIAM, 2009.
29. S. Hanhijärvi, M. Ojala, N. Vuokko, K. Puolamäki, N. Tatti, and H. Mannila. Tell me something I don't know: randomization strategies for iterative data mining. In *KDD*, pages 379–388. ACM, 2009.
30. H. Heikinheimo, J. K. Seppänen, E. Hinkkanen, H. Mannila, and T. Mielikäinen. Finding low-entropy sets and trees from binary data. In *KDD*, pages 350–359, 2007.
31. H. Heikinheimo, J. Vreeken, A. Siebes, and H. Mannila. Low-entropy set selection. In *SDM*, pages 569–580, 2009.
32. A. Henelius, J. Korpela, and K. Puolamäki. Explaining interval sequences by randomization. In *ECML PKDD*, pages 337–352. Springer, 2013.
33. IBM. *IBM Intelligent Miner User's Guide, Version 1, Release 1*, 1996.
34. S. Jaroszewicz and D. A. Simovici. Interestingness of frequent itemsets using bayesian networks as background knowledge. In *KDD*, pages 178–186. ACM, 2004.
35. E. Jaynes. On the rationale of maximum-entropy methods. *Proc. IEEE*, 70(9):939–952, 1982.
36. R. M. Karp. Reducibility among combinatorial problems. In *Proc. Compl. Comp. Comput.*, pages 85–103, New York, USA, 1972.
37. K.-N. Kontonasios and T. De Bie. An information-theoretic approach to finding noisy tiles in binary databases. In *SDM*, pages 153–164. SIAM, 2010.
38. K.-N. Kontonasios and T. De Bie. Formalizing complex prior information to quantify subjective interestingness of frequent pattern sets. In *IDA*, pages 161–171, 2012.
39. J. Lijffijt, P. Papapetrou, and K. Puolamäki. A statistical significance testing approach to mining the most informative set of patterns. *Data Min. Knowl. Disc.*, pages 1–26, 2012.
40. C. Lucchese, S. Orlando, and R. Perego. Mining top-k patterns from binary datasets in presence of noise. In *SDM*, pages 165–176, 2010.
41. M. Mampaey. Mining non-redundant information-theoretic dependencies between itemsets. In *DaWaK*, pages 130–141, 2010.
42. M. Mampaey, J. Vreeken, and N. Tatti. Summarizing data succinctly with the most informative itemsets. *TKDD*, 6:1–44, 2012.
43. H. Mannila and H. Toivonen. Multiple uses of frequent sets and condensed representations. In *KDD*, pages 189–194, 1996.
44. H. Mannila, H. Toivonen, and A. I. Verkamo. Efficient algorithms for discovering association rules. In *KDD*, pages 181–192, 1994.
45. H. Mannila, H. Toivonen, and A. I. Verkamo. Levelwise search and borders of theories in knowledge discovery. *Data Min. Knowl. Disc.*, 1(3):241–258, 1997.
46. R. Meo. Theory of dependence values. *ACM Trans. Database Syst.*, 25(3):380–406, 2000.
47. P. Miettinen and J. Vreeken. Model order selection for Boolean matrix factorization. In *KDD*, pages 51–59. ACM, 2011.
48. P. Miettinen and J. Vreeken. MDL4BMF: Minimum description length for Boolean matrix factorization. Technical Report MPI-I-2012-5-001, Max Planck Institute for Informatics, 2012.
49. P. Miettinen, T. Mielikäinen, A. Gionis, G. Das, and H. Mannila. The discrete basis problem. *IEEE TKDE*, 20(10):1348–1362, 2008.
50. F. Moerchen, M. Thies, and A. Ultsch. Efficient mining of all margin-closed itemsets with applications in temporal knowledge discovery and classification by compression. *Knowl. Inf. Sys.*, 29(1):55–80, 2011.
51. M. Ojala. Assessing data mining results on matrices with randomization. In *ICDM*, pages 959–964, 2010.

52. N. Pasquier, Y. Bastide, R. Taouil, and L. Lakhal. Discovering frequent closed itemsets for association rules. In *ICDT*, pages 398–416. ACM, 1999.
53. D. Pavlov, H. Mannila, and P. Smyth. Beyond independence: Probabilistic models for query approximation on binary transaction data. *IEEE TKDE*, 15(6):1409–1421, 2003.
54. J. Pei, A. K. Tung, and J. Han. Fault-tolerant frequent pattern mining: Problems and challenges. *Data Min. Knowl. Disc.*, 1:42, 2001.
55. R. G. Pensa, C. Robardet, and J.-F. Boulicaut. A bi-clustering framework for categorical data. In *PKDD*, pages 643–650. Springer, 2005.
56. A. K. Poernomo and V. Gopalkrishnan. Towards efficient mining of proportional fault-tolerant frequent itemsets. In *KDD*, pages 697–706, New York, NY, USA, 2009. ACM.
57. G. Rasch. *Probabilistic Models for Some Intelligence and Attainnment Tests*. Danmarks paedagogiske Institut, 1960.
58. J. Rissanen. Modeling by shortest data description. *Automatica*, 14(1):465–471, 1978.
59. G. Schwarz. Estimating the dimension of a model. *Annals Stat.*, 6(2):461–464, 1978.
60. J. K. Seppanen and H. Mannila. Dense itemsets. In *KDD*, pages 683–688, 2004.
61. A. Siebes and R. Kersten. A structure function for transaction data. In *SDM*, pages 558–569. SIAM, 2011.
62. A. Siebes, J. Vreeken, and M. van Leeuwen. Item sets that compress. In *SDM*, pages 393–404. SIAM, 2006.
63. N. Tatti. Computational complexity of queries based on itemsets. *Inf. Process. Lett.*, 98(5):183–187, 2006.
64. N. Tatti. Maximum entropy based significance of itemsets. *Knowl. Inf. Sys.*, 17(1):57–77, 2008.
65. N. Tatti and M. Mampaey. Using background knowledge to rank itemsets. *Data Min. Knowl. Disc.*, 21(2):293–309, 2010.
66. N. Tatti and F. Moerchen. Finding robust itemsets under subsampling. In *ICDM*, pages 705–714. IEEE, 2011.
67. N. Tatti and J. Vreeken. Comparing apples and oranges - measuring differences between exploratory data mining results. *Data Min. Knowl. Disc.*, 25(2):173–207, 2012.
68. N. Tatti and J. Vreeken. Discovering descriptive tile trees by fast mining of optimal geometric subtiles. In *ECML PKDD*. Springer, 2012.
69. C. Tew, C. Giraud-Carrier, K. Tanner, and S. Burton. Behavior-based clustering and analysis of interestingness measures for association rule mining. *Data Min. Knowl. Disc.*, pages 1–42, 2013.
70. J. Vreeken, M. van Leeuwen, and A. Siebes. KRIMP: Mining itemsets that compress. *Data Min. Knowl. Disc.*, 23(1):169–214, 2011.
71. C. Wang and S. Parthasarathy. Summarizing itemset patterns using probabilistic models. In *KDD*, pages 730–735, 2006.
72. G. I. Webb. Self-sufficient itemsets: An approach to screening potentially interesting associations between items. *TKDD*, 4(1):1–20, 2010.
73. G. I. Webb. Filtered-top-k association discovery. *WIREs DMKD*, 1(3):183–192, 2011.
74. Y. Xiang, R. Jin, D. Fuhry, and F. F. Dragan. Succinct summarization of transactional databases: an overlapped hyperrectangle scheme. In *KDD*, pages 758–766, 2008.
75. Y. Xiang, R. Jin, D. Fuhry, and F. Dragan. Summarizing transactional databases with overlapped hyperrectangles. *Data Min. Knowl. Disc.*, 2010.
76. M. J. Zaki. Scalable algorithms for association mining. *IEEE TKDE*, 12(3):372–390, 2000.
77. M. J. Zaki and C.-J. Hsiao. Charm: An efficient algorithm for closed itemset mining. In *SDM*, pages 457–473. SIAM, 2002.
78. M. J. Zaki, S. Parthasarathy, M. Ogihara, and W. Li. New algorithms for fast discovery of association rules. In *KDD*, Aug 1997.

Chapter 6
Negative Association Rules

Luiza Antonie, Jundong Li and Osmar Zaiane

Abstract Mining association rules associates events that took place together. In market basket analysis, these discovered rules associate items purchased together. Items that are not part of a transaction are not considered. In other words, typical association rules do not take into account items that are part of the domain but that are not together part of a transaction. Association rules are based on frequencies and count the transactions where items occur together. However, counting absences of items is prohibitive if the number of possible items is very large, which is typically the case. Nonetheless, knowing the relationship between the absence of an item and the presence of another can be very important in some applications. These rules are called negative association rules. We review current approaches for mining negative association rules and we discuss limitations and future research directions.

Keywords Negative association rules

1 Introduction

Traditional association rule mining algorithms [11] have been developed to find positive associations between items [4, 9, 26, 14]. Positive associations are associations between items existing in transactions (i. e. items that are present and observed). In market basket analysis, we are generally interested in items that were purchased, and particularly in items purchased together. The assumption is that items that appear in transactions are more important than those that do not appear. As opposed to positive associations, we call negative associations, associations that negate presence.

L. Antonie (✉)
University of Guelph, Guelph, Canada
e-mail: lantonie@uoguelph.ca

J. Li · O. Zaiane
University of Alberta, Alberta, Canada
e-mail: jundong1@cs.ualberta.ca

O. Zaiane
e-mail: aiane@cs.ualberta.ca

C. C. Aggarwal, J. Han (eds.), *Frequent Pattern Mining,*
DOI 10.1007/978-3-319-07821-2_6, © Springer International Publishing Switzerland 2014

In other words, negative association rules are rules that comprise relationships between present and absent items. Indeed, items that are not purchased when others are can be revealing and certainly important in understanding purchasing behaviour. The association "bread implies milk" indicates the purchasing behaviour of buying milk and bread together. What about the following associations: "customers who buy Coke *do not* buy Pepsi" or "customers who buy juice *do not* buy bottled water"? Associations that include negative items (i. e. items absent from the transaction) can be as valuable as positive associations in many applications, such as devising marketing strategies. Aggarwal and Yu [1] discuss some of the weaknesses and the computational issues for mining positive association rules. They observe that current methods are especially unsuitable for dealing with dense datasets, which is exactly the case when one wants to mine negative association rules.

The expensive computation part of association rule mining is the phase enumerating the frequent itemsets (i. e. a set of items). This enumeration takes place in a search space of size 2^k with k being the number of unique items in the data collection. Focusing on only positive associations significantly reduces this prohibitive search space since we only need to count the observed items in the transactions. Moreover, putting the attention on items present in transactions limits the enumeration of relevant itemsets to a depth dictated by the largest available transaction. These advantageous stratagems cannot be used if absent items are also considered.

Although interesting and potentially useful, the discovery of negative association rules is both a complex and computationally expensive problem. We consider a negative association rule either a negative association between two positive itemsets or an association rule that contains at least a negative item in the antecedent or consequent. The mining of negative association rules is a complex problem due to the increase in items when negative items are considered in the mining process. Imagine a transaction in market basket analysis where a customer buys *bread* and *milk*. When mining for positive association rules only those two items are considered (i.e. *bread* and *milk*). However, when negative items are considered (i.e. items/products not present in a basket/transaction) the search space increases exponentially because all the items in the collection, although not present in the transaction have to be considered. Not only is the problem complex, but also large numbers of negative patterns are uninteresting. The research of mining negative association patterns has to take into consideration both the complexity of the problem and the usefulness of the discovered patterns.

2 Negative Patterns and Negative Association Rules

Formally, association rules are defined as follows: Let $\mathcal{I} = \{i_1, i_2, \ldots i_m\}$ be a set of items. The total number of unique items is m, the dimensionality of the problem. Let $\mathcal{D}$ be a set of transactions, where each transaction T is a set of items such that $T \subseteq \mathcal{I}$. Each transaction is associated with a unique identifier TID. A transaction T is said to contain X, a set of items in $\mathcal{I}$, if $X \subseteq T$. X is called an itemset.

Table 6.1 Transactional
database-positive and
negative items

TID	Original TD	Augmented TD
1	A,C,D	$A, \neg B, C, D, \neg E$
2	B,C	$\neg A, B, C, \neg D, \neg E$
3	C	$\neg A, \neg B, C, \neg D, \neg E$
4	A,B,E	$A, B, \neg C, \neg D, E$
5	A,C,D	$A, \neg B, C, D. \neg E$

Definition 1 (Association Rule) An *association rule* is an implication of the form
"$X \Rightarrow Y$", where $X \subseteq \mathcal{I}, Y \subseteq \mathcal{I}$, and $X \cap Y = \emptyset$.

Definition 2 (Support) The rule $X \Rightarrow Y$ has a *support s* in the transaction set $\mathcal{D}$ if
$s\%$ of the transactions in $\mathcal{D}$ contain $X \cup Y$. In other words, the support of the rule is
the probability that X and Y hold together among all the possible presented cases.

Definition 3 (Confidence) The rule $X \Rightarrow Y$ holds in the transaction set $\mathcal{D}$ with
confidence c if $c\%$ of transactions in $\mathcal{D}$ that contain X also contain Y. In other words,
the confidence of the rule is the conditional probability that the consequent Y is true
under the condition of the antecedent X.

The problem of discovering all association rules from a set of transactions $\mathcal{D}$
consists of generating the rules that have a *support* and *confidence* greater than given
thresholds.

Definition 4 (Negative Item and Positive Item) A negative item is defined as
$\neg i_k$, meaning that item i_k is absent from a transaction T. The support of $\neg i_k$ is
$s(\neg i_k) = 1 - s(i_k)$. i_k, a positive item, is an item that is present in a transaction.

Definition 5 (Negative Association Rule) A negative association rule is an impli-
cation of the form $X \Rightarrow Y$, where $X \subseteq \mathcal{I}, Y \subseteq \mathcal{I}$, and $X \cap Y = \emptyset$ and X and/or Y
contain at least one negative item.

Definition 6 Negative Associations between Itemsets A negative association be-
tween two positive itemsets X,Y are rules of the following forms $\neg X \Rightarrow Y, X \Rightarrow \neg Y$
and $\neg X \Rightarrow \neg Y$.

Table 6.1 shows a toy transactional database with 5 transactions and 5 items. "Original TD" column shows the items present in each transaction, while "Augmented TD"
column shows both present and absent items.

Mining association rules from a transactional database that contains information
about both present and absent items is computationally expensive due to the following
reasons:

1. The number of items in the transactional database swells when their negative
   counterparts are added to a transactional database. The maximum number of
   patterns that can be found in a transactional database with d items is $2^d - 1$. The
   number of items in the "Original TD" in Table 6.1 is $n = 5$. Even for the small set
   in Table 6.1, the number of itemsets jumps dramatically from 31 to 1023 when
   the negative items are added.

Table 6.2 Example 1 data

	CM	¬CM	Σ_{row}
SM	20	60	80
¬SM	20	0	20
Σ_{col}	40	60	100

2. The length of the transactions in the database increases dramatically when negative items are considered. Picture the length of the transaction in a market basket analysis example where all products in a store have to be considered in each transaction. For example, to a basket where bread and milk are bought (i.e. milk and bread are the positive items), all the other products in the store become part of the transaction as negative items.
3. The total number of association rules that can be discovered when negative items are considered is $5^d - 2 \times 3^d + 1$. A detailed calculation for the formula can be found in [18]. The number of association rules for positive items in a transactions is $3^d - 2^{d+1} + 1$. For our small example, it means that we can find up to 180 positive rules and up to 2640 when the negative items are considered as well.
4. The number of candidate itemsets is reduced when mining positive association rules by the support based pruning. This property is no longer efficient in a transactional database that is augmented with the negative items. Given that the support of a negative item is $s(\neg i_k) = 1 - s(i_k)$, either the negative or the positive item will have a big enough support to pass the minimum support threshold.

Given the reasons above, the traditional association rule mining algorithms can not cope with mining rules when negative items are considered. This is the reason new algorithms are needed to efficiently mine association rules with negative items. Here we survey algorithms that efficiently mine some variety of negative associations from data.

3 Current Approaches

In this section we present current approaches proposed in the literature to discover negative association rules. We illustrate in Example 1 how rules discovered in the support confidence framework could be misleading sometimes and how the negative associations discovered in data can shed a new light on the discovered patterns.

Example 1 Let us consider an example from market basket data. In this example we want to study the purchase of cow's milk (CM) versus soy milk (SM) in a grocery store. Table 6.2 gives us the data collected from 100 baskets in the store. In Table 6.2 "CM" means the basket contains cow's milk and "¬ CM" means the basket does not contain cow's milk. The same applies for soy milk.

In this data, let us find the positive association rules in the "support-confidence" framework. The association rule "SM ⇒ CM" has 20 % support and 25 % confidence (support(SM ∧ CM)/support(SM)). The association rule "CM ⇒ SM" has 20 % support and 50 % confidence (support(SM ∧ CM)/support(CM)). The support is considered fairly high for both rules. Although we may reject the first rule on the confidence basis, the second rule seems a valid rule and may be considered in the data analysis. However, when a statistical significance test is considered, such as statistical correlation between the *SM* and *CM* items, one would find that the two items are actually negatively correlated. This shows that the rule "CM $\Rightarrow$ SM" is misleading. This example shows not only the importance of considering negative association rules, but also the importance of statistical significance of the patterns discovered.

The problem of finding negative association rules is complex and computationally intensive as discussed in Sect. 2. A common solution to deal with the complexity is to focus the search on special cases of interest. Some techniques employ domain knowledge to guide the search, some are focusing on a certain type of rules of interest, while others are considering interestingness measures to mine for statistically significant patterns. We give more details about some approaches that have been proposed in the literature for mining association rules with negations.

Brin et al. [8] mentioned for the first time the notion of negative relationships in the literature. They proposed to use the chi-square test between two itemsets. The statistical test verifies the independence between the two itemsets. To determine the nature (positive or negative) of the relationship, a correlation metric is used. The negative association rules that could be discovered based on these measures are the following: $\neg X \Rightarrow Y$, $X \Rightarrow \neg Y$ and $\neg X \Rightarrow \neg Y$. One limitation for this method is that the computation of the χ^2 measure can become expensive in large and dense datasets.

Aggarwal and Yu [2, 3] introduced a new method for finding interesting itemsets in data. Their method is based on mining strongly collective itemsets. The collective strength of an itemset I is defined as follows:

$$C(I) = \frac{1 - v(I)}{1 - E[v(I)]} \times \frac{E[v(I)]}{v(I)} \tag{6.1}$$

where $v(I)$ is the violation rate of an itemset I and it is the fraction of violations over the entire set of transactions and $E[v(i)]$ is its expected value. An itemset I is in a violation of a transaction if only a subset of its items appear in that transaction. The collective strength ranges from 0 to ∞, where a value of 0 means that the items are perfectly negatively correlated and a value of ∞ means that the items are perfectly positively correlated. A value of 1 indicates that the value is exactly the same as its expected value, meaning statistical independence. The advantage of mining itemsets with collective strength is that the method finds statistical significant patterns. In addition, this model has good computational efficiency, thus being a good method in mining dense datasets. This property, along with the symmetry of collective strength measure, makes this method a good candidate for mining negative association rules in data.

In [19] the authors present a new idea to mine strong negative rules. They combine positive frequent itemsets with domain knowledge in the form of a taxonomy to mine negative associations. The idea is to reduce the search space, by constraining the search to the positive patterns that pass the minimum support threshold. When all the positive itemsets are discovered, candidate negative itemsets are considered based on the taxonomy used. They are considered interesting if their support is sufficiently different than the expected support. Association rules are generated from the negative itemsets if the interestingness measure of the rule exceeds a given threshold. The type of the rules discovered with this method are implications of the form $A \Rightarrow \neg B$. The issue with this approach is that it is hard to generalize since it is domain dependant and requires a predefined taxonomy. However, it should be noted that taxonomies exist for certain applications, thus making this method useful. A similar approach is described in [25].

Wu et al. [24] derived another algorithm for generating both positive and negative association rules. The negative association discovered in this paper are of the following forms: $\neg X \Rightarrow Y$, $X \Rightarrow \neg Y$ and $\neg X \Rightarrow \neg Y$. They add on top of the support-confidence framework another measure called *mininterest* for a better pruning of the frequent itemsets generated (the argument is that a rule $A \Rightarrow B$ is of interest only if $supp(A \cup B) - supp(A)supp(B) \geq mininterest$). "Mininterest" parameter is used to assess the dependency between the two itemsets considered, A and B are not independent if they satisfy the condition. The authors consider as itemsets of interest those itemsets that exceed minimum support and minimum interest thresholds. Although [24] introduces the *"mininterest"* parameter, the authors do not discuss how to set it and what would be the impact on the results when changing this parameter.

The algorithm proposed in [20, 21], named SRM (substitution rule mining), discovers a subset of negative associations. The authors develop an algorithm to discover negative associations of the type $X \Rightarrow \neg Y$. These association rules can be used to discover which items are substitutes for others in market basket analysis. Their algorithm discovers first what they call *concrete items*, which are those itemsets that have a high chi-square value and exceed the expected support. Once these itemsets are discovered, they compute the correlation coefficient for each pair of them. From those pairs that are negatively correlated, they extract the desired rules (of the type $X \Rightarrow \neg Y$, where Y is considered as an atomic item). Although interesting for the substitution items application, SRM is limited in the kind of rules that it can discover.

Antonie and Zaïane [7] proposed an algorithm to mine strong positive and negative association rules based on the Person's ϕ correlation coefficient. For the association rule $X \Rightarrow Y$, its ϕ correlation coefficient is as follows:

$$\phi = \frac{s(XY)s(\neg X \neg Y) - s(X \neg Y)s(\neg XY)}{\sqrt{(s(X)s(\neg X)s(Y)s(\neg Y))}} \tag{6.2}$$

In their algorithm, itemset and rule generation are combined and the relevant rules are generated on-the-fly while analyzing the correlations within each candidate itemset. This avoids evaluating item combinations redundantly. For each generated candidate itemset, all possible combinations of items are computed to analyze their correlations. In the end, only those rules generated from item combinations with strong correlations are considered. The strength of the correlation is indicated by a correlation threshold, either given as input or by default set to 0.5. If the correlation between item combinations X and Y of an itemset XY, where X and Y are itemsets, is negative, negative association rules are generated when their confidence is high enough. The produced rules have either the antecedent or the consequent negated: ($\neg X \Rightarrow Y$ and $X \Rightarrow \neg Y$), even if the support is not higher than the support threshold. However, if the correlation is positive, a positive association rule with the classical support-confidence idea is generated. If the support is not adequate, a negative association rule that negates both the antecedent and the consequent is generated when its confidence and support are high enough. They define the negative associations as *confined negative association rules*. A confined negative association rule is one of the following: $\neg X \Rightarrow Y$ or $X \Rightarrow \neg Y$, where the entire antecedent or consequent is treated as an atomic entity and the entire entity is either negated or not. These rules are a subset of the entire set of generalized negative association rules.

In [22], authors extend an existing algorithm for association rule mining, GRD (generalized rule discovery), to include negative items in the rules discovered. The algorithm discovers top-k positive and negative rules. GRD does not operate in the support confidence framework, it uses leverage and the number of rules to be discovered. The limitation of the algorithm is that it mines rules containing no more than 5 items (up to 4 items in the left hand side of the rule and 1 item in the right hand side of the rule).

Cornelis et al. [10] proposed a new Apriori-based algorithm (PNAR) that exploits the upward closure property of negative association rules that if support of $\neg X$ meets the minimum support threshold, then for every $Y \subseteq \mathcal{I}$ such that $X \cap Y = \emptyset$, $\neg(XY)$ also meets the support threshold. With this upward closure property, valid positive and negative association rules are defined in the form of $C_1 \Rightarrow C_2$, $C_1 \in \{X, \neg X\}, C_2 \in \{Y, \neg Y\}$, $X, Y \subseteq \mathcal{I}$, $X \cap Y = \emptyset$, if it meets the following conditions: (1) $s(C_1 \Rightarrow C_2) \geq minsup$; (2) $s(X) \geq minsup$, $s(Y) \geq minsup$; (3) $conf(C_1 \Rightarrow C_2) \geq minconf$; (4) If $C_1 = \neg X$, then there does not exist $X' \subseteq X$ such that $s(\neg X' \Rightarrow C_2) \geq minsup$ (analogously for C_2). Then, the algorithm of mining both positive and negative valid association rules is built up around a partition of the itemset space by 4 steps: (1) generate all positive frequent itemsets $L(P_1)$; (2) for all itemsets I in $L(P_1)$, generate all negative frequent itemsets of the form $\neg(XY)$; (3) generate all negative frequent itemsets of the form $\neg X \neg Y$; (4) generate all negative frequent itemsets of the form $X \neg Y$ or $\neg XY$. The complete set of valid positive and negative association rules are derived after frequent itemsets are generated. No additional interesting measures are required in this support-confidence framework. Wang et al. [23] gave a more intuitive way to express the validity of both positive and negative association rules, the mining process is very similar to PNAR.

MINR [15] is a method that uses Fisher's exact test to identify item sets that do not occur together by chance, i.e. with a statistical significant probability. Let X and Y denote the disjoint itemsets in the antecedent and consequent part of a rule, respectively. The probability that X and Y occur together with c times by chance is:

$$Pcc(c|n, s(X), s(Y)) = \frac{\binom{s(X)}{c}\binom{n-s(X)}{s(Y)-c}}{\binom{n}{s(Y)}} \tag{6.3}$$

where n is the total number of transactions. The chance threshold is calculated independently for each candidate itemset:

$$chance(n, s(X), s(Y), p) = min \left\{ t | \sum_{i=0}^{i=t} Pcc(i|n, s(X), s(Y)) \geq p \right\} \qquad (6.4)$$

Normally, for a positive association, p-value is set to be very high (usually 0.9999), on the other hand, for a negative association, p-value is set to be very low (usually 0.001). The whole algorithm develops in an iterative way with rule generation and rule pruning. An itemset with a support greater than the positive chance threshold is considered for positive rule generation, while itemset with a support less than the negative chance threshold is considered for negative rule generation. In this way, the algorithm discovers three different types of negative association rules in the form of $X \Rightarrow \neg Y$, $\neg X \Rightarrow Y$ and $\neg X \Rightarrow \neg Y$. The first two types $X \Rightarrow \neg Y$, $\neg X \Rightarrow Y$ can be generated from the negative itemsets if the rule $X \Rightarrow Y$ satisfies the negative chance threshold and minimum confidence threshold. On the other hand, the rules in the form of $\neg X \Rightarrow \neg Y$ are derived from the positive itemsets if they meet the positive chance threshold and minimum confidence threshold.

Kingfisher [12, 13] is an algorithm developed to discover positive and negative dependency rules. The dependency rule can be formulated on the basis of association rule, that the association rule $X \Rightarrow Y$ is defined as a dependency rule if $P(X, Y) \neq P(X)P(Y)$. The dependency is positive, if $P(X, Y) > P(X)P(Y)$; and negative, if $P(X, Y) < P(X)P(Y)$. Otherwise, the rule is an independent rule. The author concentrated on a specific type of dependency rules, the rules with only one single consequent attribute. It can be noticed that the negative dependency for the rules $X \Rightarrow Y$ or $\neg X \Rightarrow \neg Y$ are the same as the positive dependency for the rules $X \Rightarrow \neg Y$ and $\neg X \Rightarrow Y$, therefore, it is enough to only consider the positive dependency rules $X \Rightarrow \neg Y$ or $\neg X \Rightarrow Y$. The statistical dependency of the rule $X \Rightarrow Y$, is measured by Fisher's exact test, the p-value, can be calculated:

$$p_F(X \Rightarrow Y = y) = \sum_{i=0}^{min\{s(XY \neq y), s(\neg X, Y = y)\}} \frac{\binom{s(X)}{s(XY=y)+i} \binom{s(\neg X)}{s(\neg XY \neq y)+i}}{\binom{n}{s(Y=y)}} \qquad (6.5)$$

where $y \in \{0, 1\}$ denotes the presence or absent of Y, and n is the total number of transactions. It can also be observed that $p_F(X \Rightarrow \neg Y) = p_F(\neg X \Rightarrow Y)$, therefore, it is enough to consider the negative rules in the form of $X \Rightarrow \neg Y$. An important task of rule mining is to find the non-redundant rules. Rules are considered as redundant when they do not add new information to the remaining rules. In order to reduce the number of discovered rules, Kingfisher focused on finding non-redundant rules. The rule $X \Rightarrow Y = y$ is non-redundant, if there does not exist any rules in the form of $X' \Rightarrow Y = y$ such that $X' \subsetneq X$ and $p_F(X' \Rightarrow Y = y) < p_F(X \Rightarrow Y = y)$, otherwise, the rule is considered as redundant. However, the statistical dependency is not a monotonic property, it is impossible to do some frequency-based pruning as Apriori-like algorithms. A straightforward solution is to list all possible negative rules in the form of $X \Rightarrow \neg Y$ in the whole search space via an enumeration tree, and then calculate their p_F-values to see if they are significant. The items are ordered in an ascending order (by frequency) in the enumeration tree and the tree is traversed by a breadth-first manner. In this way, more general rules are checked before their specializations, therefore, it is possible that redundant specializations can be pruned without checking. If the task is to search for the top K rules, the threshold of p_F-value, needs to be updated consistently, when a new K-th top rule is found with a lower p_F-value. However, in both cases, the size of the whole search space is $|\mathcal{P}(\mathcal{I})|$, where $\mathcal{P}(\mathcal{I})$ is the power set of $\mathcal{I}$, it grows exponentially with the size of attributes. In order to reduce the search space, the author fully exploits the property of p_F-value, and describes the basic branch-and-bound search by introducing three lower bounds for the measure of p_F-value, therefore, some insignificant rules can be pruned without further checking. Apart from the three lower bounds of p_F-value, anther two pruning strategies (pruning by minimality and pruning by principles of Lapis philosophorum) are also introduced to speed up the search.

4 Associative Classification and Negative Association Rules

Associative classifiers are classification models that use association rules discovered in the data to make predictions [5, 16, 17]. Training data is transformed into transactions and constrained association rules are discovered from these transactions. The constraints limit the frequent itemsets to those including a class label, and limit the rules to those with a class label as the consequent. After a pruning phase to remove noisy and redundant rules, the remaining rules, classification rules, are used as a learned classification model. Negative association rules have been used for associative classifiers [6] and it was shown that the performance of the classifiers improved when negative association rules were employed in the training and the classification process. The negative association rules generated and used in addition to the positive rules are of the form $\neg X \Rightarrow Y$ (if feature X absent then class Y) or $X \Rightarrow \neg Y$ (if feature X present then cannot be class Y), where $|Y| = 1$ and Y is a class label.

5 Conclusions

In this chapter we have surveyed some methods proposed in the literature for mining association rules with negations. Although the problem of mining these types of rules is an interesting and challenging one there is a limited body of work. None of the existing methods find all the possible negative association rules. This is due to the complexity and size of the problem. A user should choose the algorithm that is most useful for the application considered. If a taxonomy is available or substitution rules are useful, the algorithms in [19] and [20, 21] are good candidates. If a user is interested in all the negative associations between pairs of itemsets, the methods proposed in [24] and [7] should be considered. Another research direction that can be useful in some situations is the mining of top-K rules with positive and negative items. This is investigated in [22] and [12, 13] which may be of interest to users who want to investigate and use a limited number of rules.

References

1. Charu C. Aggarwal and Philip S. Yu. Mining large itemsets for association rules. *IEEE Data Eng. Bull.*, 21(1):23–31, 1998.
2. Charu C. Aggarwal and Philip S. Yu. A new framework for itemset generation. In *Proceedings of the Seventeenth ACM SIGACT-SIGMOD-SIGART Symposium on Principles of Database Systems*, PODS '98, pages 18–24, 1998.
3. C. C. Aggarwal and P. S. Yu. Mining associations with the collective strength approach. *IEEE Trans. on Knowl. and Data Eng.*, 13(6):863–873, November 2001.
4. Rakesh Agrawal, Tomasz Imieliński, and Arun Swami. Mining association rules between sets of items in large databases. In *Proc. of SIGMOD*, pages 207–216, 1993.
5. M-L Antonie and Osmar R Zaïane. Text document categorization by term association. In *Proc. of ICDM*, pages 19–26, 2002.
6. M-L Antonie and Osmar R Zaïane. An associative classifier based on positive and negative rules. In *Proc. of DMKD*, pages 64–69, 2004.
7. Maria-Luiza Antonie and Osmar R Zaïane. Mining positive and negative association rules: An approach for confined rules. In *Proc. of PKDD*, pages 27–38, 2004.
8. Sergey Brin, Rajeev Motwani, and Craig Silverstein. Beyond market basket: Generalizing association rules to correlations. In *Proc. SIGMOD*, pages 265–276, 1997.
9. Sergey Brin, Rajeev Motwani, Jeffrey D Ullman, and Shalom Tsur. Dynamic itemset counting and implication rules for market basket data. In *Proc. of SIGMOD*, pages 255–264, 1997.
10. Chris Cornelis, Peng Yan, Xing Zhang, and Guoqing Chen. Mining positive and negative association rules from large databases. In *Proc. of CIS*, pages 1–6, 2006.
11. Bart Goethals and Mohammed J Zaki. FIMI 2003: Workshop on frequent itemset mining implementations. In *Third IEEE International Conference on Data Mining Workshop on Frequent Itemset Mining Implementations*, pages 1–13, 2003.
12. Wilhelmiina Hamalainen. Efficient discovery of the top-k optimal dependency rules with fisher's exact test of significance. In *Proc. of ICDM*, pages 196–205, 2010.
13. Wilhelmiina Hamalainen. Kingfisher: an efficient algorithm for searching for both positive and negative dependency rules with statistical significance measures. *Knowl. Inf. Syst.*, 32(2):383–414, 2012.
14. Jiawei Han, Jian Pei, and Yiwen Yin. Mining frequent patterns without candidate generation. In *Proc. of SIGMOD*, pages 1–12, 2000.
15. Yun Sing Koh and Russel Pears. Efficiently finding negative association rules without support threshold. In *Proc. of Australian AI*, pages 710–714, 2007.
16. Wenmin Li, Jiawei Han, and Jian Pei. CMAR: Accurate and efficient classification based on multiple class-association rules. In *Proc. of ICDM*, pages 369–376, 2001.
17. Bing Liu, Wynne Hsu, and Yiming Ma. Integrating classification and association rule mining. In *Proc. of SIGKDD*, pages 80–86, 1998.
18. Paul David McNicholas, Thomas Brendan Murphy, and M. O'Regan. Standardising the lift of an association rule. *Comput. Stat. Data Anal.*, 52(10):4712–4721, 2008.
19. Ashok Savasere, Edward Omiecinski, and Shamkant Navathe. Mining for strong negative associations in a large database of customer transactions. In *Proc. of ICDE*, pages 494–502, 1998.
20. Wei-Guang Teng, Ming-Jyh Hsieh, and Ming-Syan Chen. On the mining of substitution rules for statistically dependent items. In *Proc. of ICDM*, pages 442–449, 2002.

21. Wei-Guang Teng, Ming-Jyh Hsieh, and Ming-Syan Chen. A statistical framework for mining substitution rules. *Knowl. Inf. Syst.*, 7(2):158–178, 2005.
22. D. R. Thiruvady and G. I. Webb. Mining negative rules using grd. In *Proc. of PAKDD*, pages 161–165, 2004.
23. Hao Wang, Xing Zhang, and Guoqing Chen. Mining a complete set of both positive and negative association rules from large databases. In *Proc. of PAKDD*, pages 777–784, 2008.
24. Xindong Wu, Chengqi Zhang, and Shichao Zhang. Efficient mining of both positive and negative association rules. *ACM Trans. on Inf. Syst.*, 22(3):381–405, 2004.
25. Xiaohui Yuan, Bill P Buckles, Zhaoshan Yuan, and Jian Zhang. Mining negative association rules. In *Proc. of ISCC*, pages 623–628, 2002.
26. Mohammed J Zaki. Parallel and distributed association mining: A survey. *IEEE Concurrency: Special Issue on Parallel Mechanisms for Data Mining*, 7(4):14–25, 1999.

Chapter 7
Constraint-Based Pattern Mining

Siegfried Nijssen and Albrecht Zimmermann

Abstract Many pattern mining systems are designed to solve one specific problem, such as frequent, closed or maximal frequent itemset mining, efficiently. Even though efficient, their specialized nature can make these systems difficult to apply in other situations than the one they were designed for. This chapter provides an overview of generic constraint-based mining systems. Constraint-based pattern mining systems are systems that with minimal effort can be programmed to find different types of patterns satisfying constraints. They achieve this genericity by providing (1) high-level languages in which programmers can easily specify constraints; (2) generic search algorithms that find patterns for any task expressed in the specification language. The development of generic systems requires an understanding of different classes of constraints. This chapter will first provide an overview of such classes constraints, followed by a discussion of search algorithms and specification languages.

Keywords Constraints · Languages · Inductive databases · Search algorithms

1 Introduction

A key component of a pattern mining system is the constraint that is used by the system. A frequent itemset mining system, for instance, is characterized by the use of a minimum support constraint; an association rule mining system, similary, is identified by a minimum confidence constraint. Constraints define to a large degree which task a pattern mining system is performing.

However, the focus of many pattern mining systems on one particular type of constraint can make their use cumbersome. As an example, consider a frequent itemset mining system that one wishes to apply in a context where the utility of the

S. Nijssen (✉)
KU Leuven, Celestijnenlaan 200 A, 3001 Leuven, Belgium

Universiteit Leiden, Niels Bohrweg 1, 2333 CA, Leiden, The Netherlands
e-mail: siegfried.nijssen@cs.kuleuven.be

A. Zimmermann
INSA Lyon, LIRIS CNRS UMR 5205 Bâtiment Blaise Pascal,
69621 Villeurbanne CEDEX, France
e-mail: albrecht.zimmermann@insa-lyon.fr

C. C. Aggarwal, J. Han (eds.), *Frequent Pattern Mining,*
DOI 10.1007/978-3-319-07821-2_7, © Springer International Publishing Switzerland 2014

items is important as well. As a basic frequent itemset mining system does not support utilities, we cannot use it directly; we either have to:

- understand the code of the frequent itemset mining algorithm to add an additional constraint to it;
- or, write a second algorithm for processing the results of the frequent itemset mining system to evaluate the additional constraint for each of the itemsets found.

Both options are cumbersome. The second option is likely to be computationally inefficient if the number of frequent itemsets is large. The first option can be efficient, provided that the programmer has a deep understanding of the code that is being modified.

These disadvantages have led researchers to develop more general systems that provide easy-to-use interfaces for specifying the constraints that the pattern mining systems need to use during the search. The development of these systems has involved several challenges:

- the identification of general classes of constraints, all of which can be processed in a generic and similar way;
- the development of languages in which constraints can be expressed, such that all expressions in the language correspond to constraints in a class of constraints supported by a system;
- the development of search algorithms that can deal with constraints in a certain class.

This chapter will provide an overview of the state-of-the-art for each of these challenges. We will first formalize the problem of constraint-based pattern mining, including a discussion of different classes of constraints. Subsequently, we will discuss the most common search algorithms for these classes of constraints. Finally, we will discuss the languages that allow for the expression of constraints in pattern mining.

2 Problem Definition

Constraint-based mining starts from the observation that many pattern mining problems can be seen as instances of the following generic problem statement:

Given
- a data language $\mathcal{L}_\mathcal{D}$
- a database $\mathcal{D} \subseteq 2^{\mathcal{L}_\mathcal{D}}$ with transactions
- a pattern language $\mathcal{L}_\pi$
- a constraint $\varphi : \mathcal{L}_\pi \times 2^{\mathcal{L}_\mathcal{D}} \mapsto \{0, 1\}$

Find all patterns $\pi \in \mathcal{L}_\pi$ for which $\varphi(\pi, \mathcal{D}) = 1$.

The *pattern language* typically describes the syntax of the patterns we wish to find in the data. *Constraints* typically describe the statistical, syntactical, or other requirements that we wish these patterns to satisfy on the data.

Frequent itemset mining (see chapter ...), for example, is an instance of this generic setting, with the following choices:

- the database has transactions that are subsets of a given set of items $\mathcal{I}$;
- the pattern language is the set of all subsets of $\mathcal{I}$: $\mathcal{L}_\pi = 2^{\mathcal{I}}$;
- the *minimum support* constraint $\varphi_{minsup}(\pi, \mathcal{D})$ is true if and only if the number of transactions of $\mathcal{D}$ that contain π is large enough, in other words, it is true if and only if:

$$|cover(\pi)| = |\{d \in \mathcal{D} | \pi \subseteq d\}| \geq \theta,$$

where θ is a user-defined threshold.

By modifying the pattern language, the data language and the constraints, different data mining problems can be formalized. The main aim of *constraint-based pattern mining* is to build generic languages in which programmers can express pattern mining problems in terms of constraints, and to develop systems that can process statements in these languages.

2.1 Constraints

Constraints can be categorized along several dimensions, for instance:

1. which information is used when evaluating the constraint? Possibilities include that the constraint only evaluates the syntax of the pattern, that the constraint requires a database of transactions, or that the constraint requires a database with labeled transactions.
2. which properties do the constraints have? The most well-known property is that of (anti-)monotonicity, but other properties have been identified as well.

In terms of constraint-based pattern mining, combining constraints from different categories of the former dimension is typically easy, whereas this proves challenging for the latter dimension. Hence, existing work has focused on the latter dimension and we will elaborate on these constraint categories below.

Anti-monotonicity Most pattern mining algorithms assume the existence of a *coverage relation* between patterns and transactions in the data. In the case of frequent itemset mining, for example, an itemset π covers a transaction $d \in \mathcal{D}$ iff $\pi \subseteq d$; hence, the subset relation is used as coverage relation. In graph mining, the subgraph isomorphism relation may be used; in sequence mining, the subsequence relation.

A second important relation is the *generality relation*. A generality relation is essentially a partial order on the set of patterns in $\mathcal{L}_\pi$. We will denote this relationship with the symbol $\succeq$: if pattern π_1 is more general than pattern π_2, we will write $\pi_1 \succeq \pi_2$.

A generality relation $\succeq$ is *compatible* with a coverage relation if it satisfies the following property for all possible transactions d: if $\pi_1 \succeq \pi_2$ and π_2 covers example d, then π_1 covers example d.

A good generality relation is usually not difficult to choose. If the coverage relation is transitive, one can always use the coverage relation as generality relation as well.

For instance, in itemset mining, the subset relation is usually used as generality relation as well: an itemset π_1 is more general than an itemset π_2 iff $\pi_1 \subseteq \pi_2$.

Based on the generality relationship, we can define the *anti-monotonicity* property of constraints[1]. A constraint $\varphi(\pi, D)$ is called anti-monotonic iff it holds for all patterns π_1, π_2 that

if $\pi_1 \succeq \pi_2$ and $\varphi(\pi_1, D)$ is false, then $\varphi(\pi_2, D)$ is false.

Minimum support is the most well-known constraint that is anti-monotonic, but several other constraints are also anti-monotonic [11, 20]. Assuming that we use the subset relation to determine the generality relation, the following constraints on itemsets are anti-monotonic:

- the maximum length constraint $|\pi| \leq \theta$ for a fixed θ;
- the maximum sum of costs constraint $c(\pi) \leq \theta$ is anti-monotonic, where $c(\pi)$ sums up the costs of the items in the itemset, $c(\pi) = \sum_{i \in \pi} c(i)$, and $c(i) \leq 0$ is a cost that is associated to each item;
- a generalization constraint, which for a given set I requires that all itemsets found satisfy $\pi \subseteq I$;
- conjunctions or disjunctions of other anti-monotonic constraints.

These constraints can be generalized to other types of patterns as well.

Monotonocity Closely related to anti-monotonicity is monotonicity. A constraint is called *monotonic* iff for all patterns π_1, π_2:

if $\pi_1 \succeq \pi_2$ and $\varphi(\pi_1, D)$ is true, then $\varphi(\pi_2, D)$ is true.

In other words, monotonicity is the "reverse" of anti-monotonicity; if a constraint $\varphi(\pi)$ is anti-monotonic, its negation $\neg\varphi(\pi)$ is monotonic. This includes constraints such as:

- the maximum support constraint $|\{d \in \mathcal{D} | \pi \subseteq d\}| \leq \theta$;
- the minimum size constraint $|\pi| \geq \theta$;
- the minimum sum of costs constraint $c(\pi) \geq \theta$;
- a negated generalization constraint $\pi \not\subseteq I$;
- a specialization constraint $\pi \supseteq I$.

This relationship between monotonic and anti-monotonic constraints, one of reversal and negation, already hints at the difficulty in using both types of constraints at the same time for efficient pattern enumeration.

Convertible (anti)-monotonicity Whether a constraint is (anti)-monotonic depends on the generality relation chosen. One of the most well-known examples is that of the maximum *average cost* of an itemset, $c(\pi) = \sum_{i \in \pi} c(i)/|\pi| \geq \theta$. If we use the subset relation to define the generality, this constraint is not anti-monotonic. Consider the following two items with their corresponding costs: $c(1) = 1$ and $c(2) = 3$. If our cost threshold is 2, the average cost of $\{1, 2\}$ is 2 and satisfies the requirement; however, itemset $\{2\}$, while a subset, does not satisfy the constraint.

---

[1] Note that in some publications, anti-monotonic constraints are called monotonic, and monotonic constraints anti-monotonic [20].

However, assume that we would use the following generality order:

$\pi_1 \succeq \pi_2$ if we can obtain π_1 from π_2 by repeatedly removing from π_2 the item with the highest cost.

Then under this order the constraint is anti-monotonic: after all, by removing the most costly items from an itemset, the average cost of the items in the itemset can only go down and it hence also must satisfy the constraint.

Note that this order is *compatible* with the use of the subset relation as coverage relation; hence, this constraint can be combined with a minimum support constraint. Such an order can be *incompatible* with another order needed to make another constraint anti-monotonic, however.

Constraints which have this property, i.e., that a different generality relation needs to be used than the coverage relation to obtain (anti-)monotonicity, are called *convertible (anti-)monotonic* in the literature [26].

Succinctness Succinctness was originally defined for itemsets [23], but we will use a slightly different definition here which is applicable to other pattern languages as well: we will call any constraint *succinct* that can be enforced by manipulating the data. Consider the following two examples:

- we want to find frequent itemsets without item i: if we remove item i from the database, we will no longer find such itemsets;
- we want to find frequent itemsets that include item i: if we remove all transactions without item i from the database, and then remove item i from the remaining transactions, we can add item i to every itemset we find in the resulting database to obtain the desired set of itemsets.

These examples can easily be generalized to require the inclusion or exclusion of an itemset $\pi \subseteq \mathcal{I}$.

Condensed representations The set of patterns satisfying the above constraints may still be large. *Condendensed representations* consitute an additional approach for reducing a set of patterns. The main idea is to determine a small set of patterns that still is sufficiently large to determine a full set of patterns. The property that a pattern is part of a condensed representation can also be seen as a constraint.

We will discuss two of the most well-known cases here.

Given a generality relation $\succeq$, a pattern π is called *closed* if there is no more specific pattern π' with $\pi \succ \pi'$ such that $cover(\pi) = cover(\pi')$.

Intuitively, closed frequent patterns [25] allow one to recover a set of frequent itemsets together with their supports.

A subtle issue is the combination of the closedness constraint with other constraints. As an example, consider the maximum size constraint. One can distinguish two settings:

- a setting in which one searches for patterns satisfying the size constraint among those patterns that are closed;
- a setting in which one searches for patterns that are closed, restricting the set of patterns that are considered in the closedness definition only to those that satisfy the constraint.

As an example, assume that {1} is not closed and that {1, 2} is closed, while we have a maximum size constraint of 1. Then itemset {1} would not be in the output in the first setting, but would be in the output of the second. Constraint-based pattern mining systems can differ in their approach for dealing with this issue.

Another condensed representation is that of maximal patterns.

Given a generality relation $\succeq$ and constraint φ, a pattern π that satisfies constraint φ is called *maximal* with respect to constraint φ if there is no more specific pattern π' with $\pi \succ \pi'$ such that π' satisfies the constraint.

Compared to closed itemsets, maximal itemsets [1, 20] no longer allow one to recover the supports of a set of patterns.

If the constraint φ is a minimum support constraint, one typically refers to maximal frequent patterns. Whereas maximal frequent patterns are the most popular, it can also be useful to study maximality with respect to other constraints. Essentially, any anti-monotonic constraint defines a *border* in the space of patterns where all patterns that satisfy the constraints are on one side of the border, while all other patterns that do not satisfy it are on the other side [11, 20].

Similarly, also a *monotonic* constraint defines a border: in this case, the border one is looking for is that of *minimal* patterns that satisfy the constraints.

Different borders can be combined. Probably the most well-known example of this is found in the analysis of supervised data. If the database consists of two classes of examples, one can ask for all patterns that are frequent in the one, but infrequent in the other; the resulting set of patterns has two borders: one of the most specific patterns in this set, the other of the most general ones.

Boundable Constraints The minimum support constraint is one example of a constraint of the kind $f(\pi) \geq \theta$. Over the years, more complex functions have been studied. One example is that of accuracy (see the chapter on supervised patterns), which calculates the accuracy of a pattern when used as a classification rule on supervised data. Many such functions no longer have the (anti-)monotonicity property. In some cases, however, one can identify an alternative function f' such that:

- it is feasible to mine all patterns with $f'(\pi) \geq \theta$;
- $f'(\pi) \geq f(\pi)$.

In this case, all patterns satisfying $f(\pi) \geq \theta$ could be determined by first mining all patterns with $f'(\pi) \geq \theta$ and then calculating $f(\pi)$ for all patterns found. Function f' can be considered a *relaxation* of function f [29]. In Chap. 17 it was discussed that for supervised data such bounds often exist.

3   Level-Wise Algorithm

Most constraint-based mining algorithms can be seen as generalized versions of frequent pattern mining algorithms. Similar to frequent itemset mining algorithms, consequently, both breadth-first (*BFS*) or level-wise, and depth-first search algorithms have been proposed. The earliest techniques typically were BFS approaches, on which later works improved. Hence, we discuss them first.

3.1 Generic Algorithm

The setting which is closest to frequent pattern mining is that of constraint-based mining under anti-monotonic constraints. In this case, we can perform a level-wise search that is mostly equal to that of the APRIORI algorithm [20]. The search starts from the empty pattern, and proceeds by specializing this pattern in a breadth-first fashion.

In Algorithm 1, a description of this algorithm is given. In this pseudo-code, we use two operators: a *downward refinement operator* ρ to specialize patterns and an *upward refinement operator* δ to generalize patterns. A downward refinement operator is an operator which for any pattern π returns a *set* of more specific patterns (i.e. for all patterns $\pi' \in \rho(\pi)$ it holds that $\pi \succ \pi'$). Typically, we assume that this operator is *globally complete*, i.e. its repeated application starting from the empty pattern will produce the complete pattern language[2]. Furthermore, this operator works in "small steps", it tries to create new patterns which are minimally more specific (*least specific specializations*).

An example of a downward refinement operator for itemset mining is $\rho(\pi) = \{\pi \cup \{i\} \mid i > \max(\pi)\}$, assuming a total order $>$ on the items; e.g., if our language is $2^{\{1,2,3,4\}}$, with the usual order of integers over the items, $\rho(\{2\}) = \{\{2,3\}, \{2,4\}\}$.

Similarly, the upward refinement operator δ returns generalizations. For a given pattern π, it is assumed to only generate patterns that should have been seen before pattern π by the level-wise algorithm.

The key property on which the algorithm relies is the anti-monotonicity of constraint φ under the chosen generality relation: by refining only patterns that satisfy the constraint in line 6 and by checking generalizations in line 7, patterns are removed from consideration that are known to specialize patterns that do not satisfy φ.

Algorithm 1 Level-Wise Search(Constraint: φ)

1: $\mathcal{C} := \{\emptyset\}$
2: $\mathcal{S} := \emptyset$
3: **while** $\mathcal{C} \neq \emptyset$ **do**
4:     $S := \{\pi \in \mathcal{C} \mid \varphi(\pi) \text{ is true}\}$
5:     $\mathcal{S} := \mathcal{S} \cup S$
6:     $\mathcal{C}' := \bigcup_{\pi \in S} \rho(\pi)$
7:     $\mathcal{C} := \{\pi \in \mathcal{C}' \mid \delta(\pi) \backslash \mathcal{S} = \emptyset\}$
8: **return** $\mathcal{F}$

Note that this algorithm can also be applied to convertible and boundable constraints [28]: in this case, a modified generality relation or constraint is used. It can also be applied in a straightforward manner if there are both anti-monotonic constraint

---

[2] More formally, let $\rho^n(\pi)$ denote the set $\bigcup_{\pi' \in \rho(\pi)} \rho^{n-1}(\pi')$, with $\rho^0(\pi) = \pi$, and let $\rho^*(\pi) = \bigcup_{i=1}^{\infty} \rho^i(\pi)$, then a refinement operator is complete if $\rho^*(\emptyset)$ equals the pattern language $\mathcal{L}$.

and non anti-monotonic constraints: in principle, we ignore the non anti-monotonic constraints during the search, and evaluate the remaining constraints for *all* found patterns *afterwards*, in a *post-processing phase*.

In the presence of *monotonic* constraints, we can improve somewhat on the need to check *all* patterns [11, 18, 20, 22]. The main idea is here to traverse the patterns in a level-wise fashion in *reverse order* by starting with the most specific patterns that satisfy the anti-monotonic constraint. Since the generalizations of a pattern that does not satisfy a monotonic constraint will not satisfy the constraint either, we can stop this reverse traversal at the point at which we no longer have patterns that satisfy the constraint. This does not change the fact that we cannot enforce the monotonic constraint in the first mining phase, however.

The level-wise algorithm is easily changed to deal with *border representations*. Assume that upward refinement operator δ generates all *least general generalizations* of a pattern π (a pattern π' is a least general generalization for a pattern π if there is no pattern π'' with $\pi' \succ \pi'' \succ \pi$). Then for each pattern π that satisfies an anti-monotonic constraint, we can essentially identify its immediate generalizations, which are clearly *not* maximal, and remove them from the solution set.

For instance, in the case of itemset mining such an operator is $\delta(\pi) = \{\pi \setminus \{i\} \mid i \in \pi\}$. It would remove all immediate subsets of an itemset from the output. As it is assumed that the upward refinement operator will always generate patterns that must have been seen already, we do not need to explicitly remove other generalizations from the output: they will have been removed at an earlier stage. With similar ideas, the minimal patterns on the border of a monotonic constraint can also be found.

4 Depth-First Algorithm

Note, however, that even though the output of these modified BFS algorithms for finding borders is correct, the running time will not be much better than that of an algorithm that generates *all* patterns satisfying the constraint. Most algorithms that are able to obtain dramatically better run times in practice are depth-first algorithms.

4.1 Basic Algorithm

The most basic depth-first constraint-based mining algorithm is given in Fig. 2 and only supports anti-monotonic constraints.

Algorithm 2 Depth-First Search(Constraint: φ , pattern π)

1: $\mathcal{F} := \emptyset$
2: **if** $\varphi(\pi)$ is true **then**
3:    **for** $\pi' \in \rho(\pi)$ **do**
4:       $\mathcal{F} := \mathcal{F} \cup$ Depth-First Search (φ, π')
5: **return** $\mathcal{F}$

Essentially, compared to the earlier level-wise algorithm, this algorithm traverses the search space in a different order in which some long patterns are already considered before some shorter patterns are evaluated. As a result, optimizations based on the fact that short patterns have been seen before long patterns are not used. In practice, however, these algorithms can be more efficient. The reason for this is that most implementations take care to maintain datastructures which allow for *incremental* constraint evaluation: for instance, to calculate the support of a pattern, they do not traverse the whole dataset, but only consider those transactions covered by the pattern's parent in the search tree. As depth-first algorithms do not need to maintain a large number of candidates, maintaining such additional data structures is feasible. A well-known datastructure in this context is the FP-Tree (see the chapter on pattern growth for more details) [27].

Note that the above algorithm works for any pattern language, including graphs, strings and trees, as long as we know that the constraint φ is anti-monotonic.

When some constraints are not anti-monotonic, a basic approach for dealing with them, as in the case of breadth-first search, is to ignore them during the search and post-process the output of the above agorithm. A similar trick can be used for boundable constraints. In this case, the anti-monotonic bound is used during the depth-first search, and each pattern found is finally evaluated using the original constraint in a post-processing step.

Many studies have explored the possibilities for deriving more efficient algorithms for more complex constraints than anti-monotonic constraints. Most of these studies have focused on the pattern language of *itemsets*, as it appears additional pruning is most easily derived for itemsets. We will discuss these approaches in a generic way in the next paragraph, inspired by work of Bucila et al. and Guns et al. [8, 13].

4.2   Constraint-based Itemset Mining

The key idea in efficient depth-first constraint-based itemset mining algorithms is to maintain four sets in each node of the search tree, which are upper- and lower-bounds on the itemsets and transaction sets that can still be found:

- I_U, the largest itemset we believe we can still find;
- I_L, the smallest itemset we believe we can still find;
- T_U, the largest transaction set we believe we can still find;
- T_L, the smallest transaction set we believe we can still find.

For some constraints, not all these 4 sets need to be maintained, but in the most generic setting all 4 sets are maintained.

During the search, any modification of any of these 4 sets may be a reason to modify another of these 4 sets as well. We will refer to this process of modifying one set based on the modification of another set as *propagation*. Different approaches differ in the algorithms and data structures used to do propagation.

An overview of the generic algorithm is given in Algorithm 3, in which $I_L = T_L = \emptyset$, $I_U = \mathcal{I}$ and $T_U = \mathcal{D}$. Line 1 performs the propagation for the constraints.

Propagation may signal that no solution can be found in the current branch of the search tree by setting *stop* to true. If the lower- and upper-bound for the itemset are identical, a pattern has been found and is added to the output. Otherwise, in line 6 an item is selected, which is recursively added to the itemset (line 7), or removed from consideration (line 8).

Note that the same itemset can never be found twice: an item which is added in line 7, will never be added to an itemset that is considered in the search tree explored in the call of line 8.

Algorithm 3 Depth-First Search(Constraint: φ , I_L, I_U, T_L, T_U)

1:    $I'_L, I'_U, T'_L, T'_U, stop :=$ Propagate $(I_L, I_U, T_L, T_U, \varphi)$
2: **if** not *stop* **then**
3:    **if** $I'_L = I'_H$ **then**
4:      **return** $\{I'_L\}$
5:    **else**
6:      Pick an item $i \in I'_U \backslash I'_L$
7:      **return** Depth-First Search(φ, $I'_L \cup \{i\}, I'_U, T'_L, T'_U$) $\cup$
8:          Depth-First Search(φ, $I'_L, I'_U \backslash \{i\}, T'_L, T'_U$)
9: **else**
10:    **return** $\emptyset$

We will now consider how this algorithm can be instantiated for different types of mining settings.

Frequent Itemset Mining This is the most simple setting. Essentially, in this case, the following propagation steps are executed:

1. T'_U is restricted to $cover(I_L)$;
2. I'_U is restricted to those items in I_U that are frequent in the database containing only the transactions of T'_U (in other words, the items that are frequent in the projected database for itemset I_L, see Chap. 3);

T'_L and I'_L are not modified by propagation.

For these choices, the search is highly similar to that of Eclat or FP-Growth; at every point in the search tree, we maintain a list of candidate items that can be added to the current itemset; the set of candidate items is reduced based on the minimum support threshold.

Attentive readers may have noticed that the search tree for the generic algorithm presented here is binary, whereas for most itemset mining algorithms the tree is not binary. This is, however, only a minor conceptual difference: the recursive calls in line 8 of our generic algorithm essentially correspond to a traversal of the candidate list in traditional frequent itemset mining algorithms, where we remember which items we may no longer consider.

The clear benefit of the non-traditional perspective is that other constraints can be added with minor effort.

Minimum Sum of Cost and Minimum Support A first approach for dealing with a monotonic minimum-sum-of-cost constraint is to add the following propagation [26]:

3 if the sum of costs of itemset I'_U is lower than the desired threshold, set *stop* to true.

The rationale for this propagation step is that we can stop a branch of the search if the most expensive itemset we can still reach is not expensive enough.

The benefit of this approach is that this propagation step is relatively easy to calculate, while for high thresholds it will already prune effectively.

A more elaborate approach was proposed by Bonchi et al. [5, 6, 7]. Its essential observation is that if an itemset needs to be both frequent and expensive, this means that a certain number of transactions in the data needs to be expensive as well. This leads to the following propagation steps:

1. T'_U is set to $T_U \cap cover(I_L)$;
2. I'_U is set to those items in I_U that are frequent in the database restricted to the transactions in T'_U;
3. from T'_U all transactions d are removed for which $c(d \cap I'_U) < \theta$, where θ is the cost threshold;
4. if T'_U was changed, go back to step 2;
5. if $I'_U \subset I'_L$, set *stop* to true.

The interesting idea in this approach is the presence of a feedback loop: the removal of items can make some transactions too cheap; when a transaction is too cheap, it will not be in the cover of an itemset, and we can remove it from consideration; this however will reduce the support of items further, potentially making them infrequent in the projected database.

The advantage of this approach is that it can reduce the size of the search tree even further. The disadvantage is that the propagation is more complex too calculate, as it involves a traversal of the data. To remedy this, Bonchi et al. [6] studied settings in which the above loop is not executed in all nodes of the seach tree.

Minimum and Maximum Support A similar idea can be used when we have a minimum support threshold on some transactions ($\mathcal{D}^+$), and a maximum support threshold on the other transactions ($\mathcal{D}^-$) [10, 18].

1. T'_U is set to $cover(I_L)$;
2. I'_U is set to those items in I_U that are frequent in the database restricted to the transactions in $T'_U \cap \mathcal{D}^+$;
3. T'_L is set to $cover(I_U)$;
4. if $|T'_L \cap \mathcal{D}^-| > \theta$, where θ is the maximum support threshold for the negative examples, then set *stop* to true.

In this approach, the main idea is that if the lowest support that can be reached on the negative transactions is not low enough, we can stop the search.

Maximal Frequent Itemsets Of particular interest in the constraint-based mining literature is the discovery of border (or boundary) representations. The simplest such setting is the discovery of maximal frequent itemsets, which can be obtained by means of the following propagations:

1. T'_U is set to $cover(I_L)$;
2. I'_U is set to those items in I_U that are frequent in the database restricted to the transactions in T'_U;
3. T'_L is set to $cover(I'_U)$;
4. if some item not in I'_U is frequent in the database restricted to the transactions in T'_L, set $stop$ to true.
5. if $|T'_L| \geq \theta$, I'_L is set to I'_U.

The arguments for these steps are the following: the set T'_L represents those transactions that will be covered by any itemset we will find in the future; if there is an item that covers a sufficiently large number of these transactions, but we cannot add this item in the current branch of the search tree, we stop traversing this branch in line 4, as elsewhere we will find itemsets that include this item.

On the other hand, if the itemset consisting of all remaining items is frequent, clearly this itemset must be maximal; we can directly include all items in the itemset.

This search strategy is embodied in the MaxMiner algorithm [1]. It was generalized to the case of finding border representations under arbitrary monotonic and anti-monotonic constraints by Bucila et al. [8].

Closed Frequent Itemsets Closed itemset mining can be achieved by another modification of the propagation for frequent itemset mining:

1. T'_U is set to $cover(I_L)$;
2. I'_U is set to those items in I_U that are frequent in the database restricted to the transactions in T'_U;
3. let I'' contain those items in $\mathcal{I}$ which are present in all transactions in T'_U;
4. if I'' contains items not in I'_U, set $stop$ to true; otherwise, let I'_L be I''.

Remember that in closed itemset mining the task is to find itemsets such that no superset has the same coverage. This propagation ensures this: in line 4, if an item can be added to the current itemset without changing the coverage, it will be added immediately if this is allowed; however, if this item may not be added, as we branched over it earlier, we stop the search, as we can no longer find closed itemsets in the current part of the search space.

This search strategy is embodied in the LCM closed itemset mining algorithm [30]. The combination of closed itemset mining with constraints was studied in more detail in the D-MINER system by Besson et al. [2, 3].

4.3 Generic Frameworks

The similarity between these depth-first search algorithms indicates that it may be possible to combine different constraints and condensed representations. Indeed, this is the key idea underlying most generic frameworks for constraint-based mining.

The DualMiner algorithm [8] essentially represents a generic depth-first algorithm for finding border representations that extends the ideas found in the MaxMiner algorithm. The authors of that work brought this development to its logical conclusion by introducing the concept of "witnesses" [17], itemsets on which constraint satisfaction is tested to derive pruning information for parts of the search space. Witness-based mining subsumes mining under (anti-)monotonic and convertible constraints, and is capable of handling additional constraint classes, and mining under conjunctions of constraints. The D-Miner system combines closed itemset mining (formal concept analysis) with constraints [2, 3].

The Constraint Programming for Itemset Mining framework [13] is built on the observation that constraint-based search, and constraint programming in particular, has been studied extensively in the general artificial intelligence literature. It shows that the mining tasks discussed earlier can be reformalized in terms of constraints present in generic constraint programming systems; furthermore, such systems provide a generic framework for constraint propagation which makes it easy to combine different constraints.

4.4 Implementation Considerations

In the above description, we intentionally left unaddressed how the indicated propagation is performed in detail. In principle, all different data structures that have been studied in the frequent itemset mining literature can be used in this context as well. For instance, the MaxMiner and DualMiner algorithms use vertical representations of the data most similar to that of Eclat; the FP-Bonsai algorithm, on the other hand, uses FP-Trees [5]. The impact of data structures in a Constraint Programming framework was studied by Nijssen et al [24]. These studies confirm that for good run times the choice of data structure is important; however, many of the above propagation procedures can be adapted to different data representations, and hence the two aspects can be considered orthogonal.

5 Languages

Most of the systems studied earlier require a language for the specification of constraints. Roughly speaking, three categories can be distinguished within these languages: special purpose languages, SQL inspired languages, and constraint programming based languages.

Special Purpose Languages Many constraint-based mining systems implement a small special purpose language. As an example, this is an expression in the language underlying the SeqLog system [19]:

```
database ca = smiles_file("molecules.ca");
database ci = smiles_file("molecules.ci");

predicate ca = minimum_frequency(ca, 10);
predicate ci = maximum_frequency(ci, 500);

mine ca and ci;
```

Essentially, this language provides a small set of built-in primitives such as smiles_file for reading a data file, minimum_frequency for specifying a minimum support constraint and maximum_frequency for specifying a maximum support constraint. For each of these primitives, the system is aware of the properties such as (anti-)monotonicity, which ensures that any conjunction or disjunction of constraints that is written is down can be processed by the system.

Similar special purpose languages were proposed by several other authors [22, 29]; they differ in the constraints that are supported and the type of patterns that can be found (itemsets [22, 29], strings [12, 19], ...).

Languages built on SQL A clear disadvantage of special purpose languages is that they are yet additional languages that the programmer has to learn. Given that many datasets are stored in databases, several projects have studied the integration of constraint-based pattern mining in database systems.

The first class of such methods aims to extend SQL with additional syntax for the formalization of data mining tasks. One early example is the MINE RULE operator [21]:

```
MINE RULE Associations AS
SELECT DISTINCT 1..n item as BODY, 1..n item AS HEAD,
               SUPPORT, CONFIDENCE
FROM Purchase
WHERE price <= 150
GROUP BY transaction
EXTRACTING RULES WITH SUPPORT: 0.1, CONFIDENCE: 0.2
```

This example mines association rules with minimum support 0.1, confidence 0.2, limiting the search to items with a price lower than $ 150, a succinct constraint. Another example is the DMQL language [15]:

```
FIND association rules
RELATED TO beer, wine, diapers
FROM products
WHERE value >= 100
WITH support threshold = 0.1
WITH confidence threshold = 0.2
```

In this example we search for association rules related to three specific products, in those transactions that have a value higher than 100; the parameters of the association rule discovery process are similar to the previous example. A third example is SPQL [7].

The advantage of these languages is that well-known syntax can be used for the expression for constraints. Furthermore, common SQL syntax can be used to specify the input of the mining task or to process its output further.

At the same time, the programmer still has to learn the additional primitives, such as the FIND or MINE RULE keywords. An alternative perspective is to avoid extending the language, but to add *mining views* to a database [4]. They are virtual tables, which, once queried, will trigger the execution of mining algorithms. This is an example:

```
SELECT R.rid, C1.*, C2.*, R.conf
FROM Sets S, Rules R, Concepts C1, Concepts C2,
WHERE R.cid = S.cid AND C1.cid = R.cida AND C2.cid = R.cidc AND
      S.supp >= 30 AND R.conf >=80
```

Here, `Sets`, `Rules` and `Concepts` are virtual mining views.

A limitation of most SQL-based approaches is however that they are limited to itemset patterns or association rules. How to specify graph mining or sequence mining tasks in this context is still an open question. Most constraint-based graph mining or sequence mining systems currently use special purpose languages.

The general idea of linking constraint-based mining to database querying has been studied in the area of *inductive databases* and *inductive querying* [9, 16].

Constraint Programming Constraint-based mining has many similarities to generic constraint satisfaction problem (CSP) solving as studied in the Artificial Intelligence (AI) community. Both areas essentially require the discovery of solutions in a space of possible solutions satisfying constraints. To deal with generic CSPs, the AI community has developed generic systems known as *constraint programming systems*. These systems provide languages in which programmers can specify constraint satisfaction problems; statements in these languages can be solved by various types of solvers, including generic propagation-based solvers. As we have seen earlier, many depth-first constraint-based itemset mining systems are also based on propagation, and hence it is not surprising that generic constraint-based itemset mining fits naturally into a constraint programming context as well.

This observation was used by Guns et al. to formalize constraint-based itemset mining tasks in generic constraint programming languages [13, 14]. This is an example in the most recent version of the MiniZinc constraint programming language:

```
int: Nr I; int: NrT; int: Freq;
array[1..NrT] of set of 1..NrI: TDB;
var set of 1..Nr I: Items;
constraint card ( cover ( Items, TDB ) ) >= Freq;
solve satisfy;
```

It specifies the task of frequent itemset mining; `cover` is a function available in a MiniZinc library, implemented in the MiniZinc language itself as well.

Statements in the MiniZinc language can be executed by a generic constraint programming system, or by a specialized data mining system, if one exists [14]. However, it was shown that generic constraint programming systems implement many types of propagation automatically, and hence that specialized systems are often not needed if a task can be modelled in the MiniZinc language.

Similar to the SQL-based languages, it is at this moment not understood how to integrate graph mining or sequence mining tasks in an elegant matter in the CP setting.

6   Conclusions

In this chapter we provided an overview of classes of constraints, algorithms for solving constraint-based mining problems and languages for specifying contraint-based mining tasks.

The trend in constraint-based mining has been to build increasingly generic systems. While initially constraint-based mining systems provided special purpose languages that only supported slightly more constraints than specialized frequent itemset mining algorithms did, in recent years the range of constraints has expanded, as well as the genericity of the languages supporting constraint-based mining, culminating in the integration with generic constraint satisfaction systems and languages.

Several open challenges remain. These include a closer integration of constraint-based mining with pattern set mining, getting a better understanding of how to integrate statistical requirements in constraint-based mining systems, and mining structured databases such as graph or sequence databases using sufficiently generic languages.

References

1. R. Bayardo. Efficiently mining long patterns from databases. In *Proceedings of ACM SIGMOD Conference on Management of Data*, 1998.
2. Jérémy Besson, Céline Robardet, and Jean-François Boulicaut. Constraint-based mining of formal concepts in transactional data. In Honghua Dai, Ramakrishnan Srikant, and Chengqi Zhang, editors, *Advances in Knowledge Discovery and Data Mining, 8th Pacific-Asia Conference*, pages 615–624, Sydney, Australia, May 2004. Springer.
3. Jérémy Besson, Céline Robardet, Jean-François Boulicaut, and Sophie Rome. Constraint-based concept mining and its application to microarray data analysis. *Intell. Data Anal.*, 9(1):59–82, 2005.
4. Hendrik Blockeel, Toon Calders, Élisa Fromont, Bart Goethals, Adriana Prado, and Céline Robardet. An inductive database system based on virtual mining views. *Data Min. Knowl. Discov.*, 24(1), 2012.
5. Francesco Bonchi, Fosca Giannotti, Claudio Lucchese, Salvatore Orlando, Raffaele Perego, and Roberto Trasarti. A constraint-based querying system for exploratory pattern discovery. *Inf. Syst.*, 34(1):3–27, 2009.
6. Francesco Bonchi, Fosca Giannotti, Alessio Mazzanti, and Dino Pedreschi. Examiner: Optimized level-wise frequent pattern mining with monotone constraint. In *ICDM*, pages 11–18. IEEE Computer Society, 2003.
7. Francesco Bonchi and Bart Goethals. Fp-bonsai: The art of growing and pruning small fp-trees. In *PAKDD*, pages 155–160, 2004.
8. Cristian Bucila, Johannes Gehrke, Daniel Kifer, and Walker White. DualMiner: A dual-pruning algorithm for itemsets with constraints. In *Proceedings of The Eight ACM SIGKDD International Conference on Knowledge Discovery and Data Mining*, Edmonton, Alberta, Canada, July 23–26 2002.
9. Luc De Raedt. A perspective on inductive databases. *SIGKDD Explorations*, 4(2):69–77, 2002.
10. L. De Raedt and S. Kramer. The level wise version space algorithm and its application to molecular fragment finding. In B. Nebel, editor, *Proceedings of the 17th International Joint Conference on Artificial Intelligence*, pages 853–862. Morgan Kaufmann, 2001.

11. Luc De Raedt, Manfred Jaeger, Sau Dan Lee, and Heikki Mannila. A theory of inductive query answering. In *ICDM*, pages 123–130. IEEE Computer Society, 2002.
12. Minos N. Garofalakis, Rajeev Rastogi, and Kyuseok Shim. Spirit: Sequential pattern mining with regular expression constraints. In *VLDB'99, Proceedings of 25th International Conference on Very Large Data Bases, September 7–10, 1999, Edinburgh, Scotland, UK*, pages 223–234. Morgan Kaufmann, 1999.
13. Tias Guns, Anton Dries, Guido Tack, Siegfried Nijssen, and Luc De Raedt. Miningzinc: A modeling language for constraint-based mining. In *IJCAI*, 2013.
14. Tias Guns, Siegfried Nijssen, and Luc De Raedt. Itemset mining: A constraint programming perspective. *Artif. Intell.*, 175(12–13):1951–1983, 2011.
15. Jiawei Han, Yongjian Fu, Krzystzof Koperski, Wei Wang, and Osmar Zaiane. Dmql: A data mining query language for relational databases. In *SIGMOD'96 Workshop. on Research Issues on Data Mining and Knowledge Discovery (DMKD'96)*, 1996.
16. T. Imielinski and H. Mannila. A database perspectivce on knowledge discovery. *Communications of the ACM*, 39(11):58–64, 1996.
17. Daniel Kifer, Johannes Gehrke, Cristian Bucila, and Walker M. White. How to quickly find a witness. In *PODS*, pages 272–283. ACM, 2003.
18. Stefan Kramer, Luc De Raedt, and Christoph Helma. Molecular feature mining in hiv data. In *KDD-2001: The Seventh ACM SIGKDD International Conference on Knowledge Discovery and Data Mining*, 2001.
19. Sau Dan Lee and Luc De Raedt. An algebra for inductive query evaluation. In *ICDM*, pages 147–154, 2003.
20. Heikki Mannila and Hannu Toivonen. Levelwise search and borders of theories in knowledge discovery. *Data Mining and Knowledge Discovery*, 1(3):241–258, 1997.
21. Rosa Meo, Giuseppe Psaila, and Stefano Ceri. A new sql-like operator for mining association rules. In *VLDB*, pages 122–133, 1996.
22. Jean-Philippe Métivier, Patrice Boizumault, Bruno Crémilleux, Mehdi Khiari, and Samir Loudni. A constraint-based language for declarative pattern discovery. In *ICDM Workshops*, pages 1112–1119, 2011.
23. Raymond T. Ng, Laks V. S. Lakshmanan, Jiawei Han, and Alex Pang. Exploratory mining and pruning optimizations of constrained associations rules. In *Proceedings of the ACM-SIGMOD Conference on Management of Data*, pages 13–24, 1998.
24. Siegfried Nijssen and Tias Guns. Integrating constraint programming and itemset mining. In *ECML/PKDD (2)*, pages 467–482, 2010.
25. Nicolas Pasquier, Yves Bastide, Rafik Taouil, and Lotfi Lakhal. Discovering frequent closed itemsets for association rules. In Catriel Beeri and Peter Buneman, editors, *ICDT*, volume 1540 of *Lecture Notes in Computer Science*, pages 398–416. Springer, 1999.
26. Jian Pei and Jiawei Han. Can we push more constraints into frequent pattern mining? In *KDD*, pages 350–354, 2000.
27. Jian Pei and Jiawei Han. Constrained frequent pattern mining: a pattern-growth view. *SIGKDD Explorations*, 4(1):31–39, 2002.
28. Jian Pei, Jiawei Han, and Laks V. S. Lakshmanan. Pushing convertible constraints in frequent itemset mining. *Data Min. Knowl. Discov.*, 8(3):227–252, 2004.
29. Arnaud Soulet and Bruno Crémilleux. Mining constraint-based patterns using automatic relaxation. *Intell. Data Anal.*, 13(1):109–133, 2009.
30. Takeaki Uno, Tatsuya Asai, Yuzo Uchida, and Hiroki Arimura. An efficient algorithm for enumerating closed patterns in transaction databases. In *Discovery Science*, pages 16–31, 2004.

Chapter 8
Mining and Using Sets of Patterns through Compression

Matthijs van Leeuwen and Jilles Vreeken

Abstract In this chapter we describe how to successfully apply the MDL principle to pattern mining. In particular, we discuss how pattern-based models can be designed and induced by means of compression, resulting in succinct and characteristic descriptions of the data.

As motivation, we argue that traditional pattern mining asks the wrong question: instead of asking for *all* patterns satisfying some interestingness measure, one should ask for a small, non-redundant, and interesting *set* of patterns—which allows us to avoid the pattern explosion. Firmly rooted in algorithmic information theory, the approach we discuss in this chapter states that the best set of patterns is that set that compresses the data best. We formalize this problem using the Minimum Description Length (MDL) principle, describe useful model classes, and briefly discuss algorithmic approaches to inducing good models from data. Last but not least, we describe how the obtained models—in addition to showing the key patterns of the data—can be used for a wide range of data mining tasks; hence showing that MDL selects *useful* patterns.

Keywords Compression · MDL · Pattern set mining · Data summarization

1 Introduction

What is the ideal outcome of pattern mining? Which patterns would we really like to find? Obviously, this depends on the task at hand, and possibly even on the user. When we are exploring the data for new insights the ideal outcome will be different than when the goal is to build a good pattern-based classifier.

There are, however, a few important general observations to be made. For starters, we are not interested in patterns that describe noise—we only want patterns that identify important associations in the data. In pattern mining, the function that usually

---

M. van Leeuwen (✉)
KU Leuven, Leuven, Belgium
e-mail: matthijs.vanleeuwen@cs.kuleuven.be

J. Vreeken
Max-Planck Institute for Informatics and Saarland University, Saarbrücken, Germany
e-mail: jilles@mpi-inf.mpg.de

C. C. Aggarwal, J. Han (eds.), *Frequent Pattern Mining*,
DOI 10.1007/978-3-319-07821-2_8, © Springer International Publishing Switzerland 2014

determines the importance of a pattern in this regard is called an *interestingness* measure.[1]

The traditional pattern mining question is to ask for *all* patterns in the data that satisfy some interestingness constraint. For example, all patterns that occur at least *n* times, or, those that are so-and-so significant according to a certain statistical test. Intuitively this makes sense, yet in practice, this approach rarely leads to satisfactory results.

The primary cause is the pattern explosion. While strict constraints only result in few patterns, these are seldom informative: they are the most obvious patterns, and hence often long-since common knowledge. However, when we loosen the constraints—to discover novel associations—the pattern explosion occurs and we are flooded with results. More often than not orders of magnitude *more* patterns are returned than there are rows in the data. In fact, even for modest amounts of data billions of patterns are discovered for non-trivial constraints. Clearly, in such numbers these patterns are impossible to consider by hand, as well as very difficult to use in any other task—therewith effectively negating the goal of mining these patterns in the first place. Not quite the ideal result.

It does, however, provide us a second observation on the ideal outcome: we do not want to have too many results. In particular, we want to avoid *redundancy*: every pattern should be interesting or useful with regard to all of the other patterns in the result.

Simply put, traditional pattern mining has been asking the wrong question. In most situations, what we really want is a small, non-redundant, and as interesting possible group of patterns. As such, instead of asking for *all* patterns that satisfy some constraint, we should ask for the *set of patterns* that is optimal with regard to a *global* interestingness criterion. This means evaluating groups of patterns indirectly, i.e. by first constructing a model using these patterns, and then scoring the quality of that model. The main questions are then how to construct such a model, and which criterion should be used? Clearly, this depends on the task at hand.

In this chapter, we focus on exploratory data analysis. That is, our goal is to explore the data for new insights, to discover any local structure in the data—in other words, to discover patterns that describe the most important associations of the data, patterns that capture the distribution of the data. As such, we are looking for a set of patterns that models the data well. To this end, we need a criterion that measures both how well the patterns capture the distribution of the data, and—to avoid overfitting and redundancy—how complex the set of patterns is. Given a global interestingness criterion we can perform model selection and identify the best model. There are a few such criteria available, including Akaike's Information Criterion (AIC) [2] and the Bayesian Information Criterion (BIC) [53]. For pattern mining, the Minimum Description Length (MDL) principle [52] is the most natural choice. It provides a principled, statistically well-founded, yet practical approach for defining an objective function for *descriptive* models—which, as patterns *describe* part of the data, fits our setting rather well.

---

[1] See Chap. 5 for a detailed overview of interestingness measures.

MDL allows us to unambiguously identify the best set of patterns as that set that provides the best lossless compression of the data. This provides us with a means to mine small sets of patterns that describe the distribution of the data very well: if the pattern set at hand would contain a pattern that describes noise, or that is redundant with regard to the rest, removing it from the set will *improve* compression. As such, the MDL optimal pattern set automatically balances the quality of fit of the data with the complexity of the model—without the user having to set any parameters, as all we have to do is minimize the encoding cost.

In this chapter we will give an overview of how MDL—or, compression—can be used towards mining informative pattern sets, as well as for how to use these patterns in a wide range of data mining tasks.

In a nutshell, we first discuss the necessary theoretical foundations in Sect. 2. In Sect. 3 we then use this theory to discuss constructing pattern-based models we can use with MDL. Section 4 covers the main approaches for mining good pattern sets, and in Sect. 5 we discuss a range of data mining tasks that pattern-based compression solves. We discuss open challenges in Sect. 6, and conclude in Sect. 7.

2   Foundations

Before we go into the specifics of MDL for pattern mining, we will have to discuss some foundational theory.

Above, we stated that intuitively our goal is to find patterns that describe *interesting structure* of the data—and want to avoid patterns that overfit, that describe noise. This raises the questions, what is significant structure, and where does structure stop and noise begin?

Statistics offers a wide range of tests to determine whether a result is significant, including via Bayesian approaches such as calculating confidence intervals, as well as frequentist approaches such as significance testing [17]. Loosely speaking, these require us to assume a background distribution or null hypothesis, and use different machinery to evaluate how likely the observed structure is under this assumption.

While a highly successful approach for confirming findings in science, in our exploratory setting this raises three serious problems. First and foremost, there are no off-the-shelf distributions for data and patterns that we can test against. Second, even if we could, by assuming a distribution we strongly influence which results will be deemed significant—a wrong choice will lead to meaningless results. Third, the choice for the significance or confidence thresholds is arbitrary, yet strongly influences the outcome. We want to avoid such far-reaching choices.

These problems were acknowledged by Ray Solomonoff, Andrey Kolmogorov, and Gregory Chaitin, whom independently invented and contributed to what is now known as algorithmic information theory [12]. In a nutshell, instead of using the probability under a distribution, in algorithmic information theory we consider the *algorithmic complexity* of the data. That is, to measure the amount of information the data contains by the amount of algorithmic 'effort' is required to generate the data using a universal Turing machine. There are different ways of formalizing such 'effort'. Here, we focus on Kolmogorov complexity.

2.1 Kolmogorov Complexity

Kolmogorov complexity measures the information content of a string s; note that any database $\mathcal{D}$ can be serialized into a string. The Kolmogorov complexity of s, $K_{\mathcal{U}}(s)$, is defined as the length in bits of the shortest program p for a Universal Turing machine $\mathcal{U}$ that generates s and then halts. Formally, we have

$$K_{\mathcal{U}}(s) = \min_{p:\mathcal{U}(p)=x} |p| \ .$$

Intuitively, program p can be regarded as the ultimate compressor of s.

Let us analyze what this entails. First of all, it is easy to see that every string s has at least one program that generates it: the program p_0 that simply prints s verbatim. Further, we know that if the string is fully random, there will be no shorter program than p_0. This gives us an upper bound. In fact, this allows us to define what structure is, and what not. Namely, any (subset of) the data for which $K(s)$ is smaller than the length of p_0 exhibits structure—and the program p is the shortest description of this structure.

Loosely speaking, the lower bound for K is zero, which will only be approximated when the data s is very simple to express algorithmically. Examples include a long series of one value, e.g., $000000000\ldots$, but also data that seems complex at first glance, such as the first n digits of π, or a fractal, have in fact a very low Kolmogorov complexity—which matches the intuition that, while the result may be complex, the process for generating this data can indeed be relatively simple.

In fact, we can regard p as two parts; the 'algorithm' that describes the compressible structure of s, and the 'input' to this algorithm that express the incompressible part of s. Separating these two components in a given dataset is exactly the goal of exploratory data analysis, and as such Kolmogorov Complexity institutes the ideal. Sadly, however, $K(s)$ is not computable. Apart from the fact that the space of possible programs is enormous, we face the problem that p has to generate s and then halt. By the halting problem we are unable to make that call.

This does not mean Kolmogorov complexity is useless. Quite the contrary, in fact. While beyond the scope of this chapter, it provides the theoretical foundations to many aspects of data analysis, statistics, data mining, and machine learning. We refer the interested reader to Li and Vitány [40] for a detailed discussion on these foundations.

Although Kolmogorov complexity itself is not computable, we can still put it to practice by approximating it. With p we have the ultimate compressor, which can exploit *any* structure present in s. The incomputability of $K(s)$ stems from this infinite 'vocabulary', as we have to consider all possible programs. We can, however, constrain the family of programs we consider to a set for which we know they halt, by limiting this vocabulary to a fixed set of regularities. In other words, by considering lossless compression algorithms.

2.2 MDL

Minimum Description Length (MDL) [20, 52], like its close cousin Minimum
Message Length (MML) [69], is in this sense a practical version of Kolmogorov
Complexity [40]. All three embrace the slogan *Induction by Compression*, but the
details on how to compress vary. For MDL, this principle can be roughly described
as follows.

Given a set of models[2] $\mathcal{M}$, the best model $M \in \mathcal{M}$ is the one that minimizes

$$L(\mathcal{D}, M) = L(M) + L(\mathcal{D}|M)$$

in which

- $L(M)$ is the length, in bits, of the description of M, and
- $L(\mathcal{D}|M)$ is the length, in bits, of the description of the data when encoded with
  M.

This is called two-part MDL, or *crude* MDL—as opposed to *refined* MDL, where
model and data are encoded together [20]. We consider two-part MDL because we
are specifically interested in the compressor: the set of patterns that yields the best
compression. Further, although refined MDL has stronger theoretical foundations, it
cannot be computed except for some special cases.

2.2.1 MDL and Kolmogorov

The MDL-optimal model M has many of the properties of the Kolmogorov optimal
program p. In fact, two-part MDL and Kolmogorov complexity have a one-to-
one connection [1, 20]. Loosely speaking, the two terms respectively express the
structure in the data, and the deviation from that structure: $L(M)$ corresponds to
the 'algorithm' part of p, which generates the structure. $L(\mathcal{D} \mid M)$, on the other
hand, does not contain any structure—as otherwise there would be a better M—and
can be seen as the 'parameter' part of p. One important difference is that $L(\mathcal{D}, M)$
happily ignores the length of the decompression algorithm—which would be needed
to reconstruct the data given the compressed representation of the model and data.
The reason is simple: its length is constant, and hence does not influence the selection
of the best model.

2.2.2 MDL and Probabilities

Any MDL-based approach encodes both the data and the models, for which *codes*
are required. It is well-known that there is a close relation between probability distributions and optimal codes. That is, Shannon's source coding theorem states that

---

[2] MDL-theorists tend to talk about *hypotheses* in this context

the optimal code length for a given symbol in a string is equal to the $-\log$ of the probability of observing it in the string [12].

As such, an alternate interpretation of MDL is to interpret $L(\mathcal{D} \mid M)$ as the (negative) log-likelihood of the data under the model, $-\log \Pr (\mathcal{D} \mid M)$, and to regard $L(M)$, as the negative log-likelihood of the model, $-\log \Pr (M)$, or, a regularization function. Hence, looking for the model that gives the best compression is similar to looking for the maximum likelihood model under a budget. As such it has a similar shape to Akaike's Information Criterion (AIC) [2] and the Bayesian Information Criterion (BIC) [53]. This of course assumes that there is a distribution for models, as well as that we have a generative model for data that can be parameterized by M. This is often not the case.

In MDL, however, we are concerned with *descriptive* models—not necessarily generative ones. As such, different from Bayesian learning, in both Kolmogorov complexity and MDL we evaluate only the data and explicit model at hand—we do not 'average' over all models, so to speak, and hence do not need access to a generative model. Moreover, MDL is different in that it requires a complete, lossless encoding of both the model and the data while BIC and AIC penalize models based only on the number of parameters.

In practice, while (refined) MDL and BIC are asymptotically the same, the two may differ (strongly) on finite data samples. Typically, MDL is a bit more conservative. For a detailed discussion on the differences between BIC and MDL we refer to Grünwald [20].

Using MDL in Practice   To use MDL in practice, one has to define the model class $\mathcal{M}$, how a single model $M \in \mathcal{M}$ describes a database, and how all of this is encoded in bits. That is, we have to define a compression scheme. In addition, we need an algorithm to mine—or approximate—the optimal model.

A key advantage of MDL is that it removes the need for user-defined parameters: the best model minimizes the total encoded size. Unfortunately, there are also disadvantages: (1) contrary to Kolmogorov Complexity, a model class needs to be defined in advance, and (2) finding the optimal model is often practically infeasible. Consequently, important design choices have to be made, and this is one of the challenges of the compression-based approach to exploratory data mining.

A standard question regarding the use of MDL concerns the requirement of a lossless encoding: if the goal is to find very short descriptions, why not use a lossy encoding? The answer is two-fold.

First, and foremost, lossless encoding ensures fair comparison between models: we know that every model is evaluated based on how well it describes the complete dataset. With lossy compression, this is not the case: two models could have the same $L(\mathcal{D}, M)$—one describing only a small part of the data in high detail, and the other describing all the data in low detail—and unlike for lossless compression, we would have no (principled) way of choosing which one is best.

Second of all, we should point out that compression is *not* the goal, but only a *means* to select the best model. By MDL, the best model provides the shortest description out of all models in the model class, and it is that model that we are interested in—in the end, the length of the description is often not of much interest. When compression is the goal, a general purpose compressor such as ZIP often provides much better compression, as it can exploit many types of statistical dependencies.

In a similar vein, is it also important to note that in MDL we are *not* concerned with materialized codes, but only interested in their *lengths*—again, as model *selection* is the goal. Although a complete overview of all useful codes—for which we can compute the optimal lengths in practice—is beyond the scope of this chapter, we will discuss a few instances in the next chapter, where we will discuss how to use MDL for pattern mining. Before we do so, however, let us quickly go into the general applicability of MDL in data mining.

2.3   MDL in Data Mining

Faloutsos and Megalooikonomou [15] argue that Kolomogorov Complexity and Minimum Description Length [20, 52] provide a powerful and well-founded approach to data mining. There exist many examples where MDL has been successfully employed in data mining, including, for example, for classification [37, 50], clustering [6, 31, 39], discretization [16, 30], defining parameter-free distance measures [11, 28, 29, 66], feature selection [48], imputation [65], mining temporally surprising patterns [8], detecting change points in data streams [36], model order selection in matrix factorization [45], outlier detection [3, 58], summarizing categorical data [43], transfer learning [54], discovering communities in matrices [9, 47, 63] and evolving graphs [60], finding sources of infection in large graphs [49], and for making sense of selected nodes in graphs [4].

We will discuss a few of these instances in Sect. 5, but first cover how to define an MDL score for a pattern based model.

3   Compression-based Pattern Models

In this section we introduce how to use the above foundations for mining small sets of patterns that capture the data distribution well. We will give both the high level picture and illustrate with concrete instances and examples. Before we go into details, let us briefly describe the basic ingredients that are required for any pattern-based model.

We assume that a dataset $\mathcal{D}$ is a bag of elements t of some data type—which we, for simplicity, will refer to as *tuples*. In the context of frequent itemset mining, each t is a transaction over a set of items $\mathcal{I}$, i.e., $t \subseteq \mathcal{I}$. Similarly, we can consider sequences, trees, graphs, time series, etc. Let us write $\mathcal{T}$ to denote the universe of possible tuples for a given data type. Clearly, all tuples in $\mathcal{D}$ are elements from $\mathcal{T}$.

Given a dataset, on of the most important choices is the *pattern language* X. A pattern language is the set of all possible patterns that we can discover for a given data type. In principle, a pattern can be any structure that describes the distribution of (a subset of) the data. Given the topic of the book, we focus on frequent patterns; e.g., when we consider itemsets, X can be the set of all possible itemsets, while for structured data, X can consist of sequential patterns, subgraphs, etc.

Clearly, the choice of X is highly important, as it determines the type of structure that we will be able to discover in the data. Another way of thinking about X is that it defines the 'vocabulary' of the compressor. If one chooses a pattern language that is highly specific, it may be impossible to find relevant structure of that type in the data. On the other hand, if a very rich, i.e., more complex pattern language is chosen, the encoding and search for the model can become rather complicated.

3.1 Pattern Models for MDL

Given a class of data and a pattern language, we can start to construct a pattern-based model. Note that by defining a model class, we essentially fix the set of possible models $\mathcal{M}$, the possible descriptions, for a given dataset $\mathcal{D}$. Given this space of possible descriptions, we can employ the MDL principle to select the best model $M \in \mathcal{M}$ for $\mathcal{D}$ simply by choosing the model that minimizes the total compressed size. In order to do so, however, we need to be able to compute $L(\mathcal{D}, M)$, however, the encoded length of the model and the data given the model.

We start with the latter, i.e., we first formally define how to compute $L(\mathcal{D} \mid M)$, the encoded length in bits of the data given the model. Generally speaking, there are many different ways to describe the same data using one model. However, by the MDL principle, our encoding should be such that we use the minimal amount of bits to do so. This helps us to make principled choices when defining the encoding scheme. Some of these may impose additional constraints and requirements on the design of the compressor, as well as determine how the score can be used. This is particularly important in light of subsequently using the pattern-based models in data mining tasks other than summarization. Here we describe three important properties that a compressor may have.

Dataset-level Compression At the highest level we need to be able to compare the encoded size of different databases. The most trivial way to do so is by comparing the total encoded size, $L(\mathcal{D}, M)$, where we induce the MDL-optimal model M for each $\mathcal{D}$.

This property alone allows us to use compression as a 'black box': without paying any attention to the contents of the models or how datasets are compressed, the MDL principle can be used to select appropriate models for a given dataset. In fact, this property does not even require datasets to consist of individual tuples that can be distinguished, nor does it require models to consist of patterns.

Moreover, it allows us to *use* compression for data mining tasks, such as for computing data dissimilarity. Note that this property generally holds for any generic compressor, and therefore compression algorithms like those in ZIP, GZIP, BZIP, etc, can also be used for such tasks. As a concrete example, the family of Normalized Compression Distance measures [41] rely on this.

As a slight variant, we can also fix the model M and see how well it compresses another dataset. That is, we require that $L(\mathcal{D} \mid M)$ is explicitly calculable for any $\mathcal{D}$ of the specified data universe $\mathcal{T}$ and any $M \in \mathcal{M}$. This allows us to calculate how well a dataset matches the distribution of another dataset. See also Sect. 5.

Tuple-level Compression In addition, a rather useful property is when each tuple $t \in \mathcal{D}$ can be compressed individually, independent of all other tuples. That is, $L(t \mid M)$ can be computed for a given M. This also implies we can simply calculate $L(\mathcal{D} \mid M)$ as

$$L(\mathcal{D} \mid M) = \sum_{t \in \mathcal{D}} L(t \mid M).$$

This property simplifies many aspects related to the induction and usage of the models. For example, as a consequence, calculating the encoded size can now be trivially parallelized. More important, though, is that common data mining tasks such as classification and clustering are now straightforward. More on this later.

Pattern-level Inspection The third and final property that we discuss here is that of *sub-tuple*, or, *pattern-level* inspection. That is, beyond computing $L(t \mid M)$ we also want to be able to inspect *how* a given tuple is encoded: what structure, in the form of a set of patterns, is used to compress it?

With this property, it becomes possible to provide explanations for certain outcomes (e.g., explain why is a certain tuple compressed better by one model than by another), but also to exploit this information to improve the model (e.g., patterns that often occur together in a tuple should probably be combined). Effectively, it is this property that makes pattern-based solutions so powerful, as it ensures that in addition to decisions, we can offer *explanations*.

3.2 Code Tables

The conceptually most simple, as well as most commonly used pattern-based model for MDL are so-called *code tables* (see e.g., [23, 56, 57, 59, 64, 68]). Informally, a code table is a dictionary, a translation table between patterns and codes. Each entry in the left column contains a pattern and corresponds to exactly one code word in the right column. Such a code table can be used to compress the data by replacing occurrences of patterns with their corresponding codes, and vice versa to decode an encoded dataset and reconstruct the original the data. Using the MDL principle, the problem can then be formulated as finding that code table that gives the best compression.

Fig. 8.1 Example code table.
The widths of the codes
represent their lengths.
$\mathcal{I} = \{A, B, C\}$. Note that the
usage column is not part of
the code table, but shown here
as illustration: for optimal
compression, codes should be
shorter the more often they
are used

Code table *CT*

Itemset	Code	Usage
A B C	▢	5
A B	▭	1
A	▬	1
B	▬	1
C	—	0

Next, we describe both the general approach, as well as cover a specific instance
for transaction data. First, we formally define a code table.

Definition 8.1 *Let $\mathcal{X}$ be a set of patterns and $\mathcal{C}$ a set of code words. A code table
CT over $\mathcal{X}$ and $\mathcal{C}$ is a two-column table such that:*

1. *The first column contains patterns, that is, elements from $\mathcal{X}$.*
2. *The second column contains elements from $\mathcal{C}$, such that each element of $\mathcal{C}$ occurs
   at most once.*

*We write $code(X \mid CT)$ for the code corresponding to a pattern $X \in CT$. Further,
we say PS for $\{X \in CT\}$, the pattern set of CT.*

Example 8.2 Throughout we will use KRIMP *[57, 68] as a running example. It was
the first pattern set mining method using code tables and MDL, and considers itemset
data. In all examples, a dataset $\mathcal{D}$ is a bag of transactions over a set of items $\mathcal{I}$, i.e.,
for each $t \in \mathcal{D}$ we have $t \subseteq \mathcal{I}$. Patterns are also itemsets and the pattern language
is the set of all possible itemsets, i.e., $\mathcal{X} = 2^{\mathcal{I}} = \{X \subseteq \mathcal{I}\}$.*

Figure 8.1 shows an example KRIMP *code table of five patterns. The left column
lists the itemsets, the second column contains the codes. Each bar represents a code,
its width represents the code length. (Note, these are obviously not real codes, but
a simplified representation; for our purposes representing code lengths suffices.)
The usage column is not part of the code table, but only used to determine the code
lengths.*

3.2.1 Encoding the Data

Given a dataset $\mathcal{D}$ and a code table CT, we need to define how to encode $\mathcal{D}$ with
CT. As already mentioned, encoding a dataset is done by replacing occurrences of
patterns in the code table by their corresponding codes. To achieve lossless encoding
of the data, we need to *cover* the complete dataset with patterns from pattern set PS.
In practice, covering a dataset is usually done on a per-tuple basis, such that each
tuple is covered by a subset of the patterns in the code table. Hence, a code table
normally has all three properties discussed in the previous subsection: it allows for
dataset-level compression, tuple-level compression, and *sub-tuple inspection.*

To encode a tuple t from database $\mathcal{D}$ with code table CT, a cover function $cover(CT, t)$ is required that identifies which elements of CT are used to encode t. The parameters are a code table CT and a tuple t, the result is a disjoint set of elements of CT that cover t. Or, more formally, a cover function is defined as follows.

Definition 8.3 *Let $\mathcal{D}$ be a database over a universe of possible tuples $\mathcal{T}$, t a tuple drawn from $\mathcal{D}$, let $\mathcal{CT}$ be the set of all possible code tables over $\mathcal{X}$, and CT a code table with $CT \in \mathcal{CT}$. Then, $\int cover : \mathcal{CT} \times \mathcal{T} \mapsto \mathcal{P}(\mathcal{X})$ is a cover function iff it returns a set of patterns such that*

1. *$cover(CT, t)$ is a subset of PS, the pattern set of CT, i.e.,*
   $X \in cover \int (CT, t) \rightarrow X \in CT$
2. *together all $X \in cover(CT, t)$ cover t completely, i.e., t can be fully reconstructed from $cover(CT, t)$*

We say that $cover(CT, t)$ covers t.

Observe that this cover function is very generic and allows many different instances. In general, finding a subset of a pattern set that covers

---

Algorithm 4 The KRIMPCOVER Algorithm

Require: Transaction $t \in \mathcal{D}$ and code table CT, both over $\mathcal{I}$.
Ensure: A cover of t using non-overlapping elements of CT.
1: $S \leftarrow$ first element X of CT for which $X \subseteq t$
2: **if** $t \setminus S = \emptyset$ **then**
3:      $Res \leftarrow \{S\}$
4: **else**
5:      $Res \leftarrow \{S\} \cup$ KRIMPCOVER$(t \setminus S, CT)$
6: **return** Res

---

a tuple can be a hard combinatorial problem. Depending on the data universe $\mathcal{T}$, pattern language $\mathcal{X}$ and requirements imposed by the task, it may therefore be beneficial to impose additional constraints to make the cover function fast and efficient to compute. Also, note that without any further requirements on code tables, it may be possible that a code table cannot cover any tuple. To remedy this, a common approach is to require that any 'valid' code table should contain at least all primitive patterns, i.e., singletons, required to cover any tuple from $\mathcal{T}$.

Example 8.4 We continue the example of KRIMP *and present its cover function in Algorithm 4. To allow for fast and efficient covering of transactions,* KRIMP *considers non-overlapping covers. Its mechanism is very simple: look for the first element in the code table that occurs in the tuple, add it to the cover and remove it from the tuple, and repeat this until the tuple is empty. Recalling that tuples and patterns are both itemsets, we have that a cover is a set of itemsets, s.t.*

$$\forall_{X,Y \in cover(t,CT)} X \cap Y = \emptyset,$$

and

$$\cup_{X \in cover(t,CT)} X = t.$$

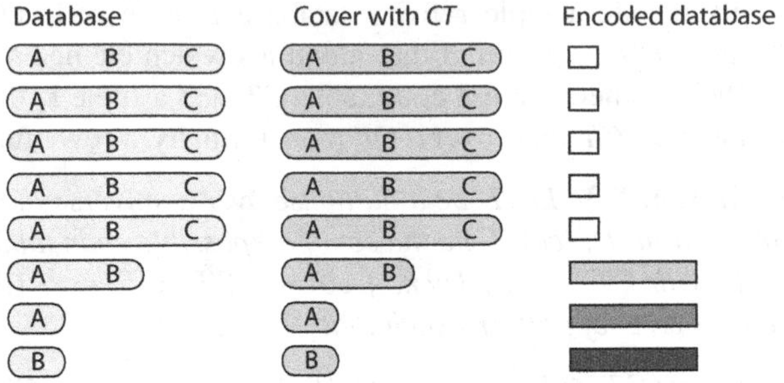

Fig. 8.2 Example database, cover and encoded database obtained by using the code table shown in Fig. 8.1. $\mathcal{I} = \{A, B, C\}$

By not allowing itemsets to overlap, it is always unambiguous what the cover of a transaction is. If overlap would be allowed, it can easily happen that multiple covers are possible and computing and testing all of them would be a computational burden.

To ensure that each code table is 'valid', each CT is required to contain at least all singleton itemsets from $\mathcal{I}$, i.e., $PS \supseteq \{\{i\} \mid i \in \mathcal{I}\}$. This way, any transaction $t \in \mathcal{P}(\mathcal{I})$ can always be covered by any $CT \in \mathcal{CT}$.

Figure 8.2 shows an example database consisting of 8 itemsets, of which 5 are identical. Also shown is the cover of this database with the example code table from Fig. 8.1. In this example, each transaction is covered by only a single itemset from the code table, resulting in very good compression. Obviously it is often not the case that complete transactions can be covered with a single itemset. For example, if itemset $\{ABC\}$ had not been in the code table, the first five transactions would have been covered by $\{AB\}$ and $\{C\}$.

To encode a database $\mathcal{D}$ using code table CT we simply replace each tuple $t \in \mathcal{D}$ by the codes of the patterns in the cover of t,

$$t \to \{code(X \mid CT) | X \in cover(CT, t)\}.$$

Note that to ensure that we can decode an encoded database uniquely we assume that $\mathcal{C}$ is a *prefix code*, in which no code is the prefix of another code [12].

Example 8.5 Figure 8.2 shows how the cover of a database can be translated into an encoded database: replace each itemset in the cover by its associated code.

3.2.2 Computing Encoded Lengths

Since MDL is concerned with the best compression, the codes in CT should be chosen such that the most often used code has the shortest length. That is, we should use *optimal* prefix codes. As there exists a nice correspondence between code lengths and probability distributions (see, e.g., [40]), the optimal code lengths can be calculated through the Shannon entropy. In MDL we are only interested in measuring complexity, and not in materialized codes. As such we do not have to require round code lengths, nor do we have to operate an actual prefix coding scheme such as Shannon-Fano or Huffman encoding.

Theorem 8.6 *Let P be a distribution on some finite set $\mathcal{D}$, there exists an optimal prefix code C on $\mathcal{D}$ such that the length of the code for $t \in \mathcal{D}$, denoted by $L(t)$ is given by*

$$L(t) = -\log(P(t)).$$

Moreover, this code is optimal in the sense that it gives the smallest expected code size for data sets drawn according to P. (For the proof, please refer to Theorem 5.4.1 in [12].)

The optimality property means that we introduce no bias using this code length. The probability distribution induced by a cover function is, of course, given by the relative usage frequency of each of the patterns.

To determine this, we need to know how often a certain code is used. We define the *usage* count of a pattern $X \in CT$ as the number of tuples t from $\mathcal{D}$ where X occurs in its cover. Normalized, this frequency represents the probability that that code is used in the encoding of an arbitrary $t \in \mathcal{D}$. The optimal code length [40] then is $-\log$ of this probability, and a code table is optimal if all its codes have their optimal length.

More formally, we have the following definition.

Definition 8.7 *Let $\mathcal{D}$ be a database drawn from a tuple universe $\mathcal{T}$, C a prefix code, cover a cover function, and CT a code table over $\mathcal{X}$ and C. The usage count of a pattern $X \in CT$ is defined as*

$$usage_{\mathcal{D}}(X) = |\{t \in \mathcal{D} | X \in cover(CT, t)\}|.$$

This implies a probability distribution over the usage of patterns $X \in CT$ in the cover of $\mathcal{D}$ by CT, which is given by

$$P(X|\mathcal{D}, CT) = \frac{usage_{\mathcal{D}}(X)}{\sum_{Y \in CT} usage_{\mathcal{D}}(Y)}.$$

The code$(X \mid CT)$ for $X \in CT$ is optimal for $\mathcal{D}$ iff

$$L(code(X \mid CT)) = |code(X \mid CT)| = -\log(P(X|\mathcal{D}, CT)).$$

A code table CT is code-optimal for $\mathcal{D}$ iff all its codes, $\{code(X \mid CT) | X \in CT\}$, are optimal for $\mathcal{D}$.

From now onward we assume that code tables are code-optimal for the database they are induced on.

Example 8.8   Figure 8.1 shows usage counts for all itemsets in the code table. For example, itemset $\{A, B, C\}$ is used 5 times in the cover of the database. These usage counts are used to compute optimal code lengths. For $X = \{A, B, C\}$:

$$P(X|\mathcal{D}, CT) = \frac{5}{8}$$

$$L(code(X \mid CT)) = -\log\left(\frac{5}{8}\right) = 0.68$$

And for $Y = \{A\}$:

$$P(Y|\mathcal{D}, CT) = \frac{1}{8}$$

$$L(code(Y \mid CT)) = -\log\left(\frac{1}{8}\right) = 3.00$$

So, $\{A, B, C\}$ *is assigned a code of length 0.68 bits, while* $\{A, B\}, \{A\}$ *and* $\{B\}$ *are assigned codes of length 3 bits each.*

Now, for any database $\mathcal{D}$ and code table CT over the same set of patterns $\mathcal{X}$ we can compute $L(\mathcal{D}|CT)$ according to the following trivial lemma.

Lemma 8.9 *Let* $\mathcal{D}$ *be a database,* CT *be a code table over* $\mathcal{X}$ *and code-optimal for* $\mathcal{D}$, cover *a cover function, and* usage *the usage function for* cover.

1. *For any* $t \in \mathcal{D}$ *its encoded length, in bits, denoted by* $L(t|CT)$, *is*

$$L(t|CT) = \sum_{X \in cover(CT, t)} L(code(X \mid CT)).$$

2. *The encoded size of* $\mathcal{D}$, *in bits, when encoded by* CT, *denoted by* $L(\mathcal{D}|CT)$, *is*

$$L(\mathcal{D}|CT) = \sum_{t \in \mathcal{D}} L(t|CT).$$

With Lemma 8.9, we can compute $L(\mathcal{D}|M)$, but we also need to know what $L(M)$ is, i.e., the encoded size of a code table.

Recall that a code table is a two-column table consisting of patterns and codes. As we know the size of each of the codes, the encoded size of the second column is easily determined: it is simply the sum of the lengths of the codes. The encoding of the first column, containing the patterns, depends on the pattern type; a lossless and succinct encoding should be chosen.

Definition 8.10 *Let* $\mathcal{D}$ *be a database,* CT *a code table over* $\mathcal{X}$ *that is code-optimal for* $\mathcal{D}$, *and* encode *an encoding for elements of* $\mathcal{X}$. *The size of* CT *in bits, denoted by* $L(CT|\mathcal{D})$, *is given by*

$$L(CT|\mathcal{D}) = \sum_{X \in CT: usage_{\mathcal{D}}(X) \neq 0} |encode(X)| + |code(X \mid CT)|.$$

Note that we do not take patterns with zero usage into account, because they are not used to code and do not get a finite code length.

With these results we have the total size of the encoded database.

Definition 8.11 *Let* $\mathcal{D}$ *be a database with tuples drawn from* $\mathcal{T}$, *let* CT *be a code table that is code-optimal for* $\mathcal{D}$ *and* cover *a cover function. The* total compressed size *of the encoded database and the code table, in bits, denoted by* $L(\mathcal{D}, CT)$ *is given by*

$$L(\mathcal{D}, CT) = L(\mathcal{D}|CT) + L(CT|\mathcal{D}).$$

3.2.3 The Problem

The overall problem is now to find the set of patterns that best describe a database $\mathcal{D}$. Given a pattern set PS, a cover function and a database, a (code-optimal) code table CT follows automatically. Therefore, each coding set has a corresponding code table and we can use this to formalize the problem.

Given a set of patterns $\mathcal{F} \subseteq \mathcal{X}$, the problem is to find a minimal subset of $\mathcal{F}$ which leads to a minimal encoded size $L(\mathcal{D}, CT)$. By requiring the smallest possible pattern set, we make sure it does not contain any unused patterns, i.e., $usage_{CT}(X) > 0$ for any pattern $X \in CT$.

More formally, in general terms, we define the problem as follows.

Problem 3.1 (Minimum Description Length Pattern Set) *Let $\mathcal{D}$ be a dataset of tuples drawn from $\mathcal{T}$, $\mathcal{F} \subseteq \mathcal{X}$ a candidate set, and enc an encoding for datasets over $\mathcal{T}$ and models over $\mathcal{X}$. Find the smallest pattern set $PS \subseteq \mathcal{F}$ such that for the corresponding model M the total compressed size with encoding enc, $L_{enc}(\mathcal{D}, M)$, is minimal.*

Naively, one might say that the solution for this problem can be found by simply enumerating all possible pattern sets given a collection of patterns $\mathcal{X}$. As such, the search space is already huge: a pattern set contains an arbitrary subset of $\mathcal{X}$, excluding only the empty set. Hence, there are

$$\sum_{k=0}^{2^{|\mathcal{X}|}-1} \binom{2^{|\mathcal{X}|} - 1}{k}$$

possible pattern sets. To determine which pattern set minimizes the objective function, we have to know the optimal cover function. Even for a greedy strategy such as covering the data using a fixed order, this explodes to having to consider all possible orders of all possible pattern sets. To make matters worse, the score typically exhibits no (weak) (anti-)monotone structure that we can exploit. As such, we relax the problem and resort to heuristics to find good models instead of the optimum.

3.3 Instances of Compression-based Models

Code tables form a generic model class that can be used with virtually any pattern and data type, given a suitable encoding. Of course there are also other compression-based model classes, and we will now discuss instances of both types.

3.3.1 Code Table Instances

The best-known instance of code tables is the one used as running example in this chapter, i.e., KRIMP code tables over itemsets and often used in conjunction with the cover function presented in Algorithm 4. As we will see in the next section, there also exist more sophisticated algorithms for inducing code tables.

In practice, we find that KRIMP returns pattern sets in the order of hundreds to a few thousand of patterns [68], which have been shown to describe the distribution of the data very well. In the next section we will discuss some of the applications in which these pattern sets have been successfully put to use.

Akoglu et al. [3] proposed the COMPREX algorithm, which describes a categorical dataset by a *set* of KRIMP code tables—by partitioning the attributes into parts that correlate strongly, and inducing a KRIMP code table for each part directly from the data.

In frequent pattern mining, and hence KRIMP, we only regard associations between 1s of the data as potentially interesting. This is mostly a matter of keeping matters computational feasible—clearly there are cases where associations between occurrences and absences are rather interesting. LESS [24] is an algorithm that describes data not using frequent itemsets, but using low-entropy sets [23]. These are itemsets for which we see the distribution its occurrences is strongly skewed. LESS code tables consist of low-entropy sets, and it uses these to identify areas of the data of where the attributes strongly interact. LESS code tables typically contain only tens to hundreds of low-entropy sets. Attribute clustering [43] provides even more succinct code tables, with the goal to provide good high-level summaries of categorical data, only up to tens of patterns are selected.

Code table instances for richer data include those for sequential patterns, i.e., serial episodes. Bathoorn et al. [5] gave a variant of KRIMP for sequential patterns without gaps, wheras the SQS [64] and GOKRIMP [34] algorithms provide fast algorithms for descriptions in terms of serial episodes where gaps are allowed. Like KRIMP, these algorithms find final selections in the order of hundreds of patterns.

Koopman and Siebes [32, 33] discussed the KRIMP framework in light of frequent patterns over multi-relational databases.

3.3.2   Other Model Classes

Like LESS, PACK [62] considers binary data symmetrically. Its patterns are itemsets, but they are modeled in a decision tree instead of a code table. This way, probabilities can be calculated more straightforwardly and refined MDL can be used for the encoding. MTV [44] also constructs a probabilistic model of the data, and aims to find that set of itemsets that best predicts the data. The framework allows both BIC and MDL to be used for model selection. Typically, between tens and hundred of itemsets are selected.

STIJL [63] describes data hierarchically in terms of dense and sparse tiles, rectangles in the data which contain surprisingly many/few 1s.

We also find compression-based models in the literature that employ lossy compression. While this contradicts MDL in principle, as long as the amount of 'lost' data is not too large, relatively fair comparisons between models can still be made.

Summarization [10] is such an approach, which identifies a group of itemsets such that each transaction is summarized by one itemset with as little loss of information as possible. Wang et al. [70] find summary sets, sets of itemsets such that each transaction is (partially) covered by the largest itemset that is frequent.

There are also model classes where the link to compression exists, but is hidden from plain sight. TILING [18] should be mentioned: a tiling is the cover of the database by the smallest set of itemsets, and is related to Set Cover [27], Minimum Entropy Set Cover [22], and matrix factorization problems [42, 45].

4   Algorithmic Approaches

So far we have discussed in detail the motivation, theoretical foundations, and models for compression-based pattern mining. Given the previous, the natural follow-up question is: *given a dataset, how can we find that model that minimizes the total compressed size?*

In this section we aim to give a brief overview of the main algorithmic strategies for inducing good code tables from data. There are two main approaches we need to discuss: candidate filtering and direct mining.

In our concise discussion on the complexity of the *Minimum Description Length Code Table* problem, we already mentioned that the search space will generally be too large to consider exhaustively. Hence, as is common with MDL-based approaches, the common solution is to resort to heuristic search strategies. This obviously implies that we usually cannot guarantee to find the best possible model, and experimental evaluation will have to reveal how useful induced models are.

In this section, we will outline common techniques. For a more in-depth discussion of the individual algorithms, we refer to the original papers; algorithmic aspects are not the main focus of this chapter.

4.1   Candidate Set Filtering

The definition of Problem 3.1 already hints at the most often used approach: candidate filtering. While the set of candidates $\mathcal{F}$ could consist of all possible patterns $\mathcal{X}$, it can also be a subset defined by some additional constraints. Typically, $\mathcal{F}$ is generated in advance and given as argument to the algorithm used for model induction.

For example, when inducing itemset-based models, it is common practice to use closed itemsets with a given minimum support threshold as candidate set. A large advantage of using smaller candidate sets, i.e., keeping $|\mathcal{F}|$ small, is that model induction can be done relatively quickly.

Given a dataset $\mathcal{D}$ and candidate set $\mathcal{F}$, a candidate set filtering method returns a model M corresponding to a pattern set $PS \subset \mathcal{F}$ for which $L(\mathcal{D}, M)$ is 'small'. (Note that we cannot claim that the compressed size is minimal due to the heuristic nature of filtering methods.)

4.1.1 Single-pass Filtering

The simplest filtering approach uses the following greedy search strategy:

1. Start with an 'empty' model M.
2. Start with an 'empty' model M.
3. Add patterns $F \in \mathcal{F}$ to M one by one. If the addition leads to better compression, keep it, otherwise, permanently discard F.

Although the basic principle of this approach is very simple, note that there are important details that need to be worked out depending on the specific model and encoding. For example, it is often impossible to start with a model that is truly empty: if a model does not contain any patterns at all, it may be impossible to encode the data at hand and hence there is no compressed size to start from. Also, adding a pattern to a model is not always straightforward: how and where in the model should it be added? Depending on design choices, there may be many possibilities and if these need all to be tested this can become a computational burden. Finally, in what order should we consider the candidates in $\mathcal{F}$? Since single-pass filtering considers every candidate only once, this choice will have a large impact on the final result.

Example 8.12 Krimp *employs single-pass filtering with several heuristic choices to ensure that it can induce good code tables from relatively large datasets and candidate sets in reasonable time.*

To ensure any transaction can be encoded, the induction process departs from the code table containing all singleton itemsets, i.e., $\{\{i\} \mid i \in \mathcal{I}\}$. *Candidate itemsets are considered in a fixed order, on frequencies and lengths, maximizing the probability that we encounter candidates that aid compression. Finally, with the same goal, we imposed an order on the itemsets in the code table. Together with the cover function, which does not allow overlap, this means that each candidate itemset can be efficiently evaluated. To further illustrate this example, the* Krimp *algorithm is given as Algorithm 5.*

Algorithm 5 The Krimp Algorithm

Require: A transaction database $\mathcal{D}$ and a candidate set $\mathcal{F}$, both over a set of items $\mathcal{I}$

Ensure: Code table CT, a heuristic solution to the MDL Pattern Set problem

1: $CT \leftarrow$ **Standard Code Table**$(\mathcal{D})$
2: $\mathcal{F}_o \leftarrow \mathcal{F}$ in **Standard Candidate Order**
3: **for all** $F \in \mathcal{F}_o \setminus \mathcal{I}$ **do**
4:     $CT_c \leftarrow (CT \cup F)$ in **Standard Cover Order**
5:     **if** $L(\mathcal{D}, CT_c) < L(\mathcal{D}, CT)$ **then**
6:        $CT \leftarrow CT_c$
7: **return** CT

Other examples of compression based pattern mining algorithms employing single-pass filtering include R-Krimp [32], RDB-Krimp [33], Less [24], Pack [62], and SQS [64].

4.1.2   Iterative Candidate Selection

Single-pass filtering is a very greedy search strategy. One particular point of concern is that it considers every candidate only once, in fixed order, deciding acceptance or rejection on the candidate's quality in relation to only the model mined up to that time. This means that unless the candidate order is perfect, we will see that candidates get rejected that would have been ideal later on, and hence that sub-optimal candidates will be accepted because we do not have access to the optimal candidate at that time.

The reason this strategy still provides good results is exactly the problem it aims to resolve: redundancy. For every rejected 'ideal' candidate we will (likely) see a good enough variant later on.

The optimal result, however, may be a much smaller set of patterns that describe the data much better. One way to approximate the optimal result better is to make the search less greedy. Smets and Vreeken [59] showed that iteratively greedily adding the locally optimal candidate leads to much better code tables.

1. Start with an 'empty' model M.
2. Select that $F \in \mathcal{F}$ that minimizes $L(\mathcal{D}, M \cup F)$.
3. Add F to M and remove it from $\mathcal{F}$.
4. Repeat steps 2-3 until compression can no longer be improved.

Naively, this entails iteratively re-ranking all candidates, and taking the best one. That is, with regard to Chap. 5, this approach can be viewed as the dynamic ranking approach to pattern set mining.

The naive implementation of this strategy is computationally much more demanding than single-pass filtering, with a complexity of $O(|\mathcal{F}|^2)$ opposed to $O(|\mathcal{F}|)$. On the upside, it is less prone to local minima. If one desires to explore even a larger parts of the search space, one could maintain the top-k best models after each iteration instead of only the single best model. Such a strategy would essentially be a beam search and is employed by the Groei algorithm, as proposed by Siebes and Kersten [56] to find good approximations to the problem of finding the best description of the data in k patterns.

Instead of exactly calculating the quality of each candidate per iteration, which requires a pass over the data and is hence expensive, we can also employ a quality estimate. To this end, the Mtv algorithm uses a convex quality estimate [44], which allows both to effectively prune a large part of the candidate space, as well as to identify the best candidate without having to calculate the actual score. Slim [59] uses an optimistic estimate, and only calculates the actual score for the top-k candidates until one is accepted by MDL.

4.1.3 Pruning

Another improvement that can be used by any candidate filtering approach is to add a *pruning* step: patterns that were added to the model before may become obsolete later during the search. That is, due to other additions previously added patterns may no longer contribute to improved compression. To remedy this, we can *prune* the model, i.e., we can test whether removing patterns from the model results in improved compression.

Again, there are many possibilities. The most obvious strategy is to check the attained compression of all valid subsets of the current pattern set and choose the corresponding model with minimal compressed size. One could even include a new candidate pattern in this process, yet this requires considerable extra amount of computation.

A more efficient alternative is to prune only directly after a candidate F is accepted. To keep the pruning search space small, one could consider each pattern in the current model for removal once after acceptance of another pattern, in a heuristic order. If pruning a pattern does not result in an increased encoded size of data and model, it apparently no longer contributes to compression. When this is the case, it is permanently removed from the model. Even simple pruning techniques like this can vastly improve the compression ratios attained by pattern-based models found by candidate filtering methods.

Pruning has been shown to be one of the key elements of KRIMP [68], as it allows for removing patterns from the model for which we have found better replacements, and which if we keep them are in the way (in terms of cost) of other patterns. Pruning practically always improves performance, both in terms of speed, compression rates, as well as in smaller pattern sets [23, 64, 68].

4.2   Direct Mining of Patterns that Compress

Candidate filtering is conceptually easy and generally applicable. It allows us to mine any set of candidate patterns, and then select a good subset. However, the reason for mining code tables, the pattern explosion, is also the Achilles heel of this two-stage approach. Mining, storing, and sorting candidate patterns is computationally demanding for non-trivial data. In particular as lower thresholds correspond to better models: larger candidate sets induce a larger model space, and hence allow for better models to be discovered. However, the vast majority of these patterns will never be selected or make it into the final model, the question is: can't we mine a good code table directly from data?

The space of models $\mathcal{M}$ is too erratic to allow direct sampling of high-quality code tables. We can, however, adapt the iterative candidate selection scheme above. In particular, instead of iteratively identifying the best candidate from $\mathcal{F}$, we use the current model M to generate candidates that are likely good additions to the model.

What makes likely a good addition to the model? A pattern that helps to reduce redundancy in the encoding. In our setting, this means correlations between code usages. If the code for pattern A and the code for pattern B often co-occur, we can gain bits using a new code C meaning 'A and B'. We can hence find good candidates by mining frequent patterns in 'encoding space'. Moreover, by employing an optimistic estimate we can prune large parts of the search space, and efficiently identify the best pattern [59]. In general terms, we have

1. Start with an 'empty' model M.
2. Find that $F \in \mathcal{X}$ that minimizes $L(\mathcal{D}, M \cup F)$.
3. Add F to M.
4. Repeat steps 2-3 until compression can no longer be improved.

Because of the strong dependence on the specific encoding and pattern type, providing a universal algorithmic strategy for step 2 is hardly possible—in itemset data correlations mean co-occurrences [59], in sequential data it means close-by occurrences [64], etc. In general, the current encoding of the data will have to be inspected to see if there are any 'patterns' in there that can be exploited to improve compression.

The SLIM algorithm [59] was the first to implement this strategy for MDL, and induces KRIMP code tables by iteratively searching for pairs of itemsets that often occur together. The union of the pair that results in the best improvement in compression is added to the code table. Although it hence considers a search space of only $O(|CT|^2)$ instead of $O(|\mathcal{F}|^2)$, its results very closely approximate the ideal local greedy strategy, or, KRAMP. In particular for dense data, SLIM can be orders of magnitude faster than KRIMP, obtaining smaller code tables that offer much more succinct descriptions of the data.

To save computation, SQS does not iteratively identify the best candidate, but instead iteratively generates a set of candidates given the current model, considers all these candidates in turn, then generates new candidates, etc, until MDL tells it to stop.

5 MDL for Data Mining

So far, we considered compression for model selection, but it has been argued [15] and shown in the literature that it can also be used for many (data mining) tasks. For example, we already referred to the Normalized Compression Distance [41]. Another concrete example is the usage of MPEG video compression for image clustering [29].

In these examples, existing compression algorithms are used as 'black boxes' to approximate Kolmogorov complexity, and usually only dataset-level compression is required (to be precise, individual strings/objects are considered as 'datasets').

In this chapter, we are particularly interested in compression-based models that allow for inspection, so that any discovered local structure can be interpreted by domain experts. For that purpose pattern-based models that can be selected by means of the MDL principle have been developed. However, we have not yet discussed if and how these models can be used for tasks other than describing and summarizing the data.

In the following we will show how many learning and mining tasks can be naturally formalized in terms of compression, using the pattern-based models and MDL formulation described in this chapter. In particular, to be able to give more concrete details we will focus on using code tables as models. Again, it is important to note that the overall approach can be applied to other compression- and pattern-based models as well.

5.1 Classification

Classification is a traditional task in machine learning and data mining. Informally, it can be summarized as follows: given a training set of tuples with class labels and an 'unseen' tuple t without class label, use the training data to infer the correct class label for t. Next, we describe a simple classification scheme based on the MDL principle [37].

5.1.1 Classification through MDL

If we assume that a database $\mathcal{D}$ is an i.i.d. sample from some underlying data distribution, we expect that the optimal model for this database, i.e., optimal in MDL sense, to compress an arbitrary tuple sampled from this distribution well. For this to work, we need a model that supports tuple-level compression.

In the context of code tables, we make this intuition more formal in Lemma 8.13. We say that the patterns in CT are *independent* if any co-occurrence of two patterns $X, Y \in CT$ in the cover of a tuple is independent. That is, $P(XY) = P(X)P(Y)$, a Naïve Bayes [71] like assumption.

Lemma 8.13 *Let $\mathcal{D}$ be a bag of tuples drawn from $\mathcal{T}$, cover a cover function, CT the optimal code table for $\mathcal{D}$ and t an arbitrary transaction from $\mathcal{T}$. Then, if the patterns $X \in cover(CT, t)$ are independent,*

$$L(t|CT) = -\log\left(P(t|\mathcal{D}, CT)\right).$$

(See [37] for the proof.)

This lemma is only valid under the Naïve Bayes like assumption, which in theory might be violated. However, by MDL, if there would be patterns $X, Y \in CT$ such that $P(XY) > P(X)P(Y)$, there will be a pattern Z in the optimal code table CT that covers both X and Y.

Now, assume that we have two databases generated from two different underlying distributions, with corresponding optimal code tables. For a new tuple that is generated under one of the two distributions, we can now decide to which distribution it most likely belongs. That is, under the Naïve Bayes assumption, we have the following lemma.

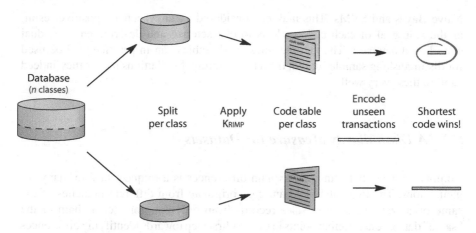

Fig. 8.3 The code table classifier in action

Lemma 8.14 *Let $\mathcal{D}_1$ and $\mathcal{D}_2$ be two bags of tuples from $\mathcal{T}$, sampled from two different distributions, CT_1 and CT_2 the optimal code tables for $\mathcal{D}_1$ and $\mathcal{D}_2$, and t an arbitrary tuple over $\mathcal{T}$. Then, by Lemma 8.13 we have*

$$L(t|CT_1) > L(t|CT_2) \;\Rightarrow\; P(t|\mathcal{D}_1) < P(t|\mathcal{D}_2).$$

Hence, the Bayes optimal choice is to assign t to the distribution that leads to the shortest code length.

5.1.2   The Code Table Classifier

The above suggests a straightforward classification algorithm based on code tables.

This classification scheme is illustrated in Fig. 8.3.

The classifier consists of a code table per class. Given a database with class labels, this database is split according to class, after which the class labels are removed from all tuples. Then, some induction method is used to obtain a code table for each single-class database. When the per-class compressors have all been constructed, classifying unseen tuples is trivial: simply assign the class label belonging to the code table that provides the minimal encoded length for the transaction.

Note that this simple yet effective scheme requires a code table to be able to compress any possible tuple, i.e., it should be possible to compute $L(t|CT)$ for any $t \in \mathcal{T}$. For this it is important to keep all 'primitive' patterns in the code table, i.e., those that are in the 'empty' code table. Further, to ensure valid codes all patterns should have non-zero usage, which can be achieved by, e.g., applying a Laplace correction: add one to the usage of each pattern in the code table.

Results of this scheme on itemset data, using KRIMP [37] and PACK [62], show this simple classifier performs *on par* with the best classifiers in the literature, including

Naïve Bayes and SVMs. This may be considered an unexpectedly positive result, as the sole goal of each code table is to characterize and describe an individual class-based database. The fact that these code tables can in practice also be used for distinguishing samples drawn from the different distributions means they indeed capture these very well.

5.2 A Dissimilarity Measure for Datasets

Comparing datasets to find and explain differences is a frequent task in many organizations. The two databases can, e.g., originate from different branches of the same organizations, such as sales records from different stores of a chain or the "same" database at different points in time. A first step towards identifying differences between datasets is to *quantify* how different two datasets are.

Although this may appear to be a simple task at first sight, in practice it turns out to be far from trivial in many cases. In particular, this is true when considering data types for which no obvious distance measures are available, such as for categorical data. In this subsection we describe a compression-based difference measure for datasets (based on [66]).

5.2.1 Code Length Differences

Let $\mathcal{D}_1$ and $\mathcal{D}_2$ be two databases with tuples drawn from the same data universe $\mathcal{T}$. The MDL principle implies that the optimal compressor induced from a database $\mathcal{D}_1$ will generally provide shorter encodings for its tuples than the optimal compressor induced from another database $\mathcal{D}_2$. This is the same principle as used by the classifier described in the previous subsection, and again we assume and exploit the tuple-level compression property.

Formally, let M_i be the optimal model induced from database $\mathcal{D}_i$, and t a transaction in $\mathcal{D}_1$. Then, the MDL principle implies that

$$|L(t|M_2) - L(t|M_1)|$$

- is small if t is equally likely under the distributions of $\mathcal{D}_1$ and $\mathcal{D}_2$;
- is large if t is more likely under the distribution of one database than under the distribution underlying the other.

Furthermore, the MDL principle implies that for the MDL-optimal models M_1 and M_2 and t from $\mathcal{D}_1$, the expected average value of $L(t|M_2) - L(t|M_1)$ is positive. The next step towards a dissimilarity measure is to aggregate these code length differences over the dataset.

If we would do this naively, the resulting aggregate would depend on the size of the data. To avoid this, we normalize by dividing by the 'native' encoded size of the database, $L(\mathcal{D}_1|M_1)$, and arrive at

$$ACLD(\mathcal{D}_1, M_2) = \frac{L(\mathcal{D}_1|M_2) - L(\mathcal{D}_1|M_1)}{L(\mathcal{D}_1|M_1)}.$$

Like Kullback-Leibler divergence, *ACLD* is asymmetric: it measures how different $\mathcal{D}_2$ is from $\mathcal{D}_1$, not vice versa. While it is reasonable to expect these to be in the same ballpark, this is not a given.

5.2.2 The Database Dissimilarity Measure

The asymmetric measure allows measuring similarity of one database to another. To make it a practical measure we would like it to be symmetric. We do this by taking the maximum value of two aggregated differences, i.e., $\max\{ACLD(\mathcal{D}_1, M_2), ACLD(\mathcal{D}_2, M_1)\}$. This can easily be rewritten in terms of compressed database sizes, as follows.

Definition 8.15 *Let $\mathcal{D}_1$ and $\mathcal{D}_2$ be two databases drawn from $\mathcal{T}$, and let M_1 and M_2 be their corresponding MDL-optimal models. Then, define the* dissimilarity *measure DS between $\mathcal{D}_1$ and $\mathcal{D}_2$ as*

$$DS(\mathcal{D}_1, \mathcal{D}_2) = \max\left(\frac{L(\mathcal{D}_1|M_2) - L(\mathcal{D}_1|M_1)}{L(\mathcal{D}_1|M_1)}, \frac{L(\mathcal{D}_2|M_1) - L(\mathcal{D}_2|M_2)}{L(\mathcal{D}_2|M_2)}\right).$$

Using this measure, we'll obtain a score of 0 iff the databases are identical, and higher scores indicate higher dissimilarity. In theory, using MDL-optimal models we find that *DS*, like NCD [11] is a metric: the symmetry axiom holds by definition, scores cannot be negative, and it holds that $DS(\mathcal{D}_1, \mathcal{D}_2) = 0$ iff $\mathcal{D}_1 = \mathcal{D}_2$. The advantage of *DS* over NCD is that we only have to induce two models, as opposed to four.

For heuristic model induction algorithms the metric property is difficult to prove. However, instantiating this measure for itemset data using KRIMP, we obtain very good results [66]: dataset pairs drawn from the same distribution have very low dissimilarities, whereas dataset pairs from different distributions have substantially larger dissimilarities.

5.3 Identifying and Characterizing Components

Though most databases are mixtures drawn from different distributions, we often assume only one distribution. Clearly, this leads to suboptimal results: the distributions need to be modeled individually.

Clustering addresses part of this problem by trying to separate the source components that make up the mixture. However, as we do not know upfront what distinguishes the different components, the appropriate distance metric is hard to define. Furthermore, in clustering we are only returned the object assignment, and not any insight in the characteristics per cluster. For example, what is typical for that cluster, and how do the different ingredients of the mixture compare to each other?

The pattern- and compression-based models described in this chapter provide all prerequisites required for data characterization, classification, and difference measurement. If a compression-based approach can be used to identify the components of a database, each represented by a pattern model, all these advantages can be obtained 'for free'.

5.3.1 MDL for Component Identification

On a high level, the goal is to discover an optimal partitioning of the database; optimal, in the sense that the characteristics of the different components are different, while the individual components are homogeneous. Translating this to MDL, the task is to partition a given database such that the total compressed size of the components is minimized—where each component is compressed by its own MDL-optimal model.

The intuition is that similar tuples of a database can be better compressed if they are assigned to the same partition and hence compressed by the same model. However, having multiple components, with corresponding models, allows models to be more specific and hence can be expected to provide better overall compression. By minimizing the total compressed size, including the sizes of the models, the different distributions of the mixture are expected to be divided over the partitions.

Following [39], we have the following problem statement:

Problem 5.1 (Identifying Database Components) *Let $\mathcal{D}$ be a bag of tuples drawn from $\mathcal{T}$. Find a partitioning $\mathcal{D}_1, \cdots, \mathcal{D}_k$ of $\mathcal{D}$ and associated models $M_1, \cdots, M_k$, such that the total compressed size of $\mathcal{D}$,*

$$\sum_{i \in \{1, \cdots, k\}} L(M_i, \mathcal{D}_i),$$

is minimized.

There are a few of observations we should make with regard to this problem. First of all, note that it is parameter-free: MDL determines the optimal number of components. Second, asking for both the partitioning and the models is in a sense redundant. For any partitioning, the best associated models are, of course, the optimal ones. The other way around, given a set of models, a database partitions naturally: each tuple goes to the model that compresses it best, as with classification.

The search space, however, is enormous, and solving the problem hard. An effective and efficient heuristic is to take an EM-like approach [14], starting with a random partitioning, and iteratively inducing models and re-assigning tuples to maximize compression, until convergence. Besides automatically determining the optimal number of components, this approach has been shown to find sound groupings [39].

5.4   Other Data Mining Tasks

So far we covered some of the most prominent tasks in data mining. However, many more tasks have been formulated in terms of MDL and pattern-based models. Below, we briefly describe five examples.

5.4.1   Data Generation—and Privacy Preservation

The MDL principle is primarily geared towards descriptive models. However, these models can also be employed as predictive models, such as in the classification example above. Furthermore, under certain conditions, compression-based models can also be used as *generative models*.

By exploiting the close relation between code lengths and probability distributions, code tables can be used for data generation. For categorical data, *synthetic* data generated from a KRIMP code table has the property that the deviation between the observed and original frequencies is very small on expectation for *all* itemsets [67]. One application is privacy preservation: the generated data has the same characteristics as the original data, yet individual details are lost and specified levels of anonymity can be obtained.

5.4.2   Missing Value Estimation

Many datasets have missing values. Under the assumption these are missing without correlation to the data, they do not affect the observed overall distribution. Consequently, despite those missing values, a model of reasonable quality can be induced given sufficient data. Given such a database and corresponding model, the best estimation for a single missing value is the one that minimizes the total compressed size. We can do so both for individual tuples, a well as for databases with many missing values: by iteratively imputing the values, and inducing the model, completed datasets with very high accuracy are obtained [65].

5.4.3   Change Detection in Data Streams

A database can be a mixture of different distributions, but in data streams concept drift is common: one distribution is 'replaced' by another distribution. In this context, it is important to detect when such change occurs. Complicating issues are that streams are usually infinite, can have high velocity, and only limited computation time is available for processing.

By first assuming that the data stream is sampled from a single distribution, a model can be induced on only few samples; how many are needed can be deduced from the attained compression ratios. Once we have a model, we can observe the compressed size of the new data; if this is considerably larger than for the earlier samples, a change has occurred and a new model should be induced. In particular for sudden distribution shifts, this scheme is highly effective [36].

5.4.4 Coherent Group Discovery

Whereas the *Identifying Database Components* problem assumes that we are interested in a partitioning of the complete database, this task aims at the discovery of coherent subsets of the data that deviate from the overall distribution. As such, it is an instance of subspace clustering. In terms of MDL, this means that the goal is to find groups that can be compressed much better by themselves than as part of the complete database.

As example application, this approach was applied to tag data obtained for different media types [38]. It was shown that using only tag information, coherent groups of media, e.g., photos, can be discovered.

5.4.5 Outlier Detection

All databases contain outliers, but defining what an outlier exactly is and detecting them are well-known to be challenging tasks. By assuming that the number of outliers is small, and given the intuition of what an outlier is this seems a safe assumption, we know that the largest part of a dataset is 'normal'. Hence, a model induced on the database should capture primarily what is normal, and not so much what is an outlier. Then, outlier detection can be formalized as a one-class classification problem: all tuples that are compressed well belong to the 'normal' distribution, while tuples that get a long encoding may be considered outliers. For transactional data, this approach performs on par with the state-of-the-art of the field [58].

5.5 The Advantage of Pattern-based Models

For each and every of these tasks, we have to point out the added benefit of using a pattern-based model. Besides obtaining competitive, state-of-the-art performance, these patterns help to characterize decisions. For example, in the case of outlier detection, we can identify *why* a tuple is identified as an anomaly by pointing out the patterns of the norm it does not comply with, as well as how strongly it is an anomaly—how much effort we have to do in order to make it 'normal'. Similar advantages hold for the classification task. For the clustering related tasks, we have the added benefit that we can offer specialized code tables, specialized descriptions per subpart of the data; we are not only told which parts of the data should go together, but also why, what patterns make these data points similar.

6 Challenges Ahead

Above we showed that compression provides a powerful approach to both mining and using patterns in a range of data mining tasks. Here we briefly identify and discuss a number of open research problems.

6.1 Toward Mining Structured Data

When compared to other data types, compressing itemset data is relatively simple. The most important reason is that the data is unordered over both rows and columns, and hence tuples can be considered as *sets* of items, and the data as a *bag* of tuples.

For 'spatial' binary data, where the order of rows and columns does matter, many tasks already become more difficult. A good example is the extension of tiling, called geometric tiling [19], which aims at finding a hierarchy of (noisy) tiles that describe the data well. Finding optimal sub-tiles is more difficult than mining itemsets, as we now also have to consider every subset of rows. STIJL efficiently finds the MDL-optimal sub-tile in order to greedily find good tilings [63].

Another possible structural constraint is time: sequences and streams are both series of data points, where sequences consist of events while data streams usually consists of complete tuples, e.g., itemsets. Initial attempts to characterize sequence data with patterns using compression include [64] and [34]. Lam et al. [35] mine sequential patterns from streams, whereas the goal of Van Leeuwen and Siebes [36] is to detect changes in data streams. All these are limited though. For example, none are suited for the high velocity of big data streams, as well as suboptimal for data consisting of shifting mixtures of distributions. Other open issues include allowing overlap between patterns, as well as allowing multiple events per time-stamp.

Adding even more structure, we have trees, graphs, as well as multi-relational data. In this area even fewer results have been published, though arguably these data types are most abundant. For graphs, SLASHBURN [26] uses compression to separate communities and hubs. For multi-relational data, two variants of KRIMP have been proposed [32, 33], yet their modeling power is limited by their restrictive pattern languages—nor are direct candidate mining strategies available.

Further, so far no pattern set mining approaches have been proposed for continuous data. Moreover, all data is assumed to be 'certain'. However, in bioinformatics, for example, many data is probabilistic in nature, e.g., representing the uncertainty of protein-protein interactions. Bonchi et al. [7] proposed an approach to model uncertain data by itemsets, yet they do so with 'certain' itemsets, i.e., without explicit probabilities. Mining pattern sets from numerical and uncertain data, as well as using them in compression-based models, are important future challenges.

6.2 Generalization

While the above challenges concern specialization for structured data types and other data primitives, another challenge concern the other direction: generalization. One of the fundamental problems in data mining is that new models, algorithms, and implementations are needed for every combination of task and data type. Though the literature flourishes, it makes the results very hard to use for non-experts.

In this chapter we have shown that patterns can actually be useful: for summarization and characterization, as well as for other tasks. One of the upcoming challenges will be to generalize compression-based data mining. Can patterns be defined in a very generic way, so that mining them and using them for modeling remains possible? For that, progress with regard to both mining and modeling needs to be made. Both are currently strongly tailored toward specific data and pattern types.

One approach may be to represent everything, both data and patterns, as queries. With such a uniform treatment, recently proposed by Siebes [55], the ideal of exploratory data mining might become reachable. Note that the high-level goal of generalizing data mining and machine learning is also pursued by De Raedt et al. [51, 21], yet with different focus: their aim is to develop declarative modeling languages for data mining, which can use existing solver technology to mine solutions.

6.3 Task- and/or User-specific Usefulness

While obtaining very good results in practice, MDL is not a magic wand. In existing approaches, the results are primarily dependent on the data and pattern languages. In other situations it may be beneficial to take specific tasks and/or users into account. In other words, one may want to keep the purpose of the patterns in mind.

As an example, the code table classifier described in the previous section works well in practice, yet it is possibly sub-optimal. It works by modeling the class distributions, not by modeling the differences between these. Although classification is hardly typical for *exploratory* data mining, similar arguments exist for other data mining tasks.

In this chapter we ignore any background knowledge the user may have. If one is interested in the optimal model *given* certain background knowledge, this entails finding MDL-optimal models given prior distributions—which reduces to the MML [69] principle. The optimal prior can be identified using the Maximum Entropy principle [25].[3]

De Bie [13] argues that the goal of the data miner in data exploration is to model the user's belief-state, so that we can algorithmically discover those results that will be most informative to the user. At the core, this reduces to compression—with the twist that the decision whether to include a pattern is made by the user.

---

[3] See Chap. 5 for a more complete discussion on MaxEnt.

6.3.1 The Optimum

A more global issue is the efficiency of the used encodings. Whereas in Kolmogorov complexity we have access to the ultimate algorithmic compressor, the MDL principle assumes that we have access to the ultimate encoding. In practice, we have to make do with an approximation. While when constructing an encoding we can make principled choices, we often have to simplify matters to allow for fast(er) induction of good models. For instance, in KRIMP it would be nice if we could encode transactions using their exact probability given the pattern set. However, calculating frequencies of an itemset given a set of itemsets and frequencies is known to be PP-hard [61]. Hence KRIMP uses a (admittedly crude) approximation of this ideal. A more efficient encoding would allow to detect more fine-grained redundancy, and hence lead to smaller and better models. Currently, however, there is very little known on how to construct a good yet practical encoding.

A second global issue we need to point out is that of complexity. Intuitively, optimizing an MDL score is rather complex. However, so far we only have hardness results for a simple encoding in Boolean matrix factorization [46]. It may be that other encodings do exhibit structure that we have not yet identified, but which may be exploited for (more) efficient search. Alternatively, so far we have no theoretical results on the quality of our greedy approximations. It may be possible to construct non-trivial MDL scores that exhibit sub-modularity, which would allow approximating the quality of the greedy strategy.

Third, for now assuming the optimization problem is hard, and there are no (useful) approximation guarantees, we need to develop smart heuristics. We described the two main approaches proposed so far, candidate filtering and direct mining. Naively, the larger part of the search space $\mathcal{M}$ we consider, the better the model M we'll be able to find. However, as the model space is too large, we have to find ways of efficiently considering what is good. The direct mining approach provides a promising direction, but is only as good as the quality estimation it employs. Improving this estimation will allow to prune away more candidates, and concentrate our effort there where it matters most.

7 Conclusions

We discussed how to apply the MDL principle for mining sets of patterns that are both informative and useful. In particular, we discussed how pattern-based models can be designed and selected by means of compression, giving us succinct and characteristic descriptions of the data.

Firmly rooted in algorithmic information theory, the approach taken in this chapter states that the best set of patterns is that set that compresses the data best. We formalized this problem using MDL, described model classes that can be used to this end, and briefly discussed algorithmic approaches to inducing good models from data. Last but not least, we described how the obtained models, which are very characteristic for the data, can be used for numerous data mining tasks, making the pattern sets practically useful.

Acknowledgments Matthijs van Leeuwen is supported by a Post-doctoral Fellowship of the Research Foundation Flanders (FWO). Jilles Vreeken is supported by the Cluster of Excellence "Multimodal Computing and Interaction" within the Excellence Initiative of the German Federal Government.

References

1. P. Adriaans and P. Vitányi. Approximation of the two-part MDL code. *IEEE TIT*, 55(1):444–457, 2009.
2. H. Akaike. A new look at the statistical model identification. *IEEE TAC*, 19(6):716–723, 1974.
3. L. Akoglu, H. Tong, J. Vreeken, and C. Faloutsos. CompreX: Compression based anomaly detection. In *CIKM*. ACM, 2012.
4. L. Akoglu, J. Vreeken, H. Tong, N. Tatti, and C. Faloutsos. Mining connection pathways for marked nodes in large graphs. In *SDM*. SIAM, 2013.
5. R. Bathoorn, A. Koopman, and A. Siebes. Reducing the frequent pattern set. In *ICDM-Workshop*, pages 1–5, 2006.
6. C. Böhm, C. Faloutsos, J.-Y. Pan, and C. Plant. Robust information-theoretic clustering. In *KDD*, pages 65–75, 2006.
7. F. Bonchi, M. van Leeuwen, and A. Ukkonen. Characterizing uncertain data using compression. In *SDM*, pages 534–545, 2011.
8. S. Chakrabarti, S. Sarawagi, and B. Dom. Mining surprising patterns using temporal description length. In VLDB, pages 606–617. Morgan Kaufmann, 1998.
9. D. Chakrabarti, S. Papadimitriou, D. S. Modha, and C. Faloutsos. Fully automatic cross-associations. In *KDD*, pages 79–88, 2004.
10. V. Chandola and V. Kumar. Summarization – compressing data into an informative representation. *Knowl. Inf. Sys.*, 12(3):355–378, 2007.
11. R. Cilibrasi and P. Vitányi. Clustering by compression. *IEEE TIT*, 51(4):1523–1545, 2005.
12. T. M. Cover and J. A. Thomas. *Elements of Information Theory*. Wiley-Interscience New York, 2006.
13. T. De Bie. An information theoretic framework for data mining. In *KDD*, pages 564–572. ACM, 2011.
14. A. Dempster, N. Laird, and D. Rubin. Maximum likelihood from incomplete data via the EM algorithm. *J. R. Statist. Soc. B*, 39(1):1–38, 1977.
15. C. Faloutsos and V. Megalooikonomou. On data mining, compression and Kolmogorov complexity. *Data Min. Knowl. Disc.*, 15(1):3–20, 2007.
16. U. Fayyad and K. Irani. Multi-interval discretization of continuous-valued attributes for classification learning. In *UAI*, pages 1022–1027, 1993.
17. R. A. Fisher. On the interpretation of χ^2 from contingency tables, and the calculation of P. *Journal of the Royal Statistical Society*, 85(1):87–94, 1922.
18. F. Geerts, B. Goethals, and T. Mielikäinen. Tiling databases. In *DS*, pages 278–289, 2004.
19. A. Gionis, H. Mannila, and J. K. Seppänen. Geometric and combinatorial tiles in 0-1 data. In *PKDD*, pages 173–184. Springer, 2004.
20. P. Grünwald. *The Minimum Description Length Principle*. MIT Press, 2007.
21. T. Guns, S. Nijssen, and L. D. Raedt. Itemset mining: A constraint programming perspective. *Artif. Intell.*, 175(12-13):1951–1983, 2011.
22. E. Halperin and R. M. Karp. The minimum-entropy set cover problem. *TCS*, 348(2-3):240–250, 2005.
23. H. Heikinheimo, J. K. Seppänen, E. Hinkkanen, H. Mannila, and T. Mielikäinen. Finding low-entropy sets and trees from binary data. In *KDD*, pages 350–359, 2007.
24. H. Heikinheimo, J. Vreeken, A. Siebes, and H. Mannila. Lowentropy set selection. In *SDM*, pages 569–580, 2009.

25. E. Jaynes. On the rationale of maximum-entropy methods. *Proc. IEEE*, 70(9):939–952, 1982.
26. U. Kang and C. Faloutsos. Beyond caveman communities: Hubs and spokes for graph compression and mining. In *ICDM*, pages 300–309. IEEE, 2011.
27. R. M. Karp. Reducibility among combinatorial problems. In *Proc. Compl. Comp. Comput.*, pages 85–103, New York, USA, 1972.
28. E. Keogh, S. Lonardi, and C. A. Ratanamahatana. Towards parameter-free data mining. In *KDD*, pages 206–215, 2004.
29. E. Keogh, S. Lonardi, C. A. Ratanamahatana, L. Wei, S.-H. Lee, and J. Handley. Compression-based data mining of sequential data. *Data Min. Knowl. Disc.*, 14(1):99–129, 2007.
30. P. Kontkanen and P. Myllymäki. A linear-time algorithm for computing the multinomial stochastic complexity. *Inf. Process. Lett.*, 103(6):227–233, 2007.
31. P. Kontkanen, P. Myllymäki, W. Buntine, J. Rissanen, and H. Tirri. An MDL framework for clustering. Technical report, HIIT, 2004. Technical Report 2004–6.
32. A. Koopman and A. Siebes. Discovering relational items sets efficiently. In *SDM*, pages 108–119, 2008.
33. A. Koopman and A. Siebes. Characteristic relational patterns. In *KDD*, pages 437–446, 2009.
34. H. T. Lam, F. Mörchen, D. Fradkin, and T. Calders. Mining compressing sequential patterns. In *SDM*, 2012.
35. H. T. Lam, T. Calders, J. Yang, F. Moerchen, and D. Fradkin.: Mining compressing sequential patterns in streams. In *IDEA*, pages 54–62, 2013.
36. M. van Leeuwen and A. Siebes. StreamKrimp: Detecting change in data streams. In *ECML PKDD*, pages 672–687, 2008.
37. M. van Leeuwen, J. Vreeken, and A. Siebes. Compression picks the item sets that matter. In *PKDD*, pages 585–592, 2006.
38. M. van Leeuwen, F. Bonchi, B. Sigurbjörnsson, and A. Siebes. Compressing tags to find interesting media groups. In *CIKM*, pages 1147–1156, 2009.
39. M. van Leeuwen, J. Vreeken, and A. Siebes. Identifying the components. *Data Min. Knowl. Disc.*, 19(2):173–292, 2009.
40. M. Li and P. Vitányi. *An Introduction to Kolmogorov Complexity and its Applications*. Springer, 1993.
41. M. Li, X. Chen, X. Li, B. Ma, and P. Vitanyi. The similarity metric. *IEEE TIT*, 50(12): 3250–3264, 2004.
42. C. Lucchese, S. Orlando, and R. Perego. Mining top-k patterns from binary datasets in presence of noise. In *SDM*, pages 165–176, 2010.
43. M. Mampaey and J. Vreeken. Summarising categorical data by clustering attributes. *Data Min. Knowl. Disc.*, 26(1):130–173, 2013.
44. M. Mampaey, J. Vreeken, and N. Tatti. Summarizing data succinctly with the most informative itemsets. *ACM TKDD*, 6:1–44, 2012.
45. P. Miettinen and J. Vreeken. Model order selection for Boolean matrix factorization. In *KDD*, pages 51–59. ACM, 2011.
46. P. Miettinen and J. Vreeken. mdl4bmf: Minimum description length for Boolean matrix factorization. *ACM TKDD. In Press*.
47. S. Papadimitriou, J. Sun, C. Faloutsos, and P. S. Yu. Hierarchical, parameter-free community discovery. In *ECML PKDD*, pages 170–187, 2008.
48. B. Pfahringer. Compression-based feature subset selection. In *Proc. IJCAI'95 Workshop on Data Engineering for Inductive Learning*, pages 109–119, 1995.
49. B. A. Prakash, J. Vreeken, and C. Faloutsos. Spotting culprits in epidemics: How many and which ones? In ICDM. IEEE, 2012.
50. J. Quinlan. *C4.5: Programs for Machine Learning*. Morgan-Kaufmann, Los Altos, California, 1993.
51. L. D. Raedt. Declarative modeling for machine learning and data mining. In *ECML PKDD*, pages 2–3, 2012.
52. J. Rissanen. Modeling by shortest data description. *Automatica*, 14(1):465–471, 1978.
53. G. Schwarz. Estimating the dimension of a model. *Annals Stat.*, 6(2):461–464, 1978.

54. H. Shao, B. Tong, and E. Suzuki. Extended MDL principle for feature-based inductive transfer learning. *Knowl. Inf. Sys.*, 35(2):365–389, 2013.
55. A. Siebes. Queries for data analysis. In *IDA*, pages 7–22, 2012.
56. A. Siebes and R. Kersten. A structure function for transaction data. In *SDM*, pages 558-569. SIAM, 2011.
57. A. Siebes, J. Vreeken, and M. van Leeuwen. Item sets that compress. In *SDM*, pages 393-404. SIAM, 2006.
58. K. Smets and J. Vreeken. The odd one out: Identifying and characterising anomalies. In *SDM*, pages 804–815. SIAM, 2011.
59. K. Smets and J. Vreeken. Slim: Directly mining descriptive patterns. In *SDM*, pages 236–247. SIAM, 2012.
60. J. Sun, C. Faloutsos, S. Papadimitriou, and P. S. Yu. Graphscope: parameter-free mining of large time-evolving graphs. In *KDD*, pages 687–696, 2007.
61. N. Tatti. Computational complexity of queries based on itemsets. *Inf. Process. Lett.*, 98(5): 183–187, 2006.
62. N. Tatti and J. Vreeken. Finding good itemsets by packing data. In *ICDM*, pages 588–597, 2008.
63. N. Tatti and J. Vreeken. Discovering descriptive tile trees by fast mining of optimal geometric subtiles. In *ECML PKDD*. Springer, 2012.
64. N. Tatti and J. Vreeken. The long and the short of it: Summarizing event sequences with serial episodes. In *KDD*. ACM, 2012.
65. J. Vreeken and A. Siebes. Filling in the blanks: Krimp minimisation for missing data. In *ICDM*, pages 1067–1072. IEEE, 2008.
66. J. Vreeken, M. van Leeuwen, and A. Siebes. Characterising the difference. In *KDD*, pages 765–774, 2007.
67. J. Vreeken, M. van Leeuwen, and A. Siebes. Preserving privacy through data generation. In *ICDM*, pages 685–690. IEEE, 2007.
68. J. Vreeken, M. van Leeuwen, and A. Siebes. Krimp: Mining itemsets that compress. *Data Min. Knowl. Disc.*, 23(1):169–214, 2011.
69. C. Wallace. *Statistical and inductive inference by minimum message length*. Springer-Verlag, 2005.
70. C. Wang and S. Parthasarathy. Summarizing itemset patterns using probabilistic models. In *KDD*, pages 730–735, 2006.
71. H. Warner, A. Toronto, L. Veasey, and R. Stephenson. A mathematical model for medical diagnosis, application to congenital heart disease. *J. Am. Med. Assoc.*, 177:177–184, 1961.

Chapter 9
Frequent Pattern Mining in Data Streams

Victor E. Lee, Ruoming Jin and Gagan Agrawal

Abstract As the volume of digital commerce and communication has exploded, the demand for data mining of streaming data has likewise grown. One of the fundamental data mining tasks, for both static and streaming data, is frequent pattern mining. The goal of pattern mining is to identity frequently occurring patterns and structures. Such patterns may indicate scientific phenomena, economic or social trends, or even security threats. Moreover, not only is pattern discovery important by itself, but it is also a building block for machine learning tasks such as association rule induction. Traditionally, algorithms for pattern discovery have processed the entire dataset as a batch, with no restriction on how many passes through the data would be taken.

However, when the data are arriving in a continuous and unending stream, our algorithm must be limited to a single pass. Moreover, the length of the stream is indeterminate, so we cannot wait for it to end. We generate an initial result after seeing a certain quantity of data, and then we periodically revise the result. A particular challenge for frequent pattern discovery is the combinatorial explosion of candidate patterns

In this chapter, we present a structured review of online frequent pattern mining techniques. We classify the methods according to the type of pattern and data, the time window being considered, and the quality of the approximation.

Keywords Frequent pattern mining · Streaming data · Lossy counting · Sliding window

V. E. Lee (✉)
John Carroll University, University Heights, OH, USA
e-mail: vlee@jcu.edu

R. Jin
Kent State University, Kent, OH, US
e-mail: jin@cs.kent.edu

G. Agrawal
Ohio State University, Columbus, OH, US
e-mail: agrawal@cse.ohio-state.edu

C. C. Aggarwal, J. Han (eds.), *Frequent Pattern Mining,*
DOI 10.1007/978-3-319-07821-2_9, © Springer International Publishing Switzerland 2014

1 Introduction

Frequent pattern mining is the search for frequently-occurring patterns within a dataset. The dataset may be loosely structured, such as a set of text documents, semistructured such as XML, or highly structured such as a graph. Each type of data may be the source of a different type of pattern. For example, for a dataset of purchase transactions, each transaction contains an itemset, so we may look for frequent (sub)itemsets. In fact, frequent itemset mining is the most common pattern mining application. Other important patterns include subsequences, subtrees, and subgraphs.

Not only do frequent patterns describe the highlights of a dataset, providing key insights into the data, but they also serve as a constituent for many other data mining and machine learning tasks, such as association rule mining, classification, clustering, and change detection [1, 36, 37, 38, 41, 51, 80].

The frequent itemset mining task gained wide attention in the data mining community with the publication of Agrawal and Srikant's Apriori algorithm [2] in 1994. The next year, the pattern space was extended from itemsets to sequential patterns in another seminal paper [3], also by Agrawal and Srikant. Since then, many efficient frequent pattern algorithms have been developed [4, 30, 32, 44, 76, 79, 78]. A popular survey of frequent pattern mining algorithms is [33]. There algorithms assume that the dataset is static, stored on disk, and that two or more passes over the dataset may be taken.

In a streaming environment, however, a mining algorithm may take only a single pass over the data [8]. The aforementioned algorithms at best only guarantee an approximate result after one pass. Thus, the need for a new class of mining techniques arose.

Compared with other stream processing tasks, frequent pattern mining presents three computational challenges. First, there is generally an exponential number of patterns to consider. For example, if we are seeking subsequences, a sequence of length N contains 2^N possible subsequences. The classic Apriori-style algorithm evaluates $O(k^2)$ candidate subpatterns in order to find one pattern of length k. However, if data are streaming in quickly, the computational complexity needs to be linear or nearly so, in order to keep up with with newly arriving data.

Second, the memory requirements can also be substantial. Because the search space is so large, the answering set itself may also be very large. To make matters worse, many stream mining algorithms produce approximate results biased towards false positive selections, so as not to miss any true positive results. Hence, a naïve streaming data algorithm could require more memory than a static data algorithm. Therefore, the mining algorithm needs to be very memory-efficient.

Third, the algorithms must balance the need for accuracy vs. the efficiency. Reducing the error of the approximate results usually requires expending more memory and more computational time, with diminishing returns. A good mining algorithm should allow the user to adjust the balance between the accuracy and computational resources.

In the last several years, researchers have introduced several new algorithms to find frequent patterns over data streams. In this chapter, we will conduct a survey of these algorithms.

2 Preliminaries

In this section, we define the general problem of frequent pattern mining in streaming data. We discuss some common variations of the task and we present popular approaches to the simplest variation: frequent item mining.

2.1 Frequent Pattern Mining: Definition

Many of the existing surveys or reviews of frequent pattern mining have focused exclusively on frequent itemset mining. We aim to offer a broader coverage, including sequences, trees, and graphs among the types of data and patterns to be considered. We start by giving a formal definition of the Frequent Pattern Mining problem.

Let $\mathcal{X} = \{x_1, x_2, \ldots, x_m\}$ be the set of all possible data items x_i. A pattern P is a sequence or set of data items, with $\mathcal{P}$ being all the possible patterns of interest. A streaming dataset $\mathcal{T}$ is a sequence of transacted patterns, i.e., $\mathcal{T}=\{T_1, T_2, T_3, \ldots\}$. The sequence is of indefinite length. At a time j, the data window $T_{i,j}$ is the finite data subsequence from some earlier time i to the present: $T_{i,j} = \{T_i, T_{i+1}, T_{i+2}, \ldots, T_j\}$. Because smaller patterns may be subsets of larger patterns, any data window may contain numerous patterns. Let $Patt(T)$ be a subpattern enumeration function which generates the *multiset* of all patterns contained within T. The *support s* of a subpattern P in dataset T is the frequency of p within T:

$$s(P) = \frac{count(P, Patt(T))}{|T|} \tag{9.1}$$

where $count(P, Patt(T))$ is the number of times that pattern P occurs in multiset $Patt(T)$. Then, for a given support threshold θ, $0 < \theta < 1$, a pattern P is a *frequent pattern* of T iff $s(P) \geq \theta$. The **Frequent Pattern Mining in Streaming Data Problem** seeks to find the set of all θ-frequent patterns $P \in \mathcal{P}$ contained with a data window T_{ij}. A variant task seeks to find the **Top-K Frequent Patterns**, regardless of support threshold.

This general model fits all the common types of patterns sought in streaming data: itemsets, subsequences, subtrees, and subgraphs. Note that we have said that $Patt$ considers the whole data subsequence $T_{i,j}$ to enumerate subpatterns. However, in the overwhelming majority of research works, we look for subpatterns only within each individual data object. With this typical restriction, $Patt(T_{i,j}) = \cup Patt(T_a) \forall T_a \in T_{i,j}$. When the patterns of interest themselves are subsequences, there are a few works [16, 72] which combine adjacent data objects from the data stream to form candidate patterns.

Let us see how this model fits the the most common application, frequent itemset mining. Each object is an itemset.

In the example in Table 9.1, each data object in the stream is an itemset. The patterns being sought are frequent (sub)itemsets contained within each data object. For example, $T_1 = \{A, B, D, E\}$ and $Patt(T_1) = \{(A, B), (A, D), (A, E), (B, D),$

Table 9.1 Example of itemset stream	Itemset ID	Contents
	1	A,B,D,E
	2	B,C
	3	A,B,C,D,E
	4	B,C,D,E
	5	A,C,E

Table 9.2 Common varieties of patterns from a data steam	Data object	Pattern
	Item or Itemset	Item
	Itemset	Subitemset
	Sequence	Subsequence
	Itemset	Sequence of items spanning a sequence of itemsets
	Tree	Subtree
	Graph	Subgraph or subtree

$(B, E), (D, E), (A, B, D), (A, B, E), (A, D, E), (B, D, E), (A, B, D, E)\}$, excluding singleton items. If we set $\theta = 0.6$, then an itemset must occur in 3 of the 5 objects to be considered frequent. The frequent itemsets are $Patt_{\theta=0.6}(T_{1,5}) = (A, E), (B, C), (B, D), (B, E), (D, E)$, and (B, D, E).

Table 9.2 lists the types of data streams and patterns that are notable in the literature.

Different types of data suggest different types of patterns. For example, a natural language text document can be considered either a bag of words, for itemset mining, or a sequence of words, for sequence mining. There are two major ways of formulating the subsequence problem. Each data object can itself be a sequence. Alternately, the data stream itself forms an unending sequence. An XML document, when read from top to bottom, is a depth-first traversal of a tree, so it may be suitable for subtree mining.

Arguably the most important frequent pattern mining task is *frequent itemset mining*, proposed by Rakesh Agrawal and Srikant in 1993 [2]. In this setting, each object in the dataset $\mathcal{T}$ is a set of items. Let $\mathcal{X}$ be the set of all possible items in the dataset $\mathcal{T}$. Then data object T_i can be represented as $T_i = \{x_{i1}, \cdots, x_{i|T_i|}\}$, where $x_{ij} \subseteq \mathcal{X}$. The pattern space $\mathcal{P}$ is the power-set of $\mathcal{X}$. Note that in this setting, the set of all possible transactional objects T is the same as $\mathcal{P}$.

Because the majority of work in mining frequent patterns over data streams focuses on frequent itemset mining, we devote the major portion of our chapter to itemset patterns. Many techniques developed in the itemset context can be transferred easily to mining other types of patterns, such as graph mining [38].

2.2 Data Windows

In the data stream setting, the sequence of data objects, $\mathcal{T} = (T_1, T_2, \cdots, T_i, \cdots)$, arrives over time with no known ending time. After some initial delay from the starting point, a mining result is generated based on the window of date received so far. As the sequence continues, the window is updated and so are the results. However, it is not necessarily true that we want to give equal consideration to all the data received from the start up to the present. Consequently, several standard data window models exist. A window is a subsequence between the i-th and j-th transactions, denoted as $T_{i,j} = (T_i, T_{i+1}, \cdots, T_j), i \leq j$. A user can ask different types of frequent pattern-mining questions over different type of window model.

Landmark Window In this model, we seek the frequent patterns contained in the window from a fixed starting timepoint s to the current time t. In other words, we are trying to find the frequent patterns over the window $T_{s,t}$. A special case of the landmark window is when $s = 1$. In this case, we are interested in the frequent patterns over the entire data stream. Clearly, the difficulty to solve the special case of $s = 1$ is essentially the same as the more general cases, and all cases require an efficient single-pass mining algorithm. For simplicity, we will focus on the case where the full data stream is the window.

Note that in this model, we treat each timepoint after the starting point as equally important. However, in many cases, we are more interested in the *recent* timepoints. The following models address this issue.

Sliding Window Given a window width w and current timepoint t, we are interested in the frequent patterns occurring in the window $[t - w + 1, t]$. As time advances, the window will keep its width and move along with the current timepoint. In this model, we are not interested in the data which arrived before the timepoint $t - w + 1$.

Damped Window Model This model assigns greater weight to more recently arrived transactions. A simple way to do that is to define a *decay rate* $\delta, 0 < \delta \leq 1$ [14]. As each new data transaction arrives, the support levels of the previously recorded patterns are multiplied by δ to reduce their significance. Thus, a pattern that occurred k time steps ago has a weight of δ^k. The total support for a pattern is the sum of its time-decayed counts.

Time-Tilted Window The time-tilted window was introduced by Giannella et al. [28]. In this model, we are interested in frequent itemsets over a set of windows of varying width. Each window corresponds to different time granularity based on their recency. In the log-time version, each window is twice as wide as it more recent neighbor. Specifically, the two most recent windows are 1 time unit wide. The one before that is 2 units wide, and the one before that is 4 units wide. Such model can allow us to pose more complicated queries over the data stream. Giannella et. al. have developed a variant of FP-tree, called FP-stream, for dynamically updating frequent patterns on streaming data and answering the approximate frequent itemsets for even arbitrary time intervals [28].

2.3   Frequent Item Mining

Before looking at the more challenging case of frequent itemset mining, we consider some algorithms for frequent item mining. Algorithms for this problem fall into two groups: oount-based and sketch-based. FREQUENT is a simple but effect count-based algorithm which addresses the Top-k Frequent Item problem. The idea was originally published in 1982 by Misra and Gries [66] and then was rediscovered in 2002 [25]. The algorithm maintains up to k counters. For each item x, if it has already been assigned a counter, then increment its count. Otherwise, if fewer than k counters are currently used, then assign a oounter to x with value of 1. Otherwise, decrement all the counts. Any count that drops to 0 is deassigned.

LOSSY COUNTING, by Manku and Motwani [61], computes an approximate answer to the query for all θ-frequent patterns. The data stream is processed in batches of size $B = 1/\epsilon, 0 < \epsilon < 1$. For the nth batch, count all the incoming items and add these to counts from previous batches. However, any item whose count is now less than n is dropped from memory. As a consequence, Lossy Counting guarantees that it tracks all items with support $s(x) \geq \epsilon$ and it undercounts the actual occurrences by no more than $\lceil \epsilon n \rceil$, so ϵ serves as an error parameter. We discuss this algorithm in greater detail in the next section.

Sketch-based algorithms use a set of hash functions to project the counts for every individual item onto a matrix of counters. By using multiple independent hash functions and recording each item arrival at several matrix locations, there is a high probability that we can retrieve a close estimate of the true frequency of a given item. COUNTMIN [24] uses d hash functions h_j and a matrix M with d rows of length w. Each hash function maps an item to one of its w columns. When an item arrives, the d different matrix elements are incremented. The estimated frequency for item x is the minimum of its d corresponding count values: $\hat{f}(x) = min_{1 \leq j \leq d} M[j, h_j(x)]$. If we set sizes $d = log 1/\delta$ and $w = 2/\epsilon$, we guarantee that $\hat{f}(x)$ has an error of at most ϵN with probability of at least $1 - \delta$. The COUNT sketch algorithm by Charikar et al. [17] is similar. It uses an additional set of hash functions to decide whether to increment or decrement, and the estimate is the median of the d values rather than the minimum.

3   Frequent Itemset Mining Algorithms

For static datasets, the classic method for finding frequent patterns is the Apriori approach [2]. However, even for static data, the Apriori method is inefficient for large datasets for two reasons: (1) it makes numerous passes over the data, and (2) it starts with a potentially large number of small candidate itemsets. Due to the single-pass constraint, most frequent pattern mining algorithms for streaming data produce settle for an appropriate set of frequent patterns. That is, we do not for certain that the result set is exactly equal to the true set of frequent patterns. The algorithms fall into two categories: those that produce false positive results and those that produce false negative results. A false positive algorithm guarantees that its result set includes every true frequent pattern, but it may include some additional ones. A false negative algorithm guarantees that every pattern it returns is frequent, but it may fail to detect some true frequent ones.

Table 9.3 Algorithms for frequent itemset mining of streaming data

Publication	Window[a]	Batch?	Accuracy	Algorithm/(D)ata structure
All itemsets				
Manku'02 [61]	L	b	False +	Lossy counting
Chang'03 [14]	D		False +	Estdec
Cheung'03 [21]	L		Exact	FELINE, CATS(D)
Giannella'03 [28]	T	b	False +	FP-stream(D)
Li'04 [52]	L	b	False +	DSM-FI, IsFI-Forest(D)
Yu'04 [77]	L	b	False −	FPDM
Jin'05 [40]	S	b	False +	StreamMining
Lin'05 [58]	S	b	False ±	PFP
Calders'07 [13]	A		Exact	
Raissi'07 [70]	T	b	False +	FIDS
Li'08 [55]	L	b	False +	DSM-FI
Ng'08 [67]	L	b	False +	CLCA
Li'09 [49]	S		Exact	MFI-transSW
Tanbeer'09 [71]	S,L,D	b	Exact	CPS-Tree(D)
Closed itemsets				
Chi'04 [23]	S		Exact	MOMENT
Jiang'06 [39]	S		Exact	CLI-Stream
Chen'07 [19]	S		exact	GC-Tree(D)
Cheng'08 [20]	S		False +	IncMine
Li'09 [56]	S		Exact	NewMoment
Liu'09 [59]	L	b	False +	FP-CLS
Gupta'10 [31]	D		Exact	CLICI
Maximal itemsets				
Lee'05 [45]	L,D		False ±	estDec+, Cp-Tree(D)
Li'05 [53]	L,S	b	False +	DSM-MSI, SFO-Forest(D)
Mao'07 [63]	L		Exact	INSTANT
Li'11 [50]	L	b	False −	FNMFIMoDS
Li'12 [57]	L		Exact	INSTANT+, FP-FOREST(D)
Frequent itemsets from uncertain data				
Leung'09 [46]	S	b	Exact w.r.t. expected supp.	SUF-Growth
Leung'11 [47, 48]	D,L	b	False +	TUF-Streaming
Hewa'12 [34]	L or D	b	False +	UHS-Stream, TFUHS-Stream
Top-K frequent itemsets				
Wong'06 [74]	L,S	b	False ±	
Patnaik'13 [68]	S	b	False ±	

[a]*L* landmark, *S* sliding window, *D* damped, *T* tilted time window, *A* all types

Tables 9.3 list a representative set of algorithms which have been proposed for mining frequent itemsets. The table is subdivided into five sections, each for a different category of pattern or data: (1) general itemsets, (2) closed itemsets, (3) maximal itemsets, (4) itemsets with uncertain data, and (5) top-K itemsets. The Window column indicates what type of window is supported: (L)andmark, (S)liding, (D)amped, (T)ilted time, or (A)ll. If the Batch column is marked, then a key step of the algorithm requires that transactions be processed in batches; unmarked columns indicate algorithms that can update after each individual transaction. The Accuracy column tells us whether the resulting set of frequent patterns will be exactly correct, may include false positive results, false negative results, or both. The last column provides the authors' names for their algorithms and data structures.

In the following, we examine a range of algorithms, to see the benefits and trade-offs involved for the different types of time windows, update intervals, and accuracy guarantees.

3.1 Mining the Full Data Stream

In the most basic version of frequent itemset mining, we seek to identify every itemset that occurs with a support level greater than θ, across the full history of the data stream. However, if we want exact results and if we are to consider the complete history, then it is necessary to record every arriving pattern, either directly on in some compressed format. If we fail to record even a few infrequent pattern occurrences, then we have miscounted. If a pattern later becomes more frequent, then the few occurrences that we skipped could make the difference between exceeding the θ threshold or not. However, to count every itemset can easily exceed the available memory. Therefore, several approximation techniques have been developed.

The approximation algorithms generate a *result set*, a set of itemsets which may or may not be exactly equal to all those whose support level exceeds θ. The algorithms fall into two categories: those that produce false positive results and those that produce false negative results. A false positive algorithm guarantees that its result set includes every true frequent pattern, but it may include some additional ones. A false negative algorithm guarantees that every pattern it returns is frequent, but it may fail to detect some true frequent ones.

Lossy Counting, A True Positive Algorithm Manku and Motwani proposed the first one-pass algorithm, LOSSY COUNTING, to find all frequent itemsets over a data stream [61]. Many of the algorithms developed since then still use the basic idea behind lossy counting. Their algorithm is *false positive oriented* in the sense that it does not allow false negatives, and it has a provable bound on false positives. It uses a user-defined error parameter ϵ to control the quality of the result set for a given support level θ. More precisely, its result set is guaranteed to have all itemsets whose frequency exceeds θ, and it contains no itemsets whose true frequency is less than $\theta - \epsilon$. In other words, the itemsets whose frequency are between $\theta - \epsilon$ and θ possibly appear in the result set and are the false positives.

The algorithm maintains a prefix tree T of potentially frequent patterns. As data are streaming as part of the kth bucket, every pattern is recorded. Patterns are recorded in tuple form: $\langle p, \hat{f}(p), err(p) \rangle$, where $\hat{f}(p)$ is the number of occurrences of pattern p since its inclusion in T, and $err(p) = k - 1$. This error is the number of buckets that have passed prior to the pattern being added to T. If a pattern is not yet in the tree, a new tuple is created. If a pattern is already in the tree, then $\hat{f}$ is incremented.

The tree is pruned at the conclusion of each bucket. A pattern is deleted from bucket i if $\hat{f}(p) < i - err(p)$. Recall that $err(p)$ relates to the bucket during which the pattern was first added to T. In other words, a pattern is pruned if $\hat{f}(p) < i - k + 1$.

The quantity $(i - k)$ is the number of buckets between when the pattern was added and the present, so the pruning rule amounts to saying that a pattern must occur on average once per bucket.

If and only if a pattern occurs on average once per bucket, LOSSY COUNTING will record it and not prune it. The rate of occurrence of a pattern is its support, and since the bucket size is chosen to be $1/\epsilon$, the minimum support level for not being pruned is exactly ϵ, the error rate. The converse rule is that if a pattern's average support is less than once per bucket, it will NOT be recorded. This is the "loss" in lossy counting and the foundation for the true positive guarantee: the estimated support undercounts the true support by no more that ϵ.

The main problem with LOSSY COUNTING is that it must record a relatively large amount of data. For example, suppose that $\theta = 0.10$ and $\epsilon = 0.01$. In order to guarantee that every reported frequent pattern has an actual support of at least 10–1 % = 9 %, LOSSY COUNTING would need to remember every pattern that occurs with only 1 % support. There may be orders of magnitude more patterns that satisfy a support level of ϵ vs. a support level of θ. Hence, numerous works have striven to reduce this burden, but using more sophisticated schemes that achieve a good error rate without imposing as large a memory requirement.

FPDM, A True Negative Algorithm In response the Manku and Motwani's work, Yu et al. propose a work which takes a much different tack [77]. Their algorithm does not allow false positives, and has a high-probability of finding itemsets which are truly frequent. In particular, they use a user-defined parameter δ to control the probability of finding frequent itemsets which satisfy support level θ. Specifically, the result set does not include any itemsets whose frequency is less than θ, whereas it includes any θ-frequent itemset with probability of at least $1 - \delta$. It utilizes the Chernoff bound to achieve such probabilistic quality guarantee.

We can model the appearance of an itemset as a binomial random variable, meaning that the pattern either appears or does not in each transaction. Our support threshold θ serves as the variable's expected value. If in n trials the actual number of appearances is $\hat{f}$, then the Chernoff bound states that

$$Pr\{|\hat{f}/n - \theta| \geq \epsilon\} \leq 2e^{\frac{-n\epsilon^2}{2\theta}} \tag{9.2}$$

The term on the left side of the inequality is the probability that the observed rate of occurrences will differ from the expected rate by at least a given error threshold. We define a parameter δ to be equal to the term on the right side of the inequality. That is, it is our target confidence level that our observed support is ϵ-close to the true support. Then by rearranging,

$$\epsilon_n = \sqrt{\frac{2\theta \ln(2/\delta)}{n}} \tag{9.3}$$

Equation 9.3 expresses the mutual dependence between these several parameters. For a fixed support threshold θ, the error ϵ_n between the true and observed support levels will diminish as n increases. The confidence factor δ is inversely proportional to the error rate and the necessary number of trials.

The FDPM algorithm uses this bound as follows. A pattern X is *potentially frequent* if $\hat{f}(X) \geq \theta - \epsilon_n$. For target levels θ and δ, Yu et al. show that the a required number of transactions per batch is

$$n_0 = \frac{2 + 2\ln(2/\delta)}{\theta} \tag{9.4}$$

For simplicity, we assume that the transaction sequence has been decomposed into an itemset sequence $I = \{I_1, I_2, I_3, \dots\}$.

Algorithm 6 Procedure FPDM-1(θ, δ) [77]

1: $n \leftarrow 0, T \leftarrow \emptyset$;
2: **while** there is another itemset I_i **do**
3:     **if** $I_i \in T$ **then**
4:         Increment $\hat{f}(I_i)$;
5:     **else**
6:         Add $\left\langle I_i, \hat{f}(I_i) \right\rangle$ to T;
7:         **if** $|T| > n_0$ **then**
8:             Calculate ϵ_n for each itemset;
9:             Prune itemsets that are not potentially frequent;
10: $n \leftarrow n + 1$;

The size of n_0 can be set by the amount of available memory. An interesting property of this algorithm is that when $n < n_0$, the result set will be exact. When n is just over n_0, then the error will be at a maximum given by Eq. 9.3. As n continues to increase, the error will tend to decrease.

Comparing Lossy Counting and FPDM Both algorithms logically partitioned the data stream into equally-sized segments and find the potentially frequent itemsets for each segment. They aggregate these locally frequent itemsets and further prune the infrequent ones. However, the number of transactions in each segment as well as the method to define potentially frequent itemsets is different for these two methods. In LOSSY COUNTING, the number of transactions in a segment is $\lceil 1/\epsilon \rceil$, and an itemset which occurs more than once in a segment is potentially frequent. In FDPM, the number of transactions in a segment is n_0, where n_0 is the required number of observations in order to achieve the Chernoff bound with the user-defined parameter δ.

To theoretically estimate the space requirement for both algorithms, we consider each transaction including only a single item, and the number of transactions in the entire data stream is $|D|$. LOSSY COUNTING will take $O(1/\epsilon log(\epsilon|D|))$ to find frequent items (1-itemsets). Thue, in order to reduce its error, Lossy Count will need to increase its memory usage. FPDM-1 (the simple version of FPDM on finding frequent items) will need $O((2 + 2ln(2/\delta))/\theta)$. If the user chooses to set the confidence level δ, then there is no direct control over the error rate. However, the error rate will decrease as n increases.

Note that different approaches have different advantages or drawbacks. For instance, for the false positive approach, if a second pass is allowed, we can easily eliminate false positives. For the false negative approach, we can have a small result set which is likely to contain almost all the frequent itemsets, but might miss some of them (with very small probability controlled by δ).

Finally, while LOSSY COUNTING works for any data stream, the probabilistic performance guarantee of FPDM only applies if the contents of each transaction is independent of each other. In real applications, this is often not true. The authors of FPDM [77] suggest random sampling from a data reservoir to alleviate this problem [25], at the cost of doubling the memory requirement.

3.2 Recently Frequent Itemsets

We now look at various ways to focus the selection of frequent itemsets on more recent data: the damped window model, the sliding window model, and the tilted-time model.

Damped Window Model Chang and Lee [14] study the problem of finding recently frequent itemsets over data streams using the damped window model. Specifically, in their model, the weight of previously recorded transactions in the data stream are periodically reduced by a decay factor d, where $0 < d \leq 1$. In their algorithm, this decayed weight is used counting the number of transactions and itemsets received. For example, the initial weight of a newly arrived transaction has weight 1. Suppose n_1 transactions arrive in the first time window and n_2 arrive in the second time window. At the end of the second window, the weighted count of transactions is $|D|_{t=2} = n_2 + n_1 d$. At the end of the third time window, we have $|D|_{t=3} = n_3 + n_2 d + n_1 d^2$. The weights of itemsets are discounted similarly. However, rather than updating the value of all stored itemsets every time period, a more efficient approach can be to update a value only when an itemset's weight needs to be read. To implement this we record both a count and the timestamp of the last update: $(\hat{f}(e), t(e))$. When a new instance of itemset e arrives at time t_k, we update the record as follows:

$$\hat{f}(e) \leftarrow d^{t_k - t(e)} \hat{f}(e) + 1 \tag{9.5}$$

$$t(e) \leftarrow t_k$$

Combining the damped window model with an method for estimating itemset counts, Chang and Lee produce the *estDec* algorithm. Their estimation method is similar to Carma [35]. Let e_m be an itemset contain m items. The Apriori principle tells us that the count for e_m cannot be greater than the count for any of its $(m - 1)$-subitemsets. Let $E_{m-1}(e)$ be the set of all these $(m - 1)$-sized subsets. This sets an upper bound:

$$f(e)^{max} = min\{f(a)|a \in E_{m-1}(e)\} \tag{9.6}$$

Then, noting relationships between unions and intersections of itemsets, they provide a lower limit bound:

$$f(e)^{min} = max\{f^{min}(a \cup b)|a, b \in E_{m-1}(e) \text{and} \neq b\} \tag{9.7}$$

They maintain an exact count of individual items and then use these bounds to estimate the count and error for itemsets that were fully recorded because they were infrequent at some timepoint. Their method, however, does not allow for the user to chose the size of the error, so it many be large. Moreover, errors can compound as *estDec* builds larger itemsets from smaller ones. The several set-based computations can also be expensive. This estimation method, however, is independent of the damped window method. Other works have also used the damped window but with different counting algorithms [31, 47, 71].

Sliding Windows Jin and Agarwal [40] introduce an algorithm that provides the same result quality guarantee as LOSSY COUNTING, but uses much less memory. Also similar to LOSSY COUNTING, it processes the data stream one batch at a time. It is presented as a sliding window method, but it could be modified to preserve results from one earlier batches to combine them with the current batch.

They propose a two-stage (single pass) *hybrid* approach. In the first stage, they employ a method for efficiently finding potentially frequent 2-itemsets that is more memory efficient that Aprori. Second, they apply the Apriori property to generate the potential i-itemsets, for $i > 2$. This approach finds a set of potential frequent itemsets, which is guaranteed to contain all the *true* frequent itemsets, in a single pass of the stream.

Stage one is an extension of the work by Karp, Papadimitriou and Shenker (KPS) on finding frequent elements (or 1-itemsets) [42]. Formally, given a sequence of length N and a threshold θ ($0 < \theta < 1$),

---

Algorithm 7 Procedure BoundedCount(Sequence S, θ, ϵ) [40]

1: Global $\mathcal{P} \leftarrow \emptyset$; // Set of potentially frequent items
2: $c \leftarrow 0$; // count of elimination rounds
3: **for all** $s \in S$ **do**
4:    **if** $s \in \mathcal{P}$ **then**
5:       s.count++;
6:    **else**
7:       $\mathcal{P} \leftarrow \{s\} \cup \mathcal{P}$;
8:       $s.count = 1$;
9:       **if** $|\mathcal{P}| \geq \lceil 1/\theta\epsilon \rceil$ **then**
10:          $c++$;
11:          **for all** $p \in \mathcal{P}$ **do**
12:             $p.count--$;
13:             **if** $p.count = 0$ **then**
14:                $\mathcal{P} \leftarrow \mathcal{P} - \{p\}$;
15: **for all** $p \in \mathcal{P}$ **do**
16:    **if** $p.count \leq (N\theta - c)$ **then**
17:       $\mathcal{P} \leftarrow \mathcal{P} - \{p\}$;
18: $Output(\mathcal{P})$;

---

the goal of their work is to determine the elements that occur with frequency greater than $N\theta$, without spending the memory to remember all N elements.

Suppose item e is frequent, meaning that it occurs at least $N\theta$ times in a dataset. The KPS method randomly removes $\lceil 1/\theta \rceil$ *unique* items from the dataset. This eliminates at most one instance of e. We can remove at most $\lceil \theta \rceil$ such sets, with the final set possibly being smaller than $\lceil 1/\theta \rceil$. It follows that the final set must contain e. However, there is no way to detect which members of the result set are frequent and which are not.

Jin and Agrawal enhance KPS to use a single pass and to implement an accuracy bound analogous to Manku and Motwani's work [61]. Besides reporting all items or itemsets that occur with frequency more than $N\theta$, they report only the items or itemsets which appear with frequency at least $N\theta(1 - \epsilon)$, where $0 < \epsilon \leq 1$. Algorithm 7 (BOUNDEDCOUNT outlines their method. In the first stage, we invoke the algorithm from Karp et al., with frequency level $\theta\epsilon$. $\mathcal{P}$ is the set of potentially frequent items. We maintain a *count* for each item in the set $\mathcal{P}$. This set is initially empty. As the algorithm processes a new item from a sequence, we check if it is in the set $\mathcal{P}$. If yes, its count is incremented, otherwise, it is inserted with a count of 1. When the size of the set $\mathcal{P}$ becomes larger than $\lceil 1/\theta\epsilon \rceil$, we decrement the count of each item in $\mathcal{P}$, and eliminate any item whose count has now become 0. This processing is equivalent to the removals we described earlier. We also record the number of elimination rounds, c, that occur. Clearly, $c \leq N\theta\epsilon$. In the second step, we remove all items whose reported frequency is less than $N\theta - c \geq N\theta(1 - \epsilon)$.

Algorithm 8 StreamMining(Stream $\mathcal{D}$, θ, ϵ) [40]

1: Global $X = 0$; // Number of 2-Itemsets
2: Global $N = 0$; // Number of transactions
3: Lattice $\mathcal{L} \leftarrow \emptyset$, Buffer $\mathcal{T} \leftarrow \emptyset$;
4: $c \leftarrow 0$; // number of elimination rounds
5: **for all** $t \in \mathcal{D}$ **do**
6:   $\mathcal{T} \leftarrow \mathcal{T} \cup \{t\}$;
7:   $Update(t, \mathcal{L}, 1)$;
8:   $Update(t, \mathcal{L}, 2)$;
9:   $f \leftarrow TwoItemsetPerTransaction(t)$;
10:   **if** $|\mathcal{L}_2| \geq \lceil 1/\theta\epsilon \rceil \cdot f$ **then**
11:     $ReducFreq(\mathcal{L}, 2)$;
12:     $c++, i \leftarrow 2$;
13:     **while** $\mathcal{L}_i \neq \emptyset$ **do**
14:       $i++$;
15:       **for all** $t \in \mathcal{T}$ **do**
16:         $Update(t, \mathcal{L}, i)$;
17:       $ReducFreq(\mathcal{L}, i)$;
18:     $\mathcal{T} \leftarrow \emptyset$;
19: **for all** $s \in \mathcal{L}$ **do**
20:   **if** $s.count \leq \theta|D| - c$ **then**
21:     $\mathcal{L}_i.delete(s)$;
22: $Output(\mathcal{L})$;

We now describe the complete hybrid algorithm. The algorithm is referred to as STREAMMINING and is illustrated in Algorithm 8 The algorithm has two inter-leaved phases. The first phase deals with 2-itemsets, and the second phase deals with k-itemsets, $k > 2$. The main procedure uses three subroutines, UPDATE, RE-DUCFREQ, and TWOITEMSETPERTRANSACTION, shown separately in Algorithm 9. The first phase extends the BOUNDEDCOUNT algorithm to deal with 2-itemsets. As we stated previously, the algorithm maintains a buffer $\mathcal{T}$ which stores the recently received transactions. Initially, the buffer is empty. When a new transaction t arrives, we put it in $\mathcal{T}$. Next, we call the UPDATE routine to increment counts in $\mathcal{L}_2$. This routine simply updates the count of 2-itemsets that are already in $\mathcal{L}_2$. If a new 2-itemset appears, it is inserted into $\mathcal{L}_2$.

Algorithm 9 Subprocedures for StreamMining

1: **Subprocedure** *Update*(Transaction t, Lattice $\mathcal{L}$, i)
2: **for all** i-subsets s of t **do**
3:    **if** $s \in \mathcal{L}_i$ **then**
4:       $s.count + +$;
5:    **else if** $i \leq 2$ or all $(i-1)$-subsets of $s \in \mathcal{L}_{i-1}$ **then**
6:       $\mathcal{L}_i.insert(s)$;

7: **Subprocedure** *ReducFreq*(Lattice $\mathcal{L}$, i)
8: **for all** i-itemsets $s \in \mathcal{L}_i$ **do**
9:    $s.count - -$;
10:    **if** $s.count = 0$ **then**
11:       $\mathcal{L}_i.delete(s)$;

12: **Function** *TwoItemsetPerTransaction*(Transaction t)
13: $N++$;
14: $X \leftarrow X + \binom{|t|}{2}$;
15: $f \leftarrow \lceil X/N \rceil$; // Average number of 2-Itemsets per tranactions
16: **if** $|\mathcal{L}_2| \geq \lceil 1/\theta\epsilon \rceil f$ **then**
17:    $N \leftarrow N - \lceil 1/\theta\epsilon \rceil$;
18:    $X \leftarrow X - \lceil 1/\theta\epsilon \rceil \cdot f$;
19: **return** f;

When the size of $\mathcal{L}_2$ is beyond the threshold $\lceil 1/\theta\epsilon \rceil f$, where f is a weighted average number of 2-itemsets per transaction, we call the procedure REDUCFREQ to reduce the count of each 2-itemsets in $\mathcal{L}_2$, and the itemsets whose count becomes zero are deleted. Invoking REDUCFREQ on $\mathcal{L}_2$ triggers the *second* phase.

The second phase of the algorithm deals with all k-itemsets, $k > 2$. This process is carried out level-wise, i.e, it proceeds from 3-itemsets to the largest potential frequent itemsets. For each transaction in the buffer $\mathcal{T}$, we enumerate all i-subsets.

For any i-subset that is already in $\mathcal{L}$, the process will be the same as for a 2-itemset, i.e, we will simply increment the count. However, an i-subset that is not in $\mathcal{L}$ will be inserted in $\mathcal{L}$ only if all of its $i - 1$ subsets are in $\mathcal{L}$ as well. Thus, we use the Apriori property.

After updating i-itemsets in $\mathcal{L}$, we again invoke the REDUCFREQ routine. Thus, the itemsets whose count is only 1 will be deleted from the lattice. This procedure will continue until there are no frequent k-itemsets in $\mathcal{L}$. At the end of this, we clear the buffer, and start processing new transactions in the stream. This will restart the first phase of our algorithm to deal with 2-itemsets.

In their experimental results, Jin and Agrawal show that STREAMMINING is more memory efficient than LOSSY COUNTING. On the T10.I4.N10K dataset used in Manku and Motwani's paper, with 1 million transactions and a support level of 1 %, LOSSY COUNTING requires an out-of-core data structure on top of a a 44 MB buffer. On the other hand, for larger datasets ranging from 4 million to 20 million transactions, Jin and Agrawal's algorithm only requires 2.5 MB in main memory.

Tilted Time Window Model Giannella et al. [28] employ a tilted time window model to record a compressed history of the entire data stream's frequent patterns. Its unique windowing structure allows it to approximately answer queries about frequent itemsets for a time window from any t_1 to t_2. The window size of the response is within 50 % of the size requested by the user.

The most unique feature are the time windows of growing width. Data are processed and recorded in batches of uniform size, but as time progresses, older records are merged together. This leverages the fact that users often want to know about recent history with fine granularity but are satisfied with a coarser granularity for longer or older time periods. Because the window sizes grow exponentially, the number of windows grows logarithmically, thus making the memory requirements tractable.

The candidate frequent patterns for each data batch are discovered using an FP-Tree with error factor ϵ, like the one in LOSSY COUNTING [61]. These candidate patterns and their counts are summarized in a table. $f(i, j)$ means the table of frequent patterns for the time interval $[t_i : t_j]$. As batches become older, they are merged with adjacent batches into windows with sizes that are powers of 2. If the total number of elapsed batches is 2^n for some integer n, then there will $n + 1$ windows of size 1, 1, 2, 4, 8, etc. When the total number of batches is not such a perfect number, there may be temporarily up to three windows covering just a single batch and up to two windows for larger sizes. For example, if 16 batches have been processed, then the windows would be sized as follows:

$$f(0, 7), f(8, 11), f(12, 13), f(14), f(15)$$

However, if only 15 batches have been processed, then these would be the sizes:

$$f(0, 3), f(4, 7), f(8, 9), f(10, 11), f(12), f(13), f(14)$$

Similar to how LOSSY COUNTING will delete an itemset if its support level fails below ϵ, the tilted time window algorithm (FP-STREAMING also drops infrequent itemsets, but the test condition is more complex. For an itemset I, its will be dropped from the time window $[0 : m]$ if

1. The support for I within each individual time window between m and n must be below θ.
2. The support for I within each subsequence of time windows between m and n must be below ϵ.

3.3 Closed and Maximal Itemsets

An alternative to mining for all frequent itemsets (FI) is to look for closed frequent itemsets (CFI) or maximal frequent itemsets (MFI). A frequent itemset I is *closed* if every proper superset of I has a lesser support; that is, $s(I') < s(I)$, for all $I' \supset I$. A frequent itemset I is *maximal* if every proper superset of I is not frequent, or $s(I') < \theta$, for all $I' \supset I$. MFIs sit at the boundary between frequent and infrequent itemsets. They are usually at least an order of magnitude fewer in number than FIs, so they offer substantial memory savings. However, one cannot recover the set of FIs from the set of MFIs, so there is a loss of information. CFIs, on the hand, are more numerous. While they do not offer as great a reduction in memory consumption, the set of CFIs and their supports completely specify the set of FIs and their supports. Thus, computing and storing only CFIs can represent a substantial memory and computational savings for frequent itemset mining.

Closed Frequent Itemsets Chi et al. produced MOMENT, the first algorithm for mining closed frequent itemsets from a data stream, choosing to use a sliding data window [23]. They utilize a *closed enumerated tree* (CET) with lexicographically ordered items, recording both CFIs and those itemsets at the boundary between CFIs and other itemsets. This boundary could change with each new transaction. In practice, however, itemsets tend to change status slowly, so the tree's structure does not change often. Figure 9.1 shows the CET that corresponds to the following data with $\theta = 0.5$: $T_1 = (C, D), T_2 = (A, B), T_3 = (A, B, C), T_4 = (A, B, C)$. To maintain the necessary information, the tree's nodes are classified into four different categories.

- Infrequent gateway node (solid circle): I is infrequent AND I is the union of I's frequent parent and another frequent itemset.
- Unpromising gateway node (dotted rectangle): I is frequent AND I has a descendant which is a CFI with the same support as I.
- Intermediate node (no border): I is frequent AND has a child with the same support AND I is not a unpromising gateway node.
- Closed node (solid rectangle): I is a closed frequent itemset.

By maintaining counts for these four types of itemsets, MOMENT is able to track the exact set of CFIs.

Fig. 9.1 Closed enumerate
tree example [23]

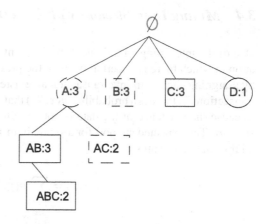

Several works [19, 20, 39, 49] have also mined for the exact set of CFIs with
a sliding window. CFI-STREAM [39] maintains a tree (DIU-tree) which records all
closed itemsets, without prejudging whether they are frequent or not. For this reason,
CFI-STREAM has two advantages: (1) It can respond to queries for any support
threshold θ, and (2) its performance is independent of θ. It outperforms MOMENT
when the support threshold is low. GC-TREE [19] exploits closure generators and a
total lexicographic ordering among all itemsets in order to compute closure more
efficiently. Each node in the tree is a 3-tuple: $\langle gen, eitem, clo \rangle$, where gen is the
closure generator, $eitem$ is the item used to extend gen to another closed itemset,
and clo is the closed itemset. In NEWMOMENT, each item's history of occurrence is
recorded as a w-bit binary vector, where w is the window width. Each time a new
transaction arrives, one simply pushes a new bit to the front and drops a bit from
the back. Bit sequences are computed for itemsets as well as items, and the itemset
histories are linked to each itemset in the frequent itemset tree. This enables more
efficient updates of support levels. As a consequence, NEWMOMENT runs faster and
uses less memory that MOMENT.

Cheng et al. [20] have introduced the notion of *semi-FCIs*, which progressively
increase the support threshold for an itemset as it resides in the window longer, to
efficiently compute an estimated set of CFIs. Furthermore, they use an inverted index
instead of a prefix tree to record semi-FCIs.

While most works have used the sliding window model, Liu et al. contribute FP-
CDS [59], which addresses the landmark window model, and Gupta et al. propose
CLICI [31], which uses the damped window model. Like LOSSY COUNTING, FP-
CDS uses batch processing and may produce false positive results within an error
threshold. CLICI uses a lattice rather than a tree structure. The lattice records all
closed itemsets, similar to FCI-STREAM.

3.4 Mining Data Streams with Uncertain Data

One of the interesting variations of the frequent itemset mining problem is the situation in which there is uncertainty about the presence of the data. We can model this by assigning each item in a transaction an existential probability. For an item x and transaction T_j, there is a probability $p(x, T_j)$ that x in fact is present in the transaction. Because the existence of the data is not certain, we can only compute an *expected support*. The expected support for a single item x is the sum of its probabilities over all transactions seem so far:

$$\hat{s}(x) = \sum_{j=1}^{n} p(x, T_j). \tag{9.8}$$

For an itemset $I = \{x_1, x_2, \ldots, x_m\}$, multiply the probabilities within each transaction:

$$\hat{s}(I) = \sum_{j=1}^{n} \prod_{x_i \in I} p(x_i, T_j). \tag{9.9}$$

There have been a small but important set of works investigating streaming uncertain data. Leung and Hao [46] extend the UF-GROWTH algorithm for static uncertain data to develop SUF-GROWTH, an exact mining algorithm. Like most exact algorithms, SUF-GROWTH employs the sliding window model to limit the memory needs. As each new transaction arrives, its probabilities are added to the pool, and those of the oldest transaction in the window are removed. The same work also presents an approximation algorithm UF-STREAMING which can produce quicker albeit less accurate results. Algorithms for the damped window and landmark window have also been developed [34, 47, 48].

A full discussion of algorithms and issues for mining uncertain data are provided in Chap. 14. In particular, we refer the reader to Sect. 8 for a presentation of algorithms for mining uncertain streaming data.

4 Mining Patterns Other than Itemsets

While the majority of research in frequent pattern mining in data streams has focused on itemsets, valuable and interesting contributions have been made for the discovery of other types of patterns. In this section, we discuss other types of patterns and introduce methods for mining them. These other patterns fall into three groups: subsequences, subtrees from trees, and subtrees and subgraphs from graphs. Selected works from each of these three categories are listed in Table 9.4.

Table 9.4 Non-itemset pattern mining algorithms for streaming data

Publication	Pattern	Window[a]	Batch?	Accuracy	Algorithm
Teng'03 [72]	Seq across window	S		False −	FTP-DS
Chang'05 [15]	Subseq	D		False +	eISeq
Raïssi'05 [69]	Subseq	T	b	False +	SPEED
Marascu'06 [62]	Subseq	T	b	False ±	SMDS
Ezeife'07 [27]	Subseq	L	b	False −	SSM
Chang'08 [16]	Closed subseq	S		Exact	SeqStream
Mendes'08 [64]	Subseq	L	b	False +	SS-BE, SS-MB
Koper'11 [43]	Subseq	L	b	False +	SS-BE2, SS-LC, SS-LC2
Asai'02 [6]	Subtree	D		False −	StreamT
Li'06 [54]	Subtree	L		False −	FQT-Stream
Bifet'08 [10]	Closed subtree	D	b	False −	AdaTreeNat
Bifet'10 [11]	Subgraph	D	b	False −	AdaGraphMiner
Aggarwal'10 [5]	Dense subgraph	L	b	False ±	

[a]*L* landmark, *S* sliding window, *D* damped, *T* tilted time window, *A* all types

4.1 Subsequences

In an early work, Teng et al. [72] defined a data stream in which each transaction is an itemset at a certain time and belonging to a particular customer. The algorithm can combine transactions belonging to the same customer to generate a sequence over time. Then, the problem is to find item sequences which occur frequently. Later authors, such as Raïssi et al. [69], assume that this sort of market basket data is preprocessed so that the algorithm is presented with a data stream in which each transaction is already an ordered sequence of items.

It is no longer necessary to merge compatible transactions to form sequences.

Raïssi et al. presented the first algorithm [69] for this data model. Their algorithm, SPEED, processes transactions in batches and uses the tilted time window model to compress older data. It maintains an item table, a sequence table, and a region tree. In the region tree, each vertex is a transaction, and the parent-child relationship is a supersequence-subsequence relationship.

SMDS [62] also uses the tilted time window mode. Its unique feature is that it incrementally computes sequence (transaction) clusters. The centroid of each cluster is the basis for frequent sequences: if a cluster has at least θ members, then the items of the centroid sequence which occur at least θ times define a frequent sequence. A prefix tree is then used to record these frequent sequences.

The damped window model is used in [15], which extends the authors' work on frequent itemsets [14]. If a pattern is not currently being monitored, its support is estimated from the support levels of its subpatterns which are being monitored.

The SSM algorithm [27] uses the landmark window model. There are three key data strutures. For each batch, SSM records all the input sequences and counts the frequency of each individual item in a hash table (D-List). Items with frequency less than ϵ are filtered out of the stored input sequences, leaving subsequences made from frequent items. These candidate sequences are used to construct a PLWAP Tree for the current batch, which in turn identifies actual frequent patterns for insertion in the FSP Tree.

Mendes et al. [64] develop two remarkably straightforward algorithms: SS-BE based on LOSSY COUNTING [61] and SS-MB based on SPACE-SAVING [65]. Like LOSSY COUNTING, they operate in the batch-processed landmark window domain. Recently, Koper and Nguyen [43] have improved the SS-BE algorithm, via changes to the pruning and other criteria. Chang et al. [16] have developed a method for discovering frequent *closed* sequences.

4.2 Subtrees and Semistructured Data

Subtrees are an interesting pattern because on one hand they are a special case of subgraphs and on the other hand they are the typical abstraction of semistructured data like XML. Mining frequent trees has many important applications, such text retrieval, web analysis, computer vision, bio- and chemical- informatics. In [6], the authors model semi-structured data and patterns as labeled ordered trees. They present an algorithm STREAMT which receives fragments of semi-structured data from a document of unknown total length through a data stream, which then returns the current set of frequent patterns upon request. This algorithm follows the damped window model, in which older items have exponentially less importance.

The data stream is assumed to be a sequence of labeled elements, each label l originating from a finite alphabet $(L) = \{1_1, 1_2, \ldots, l_k\}$. Each element may have nested within it a sequence of additional labeled elements. Each nested element may have its own nested elements. XML fits this model nicely. It is clear that this structure maps to a forest of rooted trees which are traversed depth first. Each root in the forest corresponds to the top-level element in the data stream. For convenience, The forest is treated as a single tree $\mathbf{T}$, in which the separate roots are each a child of a master root.

A pattern $\mathbf{P}$ is also a tree; every subtree of $\mathbf{T}$ is a pattern. Two patterns match if they are graph-isomorphic with matching node labels. Imagine a tree containing N nodes, $f(L)$ of which are labeled L. The pattern consisting of the single node L thus has support level $F(L)/N$. If half of the L nodes have a child with label M, then the two-node pattern $[L + child M]$ has support level $F(L)/2\,N$. When the tree becomes too large, infrequent branches may be pruned. The tree is exact up until the time of the first pruning. Because pruning removes some history, we may have false negative errors after that point.

Another work which seeks frequent trees from a XML data stream is [54]. Each transaction is a query in the form of an XML tree. We seek frequent common subtrees. Each query tree is converted to a standard linear format via a depth-first search. It is then added to a forest which records the accumulated query trees. The algorithm follows the landmark model, so the forest summarizes the full history; however, infrequent subtrees of the forest are periodically pruned.

4.3 Subgraphs

Discovering frequent subgraphs is a well-studied problem; however, there have been very few works which specifically seek frequent or important subgraphs from a stream of graphs.

In [11] Bifet et al. present what is believed to be the first algorithms for finding frequent subgraphs in a data stream of graphs. The objective is to find the *closed* subgraphs which occur at least as frequently as the minimum support threshold. To reduce the memory and computational requirements, they process the stream in batches and compute an approximate but compact *coreset* representation for the set of frequent closed graphs. Rather than recording the support of each pattern, they record the relative support, which is the difference in support between a closed graph and an its nearest subgraph. They offer three algorithms, INCGRAPHMINER, WINGRAPHMINER, and ADAGRAPHMINER. ADAGRAPHMINER is the best solution when there is concept drift in the data stream. It maintains a sliding window which adapts its width in response to the rate of concept drift, that is, whether the average value within the window has changed by more than a threshold parameter. Using the Hoeffdinger inequality, they guarantee that the number of false positive and false negative errors is bounded. Using Galois lattice theory, Bifet develops a more general methodology for identifying closed patterns in data streams [9].

In [5], the problem at hand is defined differently. Given a stream of graphs $G_1, G_2, \ldots$ which draw from a massive set of possible vertices, they detect the *frequent and significant* patterns. In particular, since the graphs will be sparse, their definition of significance draws upon two measures of density: node affinity and edge density. Given a vertex set P, let its *node affinity* $A(P)$ be the ratio of the number of graphs which contain all the vertices of P compared to the number which contain at least one vertex from P. The *edge density* $D(P)$ is defined as follows: among those graphs which fully contain P, count the average number of edges joining vertices in P and divide by the $\binom{|P|}{2}$ possible set of edges joining P. A subgraph P is (θ, γ)-significant if $A(P) \geq \theta$ and $D(P) \geq \gamma$.

The two-phase algorithm first computes patterns which have θ-affinity and then the subset of those with γ-density. Both phases of the algorithm use a min-hash approach which effectively summarizes the graph stream, estimating the probability that all members of a set have a property if some subset of them have that property. This transforms the node-affinity mining problem to a support-based mining problem. The overall algorithm maintains a set of min-hash statistics which are updated as new graphs stream in. The authors prove that their method enforces bounds on false positive and false negative errors.

5 Concluding Remarks

In this chapter, we have provided an overview of algorithms for frequent pattern mining over data streams. The streaming and unbounded nature of the data restricts us to a single pass over the data, which further complicates the already challenging task of frequent pattern mining. There are numerous variations of this mining task, including different data windowing approaches and different guarantees of the counting accuracy. Though many improvements in efficiency have been made over the years, the basic principle of LOSSY COUNTING remains one of the core approaches. To reduce the memory requirements, some algorithms seek only the maximal frequent itemsets or closed frequent itemsets. Because all or more frequent itemsets can be recovered from these, they act as a form of data compression. In these cases, it may be possible to discover an exact result set. Recently, several algorithms have considered the problem of uncertain data. In addition to frequent itemset mining, a few works have studied frequent subsequences and subgraphs. In the future, we can expect that all the forms of frequent pattern mining for fixed datasets will be investigated for streaming data as well.

One technique that has not yet been much studied is mining from a sampling of the data stream. This may be an interesting area for future work. Compared with existing sampling techniques [12, 18, 73] on disk-resident datasets for frequent itemset mining, sampling data streams brings some new issues. For example, the underlying distribution of the data stream can change from time to time. Therefore, sampling needs to adapt to the data stream. However, it is hard to accurately detect drift if we do not mine the set of frequent itemsets directly. In addition, the space requirement of the sample set can be an issue as well. As pointed by Manku and Motwani [60], methods similar to concise sampling [29] might be helpful to reduce the space and achieve better mining results.

References

1. R. Agrawal and R. Srikant. Fast algorithms for mining association rules. In *Proc. of Int. conf. Very Large DataBases (VLDB'94)*, pages 487–499, Santiago, Chile, September 1994.
2. Rakesh Agrawal and Ramakrishnan Srikant. Fast algorithms for mining association rules in large databases. In *Proceedings of the 20th International Conference on Very Large Data Bases*, VLDB '94, pages 487–499, San Francisco, CA, USA, 1994. Morgan Kaufmann Publishers Inc.
3. Rakesh Agrawal and Ramakrishnan Srikant. Mining sequential patterns. In *Data Engineering, 1995. Proceedings of the Eleventh International Conference on*, pages 3–14. IEEE, 1995.
4. R. Agrawal, H. Mannila, R. Srikant, H. Toivonent, and A. Inkeri Verkamo. Fast discovery of association rules. In U. Fayyad and et al, editors, *Advances in Knowledge Discovery and Data Mining*, pages 307–328. AAAI Press, Menlo Park, CA, 1996.
5. Charu C. Aggarwal, Yao Li, Philip S. Yu, and Ruoming Jin. On dense pattern mining in graph streams. *Proc. VLDB Endow.*, 3(1–2):975–984, September 2010.
6. Tatsuya Asai, Hiroki Arimura, Kenji Abe, Shinji Kawasoe, and Setsuo Arikawa. Online algorithms for mining semi-structured data stream. In *Data Mining, 2002. ICDM 2003. Proceedings. 2002 IEEE International Conference on*, pages 27–34. IEEE, 2002.
7. Tatsuya Asai, Kenji Abe, Shinji Kawasoe, Hiroki Arimura, and Setsuo Arikawa. Efficient algorithms for finding frequent substructures from semi-structured data streams. In *New Frontiers in Artificial Intelligence*, pages 29–45. Springer, 2007.
8. B. Babcock, S. Babu, M. Datar, R. Motwani, and J. Widom. Models and Issues in Data Stream Systems. In *Proceedings of the 2002 ACM Symposium on Principles of Database Systems (PODS 2002) (Invited Paper)*. ACM Press, June 2002.

9. Albert Bifet. Adaptive stream mining: Pattern learning and mining from evolving data streams. In *Proceedings of the 2010 conference on Adaptive Stream Mining: Pattern Learning and Mining from Evolving Data Streams*, pages 1–212, Amsterdam, The Netherlands, The Netherlands, 2010. IOS Press.

10. Albert Bifet and Ricard Gavaldà. Mining adaptively frequent closed unlabeled rooted trees in data streams. In *Proceedings of the 14th ACM SIGKDD international conference on Knowledge discovery and data mining*, KDD '08, pages 34–42, New York, NY, USA, 2008. ACM.

11. Albert Bifet, Geoff Holmes, Bernhard Pfahringer, and Ricard Gavaldà. Mining frequent closed graphs on evolving data streams. In *Proceedings of the 17th ACM SIGKDD international conference on Knowledge discovery and data mining*, KDD '11, pages 591–599, New York, NY, USA, 2011. ACM.

12. Hervé Brönnimann, Bin Chen, Manoranjan Dash, Peter Haas, and Peter Scheuermann. Efficient data reduction with ease. In *Proceedings of the Ninth ACM SIGKDD International Conference on Knowledge Discovery and Data Mining*, KDD '03, pages 59–68, New York, NY, USA, 2003. ACM.

13. Toon Calders, Nele Dexters, and Bart Goethals. Mining frequent itemsets in a stream. In *Data Mining, 2007. ICDM 2007. Seventh IEEE International Conference on*, pages 83–92. IEEE, 2007.

14. Joong Hyuk Chang and Won Suk Lee. Finding recent frequent itemsets adaptively over online data streams. In *Proceedings of the ninth ACM SIGKDD international conference on Knowledge discovery and data mining*, KDD '03, pages 487–492, New York, NY, USA, 2003. ACM.

15. Joong Hyuk Chang and Won Suk Lee. Efficient mining method for retrieving sequential patterns over online data streams. *J. Inf. Sci.*, 31(5):420–432, October 2005.

16. Lei Chang, Tengjiao Wang, Dongqing Yang, and Hua Luan. Seqstream: Mining closed sequential patterns over stream sliding windows. In *Proceedings of the 2008 Eighth IEEE International Conference on Data Mining*, ICDM '08, pages 83–92, Washington, DC, USA, 2008. IEEE Computer Society.

17. Moses Charikar, Kevin Chen, and Martin Farach-Colton. Finding frequent items in data streams. In *Automata, Languages and Programming*, pages 693–703. Springer, 2002.

18. Bin Chen, Peter Haas, and Peter Scheuermann. A new two-phase sampling based algorithm for discovering association rules. In *Proceedings of the Eighth ACM SIGKDD International Conference on Knowledge Discovery and Data Mining*, KDD '02, pages 462–468, New York, NY, USA, 2002. ACM.

19. Junbo Chen and ShanPing Li. Gc-tree:a fast online algorithm for mining frequent closed itemsets. In *Emerging Technologies in Knowledge Discovery and Data Mining*, pages 457–468. Springer, 2007.

20. James Cheng, Yiping Ke, and Wilfred Ng. Maintaining frequent closed itemsets over a sliding window. *Journal of Intelligent Information Systems*, 31(3):191–215, 2008.

21. William Cheung and Osmar R Zaiane. Incremental mining of frequent patterns without candidate generation or support constraint. In *Database Engineering and Applications Symposium, 2003. Proceedings. Seventh International*, pages 111–116. IEEE, 2003.

22. D.W. Cheung, J. Han, V. Ng, and C.Y. Wong. Maintenance of discovered association rules in large databases: An incremental updating techniques. In *Proc. 12th IEEE International Conference on Data Engineering (ICDE-96)*, New Orleans, Louisiana, U.S.A., March 1, 1996.

23. Yun Chi, Haixun Wang, Philip S. Yu, and Richard R. Muntz. Moment: Maintaining closed frequent itemsets over a stream sliding window. In *Proceedings of the Fourth IEEE International Conference on Data Mining*, ICDM '04, pages 59–66, Washington, DC, USA, 2004. IEEE Computer Society.

24. Graham Cormode and S. Muthukrishnan. An improved data stream summary: The count-min sketch and its applications. *J. Algorithms*, 55(1):58–75, April 2005.

25. Erik D. Demaine, Alejandro López-Ortiz, and J. Ian Munro. Frequency estimation of internet packet streams with limited space. In *Proceedings of the 10th Annual European Symposium on Algorithms*, ESA '02, pages 348–360, London, UK, UK, 2002. Springer-Verlag.

26. C. I. Ezeife and Yi Lu. Mining web log sequential patterns with position coded pre-order linked wap-tree. *Data Min. Knowl. Discov.*, 10(1):5–38, January 2005.
27. CI Ezeife and Mostafa Monwar. Ssm: a frequent sequential data stream patterns miner. In *Computational Intelligence and Data Mining, 2007. CIDM 2007. IEEE Symposium on*, pages 120–126. IEEE, 2007.
28. Chris Giannella, Jiawei Han, Jian Pei, Xifeng Yan, and Philip S Yu. Mining frequent patterns in data streams at multiple time granularities. *Next generation data mining*, 212:191–212, 2003.
29. Phillip B. Gibbons and Yossi Matias. New Sampling-Based Summary Statistics for Improving Approximate Query Answers. In *Proc. of the 1998 ACM SIGMOD*, pages 331–342. ACM Press, June 1998.
30. Bart Goethals and Mohammed J. Zaki. Workshop Report on Workshop on Frequent Itemset Mining Implementations (FIMI). 2003.
31. Anamika Gupta, Vasudha Bhatnagar, and Naveen Kumar. Mining closed itemsets in data stream using formal concept analysis. In *Proceedings of the 12th international conference on Data warehousing and knowledge discovery*, DaWaK'10, pages 285–296, Berlin, Heidelberg, 2010. Springer-Verlag.
32. J. Han, J. Pei, and Y. Yin. Mining frequent patterns without candidate generation. In *Proceedings of the ACM SIGMOD Conference on Management of Data*, 2000.
33. Jiawei Han, Hong Cheng, Dong Xin, and Xifeng Yan. Frequent pattern mining: current status and future directions. *Data Mining and Knowledge Discovery*, 15(1):55–86, 2007.
34. Chandima HewaNadungodage, Yuni Xia, Jaehwan John Lee, and Yi-cheng Tu. Hyper-structure mining of frequent patterns in uncertain data streams. *Knowledge and Information Systems*, 37(1):219–244, 2013.
35. C. Hidber. Online Association Rule Mining. In *Proceedings of ACM SIGMOD Conference on Management of Data*, pages 145–156. ACM Press, 1999.
36. Jochen Hipp, Ulrich Güntzer, and Gholamreza Nakhaeizadeh. Algorithms for association rule mining—a general survey and comparison. *SIGKDD Explor. Newsl.*, 2(1):58–64, June 2000.
37. Jun Huan, Wei Wang, Deepak Bandyopadhyay, Jack Snoeyink, Jan Prins, and Alexander Tropsha. Mining protein family-specific residue packing patterns from protein structure graphs. In *Eighth International Conference on Research in Computational Molecular Biology (RECOMB)*, pages 308–315, 2004.
38. Akihiro Inokuchi, Takashi Washio, and Hiroshi Motoda. An apriori-based algorithm for mining frequent substructures from graph data. In *Principles of Knowledge Discovery and Data Mining (PKDD2000)*, pages 13–23, 2000.
39. Nan Jiang and Le Gruenwald. Cfi-stream: mining closed frequent itemsets in data streams. In *Proceedings of the 12th ACM SIGKDD international conference on Knowledge discovery and data mining*, KDD '06, pages 592–597, New York, NY, USA, 2006. ACM.
40. Ruoming Jin and Gagan Agrawal. An algorithm for in-core frequent itemset mining on streaming data. In *Proceedings of the Fifth IEEE International Conference on Data Mining*, ICDM '05, pages 210–217, Washington, DC, USA, 2005. IEEE Computer Society.
41. Ruoming Jin, Chao Wang, Dmitrii Polshakov, Srini Parthasarathy, and Gagan Agrawal. Discovering frequent topological structures from graph datasets. In *KDD*, 2005.
42. Richard M. Karp, Christos H. Papadimitrious, and Scott Shanker. A Simple Algorithm for Finding Frequent Elements in Streams and Bags. Available from http://www.cs.berkeley.edu/christos/iceberg.ps, 2002.
43. Adam Koper and Hung Son Nguyen. Sequential pattern mining from stream data. In *Advanced Data Mining and Applications*, pages 278–291. Springer, 2011.
44. Michihiro Kuramochi and George Karypis. Frequent subgraph discovery. In *ICDM '01: Proceedings of the 2001 IEEE International Conference on Data Mining*, pages 313–320, 2001.
45. Daesu Lee and Wonsuk Lee. Finding maximal frequent itemsets over online data streams adaptively. In *Data Mining, Fifth IEEE International Conference on*, pages 8–pp. IEEE, 2005.
46. CK-S Leung and Boyu Hao. Mining of frequent itemsets from streams of uncertain data. In *Data Engineering, 2009. ICDE'09. IEEE 25th International Conference on*, pages 1663–1670. IEEE, 2009.

47. Carson Kai-Sang Leung and Fan Jiang. Frequent itemset mining of uncertain data streams using the damped window model. In *Proceedings of the 2011 ACM Symposium on Applied Computing*, SAC '11, pages 950–955, New York, NY, USA, 2011. ACM.
48. Carson Kai-Sang Leung and Fan Jiang. Frequent pattern mining from time-fading streams of uncertain data. In *Data Warehousing and Knowledge Discovery*, pages 252–264. Springer, 2011.
49. Hua-Fu Li and Suh-Yin Lee. Mining frequent itemsets over data streams using efficient window sliding techniques. *Expert Systems with Applications*, 36(2):1466–1477, 2009.
50. Haifeng Li and Ning Zhang. A false negative maximal frequent itemset mining algorithm over stream. In *Advanced Data Mining and Applications*, pages 29–41. Springer, 2011.
51. Wenmin Li, Jiawei Han, and Jian Pei. Cmar: Accurate and efficient classification based on multiple class-association rules. In *Proceedings of the 2001 IEEE International Conference on Data Mining*, ICDM '01, pages 369–376, Washington, DC, USA, 2001. IEEE Computer Society.
52. Hua-Fu Li, Suh-Yin Lee, and Man-Kwan Shan. An efficient algorithm for mining frequent itemsets over the entire history of data streams. In *Proc. of First International Workshop on Knowledge Discovery in Data Streams*, 2004.
53. Hua-Fu Li, Suh-Yin Lee, and Man-Kwan Shan. Online mining (recently) maximal frequent itemsets over data streams. In Research Issues in Data Engineering: Stream Data Mining and Applications, 2005. RIDE-SDMA 2005. 15th International Workshop on, pages 11–18. IEEE, 2005.
54. Hua-Fu Li, Man-Kwan Shan, and Suh-Yin Lee. Online mining of frequent query trees over xml data streams. In *Proceedings of the 15th international conference on World Wide Web*, pages 959–960. ACM, 2006.
55. Hua-Fu Li, Man-Kwan Shan, and Suh-Yin Lee. Dsm-fi: an efficient algorithm for mining frequent itemsets in data streams. *Knowledge and Information Systems*, 17(1):79–97, 2008.
56. Hua-Fu Li, Chin-Chuan Ho, and Suh-Yin Lee. Incremental updates of closed frequent itemsets over continuous data streams. *Expert Systems with Applications*, 36(2):2451–2458, 2009.
57. Haifeng Li, Ning Zhang, and Zhixin Chen. A simple but effective maximal frequent itemset mining algorithm over streams. *Journal of Software*, 7(1):25–32, 2012.
58. Chih-Hsiang Lin, Ding-Ying Chiu, Yi-Hung Wu, and Arbee L.P. Chen. Mining frequent itemsets from data streams with a time-sensitive sliding window. In *Proceedings of the Fifth SIAM International Conference on Data Mining*, volume 119, page 68. SIAM, 2005.
59. Xuejun Liu, Jihong Guan, and Ping Hu. Mining frequent closed itemsets from a landmark window over online data streams. *Comput. Math. Appl.*, 57(6):927–936, March 2009.
60. G. S. Manku and R. Motwani. Approximate Frequency Counts Over Data Streams. In *Proceedings of Conference on Very Large DataBases (VLDB)*, pages 346–357, September 2002.
61. Gurmeet Singh Manku and Rajeev Motwani. Approximate frequency counts over data streams. In *Proceedings of the 28th international conference on Very Large Data Bases*, VLDB '02, pages 346–357. VLDB Endowment, 2002.
62. Alice Marascu and Florent Masseglia. Mining sequential patterns from data streams: A centroid approach. *J. Intell. Inf. Syst.*, 27(3):291–307, November 2006.
63. Guojun Mao, Xindong Wu, Xingquan Zhu, Gong Chen, and Chunnian Liu. Mining maximal frequent itemsets from data streams. *Journal of Information Science*, 33(3):251–262, 2007.
64. Luiz F. Mendes, Bolin Ding, and Jiawei Han. Stream sequential pattern mining with precise error bounds. In *Proceedings of the 2008 Eighth IEEE International Conference on Data Mining*, ICDM '08, pages 941–946, Washington, DC, USA, 2008. IEEE Computer Society.
65. Ahmed Metwally, Divyakant Agrawal, and Amr El Abbadi. Efficient computation of frequent and top-k elements in data streams. In *Proceedings of the 10th International Conference on Database Theory*, ICDT'05, pages 398–412, Berlin, Heidelberg, 2005. Springer-Verlag.
66. Jayadev Misra and David Gries. Finding repeated elements. Technical report, Cornell University, Ithaca, NY, USA, 1982.

67. Willie Ng and Manoranjan Dash. Efficient approximate mining of frequent patterns over transactional data streams. In *Proceedings of the 10th international conference on Data Warehousing and Knowledge Discovery*, DaWaK '08, pages 241–250, Berlin, Heidelberg, 2008. Springer-Verlag.
68. Debprakash Patnaik, Srivatsan Laxman, Badrish Chandramouli, and Naren Ramakrishnan. A general streaming algorithm for pattern discovery. *Knowledge and Information Systems*, pages 1–26, 2013.
69. Chedy Raïssi, Pascal Poncelet, and Maguelonne Teisseire. Need for SPEED: Mining sequential patterns in data streams. *BDA'05: Bases de données Avanées Actes*, 2005.
70. Chedy Raïssi, Pascal Poncelet, and Maguelonne Teisseire. Towards a new approach for mining frequent itemsets on data stream. *Journal of Intelligent Information Systems*, 28(1):23–36, 2007.
71. Syed Khairuzzaman Tanbeer, Chowdhury Farhan Ahmed, Byeong-Soo Jeong, and Young-Koo Lee. Sliding window-based frequent pattern mining over data streams. *Information sciences*, 179(22):3843–3865, 2009.
72. Wei-Guang Teng, Ming-Syan Chen, and Philip S. Yu. A regression-based temporal pattern mining scheme for data streams. In *Proceedings of the 29th international conference on Very large data bases-Volume 29*, VLDB '03, pages 93–104. VLDB Endowment, 2003.
73. H. Toivonen. Sampling large databases for association rules. In *Proc. of the 22nd VLDM Conference.*, 1996.
74. Raymond Chi-Wing Wong and Ada Wai-Chee Fu. Mining top-k frequent itemsets from data streams. *Data Mining and Knowledge Discovery*, 13(2):193–217, 2006.
75. Dong Xin, Jiawei Han, Xifeng Yan, and Hong Cheng. Mining compressed frequent-pattern sets. In *VLDB*, 2005.
76. Xifeng Yan and Jiawei Han. gspan: Graph-based substructure pattern mining. In *ICDM '02: Proceedings of the 2002 IEEE International Conference on Data Mining (ICDM'02)*, page 721, 2002.
77. Jeffery Xu Yu, Zhihong Chong, Hongjun Lu, and Aoying Zhou.Li False positive or false negative: mining frequent itemsets from high speed transactional data streams. In *Proceedings of the Thirtieth international conference on Very large data bases-Volume 30*, VLDB '04, pages 204–215. VLDB Endowment, 2004.
78. Mohammed J. Zaki. Efficiently mining frequent trees in a forest. In *KDD '02: Proceedings of the eighth ACM SIGKDD international conference on Knowledge discovery and data mining*, pages 71–80, 2002.
79. M. J. Zaki, S. Parthasarathy, M. Ogihara, and W. Li. New algorithms for fast discovery of association rules. In *3rd Intl. Conf. on Knowledge Discovery and Data Mining.*, August 1997.
80. Mohammed J. Zaki and Charu C. Aggarwal. Xrules: an effective structural classifier for xml data. In *KDD '03: Proceedings of the ninth ACM SIGKDD international conference on Knowledge discovery and data mining*, pages 316–325, 2003.

Chapter 10
Big Data Frequent Pattern Mining

David C. Anastasiu, Jeremy Iverson, Shaden Smith and George Karypis

Abstract Frequent pattern mining is an essential data mining task, with a goal of discovering knowledge in the form of repeated patterns. Many efficient pattern mining algorithms have been discovered in the last two decades, yet most do not scale to the type of data we are presented with today, the so-called "Big Data". Scalable parallel algorithms hold the key to solving the problem in this context. In this chapter, we review recent advances in parallel frequent pattern mining, analyzing them through the Big Data lens. We identify three areas as challenges to designing parallel frequent pattern mining algorithms: memory scalability, work partitioning, and load balancing. With these challenges as a frame of reference, we extract and describe key algorithmic design patterns from the wealth of research conducted in this domain.

Keywords Data mining · Parallel algorithms · Frequent pattern mining · Frequent sequence mining · Frequent graph mining · Motif discovery · Memory scalability · Work partitioning · Load balancing

1 Introduction

As an essential data mining task, frequent pattern mining has applications ranging from intrusion detection and market basket analysis, to credit card fraud prevention and drug discovery. Many efficient pattern mining algorithms have been discovered in the last two decades, yet most do not scale to the type of data we are presented with

D. C. Anastasiu (✉) · J. Iverson · S. Smith · G. Karypis
Department of Computer Science and Engineering,
University of Minnesota, Minneapolis, USA
e-mail: dragos@cs.umn.edu

J. Iverson
e-mail: jiverson@cs.umn.edu

S. Smith
e-mail: shaden@cs.umn.edu

G. Karypis
e-mail: karypis@cs.umn.edu

C. C. Aggarwal, J. Han (eds.), *Frequent Pattern Mining,*   225
DOI 10.1007/978-3-319-07821-2_10, © Springer International Publishing Switzerland 2014

today, the so-called "Big Data". Web log data from social media sites such as Twitter produce over 100 TB of raw data daily [32]. Giants such as Walmart register billions of yearly transactions [5]. Today's high-throughput gene sequencing platforms are capable of generating terabytes of data in a single experiment [16]. Tools are needed that can effectively mine frequent patterns from these massive data in a timely manner.

Some of today's frequent pattern mining source data may not fit on a single machine's hard drive, let alone in its volatile memory. The exponential nature of the solution search space compounds this problem. Scalable parallel algorithms hold the key to addressing pattern mining in the context of Big Data. In this chapter, we review recent advances in solving the frequent pattern mining problem in parallel. We start by presenting an overview of the frequent pattern mining problem and its specializations in Sect. 2. In Sect. 3, we examine advantages of and challenges encountered when parallelizing algorithms, given today's distributed and shared memory systems, centering our discussion in the frequent pattern mining context. We survey existing serial and parallel pattern mining methods in Sects. 4–6. Finally, Sect. 7 draws some conclusions about the state-of-the-art and further opportunities in the field.

2 Frequent Pattern Mining: Overview

Since the well-known itemset model was introduced by Agrawal and Srikant [1] in 1994, numerous papers have been published proposing efficient solutions to the problem of discovering frequent patterns in databases. Most follow two well known paradigms, which we briefly describe in this section, after first introducing notation and concepts used throughout the paper.

2.1 Preliminaries

Let $I = \{i_1, i_2, \ldots, i_n\}$ be a set of items. An **itemset** C is a subset of I. We denote by $|C|$ its *length* or *size*, i.e. the number of items in C. Given a list of transactions $\mathcal{T}$, where each transaction $T \in \mathcal{T}$ is an itemset, $|\mathcal{T}|$ denotes the total number of transactions. Transactions are generally identified by a transaction id (*tid*). The **support** of C is the proportion of transactions in $\mathcal{T}$ that contain C, i. e., $\phi(C) = |\{T | T \in \mathcal{T}, C \subseteq T\}|/|\mathcal{T}|$. The *support count*, or *frequency* of C is the number of transactions in $\mathcal{T}$ that contain C. An itemset is said to be a **frequent itemset** if it has a support greater than some user defined minimum support threshold, σ.

The itemset model was extended to handle sequences by Srikant and Agrawal [54]. A **sequence** is defined as an ordered list of itemsets, $s = \langle C_1, C_2, \ldots, C_l \rangle$, where $C_j \subseteq I, 1 \leq j \leq l$. A sequence database $\mathcal{D}$ is a list of $|\mathcal{D}|$ sequences, in which each sequence may be associated with a customer id and elements in the sequence may have an assigned timestamp. Without loss of generality, we assume a lexicographic order of items $i \in C$, $C \subseteq I$. We assume sequence elements are ordered in non-decreasing order based on their timestamps. A sequence $s' = \langle C'_1, C'_2, \ldots, C'_m \rangle$, $m \leq l$ is a **sub-sequence** of s if there exist integers $i_1, i_2, \ldots, i_m$ s.t. $1 \leq i_1 \leq i_2 \leq \ldots \leq i_m \leq l$ and $C'_j \subseteq C_{i_j}$, $j = 1, 2, \ldots, m$. In words, itemsets in s' are subsets of those in s and follow the same list order as in s. If s' is a sub-sequence of s, we write that $s' \subseteq s$ and say that s *contains* s' and s is a **super-sequence** of s'. Similar to the itemset support, the **support** of s is defined as the proportion of sequences in $\mathcal{D}$ that contain s, i. e., $\phi(s) = |\{s'|s' \in \mathcal{D}, s \subseteq s'\}|/|\mathcal{D}|$. A sequence is said to be a **frequent sequence** if it has a support greater than σ.

A similar model extension has been proposed for mining structures, or graphs/networks. We are given a set of graphs $\mathcal{G}$ of size $|\mathcal{G}|$. Graphs in $\mathcal{G}$ typically have labelled edges and vertices, though this is not required. $V(G)$ and $E(G)$ represent the vertex and edge sets of a graph G, respectively. The graph $G = (V(G), E(G))$ is said to be a **subgraph** of another graph $H = (V(H), E(H))$ if there is a bijection from $E(G)$ to a subset of $E(H)$. The relation is noted as $G \subseteq H$. The **support** of G is the proportion of graphs in $\mathcal{G}$ that have G as a subgraph, i.e., $\phi(G) = |\{H|H \in \mathcal{G}, G \subseteq H\}|/|\mathcal{G}|$. A graph is said to be a **frequent graph** if it has a support greater than σ.

The problem of frequent pattern mining (FPM) is formally defined as follows. Its specialization for the frequent itemset mining (FIM), frequent sequence mining (FSM), and frequent graph mining (FGM) is straight-forward.

Definition 1 *Given a pattern container $\mathcal{P}$ and a user-specified parameter σ ($0 \leq \sigma \leq 1$), find all sub-patterns each of which is supported by at least $\lceil \sigma |\mathcal{P}| \rceil$ patterns in $\mathcal{P}$.*

At times, we may wish to restrict the search to only **maximal** or **closed** patterns. A maximal pattern m is not a sub-pattern of any other frequent pattern in the database, whereas a closed pattern c has no proper super-pattern in the database with the same support.

A number of variations of the frequent sequence and frequent graph problems have been proposed. In some domains, the elements in a sequence are symbols from an alphabet $\mathcal{A}$, e.g., $\mathcal{A} = \{A, C, G, T\}$ and $s = \langle TGGTGAGT \rangle$. We call these sequences *symbol sequences*. The symbol sequence model is equivalent to the general itemset sequence model where $|C| = 1$ for all $C \in s$, $s \in \mathcal{D}$. Another interesting problem, *sequence motif mining*, looks to find frequent sub-sequences within one (or a few) very long sequences. In this case, the support threshold is given as a support count, the minimum number of occurrences of the sub-sequence, rather than a value $0 \leq \sigma \leq 1$, and additional constraints may be specified, such as minimum/maximum sub-sequence length. A similar problem is defined for graphs, unfortunately also called *frequent graph mining* in the literature, where the support of G is the number of *edge-disjoint* subgraphs in a large graph $\mathcal{G}$ that are isomorphic to G. Two subgraphs are edge-disjoint if they do not share any edges. We call each appearance of G in $\mathcal{G}$ an *embedding*. Two graphs G and H are isomorphic if there exists a bijection between their vertex sets, $f : V(G) \rightarrow V(H)$, s.t. any two vertices $u, v \in V(G)$ are adjacent in G if and only if $f(u)$ and $f(v)$ are adjacent in H.

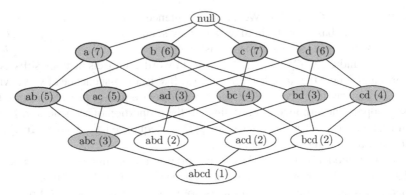

Fig. 10.1 An itemset lattice for the set of items $I = \{a, b, c, d\}$. Each node is a candidate itemset with respect to transactions in Table 10.1. For convenience, we include each itemset frequency. Given $\sigma = 0.5$, tested itemsets are shaded *gray* and frequent ones have *bold borders*

Table 10.1 Example transactions with items from the set $I = \{a, b, c, d\}$	Tid	Items
	1	a, b, c
	2	a, b, c
	3	a, b, d
	4	a, b
	5	a, c
	6	a, c, d
	7	c, d
	8	b, c, d
	9	a, b, c, d
	10	d

2.2 Basic Mining Methodologies

Many sophisticated frequent itemset mining methods have been developed over the years. Two core methodologies emerge from these methods for reducing computational cost. The first aims to prune the candidate frequent itemset search space, while the second focuses on reducing the number of comparisons required to determine itemset support. While we center our discussion on frequent itemsets, the methodologies noted in this section have also been used in designing FSM and FGM algorithms, which we describe in Sects. 5 and 6, respectively.

2.2.1 Candidate Generation

A brute-force approach to determine frequent itemsets in a set of transactions is to compute the support for every possible candidate itemset. Given the set of items I and a partial order with respect to the subset operator, one can denote all possible candidate itemsets by an *itemset lattice*, in which nodes represent itemsets and edges correspond to the subset relation. Figure 10.1 shows the itemset lattice containing candidate itemsets for example transactions denoted in Table 10.1. The brute-force approach would compare each candidate itemset with every transaction $C \in \mathcal{T}$ to check for containment. An approach like this would require $O(|\mathcal{T}| \cdot L \cdot |I|)$ item comparisons, where the number of non-empty itemsets in the lattice is $L = 2^{|I|} - 1$. This type of computation becomes prohibitively expensive for all but the smallest sets of items and transaction sets.

One way to reduce computational complexity is to reduce the number of candidate itemsets tested for support. To do this, algorithms rely on the observation that every candidate itemset of size k is the union of two candidate itemsets of size $(k - 1)$, and on the converse of the following lemma.

Lemma 10.1 (Downward Closure) *The subsets of a frequent itemset must be frequent.*

Conversely, the supersets of an infrequent itemset must be infrequent. Thus, given a sufficiently high minimum support, there are large portions of the itemset lattice that do not need to be explored. None of the white nodes in Fig. 10.1 must be tested, as they do not have at least two frequent parent nodes. This technique is often referred to as *support-based pruning* and was first introduced in the Apriori algorithm by Agrawal and Srikant [1].

Algorithm 10 Frequent itemset discovery with Apriori.

1: $k = 1$
2: $F_k = \{i | i \in I, \phi(\{i\}) \geq \sigma\}$
3: **while** $F_k \neq \emptyset$ **do**
4:    $k = k + 1$
5:    $F_k = \{C | C \in F_{k-1} \times F_{k-1}, |C| = k, \phi(C) \geq \sigma\}$
6: $Answer = \bigcup F_k$

Algorithm 10 shows the pseudo-code for Apriori-based frequent itemset discovery. Starting with each item as an itemset, the support for each itemset is computed, and itemsets that do not meet the minimum support threshold σ are removed. This results in the set $F_1 = \{i | i \in I, \phi(\{i\}) \geq \sigma\}$ (line 2). From F_1, all candidate itemsets of size two can be generated by joining frequent itemsets of size one, $F_2 = \{C | C \in F_1 \times F_1, |C| = 2, \phi(C) \geq \sigma\}$. In order to avoid re-evaluating itemsets of size one, only those sets in the Cartesian product which have size two are checked. This process can be generalized for all $F_k, 2 \leq k \leq |I|$ (line 5). When $F_k = \emptyset$, all frequent itemsets have been discovered and can be expressed as the union of all frequent itemsets of size no more than k, $F_1 \cup F_2 \cup \cdots \cup F_k$ (line 7).

In practice, the candidate generation and support computation step (line 5) can be made efficient with the use of a *subset function* and a hash tree. Instead of computing the Cartesian product, $F_{k-1} \times F_{k-1}$, we consider all subsets of size k within all transactions in $\mathcal{T}$. A subset function takes as input a transaction and returns all its subsets of size k, which become candidate itemsets. A hash tree data structure can be used to efficiently keep track of the number of times each candidate itemset is

Fig. 10.2 Prefix tree showing
prefix-based 1-length
equivalence classes in the
itemset lattice for
$I = \{a, b, c, d\}$

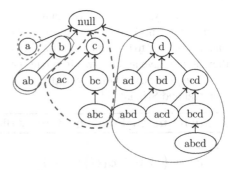

Fig. 10.3 Suffix tree showing
suffix-based 1-length
equivalence classes in the
itemset lattice for
$I = \{a, b, c, d\}$

encountered in the database, i.e. its support count. Details for the construction of the
hash tree can be found in the work of Agrawal and Srikant [1].

2.2.2 Pattern Growth

Apriori-based algorithms process candidates in a breath-first search manner, de-
composing the itemset lattice into level-wise itemset-size based equivalence classes:
k-itemsets must be processed before $(k + 1)$-itemsets. Assuming a lexicographic ordering of itemset items, the search space can also be decomposed into prefix-based
and suffix-based equivalence classes. Figures 10.2 and 10.3 show equivalence classes
for 1-length itemset prefixes and 1-length itemset suffixes, respectively, for our test
database. Once frequent 1-itemsets are discovered, their equivalence classes can
be mined independently. Patterns are *grown* by appending (prepending) appropriate
items that follow (precede) the parent's last (first) item in lexicographic order.

Zaki [63] was the first to suggest prefix-based equivalence classes as a means of
independent sub-lattice mining in his algorithm, Equivalence CLAss Transformation (ECLAT). In order to improve candidate support counting, Zaki transforms the
transactions into a *vertical database* format. In essence, he creates an inverted index,
storing, for each itemset, a list of tids it can be found in. Frequent 1-itemsets are
then those with at least $\lceil \sigma |\mathcal{T}| \rceil$ listed tids. He uses lattice theory to prove that if two

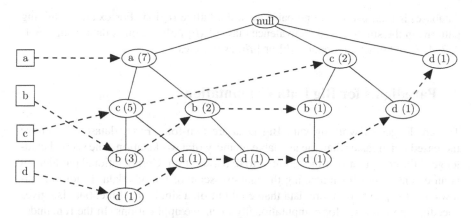

Fig. 10.4 The FP-tree built from the transaction set in Table 10.1

itemsets C_1 and C_2 are frequent, so will their intersection set $C_1 \cap C_2$ be. After creating the vertical database, each equivalence class can be processed independently, in either breath-first or depth-first order, by recursive intersections of candidate itemset tid-lists, while still taking advantage of the downward closure property. For example, assuming $\{b\}$ is infrequent, we can find all frequent itemsets having prefix a by intersecting tid-lists of $\{a\}$ and $\{c\}$ to find support for $\{ac\}$, then tid-lists of $\{ac\}$ and $\{d\}$ to find support for $\{acd\}$, and finally tid-lists of $\{a\}$ and $\{d\}$ to find support for $\{ad\}$. Note that the $\{ab\}$-rooted subtree is not considered, as $\{b\}$ is infrequent and will thus not be joined with $\{a\}$.

A similar divide-and-conquer approach is employed by Han et al. [21] in FP-growth, which decomposes the search space based on length-1 suffixes. Additionally, they reduce database scans during the search by leveraging a compressed representation of the transaction database, via a data structure called an FP-tree. The FP-tree is a specialization of a prefix-tree, storing an item at each node, along with the support count of the itemset denoted by the path from the root to that node. Each database transaction is mapped onto a path in the tree. The FP-tree also keeps pointers between nodes containing the same item, which helps identify all itemsets ending in a given item. Figure 10.4 shows an FP-tree constructed for our example database. Dashed lines show item-specific inter-node pointers in the tree.

Since the ordering of items within a transaction will affect the size of the FP-tree, a heuristic attempt to control the tree size is to insert items into the tree in non-increasing frequency order, ignoring infrequent items. Once the FP-tree has been generated, no further passes over the transaction set are necessary. The frequent itemsets can be mined directly from the FP-tree by exploring the tree from the bottom-up, in a depth-first fashion.

A concept related to that of equivalence class decomposition of the itemset lattice is that of *projected databases*. After identifying a region of the lattice that can be mined independently, a subset of $\mathcal{T}$ can be retrieved that only contains itemsets represented in that region. This subset, which may be much smaller than the original database, is then used to mine patterns in the lattice region. For example, mining patterns in the suffix-based equivalence class of $\{b\}$ only requires data from tids 1, 2, 3, 4, 8, and 9, which contain $\{b\}$ or $\{ab\}$ as prefixes.

3 Paradigms for Big Data Computation

The challenges of working with Big Data are two-fold. First, dataset sizes have increased much faster than the available memory of a workstation. The second challenge is the computation time required to find a solution. Computational parallelism is an essential tool for managing the massive scale of today's data. It not only allows one to operate on more data than could fit on a single machine, but also gives speedup opportunities for computationally intensive applications. In the remainder of this section we briefly discuss the principles of parallel algorithm design, outlining some challenges specific to the frequent pattern mining problem, and then detail general approaches for addressing these challenges in shared and distributed memory systems.

3.1 Principles of Parallel Algorithms

Designing a parallel algorithm is not an easy prospect. In addition to all of the challenges associated with serial algorithm design, there are a host of issues specific to parallel computation that must be considered. We will briefly discuss the topics of memory scalability, work partitioning, and load balancing. For a more comprehensive look at parallel algorithm design, we refer the reader to Grama et al. [17].

As one might imagine, extending the serial FIM methods to parallel systems need not be difficult. For example, a serial candidate generation based algorithm can be made parallel by replicating the list of transactions $\mathcal{T}$ at each process, and having each process compute the support for a subset of candidate itemsets in a globally accessible hash tree. These "direct" extensions however, rely on assumptions like unlimited process memory and concurrent read/concurrent write architecture, which ignore the three challenges outlined at the outset of this section.

One of the key factors in choosing to use parallel algorithms in lieu of their serial counterparts is data that is too large to fit in memory on a single workstation. Even while the input dataset may fit in memory, intermediary data such as candidate patterns and their counts, or data structures used during frequent pattern mining, may not. *Memory scalability* is essential when working with Big Data as it allows an application to cope with very large datasets by increasing parallelism. We call an algorithm memory scalable if the required memory per process is a function of $\Theta(\frac{n}{p}) + O(p)$, where n is the size of the input data and p is the number of processes executed in parallel. As the number of processes grows, the required amount of memory per process for a memory scalable algorithm decreases. A challenge in designing parallel FPM algorithms is thus finding ways to split both the input and intermediary data across all processes in such a way that no process has more data than it can fit in memory.

A second important challenge in designing a successful parallel algorithm is to decompose the problem into a set of *tasks*, where each task represents a unit of work, s.t. tasks are independent and can be executed concurrently, in parallel. Given these independent tasks, one must devise a *work partitioning*, or *static load balancing* strategy, to assign work to each process. A good work partitioning attempts to assign equal amounts of work to all processes, s.t. all processes can finish their computation at the same time. For example, given an $n \times n$ matrix, an $n \times 1$ vector, and p processes, a good work partitioning for the dense matrix-vector multiplication problem would be to assign each process n/p elements of the output vector. This assignment achieves the desired goal of equal loads for all processes. Unlike this problem, FPM is composed of inherently irregular tasks. FPM tasks depend on the type and size of objects in the database, as well as the chosen minimum support threshold σ. An important challenge is then to correctly gauge the amount of time individual tasks are likely to take in order to properly divide tasks among processes.

A parallel application is only as fast as its slowest process. When the amount of work assigned to a process cannot be correctly estimated, work partitioning can lead to a load imbalance. *Dynamic load balancing* attempts to minimize the time that processes are idle by *actively* distributing work among processes. Given their irregular tasks, FPM algorithms are prime targets for dynamic load balancing. The challenge of achieving good load balance becomes that of identifying points in the algorithm execution when work can be re-balanced with little or no penalty.

3.2   Shared Memory Systems

When designing parallel algorithms, one must be cognizant of the memory model they intend to operate under. The choice of memory model determines how data will be stored and accessed, which in turn plays a direct role in the design and performance of a parallel algorithm. Understanding the characteristics of each model and their associated challenges is key in developing scalable algorithms.

Shared memory systems are parallel machines in which processes share a single memory address space. Programming for shared memory systems has become steadily more popular in recent years due to the now ubiquitous multi-core workstations. A major advantage of working with shared memory is the ease of communication between processes. Any cooperation that needs to take place can be accomplished by simply changing values in memory. Global data structures may be constructed and accessed with ease by all processes.

Designing algorithms for shared memory systems comes with its own set of challenges. A consequence of shared memory programming is the need to acknowledge hardware details such as cache size and concurrent memory accesses. If two processes attempt to write to the same memory address at the same time, one may overwrite the other, resulting in a *race condition*. Moreover, a particular memory address may not necessarily be located on the same physical machine that is running the process, incurring large network delays.

With an appropriately sized dataset or a sufficiently large shared memory system, FPM input and intermediary data can be made to fit in memory. In this case, the challenge is that of structuring mining tasks in such a way as to minimize processes contention for system resources. To achieve this, processes could cooperate in counting support by assigning each process a distinct subset of the database and having them compute support for all candidate itemsets with respect to their portion of the database. This, however, can lead to race conditions when multiple processes increment counts for the same candidate itemset. A better approach decomposes the candidate itemset lattice into disjoint regions based on equivalence classes, as described in Sect. 2, and assigns distinct regions of the lattice to processes. In this way, no two processes will be responsible for updating the count of the same itemset, eliminating the possibility of race conditions.

3.3   Distributed Memory Systems

In a distributed (*share-nothing*) memory system, processes only have access to a local private memory address space. Sharing input data and task information must be done explicitly through inter-process communication. While processes most often communicate through network transmissions, they can also exchange information by writing and reading files on a shared file system. There are two programming paradigms commonly used for distributed memory systems. *Message passing* is a classic model that has been used very successfully in the scientific computing community for several decades. *MapReduce* is a recent paradigm, developed by Dean and Ghemawat [13] and designed specifically for working with Big Data applications.

3.3.1   Message Passing

The message passing paradigm implements the *ACTOR model of computation* [23], which is characterized by inherent computation concurrency within and among dynamically spawned actors. In the message passing paradigm, processes are actors, and they interact only through direct message passing. Messages are typically sent over a network connection and their transmission is thus affected by available inter-process network bandwidth. An advantage of the message passing paradigm is that the developer has total control over the size and frequency of process communication. Since all communication must be invoked through sending and receiving a message, the application must attempt to minimize network traffic, which is significantly slower than accessing local memory.

Parallel FPM algorithms designed for message passing based systems typically partition data such that each process holds an equal share of the input in local memory. As a consequence, subsets of the local transactions must be shared with other processes that need them. For example, in candidate generation approaches, one may choose to broadcast the local input among all processes, during each iteration of the algorithm, in a round robin fashion. The amount of communication necessary in this scenario may be a detriment to the overall execution. Moreover, the set of candidate itemsets must also be partitioned across processes, incurring additional communication overhead. An alternative approach may use a pattern growth based algorithm that mines distinct projected databases associated with an equivalence class. Once a projected database has been extracted by communicating with the other processes that store its transactions, it can be mined by a process without further communication overhead. However, care must be taken to choose small enough equivalence classes s.t. their projected databases and count data structures fit in the local process memory.

3.3.2   MapReduce

MapReduce [13] is a recent programming model for distributed memory systems that has become very popular for data-intensive computing. By using a restricted program model, MapReduce offers a simple method of writing parallel programs. While originally a proprietary software package at Google, several successful MapReduce open-source implementations have been developed, of which Hadoop [59] is currently the most popular.

Computation in MapReduce consists of supplying two routines, MAP and REDUCE. The problem input is specified as a set of *key-value pairs*. Each key is processed by an invocation of MAP, which emits another (possibly different) key-value pair. Emitted key-value pairs are grouped by key by the MapReduce framework, and then the list of grouped values in each group is processed individually by a REDUCE invocation. REDUCE in turn emits either the final program output or new key-value pairs that can be processed with MAP for another iteration.

The canonical example of a MapReduce program is word frequency counting in a set of documents. Program input is organized into key-value pairs of the form <ID, text>. A MAP process is assigned to each ID. MAP iterates over each word w_i in its assigned document and emits the pair $< w_i, 1 >$. Pairs from all MAP processes are grouped and passed to a single REDUCE process. Finally, REDUCE counts the appearances of each w_i and outputs the final counts.

Individual MAP and REDUCE tasks are independent and can be executed in parallel. Unlike the message passing paradigm, parallelism in MapReduce is *implicit*, and the program developer is not required to consider low-level details such as data placement or communication across memory address spaces. The MapReduce framework manages all of the necessary communication details and typically implements key-value pair transmissions via a networked file system such as GFS [15] or HDFS [7]. Certain MapReduce implementations provide some means to make global read-only data easily accessible by all MAP processes. In Hadoop, this is achieved through the use of a *Distributed Cache*, which is an efficient file caching system built into the Hadoop runtime.

A criticism of the model comes from its heavy reliance on disk access between the MAP and REDUCE stages when key-value pairs must be grouped and partitioned across REDUCE processes. If the computation is iterative in nature, a naïve MapReduce implementation could require shuffling the data between physical disk drives in each iteration.

Pattern growth methods are popular in the MapReduce environment, due to their ability to decompose the problem into portions that can be independently solved. After computing the support of 1-length patterns in a similar manner as the word frequency counting example previously described, equivalence classes can be mapped to different processes via MAP. A serial algorithm is used to complete the local task, then a REDUCE job gathers the output from all processes.

4   Frequent Itemset Mining

Much like its serial counterpart, there are two main approaches to solving the parallel frequent itemset mining (PFIM) problem, namely, candidate generation and pattern growth. In this section, we discuss a number of proposed PFIM algorithms, paying special attention to the algorithmic details addressing the three challenges introduced in Sect. 3. For easy reference, Table 10.2 lists the serial and parallel methods described in this section.

4.1   Memory Scalability

Serial FIM algorithms can be easily parallelized if memory constraints are ignored. Among the first to address the memory constraint issues related to PFIM were Agrawal and Shafer [3], who proposed the Count Distribution (CD) algorithm. In their method, the list of transactions $\mathcal{T}$ is distributed among the processes s.t. each process is responsible for computing the support of all candidate itemsets with respect to its local transactions. Instead of a globally accessible hash tree, each process builds a local hash tree which includes all candidate itemsets. Then, with a single pass over the local transactions, the local support for each candidate itemset can be computed. The global support for each candidate itemset can then be computed as the sum of each process' local support for the given candidate itemset, using a global reduction operation. At the same time that Agrawal and Shafer introduced the CD algorithm, Shintani and Kitsuregawa [52] introduced an identical algorithm which they called Non-Partitioned Apriori (NPA).

Since each process can build its hash tree and compute its local support for candidate itemsets independently, the only inter-process communication required is during

Table 10.2 Serial and parallel frequent itemset mining algorithms

Type	Acronym	Name	Cite
Serial	Apriori	Apriori	[1]
Serial	ECLAT	Equivalence CLAss Transformation	[63]
Serial	FP-growth	Frequent Pattern Growth	[21]
Serial	Partition	Partition	[51]
Serial	SEAR	Sequential Efficient Association Rules	[40]
Serial	TreeProjection	TreeProjection	[4]
Parallel	BigFIM	Frequent Itemset Mining for Big Data	[39]
Parallel	CCPD	Common Candidate Partitioned Database	[66]
Parallel	CD	Count Distribution	[3]
Parallel	CD TreeProjection	Count Distributed TreeProjection	[4]
Parallel	DD	Data Distribution	[3]
Parallel	Dist-Eclat	Distributed Eclat	[39]
Parallel	DPC	Dynamic Passes Combined-counting	[33]
Parallel	FPC	Fixed Passes Combined-counting	[33]
Parallel	HD	Hybrid Distribution	[19]
Parallel	HPA	Hash Partitioned Apriori	[52]
Parallel	HPA-ELD	HPA with Extremely Large Itemset Duplication	[52]
Parallel	IDD	Intelligent Data Distribution	[19]
Serial	IDD TreeProjection	Intelligent Data Distribution TreeProjection	[4]
Parallel	NPA	Non-Partitioned Apriori	[52]
Parallel	ParEclat	Parallel Eclat	[67]
Parallel	Par-FP	Parallel FP-growth with Sampling	[9]
Parallel	PCCD	Partitioned Candidate Common Database	[66]
Parallel	PEAR	Parallel Efficient Association Rules	[40]
Parallel	PPAR	Parallel PARTITION	[40]
Parallel	SPC	Single Pass Counting	[33]

the global reduction operation. However, since the hash tree data structure is built serially by each process, for all candidate itemsets, a bottleneck is realized when the process count becomes sufficiently high. Also, since each process will have a local copy of the hash tree corresponding to all candidate itemsets, if the number of candidate itemsets becomes too large, the hash tree will likely not fit in the main memory of each process. In this case, the hash tree will have to be partitioned on disk and the set of transactions scanned once for each partition to compute candidate itemset supports, a computationally prohibitive exercise in the context of Big Data. Mueller developed a serial algorithm for FIM, named SEAR [40], which is identical to Apriori with the exception of using a trie in place of the hash tree data structure. Furthermore, Mueller extended SEAR along the lines of the CD algorithm, which he called PEAR [40].

Another PFIM algorithm based on the CD algorithm is the Count Distribution TreeProjection algorithm developed by Agarwal et al. [4]. Their algorithm is a parallel extension of their TreeProjection algorithm. In this parallel version of the algorithm, identical lexicographic trees are built on each process, in place of the hash trees of the original CD algorithm. From the lexicographic trees, the support counts are computed and globally reduced. This method shares scalability characteristics with other CD based algorithms.

Lin et al. [33] introduced three variations of the CD algorithm for the MapReduce framework, namely Single Pass Counting (SPC), Fixed Passes Combined-counting (FPC), and Dynamic Passes Combined-counting (DPC). In SPC, the responsibility of the MAP function is to read the list of all candidate itemsets from the Distributed Cache and count the frequency of each of them with respect to a local partition of $\mathcal{T}$. Then, the REDUCE function computes a global summation of local support counts and outputs to the distributed cache the new set of candidate itemsets for the next iteration. SPC is a direct extension of CD to MapReduce. The FPC and DPC algorithms are optimizations of SPC which combine iterations of the classic Apriori algorithm into a single pass.

The PPAR algorithm of Mueller [40], which is the natural parallel extension of the algorithm Partition, takes a slightly different approach to address the memory scalability challenge. In Partition, Savasere et al. [51] partition $\mathcal{T}$ horizontally and each partition is processed independently. The union of the resulting frequent itemsets, which are locally frequent in at least one partition, are then processed with respect to each partition in order to obtain global support counts. Mueller splits $\mathcal{T}$ into p partitions and assigns one to each of the p processes. Once all processes have finished identifying the locally frequent itemsets, a global exchange is performed to get the union of all locally frequent itemsets. Then, each process gathers the counts for the global candidate itemsets. Finally, the counts are globally reduced and the final set of frequent itemsets is identified. Like CD, PPAR has the advantage that the entire list of transactions does not need to fit in the main memory of each process. However, the set union of locally frequent itemsets may be much larger than the largest set of candidate itemsets in CD, and may not fit in the local memory available to each process.

Agrawal and Shafer [3] proposed a second method, which they called Data Distribution (DD), to address the memory scalability limitations of CD. To this end, they suggest that the candidate itemsets, as well as $\mathcal{T}$, should be distributed, using a round-robin scheme. Since processes have access to only a subset of the candidate itemsets, each is responsible for computing the global support for its local candidate itemsets. In order to compute global support for the local candidate itemsets, each process needs to count the occurrences of each of its candidate itemsets with respect to the entire list of transactions. With each process given access to only a portion of $\mathcal{T}$, this requires processes to share their local portions of $\mathcal{T}$ with each other. This type of communication pattern, all-to-all, can be expensive depending on the capabilities of the framework/architecture on which it is implemented. In a distributed system, using a message passing framework, the exchange of local partitions of $\mathcal{T}$ can be done efficiently using a ring-based all-to-all broadcast, as implemented by Han et al. [19] in their Intelligent Data Distribution (IDD) algorithm.

Each of the methods described above address the candidate itemset count explosion, inherent in candidate generation based algorithms. Zaki et al. [67] proposed a parallel extension of their serial algorithm, ECLAT, which they called ParEclat. The parallelization strategy in ParEclat is to identify a level of the itemset lattice with a large enough number of equivalence classes s.t. each process can be assigned some subset of classes. At this point, each process can proceed to mine the classes it was assigned, independently of other processes. In the end, a reduction is performed to gather all of the frequent itemsets as the result. In order to mine the subset of classes assigned to it, each process needs access to all of the tid-lists belonging to items in its assigned classes. Because the equivalence classes of ECLAT can be processed independently, no communication is needed after the initial class distribution. In many cases, this vertical exploration of the itemset lattice, as opposed to the horizontal exploration of Apriori based approaches, leads to much smaller memory requirements. However, ParEclat makes an implicit assumption that the tid-lists associated with the subset of classes being explored fits in the main memory of a single process, which may not always be the case. Consider the case that some subset of items appears in a significant portion of $\mathcal{T}$, then the corresponding tid-lists for these items will be a large fraction of the original list of transactions. Based on our original assumption that $\mathcal{T}$ does not fit in the memory of a single process, it can be concluded that ParEclat is memory scalable for only certain classes of input.

Another method similar to ParEclat was introduced by Cong et al. [9]. Their method, called Par-FP, follows the same parallelization strategy as ParEclat, but uses standard FP-Growth as its underlying algorithm instead of ECLAT. Like ParEclat, Par-FP is also a pattern generation method and is thus formulated within similar memory assumptions. Moens et al. [39] proposed Dist-Eclat and BigFIM in the MapReduce context. Dist-Eclat follows the same approach as ParEclat, adapted to the MapReduce framework. BigFIM is a hybrid approach that uses a MapReduce equivalent of CD to compute candidate itemsets level-wise until each process' assigned set of candidate itemsets no longer fits in its memory, at which point the algorithm transitions to a pattern growth approach.

Zaki et al. [66] proposed shared memory implementations of both the CD algorithm and the DD algorithm. In their implementation of the CD algorithm, which is referred to as Common Candidate Partitioned Database (CCPD), each process computes support for all candidates with respect to a distinct partition of the list of transactions. The support counts are aggregated in a shared candidate hash tree, whose access is controlled with a locking mechanism at the leaf of each node. The Partitioned Candidate Common Database (PCCD) algorithm is the shared memory implementation of the DD algorithm. Zaki et al. observed that, although there is no locking required, the increased disk contention from all processes scanning a shared list of transactions leads to a slowdown when using more than one process, due to processes operating on disjoint candidate hash trees.

4.2   Work Partitioning

A limitation of the DD algorithm is that it leads to redundant work when compared with CD, which performs the same amount of work as its serial counterpart. Let the cost of incrementing the count for a candidate itemset that exists in a process' hash tree be a function $f(X)$ of the number of unique candidate itemsets stored in the hash tree X. Then, in the case of the CD algorithm, the function becomes $f(M_k)$, where M_k is the total number of candidate itemsets of a given size k. In contrast, for the DD algorithm, the cost is $f(M_k/p)$. Furthermore, in the CD algorithm, each process is required to process $|\mathcal{T}|/p$ transactions, while in DD, each process is responsible for $|\mathcal{T}|$. Since each transaction processed will generate the same number of k-size candidate itemsets, the computational cost for each algorithm at iteration k can be modeled as follows:

$$\text{Cost}_{CD} = \frac{|T|}{p} \times f(M_k), \quad \text{and}$$

$$\text{Cost}_{DD} = |T| \times f\left(\frac{M_k}{p}\right).$$

In order for the DD algorithm to be as computationally efficient as the CD algorithm, $f(M_k/p)$ must be less than $f(M_k)$ by a factor of p. However, Han et al. [19] showed that $f(M_k/p) > f(M_k) \times 1/p$, thus the DD algorithm introduces redundant computation.

The authors proposed a few optimizations to the distribution and processing of transactions which can appropriately address these issues. When applied to the DD algorithm, these optimizations are referred to as Intelligent Data Distribution (IDD). In IDD, candidate itemsets are distributed to processes based on their prefix, following a lexicographic ordering or items. Each process is assigned a subset of I as prefixes they are responsible for. A process can quickly check to see if a transaction will generate any candidate itemsets which start with items from its subset of I and skip non-matching transactions. Therefore, a process will only traverse its hash tree for those candidate itemsets it is responsible for.

IDD uses a bin-packing algorithm [41] to ensure that each process receives approximately the same number of candidate itemsets and all candidate itemsets assigned to a particular process begin with the items of its subset of I. During each iteration, the number of candidate itemsets starting with each frequent item are counted. Then, the candidate itemsets are partitioned into p different bins, such that the sum of the numbers of candidate itemsets starting with the items in each bin are roughly equal. Longer prefixes can be considered if initial partitioning does not lead to evenly packed bins, until the bins are evenly packed. The authors theoretically and experimentally show that IDD addresses the memory bottleneck issues of the CD algorithm, as well as the redundant work introduced by the DD algorithm.

In the Intelligent Data Distributed TreeProjection algoirthm, Agarwal et al. [4] distributed the lexicographic tree by assigning to each process subtrees associated with specific first items. The lexicographic tree is distributed s.t. the sum of the supports of the subtrees assigned to a process is as close to balanced as possible.

Shintani and Kitsuregawa [52] also recognized the redundant work inherent to the DD algorithm. Unlike the approach in IDD, the Hash Partitioned Apriori (HPA) algorithm uses a hash function to determine the process responsible for a subset of candidate itemsets. In other words, each process scans its portion of the transaction set and generates k-size candidate itemsets. The algorithm hashes the itemsets and determines the process responsible for computing their support. If an itemset is hashed to its own process id, then the process increases support count for the itemset in its own hash tree. Otherwise, the candidate itemset is sent to the process to which it was hashed, so that its support count can be incremented there.

In ParEclat, Zaki et al. [67] employed a greedy algorithm that distributes work evenly among processes. First, each equivalence class is given a weight based on its cardinality. This weight is then used to distribute classes so that all processes are assigned subsets of classes with even weight.

A more advanced work distribution scheme for pattern growth methods was proposed by Cong et al. in [9], for their Par-FP algorithm. The authors suggested the use of *selective sampling* as a preprocessing step to estimate the cost for FP-Growth execution rooted at a particular item. In selective sampling, the t most frequent items, where t is an application parameter, are discarded from all transactions, along with any infrequent items. An FP-Tree is built from the resulting set of items and mined by a single process, recording the execution time of mining the projected database of each sampled item. After the sampling step, the work of mining each item $i \in I$ is partitioned in the following way. First, of the sampled items, any identified as *large* are further split into subtasks which are assigned to processes in a round-robin fashion. A large item is defined as any item whose sample execution time is significantly higher than the expected execution time given an equal distribution of work. Then, all other sampled items are assigned using a binning algorithm based on their sample execution time. Finally, those frequent items which were not sampled are assigned to processes in a round-robin fashion.

4.3   *Dynamic Load Balancing*

The estimation of work execution cost in the context of FIM can be done in a reasonably accurate way on homogeneous systems through work partitioning, as a preprocessing step to the algorithm execution. MapReduce systems are often heterogeneous, often composed of nodes with diverse compute power or resource availability. Lin et al. [33] noted that the number of candidate itemsets is typically small during later iterations of Apriori in his SPC algorithm, leading to increased overhead due to startup and scheduling of MapReduce jobs. Therefore, the authors proposed two dynamic load balancing improvements to SPC, the FPC and DPC algorithms. FPC combines a fixed number of Apriori iterations into a single pass, which means less MapReduce overhead as well as less scans over the list of transactions, but can also lead to many false positive results. The DPC algorithm dynamically chooses how many iterations to combine during each MAP phase, via two heuristics. First, in a single MAP phase, the DPC algorithm will combine as many iterations as it can without exceeding a specified maximum number of candidate itemsets. Since MapReduce systems can be heterogeneous, it is possible that the appropriate threshold for one compute node leads to a significantly longer or shorter execution time on a different node for the same iteration. The second heuristic of DPC dynamically adjusts the threshold value based on the execution times from previous iterations. In this way, the DPC algorithm can accommodate systems with dynamically changing resource availability.

4.4 Further Considerations

As is the case with most things in parallel computing, there is no silver bullet. Han et al. [19] noted that, as the process count to number of candidate itemsets ratio increases, the number of candidate itemsets assigned to each process in the IDD algorithm is reduced, shrinking the size of each process' hash tree and the amount of computation work per transaction. For a sufficiently high number of processes or sufficiently low number of candidate itemsets, this decrease in work per transaction will mean that, in distributed systems, the communication of the transactions between processes will become the limiting component of the IDD algorithm. Therefore, the authors introduce the Hybrid Distribution algorithm, which uses both the CD and IDD algorithms. To do this, the processes are split into g equally sized groups. Each process group is responsible for computing the support for all candidate itemsets with respect to the $|\mathcal{T}|/g$ transactions assigned to it. This can be viewed conceptually as executing the CD algorithm on g pseudo-processes. Within each group, the local support for all candidate itemsets is computed by executing the IDD algorithm on the p/g processes in the group. Each process in a group computes the support of the candidate itemsets assigned to it with respect to the $|\mathcal{T}|/g$ assigned to its process group. The choice of g can be determined automatically at runtime. When the number of candidate itemsets is not sufficiently high to ensure that each process in the system is assigned enough candidate itemsets to offset the cost of communicating the transactions within each process group, then g is set to p, meaning the CD algorithm is executed with each process as its own group. Otherwise, g is chosen small enough to ensure each process gets an adequate number of candidate itemsets so that work outweighs communication.

Shintani and Kitsuregawa [52] noted that the HPA algorithm suffers a similar drawback as the IDD algorithm. Namely, that if the process count to number of candidate itemsets ratio becomes sufficiently high, then the system can become under utilized. For this reason, they also introduced the HPA with Extremely Large Itemset Duplication (HPA-ELD) algorithm. In the case of the HPA algorithm, the under utilization comes from the possibility that the number of candidate itemsets will be small enough so that the memory available to each process is not completely full. To address this, HPA-ELD duplicates the highest frequency itemsets on all processes until the system memory is full. Then, the support for these most frequent candidate itemsets is computed locally, just like in the CD algorithm, while the support for the rest of the candidate itemsets is computed in the same way as in HPA.

5 Frequent Sequence Mining

Many of the solutions to FSM problems follow algorithms developed for FIM. Furthermore, parallel FSM (PFSM) algorithms often directly extend serial ones and their parallelization strategies are heavily inspired by previously developed PFIM algorithms. We will briefly discuss some of the more prominent serial approaches to solving the FSM problem, and focus the remainder of the section on the challenges

Table 10.3 Serial and parallel frequent sequence mining algorithms

Type	Acronym	Name	Cite
Serial	AprioriAll	AprioriAll	[2]
Serial	BIDE	BI-Directional Extension	[56]
Serial	CloSpan	Closed Sequential Pattern Mining	[61]
Serial	FreeSpan	Frequent Pattern-projected Sequential Pattern Mining	[20]
Serial	GSP	General Sequential Patterns	[54]
Serial	PrefixScan	Prefix-projected Sequential Pattern Mining	[43]
Serial	SPADE	Sequential PAttern Discovery using Equivalence classes	[62]
Serial	WAP-miner	Web Access Pattern Miner	[42]
Parallel	ACME	Advanced Parallel Motif Extractor	[50]
Parallel	DGSP	Distributed GSP	[44]
Parallel	DPF	Data Parallel Formulation	[18]
Parallel	EVE	Event Distribution	[27]
Parallel	EVECAN	Event and Candidate Distribution	[27]
Parallel	HPSPM	Hash Partitioned Sequential Pattern Mining	[53]
Parallel	MG-FSM	Large-Scale Frequent Sequence Mining	[38]
Parallel	NPSPM	Non-Partitioned Sequential Pattern Mining	[53]
Parallel	Par-ASP	Parallel PrefixSpan with Sampling	[9]
Parallel	Par-CSP	Parallel CloSpan with Sampling	[10]
Parallel	PLUTE	Parallel Sequential Patterns Mining	[45]
Parallel	pSPADE	Parallel SPADE	[65]

of parallelizing these approaches. For easy reference, Table 10.3 lists the serial and parallel methods described in this section.

5.1 Serial Frequent Sequence Mining

As direct extensions of the Apriori algorithm for the FIM problem, AprioriAll [2] and General Sequential Patterns (GSP) [54], by Srikant and Agrawal, make use of the downward closure property and follow the same general multi-pass candidate generation outline as described in Sect. 2. The join operation is redefined for the sequence domain s.t. only $(k-1)$-length frequent sequences with the same $(k-2)$-prefix are joined. GSP generalizes the problem definition, introducing time constraints (minimum or maximum time period between elements in a frequent sequence), a sliding window (sequence itemset items may come from a set of transactions with timestamps within a user-specified time window), and user-defined taxonomies (the set of items I may be provided as a hierarchy and sequential patterns may include items across all levels of the hierarchy).

Pattern growth methods also exhibit direct extensions to the sequence domain. Zaki [64] developed Sequential PAttern Discovery using Equivalence classes (SPADE), patterned after their ECLAT [62] algorithm for FIM, which is introduced in Sect. 2. During the first scan of the sequence database, SPADE creates, for each discovered sequence element (itemset), an *id-list* containing <sequence id, timestamp> pairs denoting locations of the element in the database. Candidate support counts can then be computed via id-list intersections. Zaki used lattice theory [12] to show that any possible frequent sequence can be obtained as a union or join of shorter frequent sequences. He proposed both *breath-first-search* and *depth-first-search* approaches to recursively decompose the lattice into prefix-based classes, providing an alternative to the breath-first-search approach of GSP.

Han et al. [20] introduced Frequent pattern-projected Sequential pattern mining (FreeSpan), which Pei et al. [43] followed with Prefix-projected Sequential pattern mining (PrefixSpan). FreeSpan and PrefixSpan recursively partition the sequence data based on frequent items and sequence prefixes, respectively. They first scan the database and derive frequent items. Then, projected databases are constructed for each of the frequent items, or length-1 frequent sequences, eliminating those sequences that only contain infrequent items. By choosing to partition on sequence prefixes, PrefixSpan is able to find all frequent sequences by examining only prefix sub-sequences and projecting only their corresponding postfix sub-sequences into projected databases. Pei et al. also explore pseudo-projections (when data fits in main memory) and bi-level projections in PrefixSpan, using the downward closure property to prune items in projected databases.

Guralnik and Karypis [18] altered a tree projection algorithm for discovering frequent itemsets by Agarwal et al. [4] to mine frequent sequences. As in the FIM approach, a lexicographic *projection tree* is grown through bi-level candidate sequence generation (nodes at level $k - 1$ generate candidates for level $k + 1$). Each node in the tree represents a frequent sequence, which can be extended either by adding a lexicographically correct item in the last element of the sequence (*itemset extension*) or a new itemset/element to the sequence (*sequence extension*). Each active node, one that still has a possibility of being extended, maintains four sparse *count matrices* used to efficiently update frequencies during the projection phase.

In the context of Web access mining, Pei et al. [42] developed Web Access Pattern miner (WAP-miner) for mining *symbol sequences* constructed from Web logs. The algorithm uses a Web access pattern tree (*WAP-tree*) as a compressed representation of the sequence database and a *conditional search* tree-projection mechanism for growing sequence patterns. Rajimol and Raju [46] surveyed several WAP-tree based extensions which improve the data structure and projection mechanism of WAP-miner.

While finding all frequent patterns is costly, the majority of the resulting patterns may not be interesting. CloSpan (Yan et al. [61]) and BIDE (Wang and Han [56]) focus on finding *frequent closed sequences*. Though following a candidate generate-and-test methodology, CloSpan prunes more of the search space through *CommonPrefix* and *Backward Sub-Pattern pruning*. It also speeds up testing by noting that projected databases with the same number of items are equivalent. BIDE uses a paradigm called *BI-Directional Extension* to both check closure of candidate sequences and prune the search space.

As a related problem first studied by Mannila et al. [36], *frequent episode mining* finds frequent collections of events that occur relatively close to each other in a long symbol sequence, given a partial event order. Joshi et al. [26] developed a universal formulation for sequence patterns, encompassing both Mannila et al.'s definition of frequent episodes and Srikant and Agrawal's generalized sequential patters. For an in-depth survey of serial frequent pattern mining methods, the reader may consult the work of Han et al. [22].

Motif discovery (finding frequent sub-sequence patterns) is an important problem in the biological domain, especially as a sub-step in sequence alignment. Rigoutsos and Floratos [49] proposed a pattern growth based method for motif discovery with rigid gaps (sets of *don't care* items). Wang et al. [58] use a two step process for motif discovery. They first search for short patterns with no gaps, called segments, and then search for longer patterns made up of segments joined by variable length gaps. Liao and Chen [31] use a vertical-database format they call the *three-dimensional list* to speed up verification of patterns generated via direct spelling, i.e., extending patterns at each level by each of the characters in the often small genetic symbol alphabet.

5.2   Parallel Frequent Sequence Mining

In this section, we discuss a number of proposed parallel sequence mining algorithms, along the lines of their memory scalability, task partitioning, and load balancing choices.

5.2.1   Memory Scalability

Some algorithms aimed at distributed systems assume that either the set of input sequences, the set of candidate sequences, or global count data structures fit in the local memory of each process. This is an unrealistic expectation when mining Big Data. Although their focus is on balancing mining tasks, Cong et al. [9, 10] (Par-FP, Par-ASP, Par-CSP) accomplished the task using a sampling technique that requires the entire input set be available at each process. Shintani and Kitsuregawa [53] partitioned the input set in Non Partitioned Sequential Pattern Mining (NPSPM), yet they assumed that the entire candidate set can be replicated and will fit in the overall memory (random access memory and hard drive) of a process. Similar assumptions were made by Joshi et al. [27] in Event Distribution (EVE) and by Guralnik and Karypis [18] in their Data Parallel Formulation (DPF).

These straight-forward *data parallel formulations* assign work to processes based on the input data they have been assigned. This strategy does not scale well, as global candidate sequences and their counts must also fit in the local process memory. Most authors proposed alternatives that partition both input and intermediary data. Shintani and Kitsuregawa [53] used a hash function in Hash Partitioned Sequential Pattern Mining (HPSPM) to assign input and candidate sequences to specific processes. Joshi et al. [27] proposed to shard both input and candidate sequences in Event and Candidate Distribution (EVECAN) and rotate the smaller of the the two sets among the processes, in round-robin fashion. Guralnik and Karypis introduced several task-parallel formulations, which are discussed in the next section, that partition the problem into independent sub-problems via database projections.

Input partitioning is not inherently necessary for shared memory or MapReduce distributed systems. In the case of shared memory systems, the input data should fit in the aggregated system memory and is available to be read by all processes. However, care must be taken to ensure processes do not simultaneously attempt to read or write to the same block of memory. Zaki [65] extended his serial FSM algorithm (SPADE) to the shared memory parallel architecture, creating pSPADE. Input data is assumed residing on shared hard drive space, stored in the vertical-database format. The author proposed two data parallel formulations in pSPADE that partition the input space s.t. different processes are responsible for reading different sections of the input data (*id-lists*). Processes are either assigned id-lists for a subset of sequences, or portions of all id-lists associated with a range of sequences in $\mathcal{D}$. Then, processes collaborate to expand each node in the itemset lattice. The author found that these formulations lead to poor performance due to high synchronization and memory overheads. He then proposed two task distribution schemes, discussed in the following sections, which are able to avoid read/write conflicts through independent search space sub-lattice assignments.

The HDFS distributed file system provided by the Hadoop MapReduce framework ensures adequate space exists for a large input dataset. However, data elements are replicated across several processing nodes in the system and repeated scans of the data will incur severe hard drive and/or network I/O penalties. Qiao et al. [44] found that a straight-forward extension of GSP for a MapReduce distributed environment (DGSP) performed poorly due to repeated scans of the input data. Instead, the authors used the CD algorithm for 1-length candidate sequence counting in an algorithm called PartSpan, followed by a projection-based task partitioning for solving the remainder of the problem. Similar strategies were followed by Qiao et al. [45] in PLUTE and Miliaraki et al. [38] in MG-FSM.

5.2.2 Work Partitioning

The FSM solution space is inherently irregular. Some processes may be assigned subsections of the input with only short frequent sequences and in effect have less work to complete than the rest of the processes. A number of authors have introduced work partitioning schemes designed to combat these potential problems.

pSPADE [65] recursively decomposes the frequent sequence search space into suffix-based equivalence classes, similar to those shown in Sect. 2. A node's sub-forest can be computed independently using only the node's input data (id-lists). Zaki proposed a static task partitioning scheme, Static Load Balancing (SLB), in which nodes in the first level of the class-tree and their associated id-lists are assigned in a round-robin way to processes, in reverse order of the number of elements in each node's class. This scheme cannot gauge well the amount of work that is needed to mine each node's class and may still lead to load imbalance.

Solving the FSM problem via lexicographic tree projection in a distributed message-passing environment, Guralnik and Karypis proposed a Data Parallel Formulation (DPF) similar to the Count Distribution strategy described in Sect. 4. It partitions the input equally among the processes and requires each node to keep track of the lexicographic tree representing frequent candidates and the associated count matrices. Each node projects its local set of sequences unto level $(k - 1)$ of the tree to determine local supports for candidate itemsets at the $(k + 1)$th level. A reduction operation facilitates computing global support of candidate itemsets. Those meeting the minimum support threshold σ are then broadcast to all processes before continuing to the next level. For problems with large numbers of candidates, where the global tree and its associated count matrices do not fit in the local memory of a process, DPF has to partition the tree and perform multiple scans of the local database at each iteration. This added I/O overhead makes this formulation impractical in the Big Data context. The authors also showed that DPF's parallel efficiency will decrease as the number of processes increase, even when the amount of work increases, due to the overhead of maintaining the global tree at each process. To account for these limitations, they propose Static and Dynamic Task-Parallel Formulations (STPF and DTPF), which partition both the input data and the lexicographic tree during computation.

Both task-parallel formulations first use DPF to expand the tree up to a level $k + 1$, $k > 0$. Further expanding a $(k + 1)$-level node requires count data from its parent node. Therefore, nodes at level k are partitioned among the processes for further processing, along with their *projected databases*. A node's projected database contains a subset of the sequence database $\mathcal{D}$ s.t. each item in the itemsets of each transaction is still viable to extend the sequence represented by the node into a possible frequent sequence. These viable items are called *active items*. The question remains how to partition the k-level nodes among the processes. The authors proposed two static approaches. The first uses a bin-packing algorithm based on relative computation time estimates for expanding the nodes, computed as the sum of their children's corresponding sequential patterns' support. The second approach aims to minimize the overlap in the nodes' projected databases via repeated minimum-cut partitioning of the bipartite graph formed by the set of nodes and the set of itemsets at level $(k + 1)$ in the tree. The authors found that the bipartite graph partitioning approach is able to substantially reduce the overlap in projected databases and lead to smaller execution times as opposed to other static task parallel formulations. However, its workload prediction accuracy decreases as the numbers of processes increases, leading to unbalanced loads. As a result, the authors introduced dynamic load balancing in DTPF that monitors process workloads and reassigns work as necessary.

Cong et al. took a sampling approach to accomplish static task partitioning in Par-ASP [9], which they named *selective sampling*. After gathering 1-length frequent sequence counts, they separate a sample $\mathcal{S} \subset \mathcal{D}$ of k-length frequent prefixes of sequences in $\mathcal{D}$. The size of the prefix k is a function of the average length of sequences in $\mathcal{D}$. The authors then use one process to mine the sample via a pattern growth algorithm, recording execution times for the found frequent sub-sequences.

Mining each frequent sub-sequence becomes a task. Projected databases for these frequent sequences are computed during the mining time estimation. Task distribution is then done in the same way as in the Par-FP algorithm, as described in Sect. 4. The authors found that the serial sampling component of their algorithm accounted on average for 1.1 % of the serial mining time, which limits the speedup potential as the number of processes increases.

Unlike in the general sequence mining problem, infrequent items cannot be removed from projected sequences in the gap-constrained frequency mining problem. Miliaraki et al. proposed several ways to compress projected databases in MG-FSM, based on the concept of *w-equivalency*, i.e., the equivalency of projected databases with respect to a pivot *w*. Projected databases are constructed and compressed for all 1-length frequent sequences in $\mathcal{D}$ in the MAP phase of a MapReduce job. Run-length and variable-byte encoding are used to reduce the size of the projected databases before transmitting them to reducers, which provides a significant performance boost, as reported by the authors in their experiments. A serial algorithm is used in the RE-DUCE phase of the MapReduce job to mine the independent partitions of the sequence database.

5.2.3 Dynamic Load Balancing

The advantage of separating work into partitions that processes can accomplish independently seems clear. Yet it is not always clear how long each partition will take to mine. Dynamic load balancing aims to (re)-distribute the work as necessary when processes finish their assigned tasks, in such a way that processes are allotted equal amounts of work overall.

Zaki extended his static task partitioning scheme (SLB) in pSPADE by forming a task queue. First level class nodes are entered into the queue in the same way as they would have been assigned in SLB. In the inter-Class Dynamic Load Balancing (CDLB) scheme, processes pick up new tasks from the queue, one at a time, as soon as they finish their current work. An abnormally large sub-class may still lead to load imbalance. The Recursive Dynamic Load Balancing (RDLB) scheme exploits both inter and intra-class parallelism, allowing a process to share its work with free ones. Mining of a class sub-tree at the process level takes place in a breath-first manner, level by level. A process signals it is free via a global counter when it finishes its current work and the queue is empty. Another process that still has work then enters its unprocessed nodes on the level he is currently processing into the global queue, providing work for the first process. The author reported RDLB performs best among the three proposed schemes.

The static task-parallel formulation (STPF) presented by Guralnik and Karypis may suffer some load imbalance as the number of processes increases. The authors combat this through a receiver initiated load-balancing with random polling scheme [17]. Processes are initially assigned tasks according to the STPF scheme. As a process completes its currently assigned work, it sends a request for more work to a randomly chosen other *donor* process. The donor process responds positively, by sending half of the nodes on the current level of the projection tree it is expanding, along with their projected databases, if it has just started expanding the level. If it is in the middle of processing the level, it responds with a wait message, including an estimated completion time for processing the current level. Otherwise, if it is nearing the end of processing the level, it can simply refuse to share its load. The requesting process may choose another random process to request work from if receiving a neutral or negative message from a donor process. In their experiments, the authors show that DTPF achieves comparable or significantly lower run-times than STPF, demonstrating the utility of dynamically re-distributing work in parallel FSM.

In the context of mining closed patterns from symbol sequences, Cong et al. [10] proposed using pseudo-projections of the dataset $\mathcal{D}$ in Par-CSP, one for each found 1-length frequent sequences. The projection for the frequent 1-sequence $\langle d \rangle$ is made up of sequences in $\mathcal{D}$ containing d, minus any symbols before d in those sequences. Assuming each process has access to the input data file, a pseudo-projection is simply a set of file pointers noting the beginning of each sub-sequence included in the projection. Each process can then be assigned a sub-set of the 1-length frequent sequences and their associated pseudo-projections, which they mine using the BIDE [56] serial algorithm. Pseudo-projections are communicated to all processes via an all-to-all broadcast. Dynamic load balancing is then achieved through a master-worker scheme [17], where each process contacts the master process for another 1-length frequent sequence to process when they finish their current work. In an effort to ensure tasks in the work queue have similar sizes, the authors use a *selective sampling* scheme similar to that in Par-ASP [9] to sub-divide large tasks.

A key task in bioinformatics is the identification of frequently recurring patterns, called *motifs*, in long genomic sequences. Sequences are defined over a small alphabet, e.g. $\mathcal{A} = \{A, C, G, T\}$ for DNA sequences. Patterns are constrained by minimum and maximum length parameters, and are required to be highly similar, rather than identical, to count towards a frequently occurring pattern. Sahli et al. developed ACME [50], a parallel combinatorial motif miner that decomposes the problem into many fine-grained sub-tasks, dynamically schedules them to be executed by all available processes, and uses a cache aware search technique to perform each sub-task. They construct a full-text suffix-tree index of the sequence at each *worker process* in linear time and space, using Ukonnen's algorithm [55], and annotate each node in the tree with the number of leaves reachable through the node. Motif mining is then reduced to finding tree nodes that represent candidate patterns and summing up their annotated leaf count. The search space, represented as a trie data structure, is partitioned into prefix-coherent sub-tries by a *master process* s.t. the average number of sub-tries per worker process is at least 16. The authors show experimentally that this leads to a near-optimal workload balance in both distributed message passing and shared memory environments. A worker then requests a new prefix (sub-task) to search for when it finishes processing the current assignment.

Table 10.4 Serial and parallel frequent graph mining algorithms

Type	Acronym	Name	Cite
Serial	AGM	Apriori-based Graph Mining	[25]
Serial	FSG	Frequent Subgraph Mining	[29]
Serial	gSpan	Graph-based Substructure Pattern Mining	[60]
Serial	HSIGRAM	Horizontal Single Graph Miner	[30]
Serial	MoFa	Molecular Fragment Miner	[6]
Serial	SUBDUE	Substructure Discovery	[24]
Serial	VSIGRAM	Vertical Single Graph Miner	[30]
Parallel	p-MoFa	Parallel MoFa	[37]
Parallel	p-gSpan	Parallel gSpan	[37]
Parallel	d-MoFa	Distributed MoFa with Dynamic Load Balancing	[14]
Parallel	MotifMiner	MotifMiner Toolkit	[57]
Parallel	MRFSE	MapReduce-based Frequent Subgraph Extraction	[35]
Parallel	MRPF	MapReduce-based Pattern Finding	[34]
Parallel	SP-SUBDUE	Static Partitioning SUBDUE	[11]
Parallel	SP-SUBDUE-2	Static Partitioning SUBDUE 2	[47]

6   Frequent Graph Mining

Successful algorithms for frequent graph mining (FGM) are often directly related to those designed for FIM. In this section, we provide a brief overview of some key serial methods for FGM and then discuss strategies and challenges for parallelizing FGM. For easy reference, Table 10.4 lists the serial and parallel methods described in this section.

6.1   Serial Frequent Graph Mining

Frequent graph mining comes with an added computational expense that is not present in frequent itemset or sequence mining. Determining if one graph is a subgraph of another is an NP-complete problem, known as the *subgraph isomorphism problem*. The cost of pruning infrequent candidates by performing isomorphism checks is very costly and thus most FGM methods make an effort to reduce the amount of pruning necessary.

Inokuchi et al. [25] developed Apriori-based Graph Mining (AGM), which extended the Apriori algorithm to FGM. During candidate generation, subgraphs are extended by joining two frequent subgraphs and expanding by one edge in all possible ways. Kuramochi and Karypis [29] described the Frequent Subgraph mining algorithm (FSG), another candidate generation based method. Size-k candidates are generated by joining two frequent $(k-1)$-subgraphs. Subgraphs are joined only if they share the same $(k-2)$-subgraph. When candidates are joined, an edge is *grown* by either connecting two existing vertices or adding a new vertex. Infrequent subgraphs must be pruned after all size-k candidates are generated. FSG optimizes the pruning stage by only adding edges during candidate generation which are known to

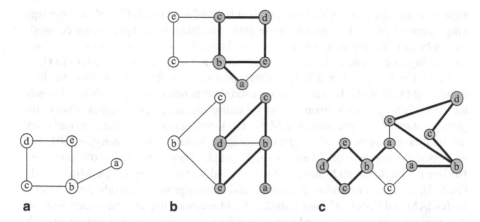

Fig. 10.5 Example of the FGM problem using both a set of graphs and a large graph as input. The embedding of the candidate subgraph in (**a**) is highlighted by *bold lines* and *gray nodes* in the example databases in (**b**) and (**c**)

be frequent. This can greatly reduce the number of infrequent subgraphs that need to be pruned.

Pattern growth methods, which traverse the pattern lattice in depth-first order, have been quite successful in solving the FGM problem. They have smaller memory footprints than candidate generation ones, because only subgraphs located on the path from the start of the lattice to the current pattern being explored need to be kept in memory. A challenge that pattern growth methods must face, however, is the added risk of duplicate generation during exploration.

Yan and Han [60] developed the first pattern growth FGM method, named graph-based Substructure pattern mining (gSpan). It avoids duplicates by only expanding subtrees which lie on the *rightmost path* in the depth-first traversal. A major speedup in gSpan is obtained by searching for the next extensions to make at the same time as executing isomorphism tests.

Molecular Fragment miner (MoFa), introduced by Borgelt and Berthold [6], is a pattern-growth method that was designed specifically for mining molecular substructures, but is also capable of general FGM. MoFa maintains *embedding lists*, which are pointers to the exact locations within G where frequent subgraphs are found. Embedding lists trade off a large amount of memory for the need to perform isomorphism checks across G when a new subgraph is explored. To avoid duplicates, MoFa maintains a hash table of all subgraphs that have been expanded and checks whether a graph exists in the hash table before considering it as an extension. A graph G is hashed by first transforming it to a string representation unique to G and graphs isomorphic to G.

Some FGM algorithms define G as a single large graph, instead of the typical set of graphs. In this formulation, the support of a graph G is defined as the number of times G appears as a subgraph in G. Figure 10.5 portrays the difference between the alternate problem definitions. A complication that appears when considering support in a single graph is that some embeddings of a subgraph G could be overlapping. Kuramochi and Karypis [30] show that, if overlapping subgraphs are counted, the problem is no longer downward closed. Thus, they only consider *edge-disjoint* embeddings, i.e., those embeddings of G that do not have any shared edges in $\mathcal{G}$.

SUBDUE, by Holder et al. [24], operates on a single graph and was one of the first developed FGM tools. It uses a greedy candidate generation approach and bounds the cost of doing isomorphism checks by doing inexact graph matches. Due to its greedy nature and inexact matches, SUBDUE is an approximate algorithm that does not find the complete set of frequent subgraphs. Kuramochi and Karypis [30] also addressed the scenario of $\mathcal{G}$ being a single graph. They presented two algorithms, Horizontal Single Graph Miner (HSIGRAM) and Vertical Single Graph Miner (VSI-GRAM), which are candidate generation and pattern growth methods, respectively. HSIGRAM and VSIGRAM are complete FGM methods that also have the ability to do approximate counting. In order to reduce the time for frequency counting on each lattice level, HSIGRAM makes use of a modified form of embedding lists: instead of storing each embedding found in $\mathcal{G}$, a single edge called the *anchor-edge* is stored. This significantly reduces the memory expense of full embedding lists at the cost of some recomputation for each embedding.

6.2 Parallel Frequent Graph Mining

The exponential cost of subgraph isomorphism inherent in FGM makes parallelism for this problem vital for Big Data input. We now describe existing parallel FGM (PFGM) methods, focusing on how they address the issues of memory scalability, work partitioning, and dynamic load balancing.

6.2.1 Memory Scalability

Many of the PFIM and PFSM techniques for achieving memory scalability are also applicable in PFGM algorithms. Methods commonly require access to the entirety of $\mathcal{G}$. This has performance implications for both shared and distributed memory systems. Meinl et al. [37] created parallel versions of gSpan and MoFa for a shared memory system. In both algorithms, $\mathcal{G}$ is globally accessible to all processes, but could be split across multiple machines, causing unexpected network delays. Additionally, MoFa must be able to keep track of the already generated subgraphs in order to avoid generating duplicates.

Cook et al. [11] developed Static Partitioning SUBDUE (SP-SUBDUE), a parallel, message-passing implementation of SUBDUE. Like its serial predecessor, SP-SUBDUE expects $\mathcal{G}$ to be a single large graph. It uses a weighted graph partitioning algorithm [28] to divide $\mathcal{G}$ into as many partitions as there are processes. Each process can then store only their local portion of $\mathcal{G}$ in memory. The consequence of partitioning $\mathcal{G}$ is that any frequent subgraphs that exist across partition boundaries will be missed. The algorithm attempts to minimize these lost subgraphs by assigning frequent edges large weights before partitioning, leading to fewer cuts across them.

MapReduce offers a framework for developing parallel programs that operate gracefully using secondary non-volatile storage. Secondary storage devices, such as hard disk drives, are almost always larger than the amount of random access memory available on a machine. MapReduce-based Pattern Finding (MRPF) by Liu et al. [34] is a MapReduce framework for the FGM problem that assumes $\mathcal{G}$ is a single graph. It employs a candidate generation scheme that uses two iterations of MapReduce to explore each level of the search lattice. In the first iteration, the MAP phase generates size-k candidates. The following REDUCE phase detects duplicates and removes them. Finally, the second MapReduce pass computes the support of new candidates. Lu et al. [35] developed MapReduce based Frequent Subgraph Extraction (MRFSE), a candidate generation MapReduce implementation for the traditional many-graph $\mathcal{G}$. On iteration k, a MAP task takes one graph from $\mathcal{G}$ and emits size-k candidates. Each REDUCE task takes all of the appearances of a candidate and computes its support. MRFSE also stores embedding lists to maintain the location of all appearances of frequent subgraphs in $\mathcal{G}$. By doing so, isomorphism tests are no longer necessary, at the cost of additional memory usage.

6.2.2 Work Partitioning

Candidate generation methods consider a different set of candidates at each iteration. Much like the parallel Apriori methods seen in Sect. 4, the candidate generation process can easily be decomposed into parallel tasks. Wang and Parthasarathy developed the MotifMiner Toolkit for distributed memory systems [57]. MotifMiner is designed for both the case when $\mathcal{G}$ is a set of graphs and when $\mathcal{G}$ is a single graph. It has a similar structure as the Data Distribution method from Sect. 4. $\mathcal{G}$ is assumed to be available to all processes and each process is assigned some subset of candidate generation to perform. Once all processes have generated their local candidate sets, each process shares its local candidates with all others.

Many parallel candidate generation methods identify the individual candidates to expand as independent tasks. MRFSE, a MapReduce algorithm by Lu et al. [35], assigns instead each graph in $\mathcal{G}$ as a task. The authors noted that the simple assignment of the same number of graphs to each MAP task is not a good work partitioning scheme because graphs in $\mathcal{G}$ can vary greatly in size. Instead, they count the edges of each graph in $\mathcal{G}$ and assign tasks such that each MAP is assigned a roughly equal number of edges.

SP-SUBDUE partitions the single graph $\mathcal{G}$ across processes in a distributed setting. Each process then runs an independent instance of SUBDUE and support counts are finalized by broadcasting the frequent subgraphs to all other processes. When a process receives a new frequent subgraph from another partition, it computes its support on the local portion of $\mathcal{G}$. Ray and Holder [47] extended this work, proposing SP-SUBDUE-2, in which each process first searches their locally discovered subgraphs before doing expensive isomorphism tests across their partition of $\mathcal{G}$.

Parallelism for pattern growth methods is commonly achieved by exploiting the independent nature of the subtrees before expansion. One of the most straight-forward ways to partition work is to assign adjacent subtrees to processes and explore each one independently. This essentially operates as if a separate FGM program is being executed on each process. Meinl et al. [37] took this approach in their shared memory implementations of gSpan and MoFa.

Within the shared memory system context, Reinhardt and Karypis [48] created a parallel implementation of VSIGRAM, which works on a single large graph. Unlike SUBDUE and its parallel extensions, it is not an approximate algorithm. Their implementation features both coarse and fine-grained work partitioning: subgraphs can be extended vertically in parallel, and parallelism can also be exploited in the support count of a single subgraph.

6.2.3 Dynamic Load Balancing

Load balancing for FGM is a notoriously difficult problem. Not only is the search space highly irregular, but accurately estimating the workload of a task is very difficult. Di Fatta and Berthold [14] showed that the time required for subtree exploration in a biological dataset follows a power-law distribution.

Shared memory systems offer good opportunities for cheap dynamic load balancing. Work queues can be easily accessed by all processes and requesting or sending work is generally an inexpensive operation. Buehrer and Parthasarathy [8] experimented with several work queue designs inside of a shared-memory parallel implementation of gSpan. In their work, a task is defined as a tuple of the subgraph to be mined and the list of graphs in $\mathcal{G}$ that it occurs in. They evaluated three dynamic load balancing schemes: (1) a global work queue, in which all processes cooperate to enqueue and dequeue tasks; (2) hierarchical queues, in which each process maintains a local private queue but enqueues to a shared one if their local queue becomes full; and (3) distributed queues, in which each process maintains a private queue and offers any additional work to idle processes if the local queue becomes full. Their experiments showed that a distributed queueing model offered the most effective load balancing.

Meinl et al. [37] evaluated another scheme for dynamic load balancing in their shared memory implementations of MoFa and gSpan. Each process maintains a local last-in-first-out (LIFO) work queue that can be shared with other processes. When work needs to be distributed to another process, the stack is split in half. The authors identify a heuristic for evenly splitting work. Since the work queue is LIFO, the tasks at the bottom of the stack will often require more work because they were discovered at the beginning of execution and therefore will have more subtrees to explore. In an effort to provide a balanced work distribution, each process is assigned every-other task. The authors also compared sender-initiated and receiver-initiated load balancing and found that there was not a significant difference relative to the rest of the computation time.

Di Fatta and Berthold [14] created a distributed implementation of MoFa using message passing. Dynamic load balance is achieved by receiver-initiated work requests. When issuing a work request, a process must choose another worker to request from. The authors present a policy called *random-ranked polling*, in which each process maintains a list of all other processes, sorted by the time when their current work unit was started. The requesting process randomly selects from the list, with a bias towards those which have spent more time on the current work unit. A process being asked for work first heuristically determines whether they should split their work based on the stack size, the support, and branching factor of the pattern currently being tested. If they are able to provide work to the requesting process, a process sends one pattern from the bottom of their work stack. The requesting process is then responsible for re-generating the embedding list of the received pattern before processing it.

7 Conclusion

Many efficient serial algorithms have been developed for solving the frequent pattern mining problem. Yet they often do not scale to the type of data we are presented with today, the so-called "Big Data". In this chapter, we gave an overview of parallel approaches for solving the problem, looking both at the initially defined frequent itemset mining problem and at its extension to the sequence and graph mining domains. We identified three areas as key challenges to parallel algorithmic design in the context of frequent pattern mining: memory scalability, work partitioning, and load balancing. With these challenges as a frame of reference, we extracted key algorithmic design patterns from the wealth of research conducted in this domain. We found that, among parallel candidate generation based algorithms, memory scalability is often the most difficult obstacle to overcome, while for those parallel algorithms based on pattern growth methods, load balance is typically the most critical consideration for efficient parallel execution.

The parallel pattern mining problem is in no way "solved". Many of the methods presented here are more than a decade old and were designed for parallel architectures very different than those that exist today. Moreover, they were not evaluated on datasets big enough to show scalability to Big Data levels. While most works included limited scalability studies, they generally did not compare their results against other existing parallel algorithms for the same problem, even those designed for the same architecture. More research is needed to validate existing methods at the Big Data scale.

Work partitioning and load balancing continue to be open problems for parallel frequent pattern mining. Better methods to estimate the cost of solving sub-problems at each process can lessen the need for dynamic load balancing and improve overall efficiency. Additionally, they can help processes intelligently decide whether to split their work with idling ones or not. Another open problem is that of mining sub-patterns in a large object, where sub-patterns can span multiple process' data. Current methods for sequence motif mining and frequent subgraph mining in a large graph either rely on maximum pattern length constraints that allow each process to store overlapping data partition boundaries or transfer data partitions amongst all processes during each iteration of the algorithm. Neither solution scales when presented with Big Data, calling for efficient methods to solve this problem exactly.

References

1. Rakesh Agrawal and Ramakrishnan Srikant. Fast algorithms for mining association rules in large databases. In *International Conference on Very Large Data Bases*, VLDB '94, pages 487–499, San Francisco, CA, USA, 1994. Morgan Kaufmann Publishers Inc.
2. Rakesh Agrawal and Ramakrishnan Srikant. Mining sequential patterns. In *International Conference on Data Engineering*, ICDE '95, pages 3–14, Washington, DC, USA, 1995. IEEE Computer Society.
3. Rakesh Agrawal and John C. Shafer. Parallel mining of association rules. *IEEE Transactions on Knowledge and Data Engineering*, 8(6):962–969, 1996.
4. Ramesh C. Agarwal, Charu C. Aggarwal, and V. V. V. Prasad. A tree projection algorithm for generation of frequent item sets. *Journal of Parallel and Distributed Computing*, 61(3):350–371, March 2001.
5. Big data meets big data analytics. http://www.sas.com/resources/whitepaper/wp_46345.pdf. Accessed: 2014-03-06.
6. Christian Borgelt and Michael R. Berthold. Mining molecular fragments: Finding relevant substructures of molecules. In *IEEE International Conference on Data Mining*, ICDM 2002, pages 51–58. IEEE, 2002.
7. Dhruba Borthakur. The hadoop distributed file system: Architecture and design. *Hadoop Project Website*, 11:21, 2007.
8. Gregory Buehrer, Srinivasan Parthasarathy, Anthony Nguyen, Daehyun Kim, Yen-Kuang Chen, and Pradeep Dubey. Parallel graph mining on shared memory architectures. Technical report, The Ohio State University, Columbus, OH, USA, 2005.
9. Shengnan Cong, Jiawei Han, Jay Hoeflinger, and David Padua. A sampling-based framework for parallel data mining. In *ACM SIGPLAN Symposium on Principles and Practice of Parallel Programming*, PPoPP '05, pages 255–265, New York, NY, USA, 2005. ACM.
10. Shengnan Cong, Jiawei Han, and David Padua. Parallel mining of closed sequential patterns. In *Proceedings of the Eleventh ACM SIGKDD International Conference on Knowledge Discovery in Data Mining*, KDD '05, pages 562–567, New York, NY, USA, 2005. ACM.
11. Diane J Cook, Lawrence B Holder, Gehad Galal, and Ron Maglothin. Approaches to parallel graph-based knowledge discovery. *Journal of Parallel and Distributed Computing*, 61(3):427–446, 2001.
12. Brian A. Davey and Hilary A. Priestley. *Introduction to lattices and order*. Cambridge University Press, Cambridge, 1990.
13. Jeffrey Dean and Sanjay Ghemawat. Mapreduce: Simplified data processing on large clusters. *Communications of the ACM*, 51(1):107–113, January 2008.
14. Giuseppe Di Fatta and Michael R. Berthold. Dynamic load balancing for the distributed mining of molecular structures. *IEEE Transactions on Parallel and Distributed Systems*, 17(8):773–785, 2006.
15. Sanjay Ghemawat, Howard Gobioff, and Shun-Tak Leung. The google file system. In *ACM SIGOPS Operating Systems Review*, volume 37, pages 29–43. ACM, 2003.

16. Carole A. Goble and David De Roure. The impact of workflow tools on data-centric research. In Tony Hey, Stewart Tansley, and Kristin M. Tolle, editors, *The Fourth Paradigm*, pages 137-145. Microsoft Research, 2009.
17. Ananth Grama, George Karypis, Vipin Kumar, and Anshul Gupta. *Introduction to Parallel Computing (2nd Edition)*. Addison Wesley, second edition, 2003.
18. Valerie Guralnik and George Karypis. Parallel tree-projection-based sequence mining algorithms. *Parallel Computing*, 30(4):443–472, April 2004.
19. Eui-Hong Han, George Karypis, and Vipin Kumar. Scalable parallel data mining for association rules. In *ACM SIGMOD International Conference on Management of Data*, SIGMOD '97, pages 277–288, New York, NY, USA, 1997. ACM.
20. Jiawei Han, Jian Pei, Behzad Mortazavi-Asl, Qiming Chen, Umeshwar Dayal, and Mei-Chun Hsu. Freespan: Frequent pattern-projected sequential pattern mining. In *ACM SIGKDD International Conference on Knowledge Discovery and Data Mining*, KDD '00, pages 355–359, New York, NY, USA, 2000. ACM.
21. Jiawei Han, Jian Pei, and Yiwen Yin. Mining frequent patterns without candidate generation. In *ACM SIGMOD International Conference on Management of Data*, SIGMOD '00, pages 1–12, New York, NY, USA, 2000. ACM.
22. Jiawei Han, Hong Cheng, Dong Xin, and Xifeng Yan. Frequent pattern mining: Current status and future directions. *Data Mining and Knowledge Discovery*, 15(1):55–86, August 2007.
23. Carl Hewitt, Peter Bishop, and Richard Steiger. A universal modular actor formalism for artificial intelligence. In *Third International Joint Conference on Artificial intelligence*, IJCAI-73, pages 235–245. Morgan Kaufmann Publishers Inc., 1973.
24. Lawrence B Holder, Diane J Cook, Surnjani Djoko, et al. Substucture discovery in the subdue system. In *AAAI Workshop on Knowledge Discovery in Databases*, KDD-94, pages 169–180, 1994.
25. Akihiro Inokuchi, Takashi Washio, and Hiroshi Motoda. An apriori-based algorithm for mining frequent substructures from graph data. In *Principles of Data Mining and Knowledge Discovery*, pages 13–23. Springer, 2000.
26. Mahesh V. Joshi, George Karypis, and Vipin Kumar. A universal formulation of sequential patterns. Technical Report 99-021, Department of Computer Science, University of Minnesota, 1999.
27. Mahesh V. Joshi, George Karypis, and Vipin Kumar. Parallel algorithms for mining sequential associations: Issues and challenges. Technical report, Department of Computer Science, University of Minnesota, 2000.
28. George Karypis and Vipin Kumar. A fast and high quality multilevel scheme for partitioning irregular graphs. *SIAM Journal of Scientific Computing*, 20(1):359–392, Dec 1998.
29. Michihiro Kuramochi and George Karypis. Frequent subgraph discovery. In *Proceedings of the 2001 IEEE International Conference on Data Mining*, ICDM 2001, pages 313–320. IEEE, 2001.
30. Michihiro Kuramochi and George Karypis. Finding frequent patterns in a large sparse graph. *Data Mining and Knowledge Discovery*, 11(3):243–271, 2005.
31. Vance Chiang-Chi Liao and Ming-Syan Chen. Dfsp: a depth-first spelling algorithm for sequential pattern mining of biological sequences. *Knowledge and Information Systems*, pages 1–17, 2013.
32. Jimmy Lin and Dmitriy Ryaboy. Scaling big data mining infrastructure: the twitter experience. *ACM SIGKDD Explorations Newsletter*, 14(2):6–19, 2013.
33. Ming-Yen Lin, Pei-Yu Lee, and Sue-Chen Hsueh. Apriori-based frequent itemset mining algorithms on mapreduce. In *Proceedings of the Sixth International Conference on Ubiquitous Information Management and Communication*, ICUIMC '12, pages 76:1–76:8, New York, NY, USA, 2012. ACM.
34. Yang Liu, Xiaohong Jiang, Huajun Chen, Jun Ma, and Xiangyu Zhang. Mapreduce-based pattern finding algorithm applied in motif detection for prescription compatibility network. In *Advanced Parallel Processing Technologies*, pages 341–355. Springer, 2009.

35. Wei Lu, Gang Chen, Anthony KH Tung, and Feng Zhao. Efficiently extracting frequent sub-graphs using mapreduce. In *2013 IEEE International Conference on Big Data*, pages 639–647. IEEE, 2013.
36. Heikki Mannila, Hannu Toivonen, and A. Inkeri Verkamo. Discovery of frequent episodes in event sequences. *Data Mining and Knowledge Discovery*, 1(3):259–289, January 1997.
37. Thorsten Meinl, Marc Worlein, Ingrid Fischer, and Michael Philippsen. Mining molecular datasets on symmetric multiprocessor systems. In *IEEE International Conference on Systems, Man and Cybernetics*, volume 2 of *SMC '06*, pages 1269–1274. IEEE, 2006.
38. Iris Miliaraki, Klaus Berberich, Rainer Gemulla, and Spyros Zoupanos. Mind the gap: Large-scale frequent sequence mining. In *ACM SIGMOD International Conference on Management of Data*, SIGMOD '13, pages 797–808, New York, NY, USA, 2013. ACM.
39. Sandy Moens, Emin Aksehirli, and Bart Goethals. Frequent itemset mining for big data. In *2013 IEEE International Conference on Big Data*, pages 111–118. IEEE, 2013.
40. Andreas Mueller. Fast sequential and parallel algorithms for association rule mining: A comparison. Technical report, University of Maryland at College Park, College Park, MD, USA, 1995.
41. Christos H. Papadimitriou and Kenneth Steiglitz. *Combinatorial optimization: algorithms and complexity*. Courier Dover Publications, 1998.
42. Jian Pei, Jiawei Han, Behzad Mortazavi-Asl, and Hua Zhu. Mining access patterns efficiently from web logs. In *Pacific-Asia Conference on Knowledge Discovery and Data Mining, Current Issues and New Applications*, PAKDD '00, pages 396–407, London, UK, UK, 2000. Springer-Verlag.
43. Jian Pei, Jiawei Han, Behzad Mortazavi-Asl, Helen Pinto, Qiming Chen, Umeshwar Dayal, and Mei-Chun Hsu. Prefixspan: Mining sequential patterns efficiently by prefix-projected pattern growth. In *International Conference on Data Engineering*, ICDE '01, pages 215–224, Washington, DC, USA, 2001. IEEE Computer Society.
44. Shaojie Qiao, Changjie Tang, Shucheng Dai, Mingfang Zhu, Jing Peng, Hongjun Li, and Yungchang Ku. Partspan: Parallel sequence mining of trajectory patterns. In *International Conference on Fuzzy Systems and Knowledge Discovery - Volume 05*, FSKD '08, pages 363–367, Washington, DC, USA, 2008. IEEE Computer Society.
45. Shaojie Qiao, Tianrui Li, Jing Peng, and Jiangtao Qiu. Parallel sequential pattern mining of massive trajectory data. *International Journal of Computational Intelligence Systems*, 3(3):343–356, 2010.
46. A. Rajimol and G. Raju. Web access pattern mining — a survey. In *International Conference on Data Engineering and Management*, ICDEM '10, pages 24–31, Berlin, Heidelberg, 2012. Springer-Verlag.
47. Abhik Ray and Lawrence B. Holder. Efficiency improvements for parallel subgraph miners. In *Florida Artificial Intelligence Research Society Conference*, FLAIRS '12, 2012.
48. Steve Reinhardt and George Karypis. A multi-level parallel implementation of a program for finding frequent patterns in a large sparse graph. In *International Symposium on Parallel and Distributed Processing*, IPDPS 2007, pages 1–8, 2007.
49. Isidore Rigoutsos and Aris Floratos. Combinatorial pattern discovery in biological sequences: The teiresias algorithm. *Bioinformatics*, 14(1):55–67, 1998.
50. Majed Sahli, Essam Mansour, and Panos Kalnis. Parallel motif extraction from very long sequences. In *ACM International Conference on Conference on Information & Knowledge Management*, CIKM '13, pages 549–558, New York, NY, USA, 2013. ACM.
51. Ashoka Savasere, Edward Omiecinski, and Shamkant B. Navathe. An efficient algorithm for mining association rules in large databases. In *International Conference on Very Large Data Bases*, VLDB '95, pages 432–444, San Francisco, CA, USA, 1995. Morgan Kaufmann Publishers Inc.
52. Takahiko Shintani and Masaru Kitsuregawa. Hash based parallel algorithms for mining association rules. In *International Conference on Parallel and Distributed Information Systems*, pages 19–30, Dec 1996.

53. Takahiko Shintani and Masaru Kitsuregawa. Mining algorithms for sequential patterns in parallel: Hash based approach. In Xindong Wu, Kotagiri Ramamohanarao, and Kevin B. Korb, editors, *Pacific-Asia Conference on Knowledge Discovery and Data Mining*, volume 1394 of *PAKDD '98*, pages 283–294. Springer, 1998.
54. Ramakrishnan Srikant and Rakesh Agrawal. Mining sequential patterns: Generalizations and performance improvements. In *International Conference on Extending Database Technology: Advances in Database Technology*, EDBT '96, pages 3–17, London, UK, UK, 1996. Springer-Verlag.
55. Esko Ukkonen. On-line construction of suffix trees. *Algorithmica*, 14(3):249–260, 1995.
56. Jianyong Wang and Jiawei Han. Bide: Efficient mining of frequent closed sequences. In *International Conference on Data Engineering*, ICDE '04, pages 79–91, Washington, DC, USA, 2004. IEEE Computer Society.
57. Chao Wang and Srinivasan Parthasarathy. Parallel algorithms for mining frequent structural motifs in scientific data. In *Annual International Conference on Supercomputing*, ICS '04, pages 31–40, New York, NY, USA, 2004. ACM.
58. Ke Wang, Yabo Xu, and Jeffrey Xu Yu. Scalable sequential pattern mining for biological sequences. In *International Conference on Information and Knowledge Management*, CIKM '04, pages 178–187, New York, NY, USA, 2004. ACM.
59. Tom White. *Hadoop: The Definitive Guide*. O'Reilly Media, 2009.
60. Xifeng Yan and Jiawei Han. gspan: Graph-based substructure pattern mining. In *IEEE International Conference on Data Mining*, ICDM 2002, pages 721–724. IEEE, 2002.
61. Xifeng Yan, Jiawei Han, and Ramin Afshar. Clospan: Mining closed sequential patterns in large databases. In Daniel Barbará and Chandrika Kamath, editors, *SIAM International Conference on Data Mining*, SDM 2003. SIAM, 2003.
62. Mohammed J. Zaki. Efficient enumeration of frequent sequences. In *Seventh International Conference on Information and Knowledge Management*, CIKM '98, pages 68–75, New York, NY, USA, 1998. ACM.
63. Mohammed J. Zaki. Scalable algorithms for association mining. *IEEE Transactions on Knowledge and Data Engineering*, 12(3):372–390, May 2000.
64. Mohammed J. Zaki. Spade: An efficient algorithm for mining frequent sequences. *Machine Learning*, 42(1–2):31-60, January 2001.
65. Mohammed J. Zaki. Parallel sequence mining on shared-memory machines. *Journal of Parallel and Distributed Computing*, 61(3):401–426, Mar 2001. Special issue on High Performance Data Mining.
66. Mohammed J. Zaki, Mitsunori Ogihara, Srinivasan Parthasarathy, and Wei Li. Parallel data mining for association rules on shared-memory multi-processors. In *ACM/IEEE Conference on Supercomputing*, pages 43–43, 1996.
67. Mohammed J. Zaki, Srinivasan Parthasarathy, Mitsunori Ogihara, and Wei Li. Parallel algorithms for discovery of association rules. *Data Mining and Knowledge Discovery*, 1(4):343–373, 1997.

Chapter 11
Sequential Pattern Mining

Wei Shen, Jianyong Wang and Jiawei Han

Abstract Sequential pattern mining, which discovers frequent subsequences as patterns in a sequence database, has been a focused theme in data mining research for over a decade. This problem has broad applications, such as mining customer purchase patterns and Web access patterns. However, it is also a challenging problem since the mining may have to generate or examine a combinatorially explosive number of intermediate subsequences. Abundant literature has been dedicated to this research and tremendous progress has been made so far. This chapter will present a thorough overview and analysis of the main approaches to sequential pattern mining.

Keywords Sequential · pattern · mining

1 Introduction

Sequential pattern mining discovers subsequences that appear in a sequence database with frequency no less than a user-specified threshold. A sequence database stores a number of records, where all records are ordered sequences of events, with or without concrete notions of time. Examples of sequences include retail customer transactions, DNA sequences, and web log data. A subsequence, such as buying first a PC, then a digital camera, and then a memory card, if it occurs frequently in a customer transaction database, is a (frequent) sequential pattern.

Sequential pattern mining is an important data mining problem with broad applications, such as mining customer purchase patterns, identifying outer membrane proteins, automatically detecting erroneous sentences, discovering block correlations in storage systems, identifying copy-paste and related bugs in large-scale software

W. Shen (✉) · J. Wang
Tsinghua University, Beijing, China
e-mail: chen-wei09@mails.tsinghua.edu.cn

J. Wang
e-mail: jianyong@tsinghua.edu.cn

J. Han
University of Illinois at Urbana-Champaign, Urbana, Illinois
e-mail: hanj@cs.uiuc.edu

C. C. Aggarwal, J. Han (eds.), *Frequent Pattern Mining,* 261
DOI 10.1007/978-3-319-07821-2_11, © Springer International Publishing Switzerland 2014

code, API specification mining and API usage mining from open source repositories, and Web log data mining [1]. Sequential pattern mining has become a focused theme in data mining research. Over the past years, various surveys on sequential pattern mining have been published and provide useful resources [1–6].

Similar to association rule mining [7], sequential pattern mining was initially motivated by the decision support problem in retail industry and was first addressed by Agrawal and Srikant in [8]. This problem was defined as follows: *Given a set of sequences, where each sequence consists of a list of elements and each element consists of a set of items, and given a user-specified min_support threshold, sequential pattern mining is to find all frequent subsequences, i.e., the subsequences whose occurrence frequency in the set of sequences is no less than min_support* [8].

Since then, abundant literature has been dedicated to this research and tremendous progress has been made. Improvements in sequential pattern mining algorithms have followed similar trend in the related area of association rule mining and have been motivated by the need to process more data at a faster speed with lower cost. Generally, sequential pattern mining algorithms can be categorized into two major classes: Apriori-based approaches [8–14] and pattern growth algorithms [15, 16].

The first class of algorithms (i.e., Apriori-based approaches) form the vast majority of algorithms proposed in the literature for sequential pattern mining. They depend mainly on the Apriori property, which states the fact that any super-pattern of an infrequent pattern cannot be frequent, and are based on a candidate generation-and-test paradigm proposed in association rule mining [7]. These methods have the disadvantage of repeatedly generating an explosive number of candidate sequences and scanning the database to maintain the support count information for these sequences during each iteration of the algorithm, which makes them computationally expensive.

To alleviate these problems, pattern growth approach for efficient sequential pattern mining adopts a divide-and-conquer, pattern growth paradigm as follows, sequence databases are recursively projected into a set of smaller projected databases based on the current sequential pattern(s), and sequential patterns are grown in each projected database by exploring only locally frequent fragments [17]. The frequent pattern growth paradigm removes the need for the candidate generation and prune steps that occur in the Apriori-based algorithms and repeatedly narrows the search space by dividing a sequence database into a set of smaller projected databases, which are mined separately.

Additionally, there are various kinds of extensions for sequential pattern mining, including (1) closed sequential pattern mining, (2) multi-level, multi-dimensional sequential pattern mining, (3) incremental methods, (4) hybrid methods, (5) approximate methods, (6) top-k closed sequential pattern mining, and (7) frequent episode mining. This chapter will discuss algorithms which fall into these categories in detail.

The remainder of this chapter is organized as follows. Section 2 defines the sequential pattern mining problem. Section 3 discusses Apriori-based approaches. In Sect. 4, pattern growth algorithms are introduced. The extensions for sequential pattern mining in different directions are discussed in Sect. 5. Finally, we conclude this chapter in Sect. 6.

Table 11.1 A sequence database

Sequence_id	Sequence
1	$\langle a(abc)(ac)d(cf) \rangle$
2	$\langle (ad)c(bc)(ae) \rangle$
3	$\langle (ef)(ab)(df)cb \rangle$
4	$\langle eg(af)cbc \rangle$

2   Problem Definition

This section presents the formal definition of the sequential pattern mining problem, and its associated notations.

Let $I = \{i_1, i_2, \ldots, i_n\}$ be a set of all *items*, which comprise the alphabet. An *itemset* (or *event*) is a subset of items and denoted by $(i_1 i_2 \cdots i_m)$, where i_k is an item. It is assumed that items in an itemset are sorted in lexicographic order. A *sequence* is an ordered list of itemsets. A sequence s is denoted by $\langle s_1 s_2 \cdots s_l \rangle$, where s_j is an itemset. For brevity, the brackets are omitted if an itemset has only one item, i.e., itemset (i) is written as i. An item can occur at most once in an itemset of a sequence, but can occur multiple times in different itemsets of a sequence. The number of instances of items in a sequence is called the *length* of the sequence. A sequence with length l is called an l-sequence. A sequence $\alpha = \langle a_1 a_2 \cdots a_n \rangle$ is called a subsequence of another sequence $\beta = \langle b_1 b_2 \cdots b_m \rangle$ and β is a super-sequence of α, denoted by $\alpha \sqsubseteq \beta$, if there exist integers $1 \leq j_1 < j_2 < \cdots < j_n \leq m$ such that $a_1 \subseteq b_{j_1}, a_2 \subseteq b_{j_2}, \ldots, a_n \subseteq b_{j_n}$.

A *sequence database* D is a set of tuples $\langle sid, s \rangle$, where sid is a sequence_id and s is a sequence. A tuple $\langle sid, s \rangle$ is said to contain a sequence α, if α is a subsequence of s. The *support* of a sequence α in a sequence database D, denoted by $support_D(\alpha)$, is the number of tuples in the database containing α. It can be denoted by $support(\alpha)$ if the sequence database is clear from the context. Given a positive integer *min_support* as the *minimum support threshold*, a sequence α is said to be *frequent* and called a *sequential pattern* in sequence database D if $support_D(\alpha) \geq min\_support$. A sequential pattern with length l is called an l-pattern. The set of frequent l-sequences is denoted by $\mathcal{F}_l$. If there exists no proper super-sequence of a sequential pattern α with the same support as α, α is called a *closed sequential pattern* (or a *frequent closed subsequence*) in sequence database D. Furthermore, a sequential pattern α is called a *maximal sequential pattern* (or a *frequent maximal subsequence*) if it is not contained in any other sequential pattern.

Formally, given a sequence database D and the minimum support threshold *min_support*, the problems of sequential pattern mining, closed sequential pattern mining, and maximal sequential pattern mining, are to find the set of all frequent subsequences, all frequent closed subsequences, and all frequent maximal subsequences from the input sequence database D, respectively. To estimate the upper bound on the number of sequential patterns given a sequence database, Raïssi and Pei proposed two novel techniques in [18].

A sequence database D is given in Table 11.1 and *min_support*=2. The set of items in the database D is $\{a, b, c, d, e, f, g\}$. A sequence $\langle a(abc)(ac)d(cf) \rangle$ has five itemsets: (a), (abc), (ac), (d), and (cf), where items a and c appear more than once respectively in different itemsets. It is a 9-sequence since there are 9 instances appearing in that sequence. Item a happens three times in this sequence, so it contributes 3 to the length of the sequence. However, the whole sequence $\langle a(abc)(ac)d(cf)\rangle$ contributes only one to the support of the sequence $\langle a \rangle$. Also, sequence $\langle a(bc)df \rangle$ is a subsequence of $\langle a(abc)(ac)d(cf)\rangle$. Since both sequences with sequence_id 1 and 3 contain subsequence $s = \langle (ab)c \rangle$, s is a sequential pattern of length 3 (i.e., 3-pattern). The sequence $\beta = \langle (ab)dc \rangle$ is closed in sequence database D as there exists no super-sequence of the sequential pattern β with the same support as β. Furthermore, this sequence β is also a maximal sequential pattern as it is not contained in any other sequential pattern in this database D.

3 Apriori-based Approaches

Sequential pattern mining can naturally be considered as association rule mining over a temporal database. While Apriori-based association rule mining algorithm [7] discovers intra-transaction patterns (itemsets), sequential pattern mining discovers inter-transaction patterns (sequences), where ordering of items and itemsets is very important. Similar to frequent patterns, sequential patterns also have the Apriori property as follows: every non-empty subsequence of a sequential pattern must also be frequent, which is anti-monotonic (or downward closed). Due to this similarity, it is therefore not surprising the earlier sequential pattern mining algorithms were derived from the Apriori algorithm [7, 19].

From the sequential pattern mining point of view, a sequence database can be represented in two data formats: (1) a horizontal data format, and (2) a vertical data format. The former uses the natural representation of the data set as $\langle sequence\_id :$ $a\_sequence\_of\_itemsets \rangle$, whereas the latter uses the vertical representation of the sequence database: $\langle item : (sequence\_id, itemset\_id) \rangle$, which can be obtained by transforming from a horizontal formatting sequence database.

Based on these data formats, there are two major categories of algorithms developed for efficient sequential pattern mining: (1) horizontal data format algorithms, represented by AprioriAll [8] and GSP [9], and (2) vertical data format algorithms, represented by SPADE [11] and SPAM [12]. We outline and analyze these two categories of Apriori-based sequential pattern mining methods in this section.

3.1 Horizontal Data Format Algorithms

3.1.1 AprioriAll

In the seminal paper on sequential pattern mining, three algorithms were introduced [8]. Among these algorithms, AprioriSome and DynamicSome were proposed for

Table 11.2 A horizontal formatting sequential database. (Adapted from [8])

Customer Id	Customer sequence
1	$\langle(30)\,(90)\rangle$
2	$\langle(10\ 20)(30)(40\ 60\ 70)\rangle$
3	$\langle(30\ 50\ 70)\rangle$
4	$\langle(30)\,(40\ 70)\,(90)\rangle$
5	$\langle(90)\rangle$

mining maximal sequential patterns, while AprioriAll was designed for mining all sequential patterns. Here, we mainly introduce AprioriAll algorithm which discoveries all sequential patterns in this section.

This problem of sequential pattern mining is solved using the following phases:

1. **Sort Phase.** This phase converts the original transaction database into the horizontal formatting sequential database (Table 11.2) by sorting the original database with customer_id as the major key and transaction_time as the minor key.
2. **Litemset Phase.** In the paper [8], the length of a sequence is the number of itemsets in the sequence. The support for an itemset i is defined as the number of customers who bought the items in i in a single transaction. Thus the itemset i and the 1-sequence $\langle i \rangle$ have the same support. An itemset with minimum support is called a large itemset or *litemset*. Note each itemset in a frequent sequence must have minimum support. Hence, any frequent sequence must be a list of litemsets.

    This phase finds the set of all litemsets L. It is straightforward to adapt any of the algorithms in [19] to find litemsets. The main difference is that the support count should be incremented only once per customer even if the customer buys the same set of items in two different transactions. With the minimum support set to 2, in the example database given in Table 11.2, the litemsets are (30), (40), (70), (40 70) and (90). Then the set of litemsets is mapped to a set of contiguous integers for the litemset equality comparison. A possible mapping for this set is (in the form of "litemset: mapped integer"): (30):1, (40):2, (70):3, (40 70):4, (90):5.
3. **Transformation Phase.** Each transaction is replaced by the set of all litemsets contained in that transaction. Transactions that do not contain any litemsets are not retained and a customer sequence that does not contain any litemsets is dropped. The customer sequences that are dropped, however, still contribute to the count of total number of customers. The transformation of the database is shown in Fig. 11.1.
4. **Sequence Phase.** AprioriAll algorithm makes multiple passes over the data. In each pass, it starts with a seed set of frequent sequences, and then uses the seed set for generating new potentially frequent sequences (through *Apriori-generate* function), called candidate sequences. The algorithm finds the support for these candidate sequences during the pass over the data. At the end of each pass, it determines which of the candidate sequences are actually frequent. These frequent candidates become the seed for the next pass. In the first pass, all frequent 1-sequences, obtained in the litemset phase, form the seed set. This algorithm terminates when either no candidates are generated or no candidates meet the minimum support.

C_Id	Original Customer Sequence	Transformed Customer Sequence	After Mapping
1	$\langle (30)\,(90) \rangle$	$\langle \{(30)\}\ \{(90)\} \rangle$	$\langle \{1\}\{5\} \rangle$
2	$\langle (10\ 20)\,(30)\,(40\ 60\ 70) \rangle$	$\langle \{(30)\}\ \{(40),\,(70),\,(40\ 70)\} \rangle$	$\langle \{1\}\{2,\,3,\,4\} \rangle$
3	$\langle (30\ 50\ 70) \rangle$	$\langle \{(30),\,(70)\} \rangle$	$\langle \{1,\,3\} \rangle$
4	$\langle (30)\,(40\ 70)\,(90) \rangle$	$\langle \{(30)\}\ \{(40),\,(70),\,(40\ 70)\}\ \{(90)\} \rangle$	$\langle \{1\}\{2,\,3,\,4\}\{5\} \rangle$
5	$\langle (90) \rangle$	$\langle \{(90)\} \rangle$	$\langle \{5\} \rangle$

Fig. 11.1 Transformed Database. (Adapted from [8])

The inefficiency of AprioriAll algorithm mainly stems from its computationally expensive data transformation phase which transforms each transaction to a set of litemsets (frequent itemsets) in order to find sequential patterns.

3.1.2 GSP

The same authors later proposed a new algorithm called GSP [9], which overcomes the drawbacks of AprioriAll and extends it by allowing time constraints, sliding time windows, and taxonomies. GSP is also a horizontal data format based sequential pattern mining algorithm. Based on the downward closure property of a sequential pattern, GSP adopts a multiple-pass, candidate-generation-and-test approach in sequential pattern mining.

The algorithm is outlined as follows. The first scan finds all of the frequent items which form the set of single item frequent sequences. Each subsequent pass starts with a seed set of sequential patterns, which is the set of sequential patterns found in the previous pass. This seed set is used to generate new potential patterns via a join phase, called candidate sequences. This join step is done in the following way. A sequence s_1 joins with s_2 if the subsequence obtained by dropping the first item of s_1 is the same as the subsequence obtained by dropping the last item of s_2. The candidate sequence generated by joining s_1 with s_2 is the sequence s_1 extended with the last item in s_2. The added item becomes a separate element if it is a separate element in s_2, and part of the last element of s_1 otherwise.

Thus, each candidate sequence contains one more item than a seed sequential pattern, where each element in the pattern may contain one or multiple items. The number of items in a sequence is called the length of the sequence. So, all the candidate sequences in a pass will have the same length. The scan of the database in one pass finds the support for each candidate sequence. All of the candidates whose support in the database is no less than *min_support* form the set of the newly found sequential patterns. This set then becomes the seed set for the next pass. The algorithm terminates when no new sequential pattern is found in a pass, or no candidate sequence can be generated.

The method is illustrated using the following example [3].

Example 1 **(GSP)** Given the sequence database D in Table 11.1 and *min_support* $= 2$, GSP first scans D, collects the support for each item, and finds the set of frequent items, i.e., frequent length-1 subsequences (in the form of

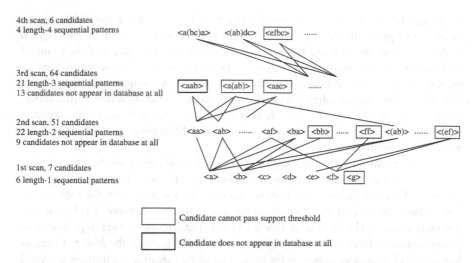

4th scan, 6 candidates
4 length-4 sequential patterns

⟨a(bc)a⟩   ⟨(ab)dc⟩   ⟨efbc⟩   ......

3rd scan, 64 candidates
21 length-3 sequential patterns
13 candidates not appear in database at all

⟨aab⟩   ⟨a(ab)⟩   ⟨aac⟩   ......

2nd scan, 51 candidates
22 length-2 sequential patterns
9 candidates not appear in database at all

⟨aa⟩ ⟨ab⟩ ...... ⟨af⟩ ⟨ba⟩ ⟨bb⟩ ...... ⟨ff⟩ ⟨(ab)⟩ ...... ⟨(ef)⟩

1st scan, 7 candidates
6 length-1 sequential patterns

⟨a⟩   ⟨b⟩   ⟨c⟩   ⟨d⟩   ⟨e⟩   ⟨f⟩   ⟨g⟩

☐ Candidate cannot pass support threshold

☐ Candidate does not appear in database at all

Fig. 11.2 Candidates, candidate generation, and sequential patterns in GSP. (Adapted from [3])

"*item: support*"): $\langle a \rangle : 4$, $\langle b \rangle : 4$, $\langle c \rangle : 4$, $\langle d \rangle : 3$, $\langle e \rangle : 3$, $\langle f \rangle : 3$, $\langle g \rangle : 1$.

By filtering the infrequent item g, we obtain the first seed set $L_1 = \{\langle a \rangle, \langle b \rangle, \langle c \rangle, \langle d \rangle, \langle e \rangle, \langle f \rangle\}$, with each member in the set representing a 1-element sequential pattern. Each subsequent pass starts with the seed set found in the previous pass and uses it to generate new potential sequential patterns, called candidate sequences.

For L_1, a set of 6 length-1 sequential patterns generates a set of $6 \times 6 + \frac{6 \times 5}{2} = 51$ candidate sequences, $C_2 = \{\langle aa \rangle, \langle ab \rangle, \ldots, \langle af \rangle, \langle ba \rangle, \langle bb \rangle, \ldots, \langle ff \rangle, \langle (ab) \rangle, \langle (ac) \rangle, \ldots, \langle (ef) \rangle\}$.

The multi-scan mining process is shown in Fig. 11.2. The set of candidates is generated by a self-join of the sequential patterns found in the previous pass. In the k-th pass, a sequence is a candidate only if each of its length-$(k-1)$ subsequences is a sequential pattern found at the $(k-1)$-th pass. A new scan of the database collects the support for each candidate sequence and finds the new set of sequential patterns. This set becomes the seed for the next pass. Clearly, the number of scans is at least the maximum length of sequential patterns. It needs one more scan if the sequential patterns obtained in the last scan still generate new candidates.

GSP, though benefits from the Apriori pruning, still generates a large number of candidates. In this example, 6 length-1 sequential patterns generate 51 length-2 candidates, 22 length-2 sequential patterns generate 64 length-3 candidates, etc. Some candidates generated by GSP may not appear in the database at all. For example, 13 out of 64 length-3 candidates do not appear in the database.

3.1.3 PSP

PSP [10] is another Apriori-based algorithm, which resumes the general procedures of GSP [9] but utilizes a different hierarchical structure for organizing candidate sequences for the purpose of retrieval efficiency. PSP organizes the candidate sequences in a prefix-tree structure according to their common elements. Any branch of the tree, from the root to a leaf stands for a candidate sequence, and the terminal node of any branch provides the support of the corresponding sequence. Adding to the support value of candidate sequence is performed by navigating to each leaf in the tree and then incrementing the value. As in this prefix-tree structure, initial subsequences common to several candidate sequences are stored only once, this structure requires less memory than the hash-tree used in the GSP approach, which fully stores all candidate sequences in the leaves.

The aforementioned Apriori-based algorithms depend largely on the Apriori property and do not exploit additional strategies to narrow the search space. During each iteration of the algorithms, they have to maintain the support count for each subsequence being mined, which makes them computationally expensive. To alleviate the problem, some Apriori-based approaches [11–14] utilize the vertical representation of the sequence database and employ simple temporal joins of the *id-lists* to calculate support for each sequence. In the remainder of this section, we introduce several vertical data format algorithms in detail.

3.2 Vertical Data Format Algorithms

3.2.1 SPADE

SPADE [11] maps a sequence database into the vertical data format which takes each item as the center of observation and takes its associated sequence and event identifiers as data sets. To find sequence of length-2 items, it just needs to join two single items if they are frequent and they share the same sequence identifier and their event identifiers (which are essentially relative timestamps) follow the sequential ordering. Similarly, SPADE can grow the sequence from length two to length three, and so on. SPADE relies on a lattice of frequent sequences generated by applying the Lattice theory [20] on frequent sequences and their subsequences. It also decomposes the original search space (lattice) into smaller pieces (sub-lattices) called *equivalence classes*, which can be loaded and processed independently in main memory. Two sequences are considered to be in the same class if they share a common k-length prefix. Each sub-lattice can be traversed via either breadth-first or depth-first search to enumerate the frequent sequences, whose counts are then calculated via simple temporal joins (or intersections) on *id-lists*. SPADE usually requires three database scans, or only a single scan with some preprocessed data. Some fragments of the SPADE mining process are illustrated using the following example [3].

Example 2 **(SPADE)** Given the sequence database D in Table 11.1 and $min\_support = 2$, SPADE first scans D, transforms the database into the vertical format by introducing EID (event_ID) which is a (local) timestamp for each event. Each single item is associated with a set of SID (sequence_id) and EID (event_id)

Fig. 11.3 Vertical format of the sequence database and fragments of the SPADE mining process. (Adapted from [3])

SID	EID	Items
1	1	a
1	2	abc
1	3	ac
1	4	d
1	5	cf
2	1	ad
2	2	c
2	3	bc
2	4	ae
3	1	ef
3	2	ab
3	3	df
3	4	c
3	5	b
4	1	e
4	2	g
4	3	af
4	4	c
4	5	b
4	6	c

a		b		···
SID	EID	SID	EID···	
1	1	1	2	
1	2	2	3	
1	3	3	2	
2	1	3	5	
2	4	4	5	
3	2			
4	3			

ab			ba		···
SID	EID(a)	EID(b)	SID	EID(b)	EID(a)···
1	1	2	2	1	3
2	1	3	3	2	4
3	2	5			
4	3	5			

aba				···
SID	EID(a)	EID(b)	EID(a)	···
1	1	2	3	
2	1	3	4	

pairs. For example, item "b" is associated with (SID, EID) pairs as follows: (1, 2), (2, 3), (3, 2), (3, 5), (4, 5), as shown in Fig. 11.3. This is because item b appears in sequence 1, event 2, and so on. Frequent single items "a" and "b" can be joined together to form a length-two subsequence by joining the same sequence_id with event_ids following the corresponding sequence order. For example, subsequence ab contains a set of triples (SID, EID(a), EID(b)), such as (1, 1, 2), and so on. Furthermore, the frequent length-2 subsequences can be joined together based on the Apriori heuristic to form length-3 subsequences, and so on. The process continues until no frequent sequences can be found or no such sequences can be formed by such joins. The detailed analysis of the method can be found in [11].

3.2.2 SPAM

SPAM [12] uses a novel search strategy that integrates a depth-first traversal of the search space with effective pruning mechanisms. The candidate sequences are stored in a lexicographic tree and each sequence in the sequence tree can be considered as either a sequence-extended sequence (via an *S-step*) or an itemset-extended sequence (via an *I-step*). A sequence-extended sequence is a sequence generated by adding a new transaction consisting of a single item to the end of its parent's sequence in the tree. An itemset-extended sequence is a sequence generated by adding an item to the last itemset in the parent's sequence, such that the item is greater than any item in that last itemset.

SPAM traverses the sequence tree described above in a standard depth-first manner. At each node, the support of each sequence-extended child and each itemset-extended child is checked. If the support of a generated sequence s is greater than or equal to *min_support*, SPAM stores that sequence and repeats DFS recursively

Fig. 11.4 Vertical format of the sequence database and fragments of the SPAM mining process. (Adapted from [6])

(a) Data sorted by CID and TID.

Customer ID (CID)	TID	Itemset
1	1	{a, b, c}
1	3	{b, c, d}
1	6	{b, c, d}
2	2	{b}
2	4	{a, b, c}
3	5	{a, b}
3	7	{b, c, d}

(b) Bitmap representation of the dataset in (a)

CID	TID	{a}	{b}	{c}	{d}
1	1	1	1	0	1
1	3	0	1	1	1
1	6	0	1	1	1
-	-	0	0	0	0
2	2	0	1	0	0
2	2	1	1	1	0
-	-	0	0	0	0
-	-	0	0	0	0
3	5	1	1	0	0
3	7	0	1	1	1
-	-	0	0	0	0
-	-	0	0	0	0

(a) S-Step processing

$(\{a\})$		$(\{a\})_s$		$\{b\}$		$(\{a\}, \{b\})$
1		0		1		0
0		1		1		1
1		1		1		1
0		1		0		0
0	S-step	0	&	1	result	0
1	→	1		1	→	0
0	process	1		0		0
0		1		0		0
1		0		1		0
0		1		1		1
0		1		0		0
0		1		0		0

(b) I-step processing

$(\{a\}, \{b\})$		$\{d\}$		$(\{a\}, \{b, d\})$
0		1		0
1		1		1
1		1		1
0		0		0
0	&	0	result	0
0		0	→	0
0		0		0
0		0		0
0		0		0
1		1		1
0		0		0
0		0		0

on *s*. Otherwise, it stops DFS on *s* by the Apriori property. The method of candidate pruning is based on Apriori (downward closure) property and is conducted at each *S-step* and *I-step*, which guarantees that all nodes corresponding to frequent sequences are visited.

To allow for efficient counting, SPAM uses a vertical bitmap representation of the data. A vertical bitmap is created for each item in the dataset, and each bitmap has a bit corresponding to each transaction in the dataset. Each bitmap partition of a sequence to be extended in the S-Step is first transformed to a *transformed bitmap*, such that all the bits less than or equal to the index of the first bit with value one (denoted by k) are set to zero, and all bits after k are set to one. Then, the resulting bitmap can be obtained by the *ANDing* operation of the transformed bitmap and the bitmap of the appended item. In the I-step, *ANDing* is performed directly without transformation of the sequence. Now support-counting becomes a simple count of how many bitmap partitions, not containing all zeros. Figure 11.4 shows the sequence database, its bitmap representation, and an example of the mining process.

3.2.3 LAPIN-SPAM

Based on SPAM, Yang and Kitsuregawa [14] proposed a new algorithm called LAst Position INduction Sequential PAttern Mining (abbreviated as LAPIN-SPAM), which can efficiently get all the frequent sequential patterns from a large database. The main difference between them is the method for candidate sequence counting and verification. While SPAM does many ANDing operations for candidate testing, LAPIN-SPAM can easily implement this process based on the following fact that if an item's last position is smaller than or equal to the current prefix position, the item can not appear behind the current prefix in the same sequence. In order to exploit this fact for candidate pruning, LAPIN-SPAM constructs an ITEM_IS_EXIST_TABLE, which is created while scanning the database for the first time. In this ITEM_IS_EXIST_TABLE, the last position information of each item in the sequence is recorded, and in each iteration, LAPIN-SPAM only needs to check this table to get information that a candidate is behind current position or not for effective candidate sequence pruning.

4 Pattern Growth Algorithms

Apriori-based algorithms generate an explosive number of candidate sequences for mining long sequential patterns, which consume a lot of memory in the mining process. To solve this problem, the pattern growth paradigm and the FP-Growth algorithm [17] emerged in the early 2000 s, firstly proposed for association rule mining. The key idea is to avoid the candidate generation and prune steps that occur in the Apriori-based algorithms and repeatedly narrow the search space by dividing a database into a set of smaller projected databases, which are mined separately.

4.1 FreeSpan

FreeSpan [15] mines sequential patterns by partitioning the search space and projecting the sequence sub-databases recursively based on the projected itemsets. The database projection can be performed as follows. At the time of deriving p's projected database from DB, the set of frequent items X of DB is already known. Only those items in X will need to be projected into p's projected database. This effectively discards irrelevant information and keeps the size of the projected database minimal. By recursively doing so, one can mine the projected databases and generate the complete set of sequential patterns in the given partition without duplication. The details are illustrated in the following example [3].

Example 3 (**FreeSpan**) Given the sequence database D in Table 11.1 and $min\_support = 2$, FreeSpan first scans D, collects the support for each item, and finds the set of frequent items. This step is similar to GSP. Frequent items are listed in support descending order (in the form of "*item: support*"), that is, f_list = $\langle a : 4, b : 4, c : 4, d : 3, e : 3, f : 3 \rangle$. They form six length-one sequential patterns: $\langle a \rangle : 4, \langle b \rangle : 4, \langle c \rangle : 4, \langle d \rangle : 3, \langle e \rangle : 3, \langle f \rangle : 3$.

According to the f_list, the complete set of sequential patterns in D can be divided into 6 disjoint subsets: (1) the ones containing only item a, (2) the ones containing item b but no item after b in f_list, (3) the ones containing item c but no item after c in f_list, and so on, and finally, (6) the ones containing item f.

The sequential patterns related to the six partitioned subsets can be mined by constructing six projected databases (obtained by one additional scan of the original database). Infrequent items, such as g in this example, are removed from the projected databases. The process for mining each projected database is detailed as follows.

Mining sequential patterns containing only item a. The $\langle a \rangle$-projected database is $\{\langle aaa \rangle, \langle aa \rangle, \langle a \rangle, \langle a \rangle\}$. By mining this projected database, only one additional sequential pattern containing only item a, i.e., $\langle aa \rangle : 2$, is found.

Mining sequential patterns containing item b but no item after b in the f_list. By mining the $\langle b \rangle$-projected database: $\{\langle a(ab)a \rangle, \langle aba \rangle, \langle (ab)b \rangle, \langle ab \rangle\}$, four additional sequential patterns containing item b but no item after b in f_list are found. They are $\{\langle ab \rangle : 4, \langle ba \rangle : 2, \langle (ab) \rangle : 2, \langle aba \rangle : 2\}$.

Mining sequential patterns containing item c but no item after c in the f_list. The mining of the $\langle c \rangle$-projected database: $\{\langle a(abc)(ac)c \rangle, \langle ac(bc)a \rangle, \langle (ab)cb \rangle, \langle acbc \rangle\}$, proceeds as follows. One scan of the projected database generates the set of length-2 frequent sequences, which are $\{\langle ac \rangle : 4, \langle (bc) \rangle : 2, \langle bc \rangle : 3, \langle cc \rangle : 3, \langle ca \rangle : 2, \langle cb \rangle : 3\}$. One additional scan of the $\langle c \rangle$-projected database generates all of its projected databases. The mining of the $\langle ac \rangle$-projected database: $\{\langle a(abc)(ac)c \rangle, \langle ac(bc)a \rangle, \langle (ab)cb \rangle, \langle acbc \rangle\}$ generates the set of length-3 patterns as follows: $\{\langle acb \rangle : 3, \langle acc \rangle : 3, \langle (ab)c \rangle : 2, \langle aca \rangle : 2\}$. Four projected database will be generated from them. The mining of the first one, the $\langle acb \rangle$-projected database: $\{\langle ac(bc)a \rangle, \langle (ab)cb \rangle, \langle acbc \rangle\}$ generates no length-4 pattern. The mining along this line terminates. Similarly, we can show that the mining of the other three projected databases terminates without generating any length-4 patterns for the $\langle ac \rangle$-projected database.

Mining other subsets of sequential patterns. Other subsets of sequential patterns can be mined similarly on their corresponding projected databases. This mining process proceeds recursively, which derives the complete set of sequential patterns.

The detailed presentation of the FreeSpan algorithm, the proof of its completeness and correctness, and the performance study of the algorithm are in [15].

4.2 PrefixSpan

PrefixSpan [16] builds upon the concept of FreeSpan but instead of projecting sequence databases it examines only the prefix subsequences and projects only their corresponding suffix subsequences into projected databases. This way, sequential patterns are grown in each projected database by exploring only local frequent sequences. For a sequence $s = \langle a(abc)(ac)d(cf) \rangle$, $\langle a \rangle$, $\langle aa \rangle$, $\langle a(ab) \rangle$ and $\langle a(abc) \rangle$ are prefixes of sequence s, but neither $\langle ab \rangle$ nor $\langle a(bc) \rangle$ is considered as a prefix if every item in the prefix $\langle a(abc) \rangle$ of sequence s is frequent in database D. Also, $\langle (abc)(ac)d(cf) \rangle$ is the suffix w.r.t. the prefix $\langle a \rangle$, $\langle (\_bc)(ac)d(cf) \rangle$ is the suffix w.r.t. the prefix $\langle aa \rangle$, and $\langle (\_c)(ac)d(cf) \rangle$ is the suffix w.r.t. the prefix $\langle a(ab) \rangle$.

The problem of mining sequential patterns can be decomposed into a set of subproblems. Let $\{\langle x_1 \rangle, \langle x_2 \rangle, \cdots, \langle x_n \rangle\}$ be the complete set of length-1 sequential patterns in a sequence database D. The complete set of sequential patterns in D can be divided into n disjoint subsets. The i-th subset ($1 \leq i \leq n$) is the set of sequential patterns with prefix $\{\langle x_i \rangle\}$. Let α be a length-l sequential pattern and $\{\beta_1, \beta_2, \cdots, \beta_m\}$ be the set of all length-$(l + 1)$ sequential patterns with prefix α. The complete set of sequential patterns with prefix α, except for α itself, can be divided into m disjoint

Prefix	Projected Database	Sequential Patterns
$\langle a \rangle$	$\langle (abc)(ac)d(cf) \rangle$, $\langle (\_d)c(bc)(ae) \rangle$, $\langle (\_b)(df)cb \rangle$, $\langle (\_f)cbc \rangle$	$\langle a \rangle$, $\langle aa \rangle$, $\langle ab \rangle$, $\langle a(bc) \rangle$, $\langle a(bc)a \rangle$, $\langle aba \rangle$, $\langle abc \rangle$, $\langle (ab) \rangle$, $\langle (ab)c \rangle$, $\langle (ab)d \rangle$, $\langle (ab)f \rangle$, $\langle (ab)dc \rangle$, $\langle ac \rangle$, $\langle aca \rangle$, $\langle acb \rangle$, $\langle acc \rangle$, $\langle ad \rangle$, $\langle adc \rangle$, $\langle af \rangle$
$\langle b \rangle$	$\langle (\_c)(ac)d(cf) \rangle$, $\langle (\_c)(ae) \rangle$, $\langle (df)cb \rangle$, $\langle c \rangle$	$\langle b \rangle$, $\langle ba \rangle$, $\langle bc \rangle$, $\langle (bc) \rangle$, $\langle (bc)a \rangle$, $\langle bd \rangle$, $\langle bdc \rangle$, $\langle bf \rangle$
$\langle c \rangle$	$\langle (ac)d(cf) \rangle$, $\langle (bc)(ae) \rangle$, $\langle b \rangle$, $\langle bc \rangle$	$\langle c \rangle$, $\langle ca \rangle$, $\langle cb \rangle$, $\langle cc \rangle$
$\langle d \rangle$	$\langle (cf) \rangle$, $\langle c(bc)(ae) \rangle$, $\langle (\_f)cb \rangle$	$\langle d \rangle$, $\langle db \rangle$, $\langle dc \rangle$, $\langle dcb \rangle$
$\langle e \rangle$	$\langle (\_f)(ab)(df)cb \rangle$, $\langle (af)cbc \rangle$	$\langle e \rangle$, $\langle ea \rangle$, $\langle eab \rangle$, $\langle eac \rangle$, $\langle eacb \rangle$, $\langle eb \rangle$, $\langle ebc \rangle$, $\langle ec \rangle$, $\langle ecb \rangle$, $\langle ef \rangle$, $\langle efb \rangle$, $\langle efc \rangle$, $\langle efcb \rangle$.
$\langle f \rangle$	$\langle (ab)(df)cb \rangle$, $\langle cbc \rangle$	$\langle f \rangle$, $\langle fb \rangle$, $\langle fbc \rangle$, $\langle fc \rangle$, $\langle fcb \rangle$

Fig. 11.5 Projected databases and sequential patterns for PrefixSpan algorithm. (Adapted from [3])

subsets. The j-th subset ($1 \le j \le m$) is the set of sequential patterns prefixed with β_j. Based on this observation, the problem can be partitioned recursively. That is, each subset of sequential patterns can be further divided when necessary. This forms a divide-and-conquer framework. In the following, let us examine how to use the prefix-based projection approach for mining sequential patterns based the following example [3].

Example 4 (**PrefixSpan**) Given the sequence database D in Table 11.1 and *min_support* $= 2$, sequential patterns in S can be mined by a prefix-projection method in the following steps.

1. *Find length-1 sequential patterns.* Scan D once to find all the frequent items in sequences. Each of these frequent items is a length-1 sequential pattern. They are $\langle a \rangle$: 4, $\langle b \rangle$: 4, $\langle c \rangle$: 4, $\langle d \rangle$: 3, $\langle e \rangle$: 3, and $\langle f \rangle$: 3, where the notation "$\langle pattern \rangle$: count" represents the pattern and its associated support count.
2. *Divide search space.* The complete set of sequential patterns can be partitioned into the following six subsets according to the six prefixes: (1) the ones with prefix $\langle a \rangle$, (2) the ones with prefix $\langle b \rangle$, ... , and (6) the ones with prefix $\langle f \rangle$.
3. *Find subsets of sequential patterns.* The subsets of sequential patterns can be mined by constructing the corresponding set of projected databases and mining each recursively. The projected databases as well as sequential patterns found in them are listed in Fig. 11.5, while the mining process is explained as follows.
   (a) *Find sequential patterns with prefix $\langle a \rangle$.* Only the sequences containing $\langle a \rangle$ should be collected. Moreover, in a sequence containing $\langle a \rangle$, only the subsequence prefixed with the first occurrence of $\langle a \rangle$ should be considered. For example, in sequence $\langle (ef)(ab)(df)cb \rangle$, only the subsequence $\langle (\_b)(df)cb \rangle$ should be considered for mining sequential patterns prefixed with $\langle a \rangle$. Notice that $(\_b)$ means that the last element in the prefix, which is a, together with b, form one element.

      The sequences in D containing $\langle a \rangle$ are projected w.r.t. $\langle a \rangle$ to form the $\langle a \rangle$-projected database, which consists of four suffix sequences: $\langle (abc)(ac)d(cf) \rangle$, $\langle (\_d)c(bc)(ae) \rangle$, $\langle (\_b)(df)cb \rangle$ and $\langle (\_f)cbc \rangle$.

By scanning the $\langle a \rangle$-projected database once, its locally frequent items are a:2, b:4, $\_b$:2, c:4, d:2, and f:2. Thus all the length-2 sequential patterns prefixed with $\langle a \rangle$ are found, and they are: $\langle aa \rangle$: 2, $\langle ab \rangle$: 4, $\langle (ab) \rangle$: 2, $\langle ac \rangle$: 4, $\langle ad \rangle$: 2, and $\langle af \rangle$: 2.

Recursively, all sequential patterns with prefix $\langle a \rangle$ can be partitioned into 6 subsets: (1) those prefixed with $\langle aa \rangle$, (2) those with $\langle ab \rangle$, ... , and finally, (6) those with $\langle af \rangle$. These subsets can be mined by constructing respective projected databases and mining each recursively in a similar way.

(b) *Find sequential patterns with prefix $\langle b \rangle$, $\langle c \rangle$, $\langle d \rangle$, $\langle e \rangle$ and $\langle f \rangle$, respectively.* This can be done by constructing the $\langle b \rangle$-, $\langle c \rangle$-, $\langle d \rangle$-, $\langle e \rangle$-, and $\langle f \rangle$- projected databases and mining them respectively. The projected databases as well as the sequential patterns found are shown in Fig. 11.5.

4. The set of sequential patterns is the collection of patterns found in the above recursive mining process.

The major advantage of PrefixSpan is that it does not generate and test any candidate sequences that do not exist in a projected database. Unlike Apriori-based algorithms, PrefixSpan only grows longer sequential patterns from the shorter frequent ones. The major cost of PrefixSpan is database projection, i.e., forming projected databases recursively. To alleviate this problem, a *pseudo-projection* method is exploited to reduce this cost. Instead of performing physical projection, one can register the index (or identifier) of the corresponding sequence and the starting position of the projected suffix in the sequence. Then, a physical projection of a sequence is replaced by registering a sequence identifier and the projected position index point. Pseudo-projection reduces the cost of projection substantially when the projected database can fit in main memory.

5 Extensions

With the successful development of the sequential pattern mining method (e.g., Apriori-based approach and pattern growth algorithm), it is interesting to explore how such a method can be extended to handle more sophisticated mining requests. In this section, we will discuss a few extensions of the sequential pattern mining approach. Specifically, extensions for sequential pattern mining include (1) closed sequential pattern mining, (2) multi-level, multi-dimensional sequential pattern mining, (3) incremental methods, (4) hybrid methods, (5) approximate methods, (6) top-k closed sequential pattern mining, and (7) frequent episode mining.

5.1 Closed Sequential Pattern Mining

The sequential pattern mining algorithms discussed above have good performance in databases consisting of short frequent sequences. However, when mining long frequent sequences, or when using very low support thresholds, the performance of such algorithms often degrades dramatically. In the field of association rule mining, algorithms such as CHARM [21], CLOSET [22], CLOSET+ [23], CARPENTER[24] are proposed for mining *frequent closed itemsets*, which overcome some of these difficulties. Similarly, in the area of sequential pattern mining, Yan et al. [25] proposed an algorithm called CloSpan, which mines only closed sequential patterns instead of mining the complete set of sequential patterns. CloSpan can produce a significantly less number of sequences than the traditional (i. e., full-set) methods while preserving the same expressive power since the whole set of frequent subsequences, together with their supports, can be derived easily from the closed sequential pattern mining results.

CloSpan [25] is developed based on the philosophy of sequential pattern growth and mines closed sequential patterns efficiently by discovery of sharing portions of the projected databases in the mining process and pruning any redundant search space, which therefore substantially enhances the mining efficiency. The algorithm first uses a lexicographic sequence tree to store the generated sequences using both I-Step and S-Step mechanisms and discovers all of the frequent sequences (closed and non-closed). During this mining process, CloSpan introduces a search space pruning condition: whenever it finds two exactly same prefix-based project databases, it can stop growing one prefix, which is called equivalent projected database pruning. To determine whether two projected databases are the same for two sequences one of which is a subsequence of another, CloSpan just needs to compare the size of the two projected databases, which has been proved in [25]. Finally, CloSpan uses a post-pruning step to filter out non-closed sequences.

BIDE [26] is an efficient algorithm for mining frequent closed sequences without candidate maintenance. It adopts a novel sequence closure checking scheme called *BI-Directional Extension*, where the forward directional extension is used to grow the prefix patterns and also checks the closure of prefix patterns, while the backward directional extension can be used to both check closure of a prefix pattern and prune the search space. Furthermore, the search space pruning is made more efficient by using the *BackScan* pruning method and the *ScanSkip* optimization technique [26]. The *BackScan* search space pruning is based on the theorem that given a prefix s_p, if $\exists i$ (i is a positive integer and is no greater than the length of s_p) and there exists any item that appears in each of its i-th semi-maximum periods, s_p can be safely pruned. The interested readers are referred to [26, 27] for more details.

Li et al. [28] proposed Gap-BIDE for mining closed sequential patterns with gap constraints from a set of input sequences. As Gap-BIDE inherits the same design philosophy as BIDE algorithm [26], it shares the same merit, that is, it does not need to maintain a candidate pattern set, which saves space consumption. Specifically, Gap-BIDE firstly establishes a framework to enumerate all the frequent gap-constrained patterns. Then, it leverages a gap-constrained Bi-Directional closure checking scheme under this enumeration framework, and thus avoids keeping a large candidate set of closed frequent gap-constrained patterns in order for checking if a newly mined pattern is closed. At last, it derives a gap-constrained BackScan pruning technique to prune the unpromising parts of the search space for closed gap-constrained sequential pattern mining.

Summarization subsequences which are used for sequence clustering are a subset of closed sequential patterns. Wang et al. [29] proposed a new algorithm, CON-TOUR, which efficiently mines summarization subsequences directly from the input sequences, and uses the set of summarization subsequences to construct a sequence clustering algorithm. CONTOUR is based on the pattern growth paradigm and leverages some effective pruning methods to prune the unpromising parts of search space by fully exploring some nice properties of the summarization subsequences.

5.2 Multi-level, Multi-dimensional Sequential Pattern Mining

Algorithms discussed so far are based on 1 or 2-dimensional spaces. In many applications, sequences are often associated with different circumstances, and such circumstances form a multiple dimensional space. For example, customer purchase sequences are associated with region, time, customer group, and others. It is interesting and useful to mine sequential patterns associated with multi-dimensional information. For example, one may find that retired customers (with age) over 60 may have very different patterns in shopping sequences from the professional customers younger than 40. Similarly, items in the sequences may also be associated with different levels of abstraction, and such multiple abstraction levels will form a multi-level space for sequential pattern mining. For example, one may not be able to find any interesting buying patterns in an electronics store by examining the concrete models of products that customers purchase. However, if the concept level is raised a little high to brand-level, one may find some interesting patterns, such as *"if one bought an IBM PC, it is likely s/he will buy a new IBM Laptop and then a Cannon digital camera within the next six months"* [3].

Pinto et al. [30] proposed a *uniform sequential* (or *Uni-Seq*) algorithm by embedding multi-dimensional information into sequences. For each sequence, a set of multi-dimensional circumstance values can be treated as one added transaction in the sequence. Similarly, for each item, its associated multi-level information can be added as additional items into the same transaction where that item resides. With such transformation, the database becomes a typical single-dimensional, single-level sequence database, and the PrefixSpan algorithm is applied to efficiently mine such transformed sequence databases. In the same work [30], Pinto et al. also proposed *Seq-Dim* and *Dim-Seq* algorithms, which divide the mining process into two steps. Seq-Dim algorithm first mines sequential patterns, and then for each sequential pattern, forms projected multi-dimensional database and finds multi-dimensional patterns within the projected databases, while Dim-Seq algorithm uses the reverse procedure.

Yu and Chen formally defined the multi-dimensional sequential pattern mining problem in [31]. In this study, they introduced two algorithms, the first of which is developed by modifying the traditional Apriori algorithm [19] and the second by modifying the PrefixSpan algorithm [16]. The first algorithm has different methods for candidate generation and support counting compared with the original Apriori algorithm. The second algorithm has different approaches for sequential pattern growth and projected database construction compared with the original PrefixSpan algorithm. For both algorithms, different dimensional scopes of each element are considered as the key factor for algorithm design. For more details, you could refer to [31].

5.3 Incremental Methods

Many real life sequence databases grow incrementally. It is undesirable to mine sequential patterns from scratch each time when a small set of sequences grow, or when some new sequences are added into the database. Incremental algorithm should be developed for sequential pattern mining so that mining can be adapted to incremental database updates.

Parthasarathy et al. [32] developed an incremental mining algorithm ISM based on the SPADE algorithm. Their goal is to minimize the I/O and computation requirements for handling incremental updates. To achieve this goal, ISM algorithm uses an efficient memory management scheme that indexes into the database efficiently, and creates an Increment Sequence Lattice (ISL), which consists of all the frequent sequences (FS) and all sequences in the negative border (NB) in the original database. FS denotes the set of all frequent sequences in the updated database, and the negative border (NB) is the collection of all sequences that are not frequent but both of whose generating subsequences are frequent. The support of each member is kept in the lattice, too. The algorithm consists of two phases. Phase 1 is for updating the supports of elements in NB and FS, and aims at pruning the sequences that become infrequent from the set of frequent sequences after the update. One scan of the database is enough to update the lattice as well as the negative border. Phase 2 is for adding to NB and FS beyond what was done in Phase 1. For the details of this algorithm, you could refer to [32].

Masseglia et al. [33] developed another incremental mining algorithm ISE for computing the frequent sequences in the updated database when new transactions and new customers are added to the original database using candidate generation-and-test approach. ISE minimizes computational costs by re-using the minimal information from the old frequent sequences, i. e. the support of frequent sequences. To find all new frequent sequences, three kinds of frequent sequences are considered. First, sequences embedded in the original database could become frequent since they have sufficient support with the incremental database, i. e. sequences similar to sequences embedded in the original database appear in the increment. Next, new frequent sequences not appearing in the original database may emerge in the incremental database. Finally, sequences of the original database might become frequent when items from the original database are added. To discover frequent sequences, the ISE algorithm executes iteratively. During the first pass on the incremental database (db), the ISE algorithm counts the support of individual items and finds the set of items occurring at least once in db. Considering the set of items embedded in the original database (DB), it determines which items of db are frequent in U (DB $\cup$ db). This set is called L_1^{db}. These frequent 1-sequences can be used as seeds to discover new frequent sequences of the following types: (1) previous frequent sequences in DB which can be extended by items of L_1^{db}; (2) frequent subsequences in DB which are predecessor items in L_1^{db}; (3) candidate sequences generated from L_1^{db}. This process is continued iteratively (since after the first step frequent 2-sequences are obtained to be used in the same manner as described before) until no more candidates are generated.

In [34], an efficient algorithm, called IncSpan, is developed, for incremental mining over multiple database increments. Several novel ideas are introduced in the algorithm development: (1) maintaining a set of "almost frequent" sequences as the candidates in the updated database, which has several nice properties and leads to efficient techniques, and (2) two optimization techniques, *reverse pattern matching* and *shared projection*, are designed to improve the performance. Reverse pattern matching is used for matching a sequential pattern in a sequence and prune some search space. Shared projection is designed to reduce the number of database projections for some sequences which share a common prefix [34].

Gao et al. [35] proposed an efficient incremental algorithm called StreamCloSeq for mining closed sequential patterns over stream data, where new data arrives continuously and the data distribution often evolves over time. StreamCloSeq stores only frequent closed sequence prefixes in the enumeration tree, used for mining and maintaining patterns in the current sliding window, to solve the frequent closed sequential pattern mining problem efficiently over stream data. Some novel effective search space pruning and pattern closure checking strategies have been also devised to accelerate the algorithm. For the details of this algorithm, you could refer to [35].

5.4 Hybrid Methods

DISC-all algorithm [36] combines the candidate sequence pruning, database partitioning and projection with a strategy called *DIrect Sequence Comparison* (abbreviated as *DISC*), which can find frequent sequences of a specific length k without having to compute the support counts of non-frequent sequences. The DISC strategy defines the order of two sequences having the same length using lexicographical ordering and temporal ordering. For example, $\langle (a)(b)(h) \rangle$ is smaller than $\langle (a)(c)(f) \rangle$ because in the second transactions, b is smaller than c owing to the lexicographical ordering. Furthermore, $\langle (a, b)(c) \rangle$ is smaller than $\langle (a)(b, c) \rangle$ because of the temporal ordering of itemsets in the first transactions.

According to this defined sequence ordering, if t_k is the smallest k-length subsequence of a customer sequence, t_k is called *k-minimum subsequence*. Then, DISC-all algorithm sorts customer sequences by the order of their associated k-minimum subsequences, which compose a *k-sorted database*. In a k-sorted database, the k-minimum subsequence at the first position is denoted by α_1 and the k-minimum subsequence at the δ-th position is denoted by α_δ, where δ is the minimum support count. These two k-minimum subsequences are compared and if they are equal then

α_1 is frequent since the first δ customer sequences in the k-sorted database take α_1 as their k-minimum subsequence, and if α_1 is smaller than α_δ, then α_1 is non-frequent and all subsequences up to and including α_δ can be skipped. This process is then repeated for the next k-minimum subsequence in the resorted k-sorted database. Thus, the DISC strategy uses a k-sorted database to find all the frequent k-sequences and skips most non-frequent k-sequences by checking only the conditional k-minimum subsequences. It uses a partitioning method similar to PrefixSpan for generating frequent 2-sequences and 3-sequences, and then employs the DISC strategy to generate the remaining frequent sequences.

5.5 Approximate Methods

Conventional sequential pattern mining methods may meet inherent difficulties in mining databases with long sequences and noise. They may generate a huge number of short and trivial patterns but fail to find interesting patterns approximately shared by many sequences. In [37], Kum et al. proposed the theme of *approximate sequential pattern mining*. The general idea is that, instead of finding exact patterns, they identify patterns approximately shared by many sequences. Instead of mining a huge set of patterns, they propose to mine consensus patterns from databases of long sequences. Intuitively, a consensus pattern is shared by many sequences and covers many short patterns. To mine consensus sequential patterns from large databases, Kum et al. developed an efficient algorithm ApproxMAP, a cluster and multiple alignment based approach, which works in two steps. First, sequences in a database are clustered based on similarity. Sequences in the same cluster may approximately follow some similar patterns. To enable the clustering of sequences, a modified version of the *hierarchical edit distance* metric is used in a density-based clustering algorithm. Then, the longest approximate sequential pattern for each cluster is generated. It is called the consensus pattern. To extract consensus patterns, a weighted sequence which records the statistics of the alignment of the sequences is derived for each cluster using multiple alignment to compress the sequential pattern information in the cluster. And then the longest consensus pattern best representing the cluster is generated from the weighted sequence.

5.6 Top-k Closed Sequential Pattern Mining

Mining closed sequential patterns may significantly reduce the number of patterns generated and is information lossless because it can be used to derive the complete set of sequential patterns. However, setting *min_support* is a subtle task: a too small value may lead to the generation of thousands of patterns, whereas a too big one may lead to no answer found. To come up with an appropriate *min_support*, one needs prior knowledge about the mining query and the task-specific data, and be able to estimate beforehand how many patterns will be generated with a particular threshold [3].

An algorithm called TSP [38] is proposed to discover top-k closed sequential patterns of length no less than min_l, where k is the desired number of closed sequential patterns to be mined and min_l is the minimal length of each pattern. TSP is a multi-pass search space traversal algorithm that finds the most frequent patterns early in the mining process and allows dynamic raising of $min\_support$ which is then used to prune unpromising branches in the search space. Also, TSP devises an efficient closed pattern verification method which guarantees that during the mining process the candidate result set consists of the desired number of closed sequential patterns. The efficiency of TSP is further improved by applying the minimum length constraint in the mining and by employing the early termination conditions developed in CloSpan [25].

5.7   Frequent Episode Mining

Introduced by Mannila et al. [39], an episode is defined as a collection of events that occur relatively close to each other in a given partial order, and the task of frequent episode mining is to find all episodes that occur frequently in an event sequence, given a class of episodes and an input sequence of events. The algorithm proposed in [39] works iteratively, alternating between building and recognition phases. First, in the building phase of an iteration i, a collection C_i of new candidate episodes of i elementary events is built, using the information available from smaller frequent episodes. Then, these candidate episodes are recognized in the event sequence and their frequencies are computed. The collection L_i consists of frequent episodes in C_i. In the next iteration $i + 1$, candidate episodes in C_{i+1} are built using the information about the frequent episodes in L_i. The algorithm starts by constructing C_i to contain all episodes consisting of single elementary events. In the end, the frequent episodes in L_i for all i are output. This is a typical Apriori-like algorithm under which the downward closure principal holds, that is, all subepisodes of a frequent episode are frequent.

Tatti and Cule [40, 41] introduced a technique for discovering closed episodes. They introduced a new subclass of general episodes, called strict episodes, which can ease the computational burden caused by the traditional closure-definition of serial episodes. An episode is strict if all nodes with the same label are connected. This class of strict episodes is large, containing all serial and parallel episodes, as well as episodes with unique labels. A natural subset relationship between episodes is introduced based on the subset relationship of sequences covering the episodes, and the subset relationship can be computed efficiently for strict episodes. In order to mine closed episodes they defined an auxiliary closure operator. This closure satisfies the needed properties so that they can use the existing framework for mining closed patterns. Discovering the true closed episodes is done via a post-processing step.

6 Conclusions and Summary

This chapter presents a comprehensive survey and analysis of main approaches to sequential pattern mining. Two main classes of algorithms (i. e., Apriori-based approaches and pattern growth algorithms) for sequential pattern mining are discussed in detail. Additionally, various kinds of extensions of sequential pattern mining are also covered in this chapter, including closed, multi-level, multi-dimensional, top-k closed sequential pattern, frequent episode mining and incremental, hybrid, approximate methods.

References

1. J. Wang, "Sequential patterns," *Encyclopedia of Database Systems. LING LIU and M. TAMER OZSU (Eds.)*, pp. 2621–2626, 2009.
2. J. Pei, J. Han, and W. Wang, "Constraint-based sequential pattern mining: the pattern-growth methods," *J. Intell. Inf. Syst.*, vol. 28, no. 2, pp. 133–160, Apr. 2007.
3. J. Han, J. Pei, and X. Yan, "Sequential pattern mining by pattern-growth: Principles and extensions," *StudFuzz*, vol. 180, pp. 183–220, 2005.
4. J. Han, H. Cheng, D. Xin, and X. Yan, "Frequent pattern mining: current status and future directions," *Data Min. Knowl. Discov.*, vol. 15, no. 1, pp. 55–86, Aug. 2007.
5. N. R. Mabroukeh and C. I. Ezeife, "A taxonomy of sequential pattern mining algorithms," *ACM Comput. Surv.*, vol. 43, no. 1, pp. 3:1–3:41, Dec. 2010.
6. C. H. Mooney and J. F. Roddick, "Sequential pattern mining—approaches and algorithms," *ACM Comput. Surv.*, vol. 45, no. 2, pp. 19:1–19:39, Mar. 2013.
7. R. Agrawal, T. Imieliński, and A. Swami, "Mining association rules between sets of items in large databases," in *ACM SIGMOD conference*, 1993, pp. 207–216.
8. R. Agrawal and R. Srikant, "Mining sequential patterns," in *ICDE Conference*, 1995, pp. 3–14.
9. R. Srikant and R. Agrawal, "Mining sequential patterns: Generalizations and performance improvements," in *EDBT Conference*, 1996, pp. 3–17.
10. F. Masseglia, F. Cathala, and P. Poncelet, "The psp approach for mining sequential patterns," in *PKDD Conference*, 1998, pp. 176–184.
11. M. J. Zaki, "Spade: An efficient algorithm for mining frequent sequences," *Mach. Learn.*, vol. 42, no. 1–2, pp. 31–60, Jan. 2001.
12. J. Ayres, J. Flannick, J. Gehrke, and T. Yiu, "Sequential pattern mining using a bitmap representation," in *ACM SIGKDD Conference*, 2002, pp. 429–435.
13. L. Savary and K. Zeitouni, "Indexed bit map (ibm) for mining frequent sequences," in *PKDD Conference*, 2005, pp. 659–666.
14. Z. Yang and M. Kitsuregawa, "Lapin-spam: An improved algorithm for mining sequential pattern," in *ICDE Workshops*, 2005.
15. J. Han, J. Pei, B. Mortazavi-Asl, Q. Chen, U. Dayal, and M.-C. Hsu, "Freespan: frequent pattern-projected sequential pattern mining," in *ACM SIGKDD Conference*, 2000, pp. 355–359.
16. J. Pei, J. Han, B. Mortazavi-asl, H. Pinto, Q. Chen, U. Dayal, and M. chun Hsu, "Prefixspan: Mining sequential patterns efficiently by prefix-projected pattern growth," in *ICDE Conference*, 2001, pp. 215–224.
17. J. Han and J. Pei, "Mining frequent patterns by pattern-growth: methodology and implications," *SIGKDD Explor. Newsl.*, vol. 2, no. 2, pp. 14–20, Dec. 2000.
18. C. Raïssi and J. Pei, "Towards bounding sequential patterns," in *ACM SIGKDD*, 2011, pp. 1379–1387.
19. R. Agrawal and R. Srikant, "Fast algorithms for mining association rules in large databases," in *VLDB Conference*, 1994, pp. 487–499.

20. B. A. Davey and H. A. Priestley, Eds., *Introduction to lattices and order*. Cambridge: Cambridge University Press, 1990.
21. M. J. Zaki and C. jui Hsiao, "Charm: An efficient algorithm for closed itemset mining," in *SDM Conference*, 2002, pp. 457–473.
22. J. Pei, J. Han, and R. Mao, "Closet: An efficient algorithm for mining frequent closed itemsets," in *ACM SIGMOD Workshop*, 2000, pp. 21–30.
23. J. Wang, J. Han, and J. Pei, "Closet+: searching for the best strategies for mining frequent closed itemsets," in *ACM SIGKDD Conference*, 2003, pp. 236–245.
24. F. Pan, G. Cong, A. K. H. Tung, J. Yang, and M. J. Zaki, "Carpenter: finding closed patterns in long biological datasets," in *ACM SIGKDD Conference*, 2003, pp. 637–642.
25. X. Yan, J. Han, and R. Afshar, "Clospan: Mining closed sequential patterns in large datasets," in *SDM Conference*, 2003, pp. 166–177.
26. J. Wang and J. Han, "Bide: Efficient mining of frequent closed sequences," in *ICDE Conference*, 2004.
27. J. Wang, J. Han, and C. Li, "Frequent closed sequence mining without candidate maintenance," *TKDE*, vol. 19, no. 8, pp. 1042–1056, 2007.
28. C. Li, Q. Yang, J. Wang, and M. Li, "Efficient mining of gap-constrained subsequences and its various applications," *ACM Trans. Knowl. Discov. Data*, vol. 6, no. 1, pp. 2:1–2:39, Mar. 2012.
29. J. Wang, Y. Zhang, L. Zhou, G. Karypis, and C. C. Aggarwal, "Contour: An efficient algorithm for discovering discriminating subsequences," *Data Min. Knowl. Discov.*, vol. 18, no. 1, pp. 1–29, Feb. 2009.
30. H. Pinto, J. Han, J. Pei, K. Wang, Q. Chen, and U. Dayal, "Multi-dimensional sequential pattern mining," in *CIKM Conference*, 2001, pp. 81–88.
31. C.-C. Yu and Y.-L. Chen, "Mining sequential patterns from multidimensional sequence data," *IEEE Trans. on Knowl. and Data Eng.*, vol. 17, no. 1, pp. 136-140, Jan. 2005.
32. S. Parthasarathy, M. J. Zaki, M. Ogihara, and S. Dwarkadas, "Incremental and interactive sequence mining," in *CIKM Conference*, 1999, pp. 251–258.
33. F. Masseglia, P. Poncelet, and M. Teisseire, "Incremental mining of sequential patterns in large databases," *Data Knowl. Eng.*, vol. 46, no. 1, pp. 97–121, 2003.
34. H. Cheng, X. Yan, and J. Han, "Incspan: incremental mining of sequential patterns in large database," in *ACM SIGKDD Conference*, 2004, pp. 527–532.
35. C. Gao, J. Wang, and Q. Yang, "Efficient mining of closed sequential patterns on stream sliding window," in *ICDM Conference*, 2011.
36. D.-Y. Chiu, Y.-H. Wu, and A. L. P. Chen, "An efficient algorithm for mining frequent sequences by a new strategy without support counting," in *ICDE Conference*, 2004, pp. 375–386.
37. H. Kum, J. Pei, W. Wang, and D. Duncan, "Approxmap: Approximate mining of consensus sequential patterns," in *SDM Conference*, 2002.
38. P. Tzvetkov, X. Yan, and J. Han, "Tsp: Mining top-k closed sequential patterns," in *ICDM Conference*, 2003.
39. H. Mannila, H. Toivonen, and A. I. Verkamo, "Discovering frequent episodes in sequences," in *KDD Conference*, 1995, pp. 210–215.
40. N. Tatti and B. Cule, "Mining closed strict episodes," in *ICDM Conference*, 2010, pp. 501–510.
41. N. Tatti and B. Cule, "Mining closed strict episodes," *Data Min. Knowl. Discov.*, vol. 25, no. 1, pp. 34–66, Jul. 2012.

Chapter 12
Spatiotemporal Pattern Mining: Algorithms and Applications

Zhenhui Li

Abstract With the fast development of positioning technology, spatiotemporal data has become widely available nowadays. Mining patterns from spatiotemporal data has many important applications in human mobility understanding, smart transportation, urban planning and ecological studies. In this chapter, we provide an overview of spatiotemporal data mining methods. We classify the patterns into three categories: (1) individual periodic pattern; (2) pairwise movement pattern and (3) aggregative patterns over multiple trajectories. This chapter states the challenges of pattern discovery, reviews the state-of-the-art methods and also discusses the limitations of existing methods.

Keywords Spatiotemporal data · Trajectory · Moving object · Data mining

1 Introduction

With the rapid development of positioning technologies, sensor networks, and online social media, spatiotemporal data is now widely collected from smartphones carried by people, sensor tags attached to animals, GPS tracking systems on cars and airplanes, RFID tags on merchandise, and location-based services offered by social media. While such tracking systems act as real-time monitoring platforms, analyzing spatiotemporal data generated from these systems frames many research problems and high-impact applications:

- Understanding animal movement is important to addressing environmental challenges such as climate and land use change, bio-diversity loss, invasive species, and infectious diseases.
- Traffic patterns help people with fastest path finding based on dynamic traffic information; automatic and early identification of traffic incidents; and safety alerts when dangerous driving behaviors are recognized.

Z. Li (✉)
Pennsylvania State University, University Park, USA
e-mail: jessieli@ist.psu.edu

C. C. Aggarwal, J. Han (eds.), *Frequent Pattern Mining,*
DOI 10.1007/978-3-319-07821-2_12, © Springer International Publishing Switzerland 2014

- Unusual vessel trajectory could be a sign of smuggling; outlying taking-off/landing patterns could be a dangerous signal for aviation; and detection of suspicious human movements could help prevent crimes and terrorism.
- Spatiotemporal interactions of human may tell the semantic relationships among them such as colleague, family or friend relationships. Different from cyber social network such as Facebook friends, spatiotemporal relationships reveal more complicated physical social network.

This book chapter discusses the state-of-art data mining methods to discover underlying patterns in movements. Various patterns, characteristics, anomalies, and actionable knowledge can be mined from massive moving object data. We will focus on following three categories of movement patterns:

- *Individual periodic pattern.* One most basic pattern in moving objects is the periodicity. Human repeat daily or weekly movement patterns. Animals have seasonal migration patterns. We will discuss how to automatically *detect the periods* in a trajectory and how to *mine frequent periodic patterns* after periods are detected. We will also describe the methods of using periodic patterns for *future movement prediction.*
- *Pairwise movement pattern.* Focusing on two moving objects only, we will discuss different trajectory *similarity measures* and the methods to mine generic, behavioral and semantic patterns. *Generic patterns* include the attraction or avoidance relationships between two moving objects. In *behavioral patterns*, we will mainly discuss how to detect the following and leadership patterns. To mine *semantic relationships*, such as colleague or friends, we will discuss the supervised learning frameworks with various spatiotemporal features.
- *Aggregate patterns over multiple trajectories.* The aggregate patterns describe a group of moving objects share similar movement patterns. *Frequent trajectory patterns* can find the frequent sequential transitions among spatial regions. *Moving object clusters*, such as flock, convoy and swarm, will detect a group of moving objects being spatially close for a relatively long period of time. *Trajectory clustering* groups similar (sub-)trajectories and reveals the popular paths shared by trajectories.

The rest of the chapter is organized as follows. Section 2 introduces the basic definitions and concepts in spatiotemporal data mining. We then study the individual periodic patterns in Sect. 3. Section 4 covers pairwise movement patterns. And we present aggregate patterns in Sect. 5. Finally, we summarize the chapter in Sect. 6.

2 Basic Concept

2.1 Spatiotemporal Data Collection

Spatiotemporal data is a broad concept. As long as the data is related to spatial and temporal information, we call it spatiotemporal data. Two most frequently seen spatiotemporal data are (1) ID-based spatiotemporal data collected from GPS and (2) location-based data collected from sensors.

Table 12.1 A sample of real moving object data showing non-constant sampling rate

Id	Timestamp	Location-long	Location-lat
2635	1997-07-24 20:50:00	− 149.007	63.809
2635	1997-07-24 21:23:35	− 148.897	63.766
2635	1997-07-27 22:30:23	− 148.967	63.824
2635	1997-07-31 02:52:48	− 149.026	63.803
2635	1997-08-03 01:47:04	− 149.046	63.795

An ID-based spatiotemporal data is essentially a trajectory. The tracking device is attached to a moving object. For example, scientists can embed sensors on animals' body and use GPS to track them; cellphone data can reveal an individual person's movement; and GPS embedded in cars can track a vehicle's movement. Suppose we have trajectories of n moving objects $\{o_1, o_2, \ldots, o_n\}$. Each trajectory is represented as a sequence of points (x_1, y_1, t_m), (x_2, y_2, t_m), $\ldots$, (x_n, y_n, t_m), where (x_i, y_i) is a location (longitude and latitude) and t_i is the time when location (x_i, y_i) is recorded. The trajectory data could contain a large set of moving objects and the tracking time for moving objects could expand several years.

A location-based spatiotemporal data is the temporal data collected from a fixed location. The tracking devices (i.e., sensors) are fixed at certain locations. For example, sensors embedded on the road can track the speed and volume of the traffic; sensors are installed at various locations to track the weather information, such as temperature, wind speed and humidity. There are a set of associated properties at location (x, y) at time t. We use $f(x, y, t, p)$ to denote the value of property p at location (x, y) at time t.

In this book chapter, we will focus on ID-based spatiotemporal data (i.e., trajectories). We will mainly discuss about the patterns of animal and human movement data.

2.2   Data Preprocessing

The raw trajectory data are unevenly sampled and could contain a long period of missing data. Table 12.1 shows a sample of raw trajectory data. As we can see that the data is sampled with uneven gaps and there could be 3–4 days missing data. Depending on different tracking scenarios, the sampling rate of movement could vary from seconds to days. For bird tracking, the data could be sampled every 3–5 days in order to save battery and make the tracking time span to several years. For vehicles, the sampling rate could be as small as seconds. For mobile phone users, there is a reported point only when the user is connecting to cellphone towers.

Most of trajectory mining methods assume the data is evenly sampled. A simple and commonly used preprocessing step is to use linear interpolation to make the data evenly gapped. If two consecutive points in a trajectory are gapped with a long time period, linear interpolation may introduce a lot of errors. For example, one data point of a human trajectory is being at home at 9 p.m. on Monday and the next point is being at home at 10 p.m. on Wednesday. If we use 1 h to linearly interpolate the missing data for these 2 days, all the points between 9 p.m. Monday to 10 p.m. Wednesday will be at home. So it is better to mark those points during the long missing period as invalid points. And when conducting pattern mining methods, we will only consider the valid points. When designing data mining methods, we should pay attention to the issue of incomplete, noisy, and unevenly sampled data. Ideally, a pattern mining method should take the raw data as input or even handle the raw data with uncertainties.

2.3   Background Information

Few moving objects move in free space. Vehicles, obviously, need to follow the road network. Planes and boats need to follow more or less the scheduled paths. Animals, which live in a more free space, are also confined to embedding landscape, such as rivers, mountains and the food resources.

When considering the background information, the mining tasks become more challenging. For example, the distance between two cars cannot be calculated simply by Euclidean distance. Similarly for animals, if there is a mountain or a big river between two animals, they could be actually far away from each other. Considering background information will result in more complex distance calculation and correspondingly require different data mining methods.

For domain experts to interpret the discovered patterns, it is important to consider the underlying geography in order to understand where, when and ultimately why the entities move the way they do. Grazing sheep, for example, may perform a certain movement pattern only when they are on a certain vegetation type. Homing pigeons may show certain flight patterns only when close to a salient landscape feature such as a rive or a highway. And, the movement patterns expressed by tracked vehicle will obviously be very dependent on the environment the vehicle is moving in, be it in a car park, in a suburb or on a highway. Thus, patterns have to be conceptualized that allow linking of the movement with the embedding environment.

3   Individual Periodic Pattern

One most common activity in moving objects is the *periodic behavior*. A periodic behavior can be loosely defined as the repeating activities at certain locations with regular time intervals. For example, bald eagles start migrating to South America in late October and go back to Alaska around mid March.

Periodic behaviors provide an insightful and concise explanation over the long moving history. For example, animal movements could be summarized using several *daily* and *yearly* periodic behaviors. Periodic behaviors are also useful for compressing movement data [3, 28, 38]. Moreover, periodic behaviors are useful in future movement prediction [17], especially for a distant querying time. At the same time,

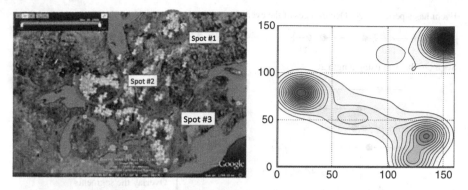

Fig. 12.1 Figure on the *left* shows the trajectory of a bald eagle over 3 years. Each *yellow pin* is a recorded GPS locations. Figure on the *right* shows the density map of all the locations in the trajectory. Periodica first detects *dense areas* as reference spots and then find periodicity for each reference spot [22]

if an object fails to follow regular periodic behaviors, it could be a signal of abnormal environment change or an accident.

In this section, we will first introduce how to automatically detect periods in a trajectory. Then, we will discuss the methods to mine frequent periodic patterns from a trajectory. Lastly, we will show how to use periodic patterns for future movement prediction.

3.1   Automatic Discovery of Periodicity in Movements

A periodic behavior can be loosely defined as the repeating activities at certain locations with *regular time intervals*. So the mining task will be, given a trajectory, find those locations and corresponding periods (i.e., regular time intervals). This is a challenging task because a real-life moving object never strictly follows a single given periodic pattern. For example, birds never follow exactly the same migration path every year. Their migration routes are strongly affected by weather conditions and thus could be substantially different from previous years. Meanwhile, even though birds generally stay in north in the summer, it is not the case that they stay at exactly the same locations on exactly the same days of the year as previous years. Therefore, "*north*" is a fairly vague geo-concept that is hard to be modeled. Moreover, birds could have multiple interleaved periodic behaviors at different spatiotemporal granularities, as a result of *daily* periodic hunting behaviors, combined with *yearly* migration behaviors.

Li et al. [22] propose Periodica to handle the aforementioned challenges. One of their key observations is that the *binary in-and-out patterns with respect to different reference spots* can reliably reveal movement periodicity. Periodica is done in two steps. In the first step, the trajectory points are clustered based on the spatial densities

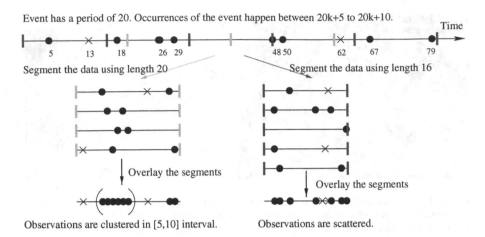

Fig. 12.2 The underlying true period is 20. The *black dots* and *cross* marks correspond to "in" and "out" events separately. When using the correct period 20 to segment and overlay the observations, as shown in the figure on the *left*, the "in" observations are clustered. Figure is from [24]

on the map to form semantic regions called *reference spots*. Figure 12.1 shows an example of an eagle movement data. There are three reference spots detected from the movement. Reference spots could be fairly large but frequently visited regions over several years, such as an area in Quebec (Spot #1 in Fig. 12.1) where birds frequently stay during the summer. In the second step, the movement is transformed into a *binary in-and-out sequence*, and then Fourier transform is applied on the sequence to detect the period.

Due to the limitations of positioning technology and data collection mechanisms, movement data collected from GPS or sensors could be *highly sparse, noisy and unsynchronized*. First, the data is often sampled at an *unsynchronized rate* (e.g., if the sampling rate of a tracking device is set to 1 h, data may be collected at 1:01, 2:08, 3:02, 4:15, and so on). Second, movement data collected can be scattered unevenly over time (e.g., collected only when the tracking device is triggered, such as the check-ins using smart phones). Third, the observations could be highly *sparse*. For example, a bird can only carry a tiny device with limited battery life. There could be only one or two reported locations in three to five days. If a sensor is not functioning or a tracking facility is turned off, it could result in a large portion of missing data. Traditional period detection methods, such as Fourier transform and auto-correlation, are known to be sensitive to such nuisances. Lomb-Scargle periodogram [27, 32] is proposed as a variation of Fourier transform to deal with unevenly spaced data, but it cannot handle the case when the data is also sparse and noisy.

Li et al. [24] develop a novel approach to detect periodicity for sparse, noisy and unsynchronized data. A *"segment-and-overlay"* idea is explored to uncover the hidden period: *Even when the observations are incomplete, the limited periodic observations will be clustered together if data is overlaid with the correct period*, as shown in Fig. 12.2. The method tries every potential periods. For a period candidate

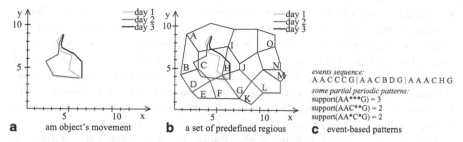

Fig. 12.3 Periodic patterns with respect to pre-defined spatial regions [28]

T, the timeline are segmented by length T and the observations are mapped to a relative timescale $[1, T]$. If T is the true period, the observations will show highly skewed distributions of the observations. Otherwise, the observations will be scattered over $[1, T]$.

3.2   Frequent Periodic Pattern Mining

Given the period, such as a day or a week, we are interested in mining the frequent regular trajectory patterns. For example, people wake up at the same time and follow more or less the same route to their work everyday. The discovery of hidden periodic patterns in spatiotemporal data, apart from unveiling important information to the data analyst, can facilitate data management substantially.

The key challenge to mine frequent pattern in movement lies in how to *transform a 2-dimensional movement sequence to 1-dimensional symbolic sequence*. As proposed by Mamoulis et al. [28], one way to handle this issue is to replace the exact locations by the regions (e.g., districts, cellphone towers, or cells of a synthetic grid) which contain them. Figure 12.3b shows an example of an area's division into such regions. By using the regions, we can transform a raw movement sequence as shown in Fig. 12.3a to an event sequence as shown in Fig. 12.3c. Now the problem becomes a traditional frequent periodic pattern mining problem [16]. In real scenario, sometimes we are interested in the *automated* discovering of descriptive regions. Mamoulis et al. [28] further propose to cluster the locations at corresponding relative timestamps, such as clustering locations at 10am over different days. They propose a top-down pattern mining method, which is more efficient than typical bottom-up method.

3.3   Using Periodic Pattern for Location Prediction

One important application of frequent periodic pattern is for future location prediction. For example, if a person repeats his periodic pattern between home and office every weekday, we could predict that this person is very likely to be in the office at

10 a.m. and to be at home at 10 p.m. Most existing techniques target at *near* future movement prediction, such as next minute or next hour. Linear motion functions [30, 31, 34, 35] have been extensively studied for movement prediction. More complicated models are studied in [36]. As pointed out by Jeung et al. [17], the actual movement of a moving object may not necessarily comply with some mathematical models. It could be more complicated than what the mathematical formulas can represent. Moreover, such models built based on recent movement are not useful for predicting distant future movement, such as next day or one month after.

Periodic patterns can help better predict future movement, especially for a *distant* query time. In [17], a prediction method based on periodic pattern is proposed. The prediction problem assumes that the period T and periodic patterns are already given. To answer predictive queries efficiently, a trajectory pattern tree is proposed to index the periodic patterns. In [17], they use *a hybrid prediction algorithm* that provides predictions for both near and distant time queries. For non-distant time queries, they use the forward query processing that treats recent movements of an object as an important parameter to predict near future locations. A set of qualified candidates will be retrieved and ranked by their premise similarities to the given query. Then they select top-k patterns and return the centers of their consequences as answers. For a distant time queries, since recent movements become less important for prediction, the backward query processing is used. Its main idea is to assign lower weights to premise similarity measure and higher weights to consequences that are closer to the query time in the ranking process of the pattern selection.

4 Pairwise Movement Patterns

In this section, we focus on pattern mining methods on two moving objects. The pairwise movement patterns are between two moving objects R and S. The trajectories of two moving objects are denoted as $R = r_1 r_2 \ldots r_n$ and $S = s_1 s_2 \ldots s_m$, where r_i and s_i are the locations of R and S at the ith timestamp.

We first introduce different similarity measures between two trajectories. Then, based on properties of patterns, we will introduce generic patterns, behavioral patterns, and semantic patterns. Generic patterns describe the overall attraction and avoidance relationship between two moving objects. Behavioral patterns describe a specific type of relationships in a (short) period of time, such as leading and following. Semantic patterns tell the semantics of a relationship (e.g., colleague and friend) in a supervised learning framework.

4.1 Similarity Measure

One way to infer the relationship strength of two moving objects is to measure the similarity of their trajectories. The simplest way of measuring the similarity between two trajectories is to use p-norm distance. The p-norm distance between trajectories of R and S is defined as:

$$L_p(R, S) = \left(\sum_{i=1}^{n} (r_i - s_i)^p \right)^{\frac{1}{p}}.$$

The p-norm distance requires the trajectory length to be the same, i.e., $n = m$. The well-known Euclidean distance and Manhattan distance are p-norm distance when $p = 2$ and $p = 1$ respectively.

The p-norm distance is easy to compute, but is sensitive to the time shift. Dynamic Time Warping (DTW) [39] can handle the local time shifting and it does not need the trajectories to be the same length. DTW is defined as:

$$DTW(R, S) = dist(r_1, s_1) + \min \left(\begin{array}{l} DTW(R[2:n], S[2:m]), \\ DTW(R[2:n], S), \\ DTW(R, S[2:m]) \end{array} \right).$$

Edit distance with Real Penalty (ERP) [4] introduces a constant value g as the gap of edit distance and uses real distance between elements as the penalty to handle local time shifting. ERP is defined as:

$$ERP(R, S) = \min \left(\begin{array}{l} ERP(R[2:n], S[2:m]) + dist(r_1, s_1), \\ ERP(R[2:n], S) + dist(r_1, g), \\ ERP(R, S[2:m]) + dist(s_1, g) \end{array} \right).$$

The Longest Common Subsequences (LCSS) [37] requires a threshold ϵ to be established. The threshold is used to determine whether or not two elements match and it allows LCSS to handle noise by quantizing the distance between two elements to two values, 0 and 1, to remove the larger distance effects caused by noise. LCSS is defined as:

$$LCSS(R, S) = \left\{ \begin{array}{ll} LCSS(R[2,n], S[2,m]) + dist(r_1, s_1) & dist(r_1, s_1) \leq \epsilon \\ \max\{LCSS(R[2,n], S), LCSS(R, S[2,m])\} & otherwise \end{array} \right.$$

Edit Distance on Real sequence (EDR) is defined similar to LCSS except EDR assigns penalties to the gaps between two matched sub-trajectories according to the lengths of gaps. EDR is defined as:

$$EDR(R, S) = \min \left(\begin{array}{l} EDR(R[2:n], S[2:m]) + subcost, \\ EDR(R[2:n], S) + 1, \\ EDR(R, S[2:m] + 1 \end{array} \right),$$

where $subcost = 0$ if $dist(r_1, s_1) \leq \epsilon$ and $subcost = 1$ otherwise.

A comparison of the similarity measures is shown in Table 12.2. All the measures except Euclidean distance can handle local time shifting. And only Euclidean distance requires the lengths of two trajectories to be the same. LCSS and EDR are more robust

Table 12.2 Summary of similarity measures [5]

Distance	Local time shifting	Noise	Metric	Computation cost
Euclidean			✓	$O(n)$
DTW	✓			$O(n^2)$
ERP	✓		✓	$O(n^2)$
LCSS	✓	✓		$O(n^2)$
EDR	✓	✓		$O(n^2)$

to noises because it does not require every point in R to be matched with a point with S. If r_i is a noise point, LCSS and EDR will skip it and assign a mismatch penalty to it. Euclidean distance and ERP are metric distances since they obey triangle inequality. Thus, efficient indexing and retrieval can be achieved by using these two distance measures.

The distance measures mentioned above are suitable to find similar trajectory with similar shapes. They can be applied on trajectories, such as hurricane trajectories and animal migration paths. To measure the similarity on human movements, it could make more sense to look at the co-locating frequency instead of trajectory shape. The meeting frequency [25] is defined as the number of timestamps that their locations are with distance ϵ:

$$freq(R, S) = \sum_{i=1}^{n} \tau(r_i, s_i),$$

where $\tau(r_i, s_i) = 1$ if $dist(r_i, s_i) \leq \epsilon$ and $\tau(r_i, s_i) = 0$ otherwise.

The similarity between two moving objects can also be measured by transitions patterns. Li et al. [20] propose to measure the similarity of two mobile users based on their location histories. The trajectory is first symbolized using the interesting locations mined from user trajectory. Given two symbolized sequences $seq_1 = r_1(k_1) \xrightarrow{\Delta t_1} r_2(k_2) \xrightarrow{\Delta t_2} \ldots r_m(k_m)$ and $seq_2 = s_1(k_1') \xrightarrow{\Delta t_1'} s_2(k_2') \xrightarrow{\Delta t_2'} \ldots s_m(k_m')$, where Δt denotes the transition time between locations and k is the number of times that the user stays in a location, seq_1 and seq_2 are similar if the following constraints are satisfied:

1. $\forall 1 \leq i \leq m, r_i = s_i$;
2. $\forall 1 \leq i \leq m, |\Delta t_i - \Delta t_i'| \leq t_{th}$, where t_{th} is a time threshold on the transition times.

4.2   Generic Pattern

Relationships between two moving objects can be classified as attraction, avoidance or neutral. In an *attraction* relationship, the presence of one individual causes the other to *approach* (i.e., reduce the distance between them). As a result, the individuals have a higher probability to be spatially close than expected based on chance. On the other hand, in an *avoidance* relationship, the presence of one individual causes the other to *move away*. So the individuals have a lower probability to be spatially close than expected. Finally, with a *neutral* relationship, individuals *do not alter* their movement patterns based on the presence (or the absence) of the other individual. So the probability that they are being spatially close is what would be expected based on independent movements.

The attraction relationship is commonly seen, for example, in animal herds or human groups (e.g., colleague and family). In addition, the avoidance relationship also naturally exists among moving objects. In animal movements, prey try to avoid predators, and different animal groups of the same species tend to avoid each other. Even in the same group, subordinate animals often avoid their more dominant group-mates. In human movements, criminals in the city try to avoid the police, whereas drug traffickers traveling on the sea try to avoid the patrol.

Intuitively, similar trajectories could be an indication of attraction relationship. The similarity can be defined by the similarity measures mentioned in the previous subsection. The assumption here is that the smaller the distance is or the higher the meeting frequency is, the stronger the attraction relationship is. Unfortunately, such assumption is often violated in real movement data. For example, two animals may be observed to be spatially close for 10 out of 100 timestamps. But is this significant enough to determine the attraction relationship? Further, another two animals are within spatial proximity for 20 out of 100 timestamps. Does this mean that the latter pair has a more significant attraction relationship than the former pair? Finally, if two animals are never being spatially close, do they necessarily have an avoidance relationship?

Li et al. [25] propose to mine significant attraction and avoidance relationships by looking into the *background territories*. The relationships are detected through the *comparison* between how frequent two objects are *expected* to meet and the *actual* meeting frequency they have. Intuitively, if the actual meeting frequency is smaller (or larger) than the expectation, the relationship is likely to be avoidance (or attraction).

Given two trajectories R and S, the probability for one point r_i in R to be spatially close to any point in S is $\frac{1}{n}\sum_{j=1}^{n} \tau(r_i, s_j)$. Then the expected meeting frequency between randomly shuffled R and S is:

$$E[freq(\sigma(R), \sigma(S))] = \sum_{i=1}^{n}\left(\frac{1}{n}\sum_{j=1}^{n}\tau(r_i, s_j)\right) = \frac{1}{n}\sum_{i=1}^{n}\sum_{j=1}^{n}\tau(r_i, s_j),$$

where $\sigma(\cdot)$ denotes a random shuffled trajectory.

However, by comparing the actual meeting frequency with the expected meeting frequency, one cannot determine a universal *degree* of the relationship. To further measure the degree, let $\mathcal{F} = \{freq(R, \sigma(S)) \mid \sigma\}$ be the multiset of all randomized meeting frequencies. The significance value of attraction (or avoidance) between to

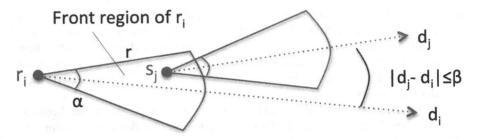

Fig. 12.4 Front region defined in [2]

moving objects R and S is defined as:

$$sig_{attract} = Pr[freq(R,S) > freq(R,\sigma(S))] +$$
$$\tfrac{1}{2}Pr[freq(R,S) = freq(R,\sigma(S))],$$
$$sig_{avoid} = Pr[freq(R,S) < freq(R,\sigma(S))] +$$
$$\tfrac{1}{2}Pr[freq(R,S) = freq(R,\sigma(S))].$$

Permutation test is conduced to get the multiset $\mathcal{F}$. Permutation test is a popular non-parametric approach, to performing hypothesis tests and constructing confidence intervals. The null hypothesis is that the movement sequences of two objects are independent. Since the total number of permutations is factorial, Monte Carlo sampling is used to approximate the significance value.

4.3 Behavioral Pattern

The behavioral patterns describe certain behaviors within a (short) period of time, such as pursuit, evasion, fighting, and play [7]. Following/leading is one interesting behavioral pattern between two moving objects. For example, animal scientists study which individual animal leads the group when animals move in order to determine the social hierarchy, whereas police and security officers look suspicious movements of a criminal who is following a victim.

Intuitively, a follower has similar trajectories as its leader but always arrives at a location with some time lag. The challenges lay in three aspects: (1) the following time lag is usually unknown and varying; (2) The follower may not have exactly the same trajectory as the leader; and (3) the following relationship could be subtle and always happens in a short period of time.

Andersson et al. [2] propose the concept of *front region*. A point s_i in the front region of r_i is defined by an apex angle α, a radius r, and an angle β restricting their difference in direction $\| d_i - d_j \|$. Figure 12.4 shows an illustration of the front region. In [2], a leader should appear in the front region of the follower(s) for at least k consecutive timestamps.

In real scenario, a leader does not necessarily appear in the *front region* of the followers for *consecutive* timestamps. Figure 12.5 illustrates a counter example. In

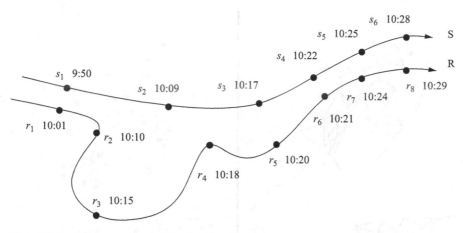

Fig. 12.5 In this example from [26], object R follows object S from *10:01* to *10:20* and moves together afterwards. Even though R follows S, s_2 is not in the front region of r_2

this example, r_2 is heading downwards at 10:10, and s_2 is apparently not in the front region of r_2. The definition of front region and the constraint on being in the front region for k consecutive timestamps are too strict to make the method applicable on real data.

Li et al. [26] propose a more relaxed definition of following pattern. Given thresholds d_{max} and l_{max}, a location pair (r_i, s_j) is said to be a *following pair* if $\| r_i - s_j \| < d_{max}$ and $0 < i - j \le l_{max}$. By considering a following pair as a matching, the problem can be mapped to local sequence alignment (LSA) problem. Smith-Waterman algorithm [33] can be applied to find the longest following interval (best local alignment).

However, experimental results show LSA is sensitive to the parameter d_{max}. To address the problem, Li et al. [26] further propose the concept of local distance minimizer. The intuition is that if object R is following S at timestamp i, then there must exit a *strictly positive integer* $\Delta(i)$ such that r_i is spatially close to $s_{i-\Delta(i)}$. In fact, the distance between r_i and S should be minimized *locally* at such $\Delta(i)$. Based on the intuition, $f(i)$ is defined as whether r_i is following s_i at timestamp i:

$$f(i) = \begin{cases} 1, & \text{if}\,\Delta(i) > 0 \text{ and } \|r_i - s_{i-\Delta(i)}\| < d_{\max} \\ 0, & \text{if}\,\Delta(i) \le 0 \text{ and } \|r_i - s_{i-\Delta(i)}\| < d_{\max} \\ \times, & \text{if}\,\|r_i - s_{i-\Delta(i)}\| \ge d_{\max} \end{cases}$$

Then the following interval $[s, t]$ should make $\sum_s^t f(i)$ maximized. The problem can be transformed to the well-known Maximum Sum Segment problem and all the following intervals can be found in linear time.

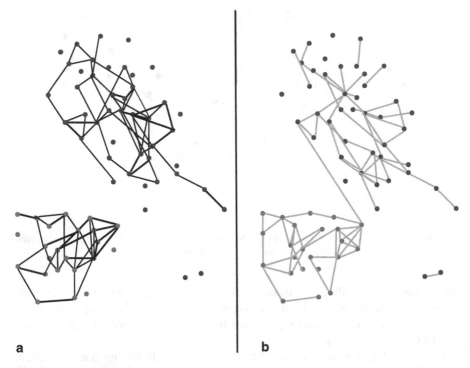

a   b

Fig. 12.6 Social networks in reality mining dataset [10]. **a** shows the inferred weighted friendship network, where the weights correspond to the relationship strengths. **b** shows the reported friendship network

4.4 Semantic Patterns

The social structure of human is one of the fundamental questions in social science. As traditional survey methods often suffer from its limited scale, Eagle and Pentland [8, 10] propose a mobile sensing framework to use human mobility data as indicators of human social network. The Reality Mining project http://reality.media.mit.edu/ tracked the movement of 94 users for one academic year and conducted survey about the relationships between those users. The studies show that human mobility patterns strongly correlate with relationships among people.

Figure 12.6 shows two social networks of all the participants in the study. Network in Fig. 12.6a is constructed from mobility data and network in Fig. 12.6b is constructed using survey data. These two networks inferred from different data show similar structure. Such observation provides strong evidence that human movement data reflects social relationship done by survey. Later in [9], Eagle and Pentland further propose to use Principle Component Analysis (PCA) to extract representative behaviors of an individual and of groups.

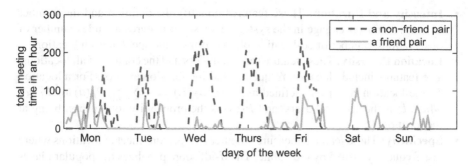

Fig. 12.7 Meeting frequency for a friend and a non-friend pair in Reality Mining dataset [23]

Category	Variables	Description	Co-location	User mobility
Intensity and Duration	NumObservations	The total number of observations of the user.		√
	NumColoc, NumColocEvening, NumColocWeekend	The number of co-location observations of the two users, in total, in the evening only, and on weekends only.	√	
	NumLocations, NumLocationsEvening, NumLocationsWeekend	The number of distinct grid boxes where the user or users were observed, in total, in the evening only, and on weekends only.	√	√
	NumHours, NumWeekdays, NumDates	The number of distinct hours of the day, days of the week, and calendar dates that the two users were observed together.	√	
	ObservationTimeSpan	The difference in seconds between the last and the first location or co-location observation.	√	√
	BoundingBoxArea	The area of the minimal axis aligned rectangle that contains the locations/co-location observations of the user/users.	√	√
Location Diversity	AvgEntropy, MedEntropy, VarEntropy, MinEntropy, MaxEntropy	The mean/median/variance/min/max of the location entropy at each location/co-location observation of the user/users.	√	√
	AvgFreq, MedFreq, VarFreq, MinFreq, MaxFreq	The mean/median/variance/min/max of the location frequency at each location/co-location observation of the user/users.	√	√
	AvgUserCount, MedUserCount, VarUserCount, MinUserCount, MaxUserCount	The mean/median/variance/min/max of the location user count at each location/co-location observation of the user/users.	√	√
Mobility Regularity	SchEntropyL, SchEntropyLH, SchEntropyLD, SchEntropyLHD	The schedule entropy of the user with respect to location, location and hour, location and day of the week, and location and hour and day of the week.		√
	SchSizeLH, SchSizeLD, SchSizeLHD	The schedule size of the user with respect to location and hour, location and day of the week, and location and hour and day of the week.		√
Specificity	AvgTFIDF, MinTFIDF, MaxTFIDF	The mean/minimum/maximum of the location TFIDF at each co-location of the two users.	√	
	PercentObservationsTogether	The total number of co-locations of the two users divided by the sum of each users total number of observations.	√	
Structural Properties	NumMutualNeighbors	The number of people who have been co-located with both users.	√	
	NeighborhoodOverlap	The number of people who have been co-located with both users divided by the number of people who have been co-located with either user.	√	
	LocationOverlap	The total number of distinct places visited by both users divided by the total number of places visited by either users.	√	

Fig. 12.8 Names and descriptions of the mobility features used in [6]

The co-locating times could be a discriminative feature to indicate the semantic relationships. Figure 12.7 shows the meeting frequency with respect to different days of the week for a friend pair and for a non-friend pair in Reality Mining dataset. It is shown in the figure that the friend pair meets more on the weekends, while the non-friend pair meets more during the weekdays. Motivated by this observation, Li et al. [23] propose to mine discriminative time intervals to classify whether two people are friends. The discriminative interval, namely T-Motif, is the time interval where there is a significant difference in meeting frequency between friend pairs and non-friend pairs.

To study how interactions in mobility data correlate with friendships on social networks, Cranshaw et al. [6] propose to build a supervised learning framework using features extracted from mobility data to predict the online relationship. They use a location sharing application based on user check-ins on Facebook to obtain the mobility data from 489 users. Using the mobility data, they propose a set of features as shown in Fig. 12.8. The features can be divided into four categories:

- **Intensity and Duration:** These features quantify the duration and the number of times that users engage in the system. This set of features includes number of observations, number of co-location observations, time spent at each location.
- **Location Diversity:** These features aim to understand the context of all locations. The features include location frequency and the location entropy. For a location L, the location entropy is defined as $Entropy(L) = -\sum_{u \in U} P_L(u) \log P_L(u)$, where U is the set of all users, and $P_L(u)$ is the probability for a user u being at the location L.
- **Specificity:** These features measure whether two persons meet at locations where less frequently visited by the public. The tf-idf score penalizes the popular places that many people frequently visit.
- **Structural Properties:** These features aim to capture network property of two users such as mutual neighbors and location overlaps.

The experimental results in [6] shows that using a variety of classification methods such as random forests and support vector machines can achieve precision above 60 % in predicting the online relationships using the mobility features.

5   Aggregate Patterns over Multiple Trajectories

The aggregate patterns describe common paths shared by a set of trajectories or a cluster of moving objects being spatially close for a long time. In this section, we first introduce the *trajectories patterns*, which is a concise description of frequent behaviors in terms of space and time. Then we will present the methods on mining moving object clusters. Finally, we discuss trajectory clustering methods.

5.1   Frequent Trajectory Pattern Mining

A frequent trajectory pattern is a popular path repeated by many trajectories. Finding frequent trajectory patterns is helpful in summarizing the historical trajectories and predicting the future movements. A trajectory pattern [14] is used to describe a set of individual trajectories visiting the same sequence of places with similar travel times. In trajectory patterns, two notions are important: (1) the geographical locations and (2) the travel time between locations.

If we assume the locations are already symbolized, frequent sequential pattern [1] can be considered as a simplified trajectory pattern. For example, if many people go from location X, to Y and then to Z, $X \to Y \to Z$ will a frequent sequential pattern. In order to enrich the sequential patterns with *transition time information* between locations, Giannotti et al. [13] propose the temporally annotated sequences (TAS). TAS has the following form:

$$T = s_0 \xrightarrow{\alpha_1} s_1 \xrightarrow{\alpha_2} \cdots \xrightarrow{\alpha_n} s_n,$$

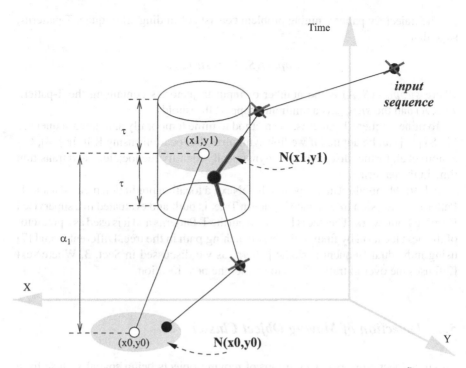

Fig. 12.9 Spatiotemporal containment of input sequence on trajectory pattern $(x_0, y_0) \xrightarrow{\alpha_1} (x_1, y_1)$
[14]

where $S = \langle s_0, \cdots, s_n \rangle$ are the elements in the sequence and $A = \langle \alpha_1, \cdots, \alpha_n \rangle$ are
annotated transition time. With TAS, the pattern could be in the format of $X \xrightarrow{30\ min}$
$Y \xrightarrow{20\ min} Z$.

Trajectory pattern [14] is defined in the same fashion of TAS where each element
in S should be a spatial location:

Definition 12.1 (T-pattern) *A Trajectory pattern, called T-pattern, is a pair (S, A),
where $S = \langle (x_0, y_0), \cdots, (x_k, y_k) \rangle$ is a sequence of points in $\mathbf{R}^2$, and $A =
\langle \alpha_1, \cdots, \alpha_k \rangle \in \mathbf{R}_+^k$ is the temporal annotation of the sequence.*

To judge whether a trajectory *contains* a trajectory pattern, Giannotti et al. [14]
propose a definition on spatiotemporal containment. In Fig. 12.9, input trajectory
sequence $S_1 \ldots S_5$ contains trajectory pattern $(x_0, y_0) \xrightarrow{\alpha_1} (x_1, y_1)$, because for each
point (x_i, y_i) in trajectory pattern, there is a point in trajectory S that is close to it. For
example, point S_3 is close to point (x_1, y_1) because it is in the spatial neighborhood
(i.e., $N(x_1, y_1)$) and also the time difference between (x_1, y_1) and S_3 is less than
threshold τ. Many approaches can be used as a neighborhood function $N(\cdot)$. One
possible neighborhood function is to use the Regions-of-Interest (RoI) to naturally
partition the space into meaning areas. If prior knowledge is not available, RoI can
also be defined as the frequently visited locations/regions mined from the trajectories.

The trajectory pattern mining problem consists of finding all frequent T-patterns, such that

$$support(S, A) \geq sup_{min},$$

where $support(S, A)$ is the number of input trajectories containing the T-pattern $T(S, A)$ and the sup_{min} is a minimum support threshold.

To mine frequent T-patterns, the method to mine temporally annotated sequences (TAS) [13] can be applied if we first symbolize the locations using RoI. In [14], Giannotti et al. further discuss how to dynamically identify the locations and transition time in the pattern.

In [29], Monreale et al. propose WhereNext, a location prediction method using T-Patterns. A decision tree, named T-pattern Tree, is built and evaluated in a supervised learning framework. The tree is learned from the T-Patterns and it is used as a predictor of the next location by finding the best matching path in the tree. Different from [17] using individual frequent periodic pattern, as we discussed in Sect. 3, WhereNext [29] uses the overall traffic flows to predict the next location.

5.2 Detection of Moving Object Cluster

Moving object clusters detect groups of moving objects being spatially close for a considerably long time. Clusters of moving objects can reveal underlying communities, such as the social groups of animals or humans, and can also indirectly identify outliers that do not conform to general group behaviors.

In this section, we will discuss patterns *flock* [15], *convoy* [18] and *swarm* [21]. A moving object cluster can be loosely defined as a set of moving objects being spatially close for k timestamps. The differences among flock, convoy and swarm lie in the definitions of "spatially close" and "k (non-)consecutive timestamps".

Gudmundsson et al. [15] first propose the concept of flock.

Definition 12.2 (Flock) *A set of moving objects O form a flock for timestamps T if (1) for every timestamp in T, there is a disc with radius r containing all the objects in O; and (2) T is consisted of at least k consecutive timestamps.*

In Fig. 12.10, o_3 and o_4 form a flock since they are in the same disc from t_1 to t_4. Since flock defines spatial constraint as a fixed-radius disc, such definition might be too strict and is independent of data distribution. For example, at timestamp t_1 in Fig. 12.10, all the objects are in a density-connected cluster but using a disc may split them into multiple clusters. To relax the rigid restriction on the disc-shape cluster, Jeung et al. [18] proposes a new concept *convoy* to discover arbitrary-shape clusters. Convoy uses DBSCAN [11] to cluster points in each timestamp. Two objects in a cluster are density-connected to each other, if only there exists a sequence of objects that connect them together. The definition of density-connected permits us to capture a group of connected points with arbitrary shape.

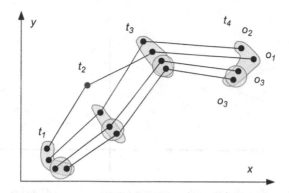

Fig. 12.10 An example of flock [15], convoy [18] and swarm [21]. If we set time constraint $k = 3$ (i.e., number of timestamps being spatially close), o_3 and o_4 form a flock since they are in a disc for four consecutive timestamps. o_1, o_3 and o_4 form a convoy since they are in the same density-connected cluster for four consecutive timestamps. Swarm considers all these four objects as a cluster since it treats o_1 at t_2 as a short deviation from the cluster

Table 12.3 Summary of moving object clusters

Pattern	Spatial constraint	Temporal constraint
Flock [15]	Disc shape	k consecutive timestamps
Convoy [18]	Arbitrary shape	k consecutive timestamps
Swarm [21]	Arbitrary shape	k (non-)consecutive timestamps

Definition 12.3 (Convoy) *A set of moving objects O form a convoy for timestamps T if (1) for every timestamp in T, all the objects in O are in the same density-connected cluster; and (2) T is consisted of at least k consecutive timestamps.*

In Fig. 12.10, three objects o_1, o_3 and o_4 are in the same density-connected cluster during the time interval $[t_1, t_4]$. Although the convoy model is much flexible than the flock, the time constraint on k *consecutive* timestamps is still too strict. The moving objects may temporarily leave the group. For example, o_1 temporarily leaves the group at t_2. If we enforce the "consecutive" time constraint, o_1 is not considered to be in the same group with other objects. Motivated by this important observation, Li et al. [21] propose the concept of *swarm* to relax the time constraint. Instead of requiring the objects being in the same cluster for *consecutive* timestamps, swarm allows the timestamps to be *non-consecutive*.

Definition 12.4 (Swarm) *A set of moving objects O form a swarm for timestamps T if (1) for every timestamp in T, all the objects in O are in the same density-connected cluster; and (2) T is consisted of at least k timestamps that are not necessarily consecutive.*

In Fig. 12.10, we can see that all the objects form a group even though o_1 temporarily leaves the cluster at t_2. Swarm is able to capture $\{o_1, o_2, o_3, o_4\}$ as one cluster.

Table 12.3 summarizes the three different patterns: flock, convoy and swarm. The definition of the swarm is the most flexible one in terms of the spatial and

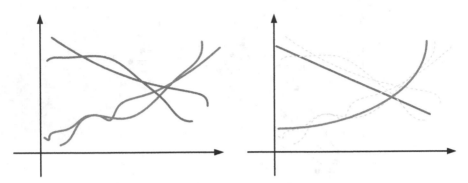

Fig. 12.11 Trajectory clustering example. Four trajectories are clustered into two clusters based on trajectory similarity

temporal constraint. The time complexity of swarm is the also highest among three patterns. Since it needs to enumerate every possible combination of objects, the time complexity is $O(2^n)$, where n is the number of moving objects in dataset. But by applying pruning rules on the search algorithm [21], swarm pattern mining is quite efficient in real scenario.

5.3 Trajectory Clustering

Different from moving object clusters that detect clusters of objects and the corresponding time intervals that they are being together, trajectory clustering will group (sub-)trajectories based on the overall trajectory similarity. Moving object cluster mining is more suitable to answer questions such as "find a group of people staying together for more than 2 hours", whereas trajectory clustering can answer questions like "group hurricane paths over years based on the trajectory similarity". Figure 12.11 illustrates an example of trajectory clustering. There are two clusters based on trajectory similarity.

A typical clustering framework needs to consider two factors: (1) similarity measure and (2) clustering methods. As we discuss earlier in Sect. 4, the typical similarity measures between two trajectories include Euclidean distance, Dynamic Time Warping and Longest Common Subsequence. And typical clustering methods include K-Means, Hierarchical clustering and Gaussian Mixture Model.

Gaffney and Smyth [12] propose to cluster trajectories based on a probabilistic modeling of trajectories. In probabilistic clustering, we assume that the data are being generated in the following "generative" manner:

- An individual is drawn randomly from the population of interest.
- The individual has been assigned to cluster k with probability w_k, $sum_{k=1}^{K} w_k = 1$. These are the prior weights on the K clusters.

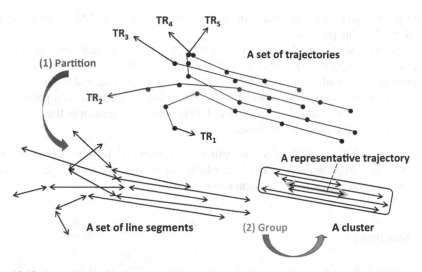

Fig. 12.12 An example of trajectory clustering in the partition-and-group framework [19]

- Given that an individual belongs to cluster k, there is a density function $f_k(y_j|\theta_k)$ which generates observed data y_i for individual j.

From this generative model, the observed density on the y's should be be a mixture model, i.e., a linear combination of the component models:

$$P(y_j|\theta) = \sum_k^K f_k(y_j|\theta_k)w_k.$$

In Gaussian mixture model, we will assume the generative models θ_k as Gaussian models. In Gaffney et al. [12], they assume the data is generated as mixtures of *regression models*, where we have measurements y which are a function of x and the density function becomes $f_k(y|x, \theta_k)$. Here x represents time, y represents the locations of object and θ_k is the regression model of y on x. The parameters in generative models can be estimated using the Expectation-Maximization (EM) algorithm. Experimental results [12] show that the proposed linear regression model performs slightly better than Gaussian mixture model. The difference becomes more obvious with higher standard deviation in data generation. Both mixture model methods perform much better than K-means.

In some applications, people are interested in discovering *similar portions* of trajectories. For example, meteorologists will be interested in the common behaviors of hurricanes near the coastline (i.e., at the time of landing) or at sea (i.e., before landing). To cluster sub-trajectories, Lee et al. [19] propose a partition-and-group framework named as TRACLUS as shown in Fig. 12.12. There are three steps in TRACLUS.

1. Partitioning: in this step, each trajectory is partitioned into a set of line segment based on characteristic points. A characteristic point is a point where the behavior

of a trajectory changes. The minimum description length (MDL) principle is adopted in this process.

2. Grouping: using given distance measure trajectory segments that are close to each other are grouped into a cluster. Density-based clustering algorithm is used in this process, which allows clusters in TRACULAS have any size and shape.

3. Representing: derive a representative trajectory for each cluster. The purpose of this representative trajectory is to describe the overall movement of the trajectory partitions that belong to the cluster.

An important step in TRACLUS is to partition a trajectory into sub-trajectories. By clustering sub-trajectories instead of the whole trajectories, we are able to discover the common paths shared by different sub-trajectories.

6 Summary

This chapter discusses many interesting state-of-the-art methods of spatiotemporal pattern mining. Discovery of spatiotemporal patterns can benefit various applications, such as ecological studies, traffic planning and social network analysis. We categorize the patterns as individual periodic patterns, pairwise patterns, and aggregate patterns over multiple trajectories.

As the collection of spatiotemporal data becomes easier and popular, spatiotemporal data mining is a promising research area with a lot of potential interesting research topics. There are still many challenging issues have not been well addressed by current methods, such as sparsity, uncertainties and noises in the data. It is also important to consider the spatial semantics (e.g., point of interest information) and constraints (e.g., road network and landscapes). So we could better understand the semantic meanings of the patterns. Finally, it will be interesting to consider human factor in the mining process and make the mining process more interactive and informative.

References

1. R. Agrawal and R. Srikant. Mining sequential patterns. In *Proc. 1995 Int. Conf. Data Engineering (ICDE'95)*, pages 3–14. IEEE, 1995.
2. M. Andersson, J. Gudmundsson, P. Laube, and T. Wolle. Reporting leaders and followers among trajectories of moving point objects. *GeoInformatica*, 12(4):497–528, 2008.
3. H. Cao, N. Mamoulis, and D. W. Cheung. Discovery of periodic patterns in spatiotemporal sequences. *Knowledge and Data Engineering, IEEE Transactions on*, 19(4):453–467, 2007.
4. L. Chen and R. T. Ng. On the marriage of lp-norms and edit distance. In *Proc. 2004 Int. Conf. Very Large Data Bases (VLDB'04)*, 2004.
5. L. Chen, M. T. Özsu, and V. Oria. Robust and fast similarity search for moving object trajectories. In *Proc. 2005 ACM-SIGMOD Int. Conf. Management of Data (SIGMOD'05)*, 2005.
6. J. Cranshaw, E. Toch, J. I. Hong, A. Kittur, and N. Sadeh. Bridging the gap between physical location and online social networks. In *Proc. 2010 Int. Conf. Ubiquitous Computing (Ubicomp'10)*, 2010.

7. S. Dodge, R. Weibel, and A.-K. Lautenschütz. Towards a taxonomy of movement patterns. *Information visualization*, 7(3–4):240–252, 2008.

8. N. Eagle and A. Pentland. Reality mining: sensing complex social systems. *Personal and ubiquitous computing*, 10(4):255–268, 2006.

9. N. Eagle and A. S. Pentland. Eigenbehaviors: Identifying structure in routine. *Behavioral Ecology and Sociobiology*, 63(7):1057–1066, 2009.

10. N. Eagle, A. Pentland, and D. Lazer. Inferring friendship network structure by using mobile phone data. In *Proceedings of the National Academy of Sciences (PNAS'09)*, pages 15274–15278, 2009.

11. M. Ester, H.-P. Kriegel, J. Sander, and X. Xu. A density-based algorithm for discovering clusters in large spatial databases. In *Proc. 1996 Int. Conf. Knowledge Discovery and Data Mining (KDD'96)*, pages 226–231, Portland, OR, Aug. 1996.

12. S. Gaffney and P. Smyth. Trajectory clustering with mixtures of regression models. In *Proc. 1999 Int. Conf. Knowledge Discovery and Data Mining (KDD'99)*, pages 63–72, San Diego, CA, Aug. 1999.

13. F. Giannotti, M. Nanni, and D. Pedreschi. Efficient mining of temporally annotated sequences. In *Proc. 2006 SIAM Int. Conf. on Data Mining (SDM'06)*, 2006.

14. F. Giannotti, M. Nanni, F. Pinelli, and D. Pedreschi. Trajectory pattern mining. In *Proc. 2007 ACM SIGKDD Int. Conf. on Knowledge Discovery and Data Mining (KDD'07)*, pages 330–339. ACM, 2007.

15. J. Gudmundsson and M. van Kreveld. Computing longest duration flocks in spatio-temporal data. In *Proc. 2006 ACM Int. Symp. Advances in Geographic Information Systems (GIS'06)*, 2006.

16. J. Han, G. Dong, and Y. Yin. Efficient mining of partial periodic patterns in time series database. In *Proc. 1999 Int. Conf. Data Engineering (ICDE'99)*, pages 106–115, Sydney, Australia, April 1999.

17. H. Jeung, Q. Liu, H. T. Shen, and X. Zhou. A hybrid prediction model for moving objects. In *Proc. 2008 Int. Conf. Data Engineering (ICDE'08)*, 2008.

18. H. Jeung, M. L. Yiu, X. Zhou, C. S. Jensen, and H. T. Shen. Discovery of convoys in trajectory databases. In *Proc. 2008 Int. Conf. Very Large Data Bases (VLDB'08)*, 2008.

19. J.-G. Lee, J. Han, and K. Whang. Clustering trajectory data. In *Proc. 2007 ACM-SIGMOD Int. Conf. Management of Data (SIGMOD'07)*, Beijing, China, June 2007.

20. Q. Li, Y. Zheng, X. Xie, Y. Chen, W. Liu, and W.-Y. Ma. Mining user similarity based on location history. In *Proceedings of the 16th ACM SIGSPATIAL international conference on Advances in geographic information systems*, page 34. ACM, 2008.

21. Z. Li, B. Ding, J. Han, and R. Kays. Swarm: Mining relaxed temporal moving object clusters. In *Proc. 2010 Int. Conf. Very Large Data Bases (VLDB'10)*, Singapore, Sept. 2010.

22. Z. Li, B. Ding, J. Han, R. Kays, and P. Nye. Mining periodic behaviors for moving objects. In *Proc. 2010 ACM SIGKDD Conf. Knowledge Discovery and Data Mining (KDD'10)*, Washington D.C., July 2010.

23. Z. Li, C. X. Lin, B. Ding, and J. Han. Mining significant time intervals for relationship detection. In *Proc. 2011 Int. Symp. Spatial and Temporal Databases (SSTD'11)*, pages 386–403, 2011.

24. Z. Li, J. Wang, and J. Han. Mining periodicity for sparse and incomplete event data. In *Proc. of 2012 ACM SIGKDD Int. Conf. on Knowledge Discovery and Data Mining (KDD'12)*, Beijing, China, Aug. 2012.

25. Z. Li, B. Ding, F. Wu, T. K. H. Lei, R. Kays, and M. C. Crofoot. Attraction and avoidance detection from movements. *Proceedings of the VLDB Endowment*, 7(3), 2013.

26. Z. Li, F. Wu, and M. C. Crofoot. Mining following relationships in movement data. In *Proc. 2013 Int. Conf. Data Mining (ICDM'13)*, 2013.

27. N. R. Lomb. Least-squares frequency analysis of unequally spaced data. In *Astrophysics and Space Science*, 1976.

28. N. Mamoulis, H. Cao, G. Kollios, M. Hadjieleftheriou, Y. Tao, and D. Cheung. Mining, indexing, and querying historical spatiotemporal data. In *Proc. 2004 ACM SIGKDD Int. Conf. Knowledge Discovery in Databases (KDD'04)*, pages 236–245, Seattle, WA, Aug. 2004.

29. A. Monreale, F. Pinelli, R. Trasarti, and F. Giannotti. Wherenext: a location predictor on trajectory pattern mining. In *Proc. 2009 ACM SIGKDD Int. Conf. on Knowledge Discovery and Data Mining (KDD'09)*, pages 637–646, 2009.

30. J. M. Patel, Y. Chen, and V. P. Chakka. Stripes: An efficient index for predicted trajectories. In *Proc. 2004 ACM-SIGMOD Int. Conf. Management of Data (SIGMOD'04)*, Paris, France, June 2004.

31. S. Saltenis, C. Jensen, S. Leutenegger, and M. Lopez. Indexing the positions of continuously moving objects. In *Proc. 2003 ACM-SIGMOD Int. Conf. Management of Data (SIGMOD'03)*, pages 331–342, San Diego, CA, June 2003.

32. J. D. Scargle. Studies in astronomical time series analysis. ii - statistical aspects of spectral analysis of unevenly spaced data. In *Astrophysical Journal*, 1982.

33. T. F. Smith and M. S. Waterman. Comparison of biosequences. *Advances in Applied Mathematics*, 2(4):482–489, 1981.

34. Y. Tao and D. Papadias. Spatial queries in dynamic environments. *ACM Trans. Database Systems*, 28:101–139, 2003.

35. Y. Tao, D. Papadias, and J. Sun. The tpr*-tree: An optimized spatio-temporal access method for predictive queries. In *Proc. 2003 Int. Conf. Very Large Data Bases (VLDB'03)*, pages 790–801, Berlin, Germany, Sept. 2003.

36. Y. Tao, C. Faloutsos, D. Papadias, and B. Liu. Prediction and indexing of moving objects with unknown motion patterns. In *Proc. 2004 ACM-SIGMOD Int. Conf. Management of Data (SIGMOD'04)*, Paris, France, June 2004.

37. M. Vlachos, D. Gunopulos, and G. Kollios. Discovering similar multidimensional trajectories. In *Proc. 2002 Int. Conf. Data Engineering (ICDE'02)*, pages 673–684, San Francisco, CA, April 2002.

38. Y. Xia, Y. Tu, M. Atallah, and S. Prabhakar. Reducing data redundancy in location-based services. In *GeoSensor*, 2006.

39. B.-K. Yi, H. V. Jagadish, and C. Faloutsos. Efficient retrieval of similar time sequences under time warping. In *Proc. 1998 Int. Conf. Data Engineering (ICDE'98)*, pages 201–208, Orlando, FL, Feb. 1998.

Chapter 13
Mining Graph Patterns

Hong Cheng, Xifeng Yan and Jiawei Han

Abstract Graph pattern mining becomes increasingly crucial to applications in a variety of domains including bioinformatics, cheminformatics, social network analysis, computer vision and multimedia. In this chapter, we first examine the existing frequent subgraph mining algorithms and discuss their computational bottleneck. Then we introduce recent studies on mining various types of graph patterns, including significant, representative and dense subgraph patterns. We also discuss the mining tasks in new problem settings such as a graph stream and an uncertain graph model. These new mining algorithms represent the state-of-the-art graph mining techniques: they not only avoid the exponential size of mining result, but also improve the applicability of graph patterns significantly.

Keywords Apriori · Frequent subgraph · Graph pattern · Significant pattern · Representative pattern · Dense pattern · Graph stream · Uncertain graph

1 Introduction

Frequent pattern mining has been a focused theme in data mining research for over a decade. Abundant literature has been dedicated to this research area and tremendous progress has been made, including efficient and scalable algorithms for frequent itemset mining, frequent sequential pattern mining, frequent subgraph mining, as well as their broad applications.

H. Cheng (✉)
Department of Systems Engineering and Engineering Management,
The Chinese University of Hong Kong, Hong Kong, China
e-mail: hcheng@se.cuhk.edu.hk

X. Yan
Department of Computer Science, University of California at Santa Barbara,
Santa Barbara, USA
e-mail: xyan@cs.ucsb.edu

J. Han
Department of Computer Science, University of Illinois at Urbana-Champaign,
Champaign, USA
e-mail: hanj@cs.uiuc.edu

C. C. Aggarwal, J. Han (eds.), *Frequent Pattern Mining,*
DOI 10.1007/978-3-319-07821-2_13, © Springer International Publishing Switzerland 2014

Frequent graph patterns are subgraphs that are found from a collection of graphs or a single massive graph with a frequency no less than a user-specified support threshold. Frequent subgraphs are useful at characterizing graph sets, discriminating different groups of graphs, classifying and clustering graphs, and building graph indices. Borgelt and Berthold [7] illustrated the discovery of active chemical structures in an HIV-screening dataset by contrasting the support of frequent graphs between different classes. Deshpande et al. [15] used frequent structures as features to classify chemical compounds. Huan et al. [22] successfully applied the frequent graph mining technique to study protein structural families. Frequent graph patterns were also used as indexing features by Yan et al. [48] to perform fast graph search. Their method outperforms the traditional path-based indexing approach significantly. Koyuturk et al. [27] proposed a method to detect frequent subgraphs in biological networks, where considerably large frequent sub-pathways in metabolic networks are observed.

In this chapter, we will first review the existing graph pattern mining methods and identify the combinatorial explosion problem in these methods—the graph pattern search space grows exponentially with the pattern size. It causes two serious problems: (1) the computational bottleneck, i.e., it takes very long, or even forever, for the algorithms to complete the mining process, and (2) patterns' applicability, i.e., the huge mining result set hinders the potential usage of graph patterns in many real-life applications. We will then introduce scalable graph pattern mining paradigms which mine *significant* subgraphs [19, 28, 33, 34, 37, 50], *representative* subgraphs [2] and *dense* subgraphs [10–12, 18, 36, 39, 41, 42, 44, 52]. In addition, we also introduce the state-of-the-art graph mining algorithms under new application settings, such as in a graph stream [1, 3, 5, 6] and in uncertain graphs [26, 53].

2 Frequent Subgraph Mining

2.1 Problem Definition

The vertex set of a graph g is denoted by $V(g)$ and the edge set by $E(g)$. A label function, l, maps a vertex or an edge to a label. A graph g is a subgraph of another graph g' if there exists a subgraph isomorphism from g to g', denoted by $g \subseteq g'$. g' is called a supergraph of g.

Definition 13.1 (Subgraph Isomorphism) *For two labeled graphs g and g', a subgraph isomorphism is an injective function $f : V(g) \rightarrow V(g')$, s.t., (1), $\forall v \in V(g), l(v) = l'(f(v))$; and (2), $\forall (u, v) \in E(g), (f(u),$
$f(v)) \in E(g')$ and $l(u, v) = l'(f(u), f(v))$, where l and l' are the labeling functions of g and g', respectively. f is called an embedding of g in g'.*

Definition 13.2 (Frequent Graph) *Given a labeled graph dataset $D = \{G_1, G_2, \ldots, G_n\}$ and a subgraph g, the supporting graph set of g is $D_g = \{G_i | g \subseteq G_i, G_i \in D\}$. The support of g is $support(g) = \frac{|D_g|}{|D|}$. A frequent graph is a graph whose support is no less than a minimum support threshold, min_sup.*

An important property, called *anti-monotonicity*, is crucial to confine the search space of frequent subgraph mining.

Definition 13.3 (Anti-Monotonicity) *Anti-monotonicity means that a size-k subgraph is frequent only if all of its subgraphs are frequent.*

Many frequent graph pattern mining algorithms [7, 8, 14, 16, 20, 21, 23, 24, 29, 30, 31, 38, 45] have been proposed. Holder et al. [20] developed SUBDUE to do approximate graph pattern discovery based on minimum description length and background knowledge. Dehaspe et al. [14] applied inductive logic programming to predict chemical carcinogenicity by mining frequent subgraphs. Besides these studies, there are two basic approaches to the frequent subgraph mining problem: the Apriori-based approach and the pattern-growth approach.

2.2 Apriori-Based Approach

Apriori-based frequent subgraph mining algorithms share similar characteristics with Apriori-based frequent itemset mining algorithms. The search for frequent subgraphs starts with small-size subgraphs, and proceeds in a bottom-up manner. At each iteration, the size of newly discovered frequent subgraphs is increased by one. These new subgraphs are generated by joining two similar but slightly different frequent subgraphs that were discovered already. The frequency of the newly formed graphs is then checked. The framework of Apriori-based methods is outlined in Algorithm 11.

Algorithm 11 Apriori(D, min_sup, S_k)

Input: Graph dataset D, minimum support threshold min_sup,
        size-k frequent subgraphs S_k
Output: The set of size-$(k + 1)$ frequent subgraphs S_{k+1}

1: $S_{k+1} \leftarrow \varnothing$;
2: **for** each frequent subgraph $g_i \in S_k$ **do**
3:   **for** each frequent subgraph $g_j \in S_k$ **do**
4:     **for** each size-$(k + 1)$ graph g formed by joining g_i and g_j **do**
5:       **if** g is frequent in D and $g \notin S_{k+1}$ **then**
6:         insert g to S_{k+1};
7: **if** $S_{k+1} \neq \varnothing$ **then**
8:   call Apriori(D, min_sup, S_{k+1});
9: **return**;

Typical Apriori-based frequent subgraph mining algorithms include AGM by Inokuchi et al. [24], FSG by Kuramochi and Karypis [29], and an edge-disjoint path-join algorithm by Vanetik et al. [38].

Fig. 13.1 AGM: Two
candidate patterns formed by
two chains

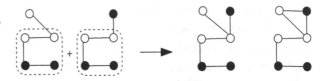

The **AGM** algorithm uses a *vertex-based candidate generation* method that increases the subgraph size by one vertex in each iteration. Two size-$(k + 1)$ frequent subgraphs are joined only when the two graphs have the same size-k subgraph. Here, *graph size* means the number of vertices in a graph. The newly formed candidate includes the common size-k subgraph and the additional two vertices from the two size-$(k + 1)$ patterns. Figure 13.1 depicts the two subgraphs joined by two chains.

The **FSG** algorithm adopts an *edge-based candidate generation* strategy that increases the subgraph size by one edge in each iteration. Two size-$(k + 1)$ patterns are merged if and only if they share the same subgraph having k edges. In the *edge-disjoint path* method [38], graphs are classified by the number of disjoint paths they have, and two paths are edge-disjoint if they do not share any common edge. A subgraph pattern with $k+1$ disjoint paths is generated by joining subgraphs with k disjoint paths.

The **Apriori**-based algorithms mentioned above have considerable overhead when two size-k frequent subgraphs are joined to generate size-$(k+1)$ candidate patterns. In order to avoid this kind of overhead, non-**Apriori**-based algorithms were developed, most of which adopt the pattern-growth methodology, as discussed below.

2.3 Pattern-Growth Approach

Pattern-growth graph mining algorithms include **gSpan** by Yan and Han [45], **MoFa** by Borgelt and Berthold [7], **FFSM** by Huan et al. [21], **SPIN** by Huan et al. [23], and **Gaston** by Nijssen and Kok [31]. These algorithms are inspired by **PrefixSpan** [32], **TreeMinerV** [51], and **FREQT** [4] in mining sequences and trees, respectively.

The pattern-growth algorithm extends a frequent graph directly by adding a new edge, in every possible position. It does not perform expensive join operations. A potential problem with the edge extension is that the same graph can be discovered multiple times. The **gSpan** algorithm helps avoiding the discovery of duplicates by introducing a *right-most extension* technique, where the only extensions take place on the *right-most path* [45]. A right-most path for a given graph is the straight path from the starting vertex v_0 to the last vertex v_n, according to a depth-first search on the graph.

Besides the frequent subgraph mining algorithms, constraint-based subgraph mining algorithms have also been proposed. Mining closed graph patterns was studied by Yan and Han [46]. Mining coherent subgraphs was studied by Huan et al. [22]. Chi et al. proposed CMTreeMiner to mine closed and maximal frequent subtrees [13]. For relational graph mining, Yan et al. [49] developed two algorithms, **CloseCut** and Splat, to discover exact dense frequent subgraphs in a set of relational graphs. For large-scale graph database mining, a disk-based frequent graph mining method was introduced by Wang et al. [40]. Jin et al. [25] proposed an algorithm, TSMiner, for mining frequent large-scale structures (defined as topological structures) from graph datasets.

For a comprehensive introduction on basic graph pattern mining algorithms including Apriori-based and pattern-growth approaches, readers are referred to the survey written by Washio and Motoda [43] and Yan and Han [47].

2.4   Closed and Maximal Subgraphs

A major challenge in mining frequent subgraphs is that the mining process often generates a huge number of patterns. This is because if a subgraph is frequent, all of its subgraphs are frequent as well. A frequent graph pattern with n edges can potentially have 2^n frequent subgraphs, which is an exponential number. To overcome this problem, *closed subgraph mining* and *maximal subgraph mining* algorithms were proposed.

Definition 13.4 (Closed Subgraph) *A subgraph g is a closed subgraph in a graph set D if g is frequent in D and there exists no proper supergraph g' such that g $\subset$ g' and g' has the same support as g in D.*

Definition 13.5 (Maximal Subgraph) *A subgraph g is a maximal subgraph in a graph set D if g is frequent, and there exists no supergraph g' such that g $\subset$ g' and g' is frequent in D.*

The set of closed frequent subgraphs contains the complete information of frequent patterns; whereas the set of maximal subgraphs, though more compact, usually does not contain the complete support information regarding to its corresponding frequent sub-patterns. Close subgraph mining methods include CloseGraph [46]. Maximal subgraph mining methods include SPIN [23] and MARGIN [35].

2.5   Mining Subgraphs in a Single Graph

While most frequent subgraph mining algorithms assume the input graph data is a set of graphs $D = \{G_1, \ldots, G_n\}$, there are some studies [8, 16, 30] on mining graph patterns from a single large graph. Defining the support of a subgraph in a set of graphs is straightforward, which is the number of graphs in the database that contain the subgraph. However, it is much more difficult to find an appropriate support definition in a single large graph since multiple embeddings of a subgraph may have overlaps. If arbitrary overlaps between non-identical embeddings are allowed, the resulting support does not satisfy the anti-monotonicity property, which is essential for most frequent pattern mining algorithms. Therefore, [8, 16, 30] investigated appropriate support measures in a single graph.

Kuramochi and Karypis [30] proposed two efficient algorithms that can find frequent subgraphs within a large sparse graph. The first algorithm, called HSIGRAM, follows a horizontal approach and finds frequent subgraphs in a breadth-first fashion. The second algorithm, called VSIGRAM, follows a vertical approach and finds the frequent subgraphs in a depth-first fashion. For the support measure defined in [30], all possible occurrences φ of a pattern p in a graph g are calculated. An *overlap-graph* is constructed where each occurrence φ corresponds to a node and there is an edge between the nodes of φ and φ' if they overlap. This is called *simple overlap* as defined below.

Definition 13.6 (Simple Overlap) *Given a pattern* $p = (V(p), E(p))$, *a simple overlap of occurrences* φ *and* φ' *of pattern* p *exists if* $\varphi(E(p)) \cap \varphi'(E(p)) \neq \varnothing$.

The support of p is defined as the size of the maximum independent set (MIS) of the overlap-graph. A later study [16] proved that the MIS-support is anti-monotone.

Fiedler and Borgelt [16] suggested a definition that relies on the non-existence of equivalent ancestor embeddings in order to guarantee that the resulting support is anti-monotone. The support is called *harmful overlap support*. The basic idea of this measure is that some of the simple overlaps (in [30]) can be disregarded without harming the anti-monotonicity of the support measure. As in [30], an overlap graph is constructed and the support is defined as the size of the MIS. The major difference is the definition of the overlap.

Definition 13.7 (Harmful Overlap) *Given a pattern* $p = (V(p), E(p))$, *a harmful overlap of occurrences* φ *and* φ' *of pattern* p *exists if* $\exists v \in V(p) : \varphi(v), \varphi'(v) \in \varphi(V(p)) \cap \varphi'(V(p))$.

Bringmann and Nijssen [8] examined the existing studies [16, 30] and identified the expensive operation of solving the MIS problem. They defined a new support measure.

Definition 13.8 (Minimum Image Based Support) *Given a pattern* $p = (V(p), E(p))$, *the minimum image based support of* p *in* g *is defined as*

$$\sigma_\wedge(p, g) = \min_{v \in V(p)} |\{\varphi_i(v) : \varphi_i \text{ is an occurrence of } p \text{ in } g\}|.$$

It is based on the number of unique nodes in the graph g to which a node of the pattern p is mapped. This measure avoids the MIS computation. Therefore it is computationally less expensive and often closer to intuition than measures proposed in [16, 30].

By taking the node in p which is mapped to the least number of unique nodes in g, the anti-monotonicity of $\sigma_\wedge$ can be guaranteed. For the definition of support, several computational benefits could be identified: (1) instead of $O(n^2)$ potential overlaps, where n is the possibly exponential number of occurrences, the method only needs to maintain a set of vertices for every node in the pattern, which can be done in $O(n)$; (2) the method does not need to solve an NP complete MIS problem; and (3) it is not necessary to compute all occurrences: it is sufficient to determine for every pair of $v \in V(p)$ and $v' \in V(g)$ if there is one occurrence in which $\varphi(v) = v'$.

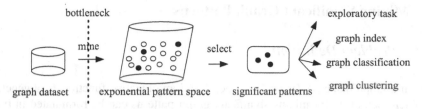

Fig. 13.2 Graph pattern application pipeline

2.6   The Computational Bottleneck

Most graph mining methods follow the combinatorial pattern enumeration paradigm. In real world applications including bioinformatics and social network analysis, the complete enumeration of patterns is practically infeasible. It often turns out that the mining results, even those for closed graphs [46] or maximal graphs [23], are explosive in size.

Figure 13.2 depicts the pipeline of graph applications built on frequent subgraphs. In this pipeline, frequent subgraphs are mined first; then significant patterns are selected based on user-defined objective functions for different applications. Unfortunately, the potential of graph patterns is hindered by the limitation of this pipeline, due to a scalability issue. For instance, in order to find subgraphs with the highest statistical significance, one has to enumerate all the frequent subgraphs first, and then calculate their p value one by one. Obviously, this two-step process is not scalable due to the following two reasons: (1) for many objective functions, the minimum frequency threshold has to be set very low so that none of significant patterns will be missed—a low-frequency threshold often means an exponential pattern set and an extremely slow mining process; and (2) there is a lot of redundancy in frequent subgraphs; most of them are not worth computing at all. When the complete mining results are prohibitively large, yet only the significant or representative ones are of real interest. It is inefficient to wait forever for the mining algorithm to finish and then apply post-processing to the huge mining result. In order to complete mining in a limited period of time, a user usually has to sacrifice patterns' quality. In short, the frequent subgraph mining step becomes the bottleneck of the whole pipeline in Fig. 13.2.

In the following discussion, we will introduce recent graph pattern mining methods that overcome the scalability bottleneck. The first series of studies [19, 28, 33, 34, 37, 50] focus on mining the optimal or significant subgraphs according to user-specified objective functions in a timely fashion by accessing only a small subset of promising subgraphs. The second study by Hasan et al. [2] generates an orthogonal set of graph patterns that are representative. All these studies avoid generating the complete set of frequent subgraphs while presenting only a compact set of interesting subgraph patterns, thus solving the scalability and applicability issues.

3 Mining Significant Graph Patterns

3.1 Problem Definition

Given a graph database $D = \{G_1, \ldots, G_n\}$ and an objective function F, a general problem definition for mining significant graph patterns can be formulated in two different ways: (1) find all subgraphs g such that $F(g) \geq \delta$ where δ is a significance threshold; or (2) find a subgraph g^* such that $g^* = \mathrm{argmax}_g F(g)$. No matter which formulation or which objective function is used, an efficient mining algorithm shall find significant patterns directly without exhaustively generating the whole set of graph patterns. There are several algorithms [19, 28, 33, 34, 37, 50] proposed with different objective functions and pruning techniques. We are going to discuss four recent studies: gboost [28], gPLS [34], LEAP [50] and GraphSig [33].

3.2 gboost: A Branch-and-Bound Approach

Kudo et al. [28] presented an application of boosting for classifying labeled graphs, such as chemical compounds, natural language texts, *etc.* A weak classifier called decision stump uses a subgraph as a classification feature. Then a boosting algorithm repeatedly constructs multiple weak classifiers on weighted training instances. A gain function is designed to evaluate the quality of a decision stump, *i.e.*, how many weighted training instances can be correctly classified. Then the problem of finding the optimal decision stump in each iteration is formulated as mining an "optimal" subgraph pattern. gboost designs a branch-and-bound mining approach based on the gain function and integrates it into gSpan to search for the "optimal" subgraph pattern.

A Boosting Framework gboost uses a simple classifier, *decision stump*, for prediction according to a single feature. The subgraph-based decision stump is defined as follows.

Definition 13.9 (Decision Stumps for Graphs) *Let t and x be labeled graphs and* $y \in \{\pm 1\}$ *be a class label. A decision stump classifier for graphs is given by*

$$h_{\langle t,y \rangle}(\mathbf{x}) = \begin{cases} y, & t \subseteq \mathbf{x} \\ -y, & otherwise \end{cases}.$$

The decision stumps are trained to find a rule $\langle \hat{t}, \hat{y} \rangle$ that minimizes the error rate for the given training data $T = \{\langle \mathbf{x}_i, y_i \rangle\}_{i=1}^{L}$,

$$\langle \hat{t}, \hat{y} \rangle = \arg \min_{t \in \mathcal{F}, y \in \{\pm 1\}} \frac{1}{L} \sum_{i=1}^{L} I(y_i \neq h_{\langle t,y \rangle}(\mathbf{x}_i))$$

$$= \arg\min_{t \in \mathcal{F}, y \in \{\pm 1\}} \frac{1}{2L} \sum_{i=1}^{L} (1 - y_i h_{\langle t, y \rangle}(\mathbf{x}_i)), \tag{13.1}$$

where $\mathcal{F}$ is a set of candidate graphs or a feature set (*i.e.*, $\mathcal{F} = \bigcup_{i=1}^{L} \{t | t \subseteq \mathbf{x}_i\}$) and $I(\cdot)$ is the indicator function. The gain function for a rule $\langle t, y \rangle$ is defined as

$$gain(\langle t, y \rangle) = \sum_{i=1}^{L} y_i h_{\langle t, y \rangle}(\mathbf{x}_i). \tag{13.2}$$

Using the gain, the search problem in Eq. (13.1) becomes equivalent to the problem: $\langle \hat{t}, \hat{y} \rangle = \arg\max_{t \in \mathcal{F}, y \in \{\pm 1\}} gain(\langle t, y \rangle)$. Then the gain function is used instead of error rate.

gboost applies AdaBoost [17] by repeatedly calling the decision stumps and finally produces a hypothesis f, which is a linear combination of K hypotheses produced by the decision stumps $f(\mathbf{x}) = sgn\left(\sum_{k=1}^{K} \alpha_k h_{\langle t_k, y_k \rangle}(\mathbf{x})\right)$. In the kth iteration, a decision stump is built with weights $\mathbf{d}^{(k)} = \left(d_1^{(k)}, \ldots, d_L^{(k)}\right)$ on the training data, where $\sum_{i=1}^{L} d_i^{(k)} = 1$, $d_i^{(k)} \geq 0$. The weights are calculated to concentrate more on hard examples than easy ones. In the boosting framework, the gain function is redefined as

$$gain(\langle t, y \rangle) = \sum_{i=1}^{L} y_i d_i h_{\langle t, y \rangle}(\mathbf{x}_i). \tag{13.3}$$

A Branch-and-Bound Search Approach According to the gain function in Eq. (13.3), the problem of finding the optimal rule $\langle \hat{t}, \hat{y} \rangle$ from the training dataset is defined as follows.

Problem 1 [Find Optimal Rule] Let $T = \{\langle \mathbf{x}_1, y_1, d_1 \rangle, \ldots, \langle \mathbf{x}_L, y_L, d_L \rangle\}$ be a training data set where $\mathbf{x}_i$ is a labeled graph, $y_i \in \{\pm 1\}$ is a class label associated with $\mathbf{x}_i$ and d_i ($\sum_{i=1}^{L} d_i = 1$, $d_i \geq 0$) is a normalized weight assigned to $\mathbf{x}_i$. Given T, find the optimal rule $\langle \hat{t}, \hat{y} \rangle$ that maximizes the gain, *i.e.*, $\langle \hat{t}, \hat{y} \rangle = \arg\max_{t \in \mathcal{F}, y \in \{\pm 1\}} y_i d_i h_{\langle t, y \rangle}$, where $\mathcal{F} = \bigcup_{i=1}^{L} \{t | t \subseteq \mathbf{x}_i\}$.

A naive method is to enumerate all subgraphs $\mathcal{F}$ and then calculate the gains for all subgraphs. However, this method is impractical since the number of subgraphs is exponential to their size. To avoid such exhaustive enumeration, the method to find the optimal rule is modeled as a branch-and-bound algorithm based on the upper bound of the gain function which is defined as follows.

Lemma 13.10 *(Upper bound of the gain) For any $t' \supseteq t$ and $y \in \{\pm 1\}$, the gain of $\langle t', y \rangle$ is bounded by $\mu(t)$ (i.e., $gain(\langle t', y \rangle) \leq \mu(t)$), where $\mu(t)$ is given by*

$$\mu(t) = max\left(2 \sum_{\{i | y_i = +1, t \subseteq x_i\}} d_i - \sum_{i=1}^{L} y_i \cdot d_i, 2 \sum_{\{i | y_i = -1, t \subseteq x_i\}} d_i + \sum_{i=1}^{L} y_i \cdot d_i\right). \tag{13.4}$$

Fig. 13.3 Branch-and-bound
search

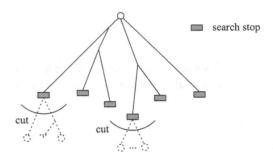

Figure 13.3 depicts a graph pattern search tree where each node represents a graph. A graph g' is a child of another graph g if g' is a supergraph of g with one more edge. g' is also written as $g' = g \diamond e$, where e is the extra edge. In order to find an optimal rule, the branch-and-bound search estimates the upper bound of the gain function for all descendants below a node g. If it is smaller than the value of the best subgraph seen so far, it cuts the search branch of that node. Under the branch-and-bound search, a tighter upper bound is always preferred since it means faster pruning.

Algorithm 12 outlines the framework of branch-and-bound for searching the optimal graph pattern. In the initialization, all the subgraphs with one edge are enumerated first and these seed graphs are then iteratively extended to large subgraphs. Since the same graph could be grown in different ways, Line 5 checks whether it has been discovered before; if it has, then there is no need to grow it again. The optimal $gain(\langle \hat{t}, \hat{y} \rangle)$ discovered so far is maintained. If $\mu(t) \leq gain(\langle \hat{t}, \hat{y} \rangle)$, the branch of t can safely be pruned.

Algorithm 12 Branch-and-Bound

Input: Graph dataset D
Output: Optimal rule $\langle \hat{t}, \hat{y} \rangle$

1: $S = \{$1-edge graph$\}$;
2: $\langle \hat{t}, \hat{y} \rangle = \varnothing$; $gain(\langle \hat{t}, \hat{y} \rangle) = -\infty$;
3: **while** $S \neq \varnothing$ **do**
4:     choose t from S, $S = S \setminus \{t\}$;
5:     **if** t was examined **then**
6:         continue;
7:     **if** $gain(\langle t, y \rangle) > gain(\langle \hat{t}, \hat{y} \rangle)$ **then**
8:         $\langle \hat{t}, \hat{y} \rangle = \langle t, y \rangle$;
9:     **if** $\mu(t) \leq gain(\langle \hat{t}, \hat{y} \rangle)$ **then**
10:        continue;
11:    $S = S \cup \{t' | t' = t \diamond e\}$;
12: **return** $\langle \hat{t}, \hat{y} \rangle$;

3.3 gPLS: A Partial Least Squares Regression Approach

Saigo et al. [34] proposed gPLS, an iterative mining method based on partial least squares regression (PLS). To apply PLS to graph data, a sparse version of PLS is developed first and then it is combined with a weighted pattern mining algorithm. The mining algorithm is iteratively called with different weight vectors, creating one latent component per one mining call. Branch-and-bound search is integrated into graph mining with a designed gain function and a pruning condition. In this sense, gPLS is very similar to the branch-and-bound mining approach in gboost.

Partial Least Squares Regression This part is a brief introduction to partial least squares regression (PLS). Assume there are n training examples $(x_1, y_1), \ldots, (x_n, y_n)$. The output y_i is assumed to be centralized $\sum_i y_i = 0$. Denote by X the design matrix, where each row corresponds to x_i^T. The regression function of PLS is

$$f(x) = \sum_{i=1}^{m} \alpha_i w_i^T x,$$

where m is the pre-specified number of components that form a subset of the original space, and w_i are weight vectors that reduce the dimensionality of x, satisfying the following orthogonality condition,

$$w_i^T X^T X w_j = \begin{cases} 1 & (i = j) \\ 0 & (i \neq j) \end{cases}.$$

Basically w_i are learned in a greedy way first, then the coefficients α_i are obtained by least squares regression without any regularization. The solutions to α_i and w_i are

$$\alpha_i = \sum_{k=1}^{n} y_k w_i^T x_k, \tag{13.5}$$

and

$$w_i = \arg\max_{w} \frac{\left(\sum_{k=1}^{n} y_k w^T x_k\right)^2}{w^T w},$$

subject to $w^T X^T X w = 1$, $w^T X^T X w_j = 0$, $j = 1, \ldots, i - 1$.

Next we present an alternative derivation of PLS called *non-deflation sparse PLS*. Define the ith latent component as $t_i = X w_i$ and T_{i-1} as the matrix of latent components obtained so far, $T_{i-1} = (t_1, \ldots, t_{i-1})$. The residual vector is computed by

$$r_i = \left(I - T_{i-1} T_{i-1}^T\right) y.$$

Then multiply it with X^T to obtain

$$v = \frac{1}{\eta} X^T \left(I - T_{i-1} T_{i-1}^T\right) y.$$

The non-deflation sparse PLS follows this idea.

In graph mining, it is useful to have sparse weight vectors w_i such that only a limited number of patterns are used for prediction. To this aim, we introduce the sparseness to the pre-weight vectors v_i as

$$v_{ij} = 0, \ if \ |v_{ij}| \leq \epsilon, \ j = 1, .., d.$$

Due to the linear relationship between v_i and w_i, w_i becomes sparse as well. Then we can sort $|v_{ij}|$ in the descending order, take the top-k elements and set all the other elements to zero.

It is worthwhile to notice that the residual of regression up to the $(i-1)$-th features,

$$r_{ik} = y_k - \sum_{j=1}^{i-1} \alpha_j w_j^T x_k, \tag{13.6}$$

is equal to the k-th element of r_i. It can be verified by substituting the definition of α_j in Eq. (13.5) into Eq. (13.6). So in the non-deflation algorithm, the pre-weight vector v is obtained as the direction that maximizes the covariance with residues. This observation highlights the resemblance of PLS and boosting algorithms.

Graph PLS: Branch-and-Bound Search   In this part, we discuss how to apply the non-deflation PLS algorithm to graph data. The set of training graphs is represented as $(G_1, y_1), \dots, (G_n, y_n)$. Let $\mathcal{P}$ be the set of all patterns, then the feature vector of each graph G_i is encoded as a $|\mathcal{P}|$-dimensional vector x_i. Since $|\mathcal{P}|$ is a huge number, it is infeasible to keep the whole design matrix. So the method sets X as an empty matrix first, and grows the matrix as the iteration proceeds. In each iteration, it obtains the set of patterns p whose pre-weight $|v_{ip}|$ is above the threshold, which can be written as

$$P_i = \left\{ p \middle| |\sum_{j=1}^{n} r_{ij} x_{jp}| \geq \epsilon \right\}. \tag{13.7}$$

Then the design matrix is expanded to include newly introduced patterns. The pseudo code of gPLS is described in Algorithm 13.

The pattern search problem in Eq. (13.7) is exactly the same as the one solved in gboost through a branch-and-bound search. In this problem, the gain function is defined as $s(p) = |\sum_{j=1}^{n} r_{ij} x_{jp}|$. The pruning condition is described as follows.

Theorem 13.11 *Define $\tilde{y}_i = sgn(r_i)$. For any pattern p' such that $p \subseteq p'$, $s(p') < \epsilon$ holds if*

$$\max \left\{ s^+(p), s^-(p) \right\} < \epsilon, \tag{13.8}$$

where

$$s^+(p) = 2 \sum_{\{i|\tilde{y}_i=+1, x_{i,j}=1\}} |r_i| - \sum_{i=1}^{n} r_i,$$

$$s^-(p) = 2 \sum_{\{i|\bar{y}_i=-1, x_{i,j}=1\}} |r_i| + \sum_{i=1}^{n} r_i.$$

Algorithm 13 gPLS

Input: Training examples $(G_1, y_1), (G_2, y_2), ..., (G_n, y_n)$
Output: Weight vectors w_i, $i = 1, ..., m$

1: $r_1 = y$, $X = \varnothing$;
2: **for** $i = 1, ..., m$ **do**
3:     $P_i = \{p || \sum_{j=1}^{n} r_{ij} x_{jp}| \geq \epsilon\}$;
4:     X_{P_i}: design matrix restricted to P_i;
5:     $X \leftarrow X \cup X_{P_i}$;
6:     $v_i = X^T r_i / \eta$;
7:     $w_i = v_i - \sum_{j=1}^{i-1}(w_j^T X^T X v_i) w_j$;
8:     $t_i = X w_i$;
9:     $r_{i+1} = r_i - (y^T t_i) t_i$;

3.4   LEAP: A Structural Leap Search Approach

Yan et al. [50] proposed an efficient algorithm which mines the most significant subgraph pattern with respect to an objective function. A major contribution of this study is the proposal of a general approach for significant graph pattern mining with non-monotonic objective functions. The mining strategy, called LEAP (Descending Leap Mine), explored two new mining concepts: (1) *structural leap search*, and (2) *frequency-descending mining*, both of which are related to specific properties in pattern search space. The same mining strategy can also be applied to searching other simpler structures such as itemsets, sequences and trees.

Structural Leap Search   Figure 13.4 shows a search space of subgraph patterns. If we examine the search structure horizontally, we find that the subgraphs along the neighbor branches likely have similar compositions and frequencies, hence similar objective scores. Take the branches A and B as an example. Suppose A and B split on a common subgraph pattern g. Branch A contains all the supergraphs of $g \diamond e$ and B contains all the supergraphs of g except those of $g \diamond e$. For a graph g' in branch B, let $g'' = g' \diamond e$ in branch A.

LEAP assumes each input graph is assigned either a positive or a negative label (*e.g.*, compounds active or inactive to a virus). One can divide the graph dataset into two subsets: a positive set D_+ and a negative set D_-. Let $p(g)$ and $q(g)$ be the frequency of a graph pattern g in positive graphs and negative graphs. Many objective

Fig. 13.4 Structural
proximity

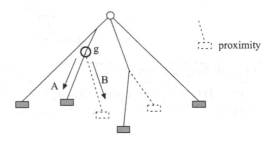

functions can be represented as a function of p and q for a subgraph pattern g, as
$F(g) = f(p(g), q(g))$.

If in a graph dataset, $g \diamond e$ and g often occur together, then g'' and g' might also
occur together. Hence, likely $p(g'') \sim p(g')$ and $q(g'') \sim q(g')$, which means similar
objective scores. This is resulted by the structural and embedding similarity between
the starting structures $g \diamond e$ and g. We call it *structural proximity*: Neighbor branches
in the pattern search tree exhibit strong similarity not only in pattern composition,
but also in their embeddings in the graph datasets, thus having similar frequencies
and objective scores. In summary, a conceptual claim can be drawn,

$$g' \sim g'' \Rightarrow F(g') \sim F(g''). \tag{13.9}$$

According to structural proximity, it seems reasonable to skip the whole search
branch once its nearby branch is searched, since the best scores between neighbor
branches are likely similar. Here, we would like to emphasize "likely" rather than
"surely". Based on this intuition, if the branch A in Fig. 13.4 has been searched, B
could be "leaped over" if A and B branches satisfy some similarity criterion. The
length of leap can be controlled by the frequency difference of two graphs g and
$g \diamond e$. The leap condition is defined as follows.

Let $I(G, g, g \diamond e)$ be an indicator function of a graph G: $I(G, g, g \diamond e) = 1$, for
any supergraph g' of g, if $g' \subseteq G$, $\exists g'' = g' \diamond e$ such that $g'' \subseteq G$; otherwise
0. When $I(G, g, g \diamond e) = 1$, it means if a supergraph g' of g has an embedding
in G, there must be an embedding of $g' \diamond e$ in G. For a positive dataset D_+, let
$D_+(g, g \diamond e) = \{G | I(G, g, g \diamond e) = 1, g \subseteq G, G \in D_+\}$. In $D_+(g, g \diamond e)$, $g' \supset g$
and $g'' = g' \diamond e$ have the same frequency. Define $\Delta_+(g, g \diamond e)$ as follows,

$$\Delta_+(g, g \diamond e) = p(g) - \frac{|D_+(g, g \diamond e)|}{|D_+|}.$$

$\Delta_+(g, g \diamond e)$ is actually the maximum frequency difference that g' and g'' could
have in D_+. If the difference is smaller than a threshold σ, then leap,

$$\frac{2\Delta_+(g, g \diamond e)}{p(g \diamond e) + p(g)} \leq \sigma \text{ and } \frac{2\Delta_-(g, g \diamond e)}{q(g \diamond e) + q(g)} \leq \sigma. \tag{13.10}$$

σ controls the leap length. The larger σ is, the faster the search is. Structural leap
search will generate an optimal pattern candidate and reduce the need for thoroughly

searching similar branches in the pattern search tree. Its goal is to help program search significantly distinct branches, and limit the chance of missing the most significant pattern.

Algorithm 14 Structural Leap Search: sLeap(D, σ, $g^\star$)

Input: Graph dataset D, difference threshold σ
Output: Optimal graph pattern candidate $g^\star$

1: $S = \{1 - \text{edge graph}\}$;
2: $g^\star = \varnothing$; $F(g^\star) = -\infty$;
3: **while** $S \neq \varnothing$ **do**
4:       $S = S \setminus \{g\}$;
5:       **if** g was examined **then**
6:           continue;
7:       **if** $\exists g \diamond e, g \diamond e \prec g$, $\frac{2\Delta_+(g,g\diamond e)}{p(g\diamond e)+p(g)} \leq \sigma$, $\frac{2\Delta_-(g,g\diamond e)}{q(g\diamond e)+q(g)} \leq \sigma$ **then**
8:           continue;
9:       **if** $F(g) > F(g^\star)$ **then**
10:          $g^\star = g$;
11:      **if** $\widehat{F}(g) \leq F(g^\star)$ **then**
12:          continue;
13:      $S = S \cup \{g' | g' = g \diamond e\}$;
14: **return** $g^\star$;

Algorithm 14 outlines the pseudo code of structural leap search (sLeap). The leap condition is tested on Lines 7–8. Note that sLeap does not guarantee the optimality of result.

Frequency Descending Mining Structural leap search takes advantages of the correlation between structural similarity and significance similarity. However, it does not exploit the possible relationship between patterns' frequency and patterns' objective scores. Existing solutions have to set the frequency threshold very low so that the optimal pattern will not be missed. Unfortunately, low-frequency threshold could generate a huge set of low-significance redundant patterns with long mining time.

Although most of objective functions are not correlated with frequency monotonically or anti-monotonically, they are not independent of each other. Cheng et al. [9] derived a frequency upper bound of discriminative measures such as information gain and Fisher score, showing a relationship between frequency and discriminative measures. According to this analytical result, if all frequent subgraphs are ranked in increasing order of their frequency, significant subgraph patterns are often in the high-end range, though their real frequency could vary dramatically across different datasets.

Figure 13.5 illustrates the relationship between frequency and G-test score for an AIDS Anti-viral dataset [50]. It is a contour plot displaying isolines of G-test score in two dimensions. The X axis is the frequency of a subgraph g in the positive

Fig. 13.5 Frequency vs.
G-test score

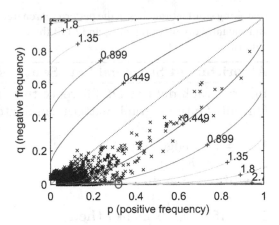

dataset, *i.e.*, $p(g)$, while the Y axis is the frequency of the same subgraph in the negative dataset, $q(g)$. The curves depict G-test score. Left upper corner and right lower corner have the higher G-test scores. The "circle" marks the highest G-score subgraph discovered in this dataset. As one can see, its positive frequency is higher than most of subgraphs.

[Frequency Association] *Significant patterns often fall into the high-quantile of frequency.*

To profit from frequency association, an iterative frequency-descending mining method is proposed in [50]. Rather than performing mining with very low frequency, the method starts the mining process with high frequency threshold $\theta = 1.0$, calculates an optimal pattern candidate g^* whose frequency is at least θ, and then repeatedly lowers down θ to check whether g^* can be improved further. Here, the search leaps in the frequency domain, by leveling down the minimum frequency threshold exponentially.

Algorithm 15 Frequency-Descending Mine: fLeap(D, ε, g^*)

Input: Graph dataset D, converging threshold ε
Output: Optimal graph pattern candidate g^*

1: $\theta = 1.0$;
2: $g = \varnothing$; $F(g) = -\infty$;
3: **do**
4:     $g^* = g$;
5:     g=fpmine(D, θ);
6:     $\theta = \theta/2$;
7: **while** $(F(g) - F(g^*) \geq \varepsilon)$
8: **return** $g^* = g$;

Algorithm 15 (fLeap) outlines the frequency-descending strategy. It starts with the highest frequency threshold, and then lowers the threshold down till the objective score of the best graph pattern converges. Line 5 executes a frequent subgraph mining routine, fpmine, which could be FSG [29], gSpan [45] *etc.* fpmine selects the most significant graph pattern g from the frequent subgraphs it mined. Line 6 implements a simple frequency descending method.

Descending Leap Mine   With structural leap search and frequency-descending mining, a general mining pipeline is built for mining significant graph patterns in a complex graph dataset. It consists of three steps as follows.

1. perform structural leap search with threshold $\theta = 1.0$, generate an optimal pattern candidate $g^\star$.
2. repeat frequency-descending mining with structural leap search until the objective score of $g^\star$ converges.
3. take the best score discovered so far; perform structural leap search again (leap length σ) without frequency threshold; output the discovered pattern.

3.5   GraphSig: A Feature Representation Approach

Ranu and Singh [33] proposed GraphSig, a scalable method to mine significant (measured by p value) subgraphs based on a feature vector representation of graphs. The first step is to convert each graph into a set of feature vectors where each vector represents a region within the graph. Prior probabilities of features are computed empirically to evaluate statistical significance of patterns in the feature space. Following the analysis in the feature space, only a small portion of the exponential search space is accessed for further analysis. This enables the use of existing frequent subgraph mining techniques to mine significant patterns in a scalable manner even when they are infrequent. The major steps of GraphSig are described as follows.

Sliding Window Across Graphs   As the first step, random walk with restart (abbr. RWR) is performed on each node in a graph to simulate sliding a window across the graph. RWR simulates the trajectory of a random walker that starts from the target node and jumps from one node to a neighbor. Each neighbor has an equal probability of becoming the new station of the walker. At each jump, the feature traversed is updated which can either be an edge label or a node label. A restart probability α brings the walker back to the starting node within approximately $\frac{1}{\alpha}$ jumps. The random walk iterates till the feature distribution converges. As a result, RWR produces a continuous distribution of features for each node where a feature value lies in the range [0, 1], which is further discretized into ten bins. RWR can therefore be visualized as placing a window at each node of a graph and capturing a feature vector representation of the subgraph within it. A graph of m nodes is represented by m feature vectors. RWR inherently takes proximity of features into account and preserves more structural information than simply counting occurrence of features inside the window.

Calculating p value of a Feature Vector To calculate p value of a feature vector, we model the occurrence of a feature vector $\underline{x}$ in a feature vector space formulated by a random graph. The frequency distribution of a vector is generated using the prior probabilities of features obtained empirically. Given a feature vector $\underline{x} = [x_1, \ldots, x_n]$, the probability of $\underline{x}$ occurring in a random feature vector $\underline{y} = [y_1, \ldots, y_n]$ can be expressed as a joint probability

$$P(\underline{x}) = P(y_1 \geq x_1, \ldots, y_n \geq x_n). \tag{13.11}$$

To simplify the calculation, we assume independence of the features. As a result, Eq. (13.11) can be expressed as a product of the individual probabilities, where

$$P(\underline{x}) = \prod_{i=1}^{n} P(y_i \geq x_i). \tag{13.12}$$

Once $P(\underline{x})$ is known, the support of $\underline{x}$ in a database of random feature vectors can be modeled as a binomial distribution. To illustrate, a random vector can be viewed as a trial and $\underline{x}$ occurring in it as "success". A database consisting m feature vectors will involve m trials for $\underline{x}$. The support of $\underline{x}$ in the database is the number of successes. Therefore, the probability of $\underline{x}$ having a support μ is

$$P(\underline{x}; \mu) = C_m^{\mu} P(\underline{x})^{\mu} (1 - P(\underline{x}))^{m-\mu}. \tag{13.13}$$

The probability distribution function (abbr. pdf) of $\underline{x}$ can be generated from Eq. (13.13) by varying μ in the range $[0, m]$. Therefore, given an observed support μ_0 of $\underline{x}$, its p value can be calculated by measuring the area under the pdf in the range $[\mu_0, m]$, which is

$$p\text{-}value\,(x, \mu_0) = \sum_{i=\mu_0}^{m} P(\underline{x}; i). \tag{13.14}$$

Identifying Regions of Interest With the conversion of graphs into feature vectors, and a model to evaluate significance of a graph region in the feature space, the next step is to explore how the feature vectors can be analyzed to extract the significant regions. Based on the feature vector representation, the presence of a "common" sub-feature vector among a set of graphs points to a common subgraph. Similarly, the absence of a "common" sub-feature vector indicates the non-existence of any common subgraph. Mathematically, the *floor* of the feature vectors produces the "common" sub-feature vector.

Definition 13.10 (Floor of Vectors) *The floor of a set of vectors $\{v_1, \ldots, v_m\}$ is a vector v_f where $v_{f_i} = min(v_{1_i}, \ldots, v_{m_i})$ for $i = 1, \ldots, n$, n is the number of dimensions of a vector. Ceiling of a set of vectors is defined analogously.*

The next step is to mine common sub-feature vectors that are also significant. Algorithm 16 presents the FVMine algorithm which explores closed sub-vectors in a bottom-up, depth-first manner. FVMine explores all possible common vectors satisfying the significance and support constraints.

Algorithm 16 FVMine($\underline{x}$, S, b)

Input: Current sub-feature vector $\underline{x}$, supporting set S of $\underline{x}$,
       current starting position b
Output: The set of all significant sub-feature vectors A

1: **if** p-$value(\underline{x}) \leq maxPvalue$ **then**
2:    $A \leftarrow A + x$;
3: **for** $i = b$ to m **do**
4:      $S' \leftarrow \{\underline{y} | y \in S, y_i > x_i\}$;
5:      **if** $|S'| < min\_sup$ **then**
6:          **continue**;
7:      $\underline{x}' = floor(S')$;
8:      **if** $\exists j < i$ such that $x'_j > x_j$ **then**
9:          **continue**;
10:     **if** p-$value(ceiling(S'), |S'|) \geq maxPvalue$ **then**
11:         **continue**;
12:     $FVMine(\underline{x}', S', i)$;

With a model to measure the significance of a vector, and an algorithm to mine closed significant sub-feature vectors, we integrate them to build the significant graph mining framework. The idea is to mine significant sub-feature vectors and use them to locate similar regions which are significant. Algorithm 17 outlines the GraphSig algorithm.

The algorithm first converts each graph into a set of feature vectors and puts all vectors together in a single set D' (lines 3–4). D' is divided into sets, such that D'_a contains all vectors produced from RWR on a node labeled a. On each set D'_a, FVMine is performed with a user-specified support and p value thresholds to retrieve the set of significant sub-feature vectors (line 7). Given that each sub-feature vector could describe a particular subgraph, the algorithm scans the database to identify the regions where the current sub-feature vector occurs. This involves finding all nodes labeled a and described by a feature vector such that the vector is a super-vector of the current sub-feature vector $\underline{v}$ (line 9). Then the algorithm isolates the subgraph centered at each node by using a user-specified radius (line 12). This produces a set of subgraphs for each significant sub-feature vector. Next, maximal subgraph mining is performed with a high frequency threshold since it is expected that all of graphs in the set contain a common subgraph (line 13). The last step also prunes out false positives where dissimilar subgraphs are grouped into a set due to the vector representation. For the absence of a common subgraph, when frequent subgraph mining is performed on the set, no frequent subgraph will be produced and as a result the set is filtered out.

Algorithm 17 GraphSig(D, min_sup, $maxPvalue$)

Input: Graph dataset D, support threshold min_sup,
      p-value threshold $maxPvalue$
Output: The set of all significant sub-feature vectors A

1: $D' \leftarrow \varnothing$;
2: $A \leftarrow \varnothing$;
3: **for** each $g \in D$ **do**
4:     $D' \leftarrow D' + RWR(g)$;
5: **for** each node label a in D **do**
6:     $D'_a \leftarrow \{\underline{v}|\underline{v} \in D', label(\underline{v}) = a\}$;
7:     $S \leftarrow FVMine(floor(D'_a), D'_a, 1)$;
8:     **for** each vector $\underline{v} \in S$ **do**
9:         $V \leftarrow \{u|u \text{ is a node of label } a, \underline{v} \subseteq vector(u)\}$;
10:         $E \leftarrow \varnothing$;
11:         **for** each node $u \in V$ **do**
12:             $E \leftarrow E + CutGraph(u, radius)$;
13:             $A \leftarrow A + Maximal\_FSM(E, freq)$;

---

4 Mining Representative Orthogonal Graphs

In this section we will discuss ORIGAMI, an algorithm proposed by Hasan et al. [2], which mines a set of α-orthogonal, β-representative graph patterns. Intuitively, two graph patterns are α-orthogonal if their similarity is bounded by a threshold α. A graph pattern is a β-representative of another pattern if their similarity is at least β. The orthogonality constraint ensures that the resulting pattern set has controlled redundancy. For a given α, more than one set of graph patterns qualify as an α-orthogonal set. Besides redundancy control, representativeness is another desired property, *i.e.*, for every frequent graph pattern not reported in the α-orthogonal set, we want to find a representative of this pattern with a high similarity in the α-orthogonal set.

The set of representative orthogonal graph patterns is a compact summary of the complete set of frequent subgraphs. Given user specified thresholds $\alpha, \beta \in [0, 1]$, the goal is to mine an α-orthogonal, β-representative graph pattern set that minimizes the set of unrepresented patterns.

4.1 Problem Definition

Given a collection of graphs D and a similarity threshold $\alpha \in [0, 1]$, a subset of graphs $\mathcal{R} \subseteq D$ is α-orthogonal with respect to D iff for any $G_a, G_b \in \mathcal{R}$, $sim(G_a, G_b) \leq \alpha$ and for any $G_i \in D \backslash \mathcal{R}$ there exists a $G_j \in \mathcal{R}$, $sim(G_i, G_j) > \alpha$.

Given a collection of graphs D, an α-orthogonal set $\mathcal{R} \subseteq D$ and a similarity threshold $\beta \in [0, 1]$, $\mathcal{R}$ represents a graph $G \in D$, provided that there exists some $G_a \in \mathcal{R}$, such that $sim(G_a, G) \geq \beta$. Let $\Upsilon(\mathcal{R}, D) = \{G | G \in D \text{ s.t. } \exists G_a \in \mathcal{R}, sim(G_a, G) \geq \beta\}$, then $\mathcal{R}$ is a β-representative set for $\Upsilon(\mathcal{R}, D)$.

Given D and $\mathcal{R}$, the residue set of $\mathcal{R}$ is the set of unrepresented patterns in D, denoted as $\triangle(\mathcal{R}, D) = D \backslash \{\mathcal{R} \cup \Upsilon(\mathcal{R}, D)\}$.

The problem defined in [2] is to find the α-orthogonal, β-representative set for the set of all maximal frequent subgraphs $\mathcal{M}$ which minimizes the residue set size. The mining problem can be decomposed into two subproblems of *maximal subgraph mining* and *orthogonal representative set generation*, which are discussed separately. Algorithm 18 shows the algorithm framework of ORIGAMI.

Algorithm 18 ORIGAMI(D, min_sup, α, β)

Input: Graph dataset D, minimum support min_sup, α, β
Output: α-orthogonal, β-representative set $\mathcal{R}$

1: EM=Edge-Map(D);
2: $\mathcal{F}_1$=Find-Frequent-Edges(D, min_sup);
3: $\widehat{\mathcal{M}} = \phi$;
4: **while** stopping_condition() $\neq$ **true do**
5:     M=Random-Maximal-Graph(D, $\mathcal{F}_1$, EM, min_sup);
6:     $\widehat{\mathcal{M}} = \widehat{\mathcal{M}} \cup M$;
7: $\mathcal{R}$=Orthogonal-Representative-Sets($\widehat{\mathcal{M}}$, α, β);
8: **return** $\mathcal{R}$;

4.2 Randomized Maximal Subgraph Mining

As the first step, ORIGAMI mines a set of maximal subgraphs, on which the α-orthogonal, β-representative graph pattern set is generated. This is based on the observation that the number of maximal frequent subgraphs is much fewer than that of frequent subgraphs, and the maximal subgraphs provide a synopsis of the frequent ones to some extent. Thus it is reasonable to mine the representative orthogonal pattern set based on the maximal subgraphs rather than the frequent ones. However, even mining all of maximal subgraphs could be infeasible in some real world applications. To avoid this problem, ORIGAMI first finds a sample $\widehat{\mathcal{M}}$ of the complete set of maximal frequent subgraphs $\mathcal{M}$.

The goal is to find a set of maximal subgraphs, $\widehat{\mathcal{M}}$, which is as diverse as possible. To achieve this goal, ORIGAMI avoids using combinatorial enumeration to mine maximal subgraph patterns. Instead, it adopts a random walk approach to enumerate a diverse set of maximal subgraphs from the positive border of such maximal patterns. The randomized mining algorithm starts with an empty pattern and iteratively adds a random edge during each extension, until a maximal subgraph M is generated and no more edges can be added. This process walks a random chain in the partial order of frequent subgraphs. To extend an intermediate pattern, $S_k \subseteq M$, it chooses a random vertex v from which the extension will be attempted. Then a random edge e incident on v is selected for extension. If no such edge is found, no extension is possible from the vertex. When no vertices can have any further extension in S_k, the random walk terminates and $S_k = M$ is the maximal graph. On the other hand, if a random edge e is found, the other endpoint v' of this edge is randomly selected. By adding the edge e and its endpoint v', a candidate subgraph pattern S_{k+1} is generated and its support is computed. This random walk process repeats until no further extension is possible on any vertex. Then the maximal subgraph M is returned.

Ideally, the random chain walks would cover different regions of the pattern space, thus would produce dissimilar maximal patterns. However, in practice, this may not be the case, since duplicate maximal subgraphs can be generated in the following ways: (1) multiple iterations following overlapping chains, or (2) multiple iterations following different chains but leading to the same maximal pattern. Let's consider a maximal subgraph M of size n. Let $e_1 e_2 \ldots e_n$ be a sequence of random edge extensions, corresponding to a random chain walk leading from an empty graph ϕ to the maximal graph M. The probability of a particular edge sequence leading from ϕ to M is given as

$$P[(e_1 e_2 \ldots e_n)] = P(e_1) \prod_{i=2}^{n} P(e_i | e_1 \ldots e_{i-1}). \tag{13.15}$$

Let $ES(M)$ denote the set of all valid edge sequences for a graph M. The probability that a graph M is generated in a random walk is proportional to

$$\sum_{e_1 e_2 \ldots e_n \in ES(M)} P[(e_1 e_2 \ldots e_n)]. \tag{13.16}$$

The probability of obtaining a specific maximal pattern depends on the number of chains or edge sequences leading to that pattern and the size of the pattern. According to Eq. (13.15), as a graph grows larger, the probability of the edge sequence becomes smaller. So this random walk approach in general favors a maximal subgraph of smaller size than one of larger size. To avoid generating duplicate maximal subgraphs, a termination condition is designed based on an estimate of the collision rate of the generated patterns. Intuitively the collision rate keeps track of the number of duplicate patterns seen within the same or across different random walks. As a random walk chain is traversed, ORIGAMI maintains the signature of the intermediate patterns in a bounded size hash table. As an intermediate or maximal subgraph is generated, its signature is added to the hash table and the collision rate is updated. If the collision rate exceeds a threshold ϵ, the method could (1) abort further extension along the current path and randomly choose another path; or (2) trigger the termination condition across different walks, since it implies that the same part of the search space is being revisited.

4.3 Orthogonal Representative Set Generation

Given a set of maximal subgraphs $\widehat{\mathcal{M}}$, the next step is to extract an α-orthogonal β-representative set from it. We can construct a meta-graph $\Gamma(\widehat{\mathcal{M}})$ to measure similarity between graph patterns in $\widehat{\mathcal{M}}$, in which each node represents a maximal subgraph pattern, and an edge exists between two nodes if their similarity is bounded by α. Then the problem of finding an α-orthogonal pattern set can be modeled as finding a maximal clique in the similarity graph $\Gamma(\widehat{\mathcal{M}})$.

For a given α, there could be multiple α-orthogonal pattern sets as feasible solutions. We could use the size of the residue set to measure the goodness of an α-orthogonal set. An optimal α-orthogonal β-representative set is the one which minimizes the size of the residue set. [2] proved that this problem is NP-hard.

Given the hardness result, ORIGAMI resorts to approximate algorithms to solve the problem which guarantees local optimality. The algorithm starts with a random maximal clique in the similarity graph $\Gamma(\widehat{\mathcal{M}})$ and tries to improve it. At each state transition, another maximal clique which is a local neighbor of the current maximal clique is chosen. If the new state has a better solution, the new state is accepted as the current state and the process continues. The process terminates when all neighbors of the current state have equal or larger residue sizes. Two maximal cliques of size m and n (assume $m \geq n$) are considered neighbors if they share exactly $n - 1$ vertices. The state transition procedure selectively removes one vertex from the maximal clique of the current state and then expands it to obtain another maximal clique which satisfies the neighborhood constraints.

5 Mining Dense Graph Patterns

Mining dense subgraphs is an important graph mining task. Dense subgraphs are useful patterns for many applications, such as detecting communities in social networks, detecting link spams on the Web, finding motifs in biological networks, visualizing complex networks, and so on. There are different definitions of dense subgraph patterns, including clique [10, 12, 42, 44], quasi-clique [36], k-core [11], k-truss [39, 52], dense neighborhood graph [41], dense bipartite subgraph [18], etc. Such different definitions lead to different properties of various dense subgraph patterns, and thus different algorithm design. While some dense subgraph patterns can be discovered in polynomial time, mining other types of dense subgraphs has been proved to be NP-complete.

5.1 Cliques and Quasi-Cliques

In an undirected graph $G = (V, E)$, a clique is a subset of vertices, $C \subseteq V$, such that every two vertices in C are connected by an edge. C is called a maximal clique if there exists no clique C' in G such that $C \subset C'$. A set of vertices C is an α-quasi-clique, if the number of edges in the induced subgraph by C is no less than $\alpha \binom{|C|}{2}$, where $\alpha \in (0, 1)$.

Cheng et al. [10] proposed an external-memory algorithm, called ExtMCE, for maximal clique enumeration from massive scale-free graphs which cannot fit into the main memory. Given an input graph G, ExtMCE recursively computes a portion of G at a time which can be fit into the main memory for MCE computation. To create portions of G that can be used for MCE computation effectively, a novel concept of H^*-graph is proposed, which is inspired by the *h-index* concept. The core part of the H^*-graph is the largest set of h vertices in G that have degree at least h, called the *h-vertices*. The induced subgraph of G by the h-vertices is further extended to their neighborhood to form the H^*-graph. Theoretical analysis is provided to show that the H^*-graph is a small portion of a large scale-free network, thus it is feasible to perform MCE computation based on the H^*-graph in memory.

Given the H^*-graph G_{H^*}, the local maximal cliques are first extracted from G_{H^*}, based on which the global maximal cliques are identified by linking to the remaining part of G. After that, G_{H^*} is removed from G. In the following steps, another subgraph having similar structure as G_{H^*} is extracted from G for local and global maximal clique enumeration in memory, in a similar way as described above. The recursive steps continue until G becomes empty.

Recently, Wang et al. [42] studied redundancy-aware maximal clique computation by generating a subset of maximal cliques as a concise and complete summary of the whole set of maximal cliques $\mathcal{M}(G)$ in a graph G. A notion of *visibility* is introduced to measure how well a maximal clique is covered by the subset. Then the problem is to find a subset of maximal cliques $\mathcal{S} \subseteq \mathcal{M}(G)$, such that the visibility of each maximal clique $C \in \mathcal{M}(G)$ is at least τ, a user-specified parameter. A randomized algorithm and its deterministic version are proposed for the redundancy-aware maximal clique computation.

Xiang et al. [44] proposed an algorithm for computing the maximum clique, which is a clique of the largest possible size in a graph G, based on the MapReduce framework. The main idea is to recursively partition the graph into smaller, possibly overlapping subgraphs so that each node in the MapReduce cluster can independently compute the maximum clique for its partition. When two vertices connected by an edge are placed in two different partitions, one of the vertices has to be replicated in the other's partition so that the edge is preserved. The partitioning algorithm removes one vertex at a time from the graph and puts the subgraph consisting of this vertex plus all its neighbors in one partition. By repeatedly partitioning the graph, it generates multiple subgraphs on which the maximum clique can be computed independently. In addition, the algorithm uses a branch and bound approach to avoid computing the maximum clique of a partition if the size of the maximum clique is smaller than the largest clique found so far.

Tsourakakis et al. [36] proposed to mine the optimal quasi-clique. The density function defined as $f_\alpha(C) = e[C] - \alpha\binom{|C|}{2}$ ($e[C]$ is the number of edges in the induced subgraph by C) measures the edge surplus of a vertex set C over the expected number of edges under the random-graph model. An optimal quasi-clique is a vertex set C that maximizes the function $f_\alpha(C)$. The optimal quasi-cliques are shown to be subgraphs with a large edge density edge density and a small diameter.

5.2   K-Core and K-Truss

Given a graph $G = (V, E)$, the k-core of G is the largest subgraph of G in which every vertex has degree of at least k within the subgraph. The k-truss of G is the largest subgraph of G in which every edge is contained in at least $(k - 2)$ triangles within the subgraph.

The problem of core decomposition in a graph G is to find the k-cores of G for all k. There is a simple and efficient algorithm for core decomposition, by recursively removing the lowest degree vertices and their incident edges. Cheng et al. [11] proposed an external-memory algorithm, called EMcore, for core decomposition in massive graphs which cannot fit into the main memory. EMcore first partitions the original graph into small blocks, and then loads the relevant blocks into main memory. It takes a top-down approach that recursively computes the k-cores from larger values of k to smaller ones, and progressively reduces search space and disk I/O cost by removing the vertices in each computed k-core. The algorithm requires only $O(k_{max})$ scans of the input graph, where k_{max} is the largest core number of the graph.

k-truss (or called *triangle k-core* in [52]) is a type of more cohesive subgraph pattern than k-core, as it is defined based on triangles instead of degrees. A k-truss is a $(k - 1)$-core, but not vice versa. The problem of truss decomposition in a graph G is to find the k-trusses of G for all k. There exists a polynomial time algorithm for computing k-truss. Wang and Cheng [39] proposed an efficient in-memory algorithm for truss decomposition in $O(m^{1.5})$ time, where m is the number of edges in G. They also developed two I/O-efficient algorithms for truss decomposition in massive networks that cannot fit in memory. The first one is a bottom-up approach that employs an effective pruning strategy by removing a large portion of edges before computing each k-truss. The second one takes a top-down approach to find the k-trusses of larger values of k. Zhang and Parthasarathy [52] proposed algorithms to identify triangle k-cores in both static and dynamic graphs.

5.3 Other Dense Subgraph Patterns

Wang et al. [41] defined another type of dense subgraph pattern, called dense neighborhood graph (DN-graph), based on the common neighbors. A DN-graph with parameter λ, denoted as $G'(V', E', \lambda)$, is a connected subgraph of graph $G(V, E)$ that satisfies the following conditions: (1) every connected pair of vertices in G' share at least λ common neighbors; and (2) for any $v \in V \setminus V'$, $\lambda(V' \cup \{v\}) < \lambda$, and for any $v \in V'$, $\lambda(V' - \{v\}) \leq \lambda$. Here, $\lambda(V)$ measures the minimal joint neighborhood size between any connected vertex pair inside the vertex set V. It is proved that mining DN-graphs is NP-complete, by making a connection with the maximal clique mining problem. As the joint neighborhood size of two vertices u, v is in fact the number of triangles the edge $e(u, v)$ participates in a graph, approximate algorithms are proposed which iteratively generate triangles to refine the λ values of graph edges. Another algorithm StreamDN is also proposed for semi-streaming graph setting, which assumes that the graph vertices can be fitted into main memory, while the edges are stored in an ordered manner in the secondary storage.

Gibson et al. [18] proposed an algorithm for mining dense bipartite subgraphs in massive graphs. Informally, a dense bipartite subgraph in a graph $G(V, E)$ is a pair of subset $A, B \subseteq V$ of nodes such that $(a, b) \in E$ for 'most' $a \in A, b \in B$. Here, 'most' parameterizes the density of the subgraph. An efficient algorithm is proposed to mine the dense bipartite subgraphs by recursively fingerprinting the graph. Specifically, for each node in a graph, the algorithm applies a shingling algorithm to the set of destinations linked-to from that node. An s-element subset of the destination set forms a shingle. Nodes that share a sufficiently large number of shingles are clustered together. In the next phase, shingles that tend to occur on the same nodes are grouped together, as the commonly co-occurring shingles lead to the identification of a dense bipartite subgraph. The recursive shingling process can convert dense subgraphs of arbitrary size into small-size fingerprints, which allow identifying the dense subgraphs in a straightforward manner.

6 Mining Graph Patterns in Streams

Recently, there have been studies on mining graph patterns in a streaming environment. In the streaming model, one assumes that the input can be read sequentially in a number of passes over the data, while the total amount of random access memory (RAM) for computation is sublinear in the size of the input. The goal is to reduce the number of passes on the data, while minimizing the amount of RAM for storing the intermediate results.

Aggarwal et al. [1] studied mining dense graph patterns in graph streams. The stream S is defined as the sequence $G_1, \ldots, G_r, \ldots$, where each graph G_i is a set of edges. The mined patterns should have two desired properties: node co-occurrences and edge density. That is, a set of nodes co-occur frequently in the graphs, and the edges among these nodes are dense. The node co-occurrence over a set of nodes

P is defined by a parameter called node affinity $A(P)$, and the edge density is defined by $D(P)$. A set of nodes P is said to be (θ, γ)-significant, if $A(P) \geq \theta$ and $D(P) \geq \gamma$. The dense pattern mining algorithm works in two phases. The first phase uses min-hash to identify all possible groups of nodes satisfying the affinity property. Specifically, the algorithm scans the graphs one by one and generates k independent minimum hash values together with the corresponding indices for each node. A $k \times N$ matrix is formed to store these hash values, where N is the number of nodes in the whole graph set. Then the problem of affinity-based node pattern mining on the original graph set can be transformed to a support-based mining problem on the transformed table. In the second phase, a new graph fragment database is created from the transformed table in the first phase for computing the edge density estimates for the identified node patterns. It is also discussed how to consolidate the two phases into a single pass, so that the mining technique can be used for a data stream.

Bifet et al. [6] studied the problem of mining frequent closed graphs on evolving data streams. First, an incremental mining algorithm is proposed, which assumes that the data arrives in batches of graphs. Every time a new batch of graphs arrives, a closed subgraph mining algorithm is applied on the new batch and the set of frequent closed graphs is updated. The second algorithm focuses on mining frequent subgraphs with a sliding window. The difference to the incremental mining algorithm lies in the management of the items in the sliding window. When the window is full, the oldest batch on the sliding window is deleted. The third algorithm is an adaptive mining one, which can adapt to the changes on the stream, maintaining only the currently frequent closed graphs.

Bahmani et al. [5] studied the problem of mining the densest subgraph in a streaming environment, and showed how to parallelize their algorithms in the MapReduce model. They adopt the semi-streaming model, which assumes that the set of nodes is known in advance, and the edges are streamed. For an undirected graph $G = (V, E)$, given a subset of vertices $S \subseteq V$, its density is defined as $\rho(S) = \frac{|E(S)|}{|S|}$, where $E(S)$ is the set of edges in the induced subgraph by S. For a directed graph $G = (V, E)$, given two subsets of vertices $S, T \subseteq V$, their density is defined as $\rho(S, T) = \frac{|E(S,T)|}{\sqrt{|S||T|}}$, where $E(S, T)$ is the set of edges in the induced subgraph by S, T. The problem of mining the densest subgraph is to find S which maximizes $\rho(S)$ in an undirected graph, or to find S, T which maximize $\rho(S, T)$ in a directed graph. Streaming algorithms are proposed for finding approximately densest subgraphs. Specifically, a $(2 + 2\epsilon)$-approximation algorithm is proposed for both undirected graphs and directed graphs. A $(3 + 3\epsilon)$-approximation algorithm is proposed when there is a size constraint of k nodes on the densest subgraph. All these algorithms make $O(\log n)$ passes over the input graph and use $O(n)$ main memory. As an example, the approximate algorithm on undirected graphs works as follows. Starting with the input graph G, the algorithm computes the current density, $\rho(G)$, and then removes all the nodes (and their incident edges) whose degree is less than $(2 + 2\epsilon) \cdot \rho(G)$. If the resulting graph is non-empty, then the algorithm recurses on the remaining graph with the node set denoted by S, until the graph becomes empty. The node subset S which achieves the highest density in the process is returned as the approximately densest subgraph.

Angel et al. [3] studied how to maintain dense subgraphs under streaming edge weight updates, with an application of real-time story identification on social media stream. The intuition is that stories can be identified via groups of tightly-coupled real-world entities that are involved in the story. In this work, an entity graph is built whose vertices correspond to real-world entities, and edge weights correspond to their pairwise association strengths in posts. Given a stream of user generated posts, a corresponding stream of edge weight updates between the mentioned entities can be obtained. The problem is, given the edge weight updates, how to maintain the dense subgraphs, where the density is calculated based on the total edge weights in a subgraph. An algorithm, called DYNDENS is proposed for this purpose. Specifically, given an edge weight update on edge (a, b), it will explore dense subgraphs containing a or b, or containing both a and b by augmenting them with neighborhood vertices. A dense subgraph index is designed to support easy access to dense subgraphs, insertion, update and deletion of dense subgraphs from the index. Theoretical analysis of DYNDENS is provided to bound the number of exploration iterations that are required, as a function of the magnitude of the edge weight update performed.

7 Mining Graph Patterns in Uncertain Graphs

There are also some studies on mining graph patterns in uncertain graphs in the literature. In the uncertain graph model [26], each edge of a graph is associated with a probability to quantify the likelihood that this edge exists in the graph. The existence of edges is mutually independent. Some other studies [53] assume that each vertex of a graph is also associated with an existence probability, and each edge is associated with a conditional existence probability given the two endpoints. Uncertain graphs have practical importance in the real world, for example, for modeling the reliability of a link in the telecommunication or electrical networks, or modeling the interaction in a protein interaction network.

Jin et al. [26] studied the problem of discovering highly reliable subgraphs in uncertain graphs. Given an uncertain graph $\mathcal{G} = (V, E, P)$, its network reliability $\mathbf{R}[\mathcal{G}]$ is defined as the probability that its sampled realizations remain connected. Mathematically, it is defined as $\mathbf{R}[\mathcal{G}] = \sum_{G \sqsubseteq \mathcal{G}} \mathbf{I}(G) \cdot \mathbf{Pr}[G]$, where G is a deterministic graph as a possible outcome of the edge probabilities in $\mathcal{G}$, $\mathbf{Pr}[G]$ is the probability of sampling G from $\mathcal{G}$, $\mathbf{I}(G)$ is an indicator function and $\mathbf{I}(G) = 1$ if G is connected, and $\mathbf{I}(G) = 0$ otherwise. The network reliability definition can be easily generalized to the induced subgraph $\mathcal{G}[V_s]$ for a subset of vertices $V_s \subseteq V$. The problem of discovering highly reliable subgraphs is, given an uncertain graph $\mathcal{G}$ and a reliability threshold $\alpha \in [0, 1]$, determine all induced subgraphs whose network reliability is at least α. The resulting set is denoted by $\mathcal{S}_\alpha$. As the network reliability problem is #P-complete, a sampling approach for subgraph reliability estimation is proposed. First, N possible graphs $G_1, G_2, \ldots, G_N$ are sampled from $\mathcal{G}$. For any subset of vertices V_s, compute the indicator function $\mathbf{I}(G_i[V_s])$. Then the sampling estimator of the subgraph reliability is $\mathbf{R}[\mathcal{G}[V_s]] \approx \widehat{\mathbf{R}}[\mathcal{G}[V_s]] = \frac{\sum_{i=1}^{N} \mathbf{I}(G_i[V_s])}{N}$. In order to control the number of false positives or negatives, two sets are used to approximate $\mathcal{S}_\alpha$ by samling: (1) the first set $\overline{S}$ which tries to maximize the recall of discovering highly reliable subgraphs; and (2) the second set $\underline{S}$ which tries to maximize the precision of discovery.

Under this sampling framework, a subproblem is to determine the vertex subset V_s in a graph G for discovering the induced subgraph. This is formulated as the frequent cohesive set mining problem. Specifically, given a set of graphs $D = \{G_1, G_2, \ldots, G_N\}$ with vertices $V(G_1) = V(G_2) = \ldots = V(G_N) = V$ and a minimal support threshold θ, a frequent cohesive set is any subset of vertices $V_s \subseteq V$ taht is a cohesive set in at least $\theta \cdot N$ graphs. A two-stage mining algorithm is proposed for mining the frequent cohesive sets. In the first stage, a top-down peeling process is employed to iteratively refine patterns to make them converge into maximal frequent cohesive sets. In the second stage, a DFS mining process is employed which utilizes the maximal frequent cohesive sets as the boundary to prune the search space and discover all the non-maximal frequent cohesive sets.

Zou et al. [53] studied the problem of mining frequent subgraphs on uncertain graph data under probabilistic semantics. An uncertain graph database is denoted as $D = \{G_1, G_2, \ldots, G_n\}$ where each G_i is an uncertain graph. An implicated graph database is denoted as $D' = \{G'_1, G'_2, \ldots, G'_m\}$ where each G'_i is an implicated certain graph of an uncertain graph in D. For a certain subgraph S, its support in D' is $sup(S; D') = \frac{|\{G' \in D' | S \subseteq G'\}|}{|D'|}$. Then the probability that the support of S is no less than $0 \leq \varphi 1$ across all implicated graph databases of D is $Pr(S; D, \varphi) = \sum_{D' \in Imp(D), sup(S; D') \geq \varphi} Pr(D \Rightarrow D')$, where $Pr(D \Rightarrow D')$ is the probability of D implicating D', and $Imp(D)$ is the set of all implicated graph databases of D. A subgraph S is called (φ, τ)-probabilistic frequent if the φ-frequent probability of S is no less than a user-specified confidence threshold $0 \leq \tau \leq 1$. The problem of mining frequent subgraphs in an uncertain graph database under probabilistic semantics is, given an uncertain graph database D, a support threshold $0 \leq \varphi \leq 1$, and a confidence threshold $0 \leq \tau \leq 1$, find all (φ, τ)-probabilistic frequent subgraphs in D.

It is proven that it is #P-hard to compute the φ-frequent probability of a subgraph S in an uncertain graph database and to count the number of frequent subgraphs in an uncertain graph database. Then an approximate mining algorithm is proposed to find a broader set of subgraphs including all frequent subgraphs and a fraction of infrequent subgraphs but with φ-frequent probability at least $\tau - \varepsilon$, where $0 \leq \varepsilon \leq \tau$ is an error tolerance. The main steps of the algorithm are as follows. First, organize all subgraphs in D into a search tree, where nodes represent subgraphs, and each node is subgraph isomorphic to all its children if it has any, and has one less edge than all of them. Second, examine the subgraphs in the search tree in depth-first order. For each examined subgraph S, determine in polynomial time whether S has φ-frequent probability at least $\tau - \varepsilon$ and probably at least τ. If the answer is "yes", then output S and proceed to examine the descendants of S in depth-first order. Otherwise, all descendants of S are infrequent and can be pruned due to the apriori property. A dynamic programming algorithm is proposed for approximating the φ-frequent probability, $Pr(S; D, \varphi)$, of a subgraph S by an interval $[p_l, p_u]$ such that $|p_u - p_l| \leq \varepsilon$ and $Pr(S; D, \varphi) \in [p_l, p_u]$.

8 Conclusions

Frequent subgraph mining is one of the fundamental tasks in graph data mining. The inherent complexity in graph data causes the combinatorial explosion problem. As a result, a mining algorithm may take a long time or even forever to complete the mining process on some real graph datasets.

In this chapter, we introduced several state-of-the-art methods that mine a compact set of significant or representative subgraphs without generating the complete set of graph patterns. The proposed mining and pruning techniques were discussed in details. These methods greatly reduce the computational cost, while at the same time, increase the applicability of the generated graph patterns. Other variations of frequent subgraph patterns include dense subgraphs, reliable subgraphs and so on. The new application scenarios also call for new graph pattern mining algorithms, for example, mining graphs patterns in a graph stream or in uncertain graphs. We also introduced the state-of-the-art research results under such settings. These research results have made significant progress on graph mining research with many new applications.

References

1. C. C. Aggarwal, Y. Li, P. S. Yu, and R. Jin. On dense pattern mining in graph streams. *PVLDB*, 3(1):975–984, 2010.
2. M. Al Hasan, V. Chaoji, S. Salem, J. Besson, and M. J. Zaki. ORIGAMI: Mining representative orthogonal graph patterns. In *Proc. 2007 Int. Conf. Data Mining (ICDM'07)*, pages 153–162, 2007.
3. A. Angel, N. Koudas, N. Sarkas, and D. Srivastava. Dense subgraph maintenance under streaming edge weight updates for real-time story identification. *PVLDB*, 3(5):574–585, 2012.
4. T. Asai, K. Abe, S. Kawasoe, H. Arimura, H. Satamoto, and S. Arikawa. Efficient substructure discovery from large semi-structured data. In *Proc. 2002 SIAM Int. Conf. Data Mining (SDM'02)*, pages 158–174, 2002.
5. B. Bahmani, R. Kumar, and S. Vassilvitskii. Densest subgraph in streaming and MapReduce. *PVLDB*, 5(5):454–465, 2012.
6. A. Bifet, G. Holmes, B. Pfahringer, and R. Gavalda. Mining frequent closed graphs on evolving data streams. In *KDD*, pages 591–599, 2011.
7. C. Borgelt and M. R. Berthold. Mining molecular fragments: Finding relevant substructures of molecules. In *Proc. 2002 Int. Conf. Data Mining (ICDM'02)*, pages 211–218, 2002.
8. B. Bringmann and S. Nijssen. What is frequent in a single graph? In *Proc. 2008 Pacific-Asia Conf. Knowledge Discovery and Data Mining (PAKDD'08)*, pages 858–863, 2008.
9. H. Cheng, X. Yan, J. Han, and C.-W. Hsu. Discriminative frequent pattern analysis for effective classification. In *Proc. 2007 Int. Conf. Data Engineering (ICDE'07)*, pages 716–725, 2007.
10. J. Cheng, Y. Ke, A. Fu, J. X. Yu, and L. Zhu. Finding maximal cliques in massive networks by H*-graph. In *SIGMOD*, pages 447–458, 2010.
11. J. Cheng, Y. Ke, S. Chu, and M. T. Ozsu. Efficient core decomposition in massive networks. In *ICDE*, pages 51–62, 2011.
12. J. Cheng, L. Zhu, Y. Ke, and S. Chu. Fast algorithms for Maximal Clique Enumeration with Limited Memory. In *KDD*, pages 1240–1248, 2012.
13. Y. Chi, Y. Xia, Y. Yang, and R. Muntz. Mining closed and maximal frequent subtrees from databases of labeled rooted trees. *IEEE Trans. Knowledge and Data Eng.*, 17:190–202, 2005.

14. L. Dehaspe, H. Toivonen, and R. King. Finding frequent substructures in chemical compounds. In *Proc. 1998 Int. Conf. Knowledge Discovery and Data Mining (KDD'98)*, pages 30–36, 1998.

15. M. Deshpande, M. Kuramochi, N. Wale, and G. Karypis. Frequent substructure-based approaches for classifying chemical compounds. *IEEE Trans. on Knowledge and Data Engineering*, 17:1036–1050, 2005.

16. M. Fiedler and C. Borgelt. Support computation for mining frequent subgraphs in a single graph. In *Proc. 5th Int. Workshop on Mining and Learning with Graphs (MLG'07)*, 2007.

17. Y. Freund and R. Schapire. A decision-theoretic generalization of on-line learning and an application to boosting. In *Proc. 2nd European Conf. Computational Learning Theory*, pages 23–27, 1995.

18. D. Gibson, R. Kumar, and A. Tomkins. Discovering large dense subgraphs in massive graphs. In *VLDB*, pages 721–732, 2005.

19. H. He and A. K. Singh. Efficient algorithms for mining significant substructures in graphs with quality guarantees. In *Proc. 2007 Int. Conf. Data Mining (ICDM'07)*, pages 163–172, 2007.

20. L. B. Holder, D. J. Cook, and S. Djoko. Substructure discovery in the subdue system. In *Proc. AAAI'94 Workshop Knowledge Discovery in Databases (KDD'94)*, pages 169–180, 1994.

21. J. Huan, W. Wang, and J. Prins. Efficient mining of frequent subgraph in the presence of isomorphism. In *Proc. 2003 Int. Conf. Data Mining (ICDM'03)*, pages 549–552, 2003.

22. J. Huan, W. Wang, D. Bandyopadhyay, J. Snoeyink, J. Prins, and A. Tropsha. Mining spatial motifs from protein structure graphs. In *Proc. 8th Int. Conf. Research in Computational Molecular Biology (RECOMB)*, pages 308–315, 2004.

23. J. Huan, W. Wang, J. Prins, and J. Yang. SPIN: Mining maximal frequent subgraphs from graph databases. In *Proc. 2004 ACM SIGKDD Int. Conf. Knowledge Discovery in Databases (KDD'04)*, pages 581–586, 2004.

24. A. Inokuchi, T. Washio, and H. Motoda. An apriori-based algorithm for mining frequent substructures from graph data. In *Proc. 2000 European Symp. Principle of Data Mining and Knowledge Discovery (PKDD'00)*, pages 13–23, 1998.

25. R. Jin, C. Wang, D. Polshakov, S. Parthasarathy, and G. Agrawal. Discovering frequent topological structures from graph datasets. In *Proc. 2005 ACM SIGKDD Int. Conf. Knowledge Discovery in Databases (KDD'05)*, pages 606–611, 2005.

26. R. Jin, L. Liu, and C. C. Aggarwal. Discovering highly reliable subgraphs in uncertain graphs. In *KDD*, pages 992–1000, 2011.

27. M. Koyuturk, A. Grama, and W. Szpankowski. An efficient algorithm for detecting frequent subgraphs in biological networks. *Bioinformatics*, 20:I200–I207, 2004.

28. T. Kudo, E. Maeda, and Y. Matsumoto. An application of boosting to graph classification. In *Advances in Neural Information Processing Systems 18 (NIPS'04)*, 2004.

29. M. Kuramochi and G. Karypis. Frequent subgraph discovery. In *Proc. 2001 Int. Conf. Data Mining (ICDM'01)*, pages 313–320, 2001.

30. M. Kuramochi and G. Karypis. Finding frequent patterns in a large sparse graph. *Data Mining and Knowledge Discovery*, 11:243–271, 2005.

31. S. Nijssen and J. Kok. A quickstart in frequent structure mining can make a difference. In *Proc. 2004 ACM SIGKDD Int. Conf. Knowledge Discovery in Databases (KDD'04)*, pages 647–652, 2004.

32. J. Pei, J. Han, B. Mortazavi-Asl, H. Pinto, Q. Chen, U. Dayal, and M.-C. Hsu. PrefixSpan: Mining sequential patterns efficiently by prefix-projected pattern growth. In *Proc. 2001 Int. Conf. Data Engineering (ICDE'01)*, pages 215–224, 2001.

33. S. Ranu and A. K. Singh. GraphSig: A scalable approach to mining significant subgraphs in large graph databases. In *Proc. 2009 Int. Conf. Data Engineering (ICDE'09)*, pages 844–855, 2009.

34. H. Saigo, N. Krämer, and K. Tsuda. Partial least squares regression for graph mining. In *Proc. 2008 ACM SIGKDD Int. Conf. Knowledge Discovery in Databases (KDD'08)*, pages 578–586, 2008.

35. L. Thomas, S. Valluri, and K. Karlapalem. MARGIN: Maximal frequent subgraph mining. In *Proc. 2006 Int. Conf. on Data Mining (ICDM'06)*, pages 1097–1101, 2006.

36. C. E. Tsourakakis, F. Bonchi, A. Gionis, F. Gullo, and M. A. Tsiarli. Denser than the densest subgraph: Extracting optimal quasi-cliques with quality guarantees. In *KDD*, pages 104–112, 2013.
37. K. Tsuda. Entire regularization paths for graph data. In *Proc. 2007 Int. Conf. Machine Learning (ICML'07)*, pages 919–926, 2007.
38. N. Vanetik, E. Gudes, and S. E. Shimony. Computing frequent graph patterns from semistructured data. In *Proc. 2002 Int. Conf. on Data Mining (ICDM'02)*, pages 458–465, 2002.
39. J. Wang and J. Cheng. Truss decomposition in massive networks. *PVLDB*, 5(9):812–823, 2012.
40. C. Wang, W. Wang, J. Pei, Y. Zhu, and B. Shi. Scalable mining of large disk-base graph databases. In *Proc. 2004 ACM SIGKDD Int. Conf. Knowledge Discovery in Databases (KDD'04)*, pages 316–325, 2004.
41. N. Wang, J. Zhang, K. L. Tan, A. K. H. Tung. On Triangulation-based Dense Neighborhood Graphs Discovery. *PVLDB*, 4(2):58–68, 2010.
42. J. Wang, J. Cheng, and A. Fu. Redundancy-aware maximal cliques Cliques. In *KDD*, pages 122–130, 2013.
43. T. Washio and H. Motoda. State of the art of graph-based data mining. *SIGKDD Explorations*, 5:59–68, 2003.
44. J. Xiang, C. Guo, and A. Aboulnaga. Scalable maximum clique computation using MapReduce. In *ICDE*, pages 74–85, 2013.
45. X. Yan and J. Han. gSpan: Graph-based substructure pattern mining. In *Proc. 2002 Int. Conf. Data Mining (ICDM'02)*, pages 721–724, 2002.
46. X. Yan and J. Han. CloseGraph: Mining closed frequent graph patterns. In *Proc. 2003 ACM SIGKDD Int. Conf. Knowledge Discovery and Data Mining (KDD'03)*, pages 286–295, 2003.
47. X. Yan and J. Han. Discovery of frequent substructures. In D. Cook and L. Holder (eds.), Mining Graph Data, pages 99–115, John Wiley Sons, 2007.
48. X. Yan, P. S. Yu, and J. Han. Graph indexing: A frequent structure-based approach. In *Proc. 2004 ACM-SIGMOD Int. Conf. Management of Data (SIGMOD'04)*, pages 335–346, 2004.
49. X. Yan, X. J. Zhou, and J. Han. Mining closed relational graphs with connectivity constraints. In *Proc. 2005 ACM SIGKDD Int. Conf. Knowledge Discovery in Databases (KDD'05)*, pages 324–333, 2005.
50. X. Yan, H. Cheng, J. Han, and P. S. Yu. Mining significant graph patterns by scalable leap search. In *Proc. 2008 ACM SIGMOD Int. Conf. on Management of Data (SIGMOD'08)*, pages 433–444, 2008.
51. M. J. Zaki. Efficiently mining frequent trees in a forest. In *Proc. 2002 ACM SIGKDD Int. Conf. Knowledge Discovery in Databases (KDD'02)*, pages 71–80, 2002.
52. Y. Zhang and S. Parthasarathy. Extracting analyzing and visualizing triangle k-core motifs within networks. In *ICDE*, pages 1049–1060, 2012.
53. Z. Zou, H. Gao, and J. Li. Discovering frequent subgraphs over uncertain graph databases under probabilistic semantics. In *KDD*, pages 633–642, 2010.

Chapter 14
Uncertain Frequent Pattern Mining

Carson Kai-Sang Leung

Abstract Frequent pattern mining aims to discover implicit, previously unknown and potentially useful knowledge—in the form of frequently occurring sets of items—that are embedded in data. Many of the models and algorithms developed in the early days mine frequent patterns from traditional transaction databases of precise data such as shopper market basket data, in which the contents of databases are known. However, we are living in an uncertain world, in which uncertain data can be found in various real-life applications. Hence, in recent years, researchers have paid more attention to frequent pattern mining from probabilistic datasets of uncertain data. This chapter covers key models, algorithms and topics about uncertain frequent pattern mining.

Keywords Data mining · Knowledge discovery from uncertain data · Association rule mining · Frequent patterns · Frequent itemsets · Probabilistic approach · Uncertain data

1 Introduction

As an important data mining task, *frequent pattern mining* [8, 12] aims to discover implicit, previously unknown and potentially useful knowledge—revealing patterns on collections of frequently co-occurring items, objects or events—that are embedded in data. Nowadays, frequent pattern mining is commonly used in various real-life business, government, and science applications (e.g., banking, bioinformatics, environmental modeling, epidemiology, finance, marketing, medical diagnosis, meteorological data analysis). Uncertain data are present in many of these applications. Uncertainty can be caused by (i) our limited perception or understanding of reality; (ii) limitations of the observation equipment; or (iii) limitations of available resources for the collection, storage, transformation, or analysis of data. It can also be inherent in nature (say, due to prejudice). Data collected by acoustic, chemical, electromagnetic, mechanical, optical radiation, thermal sensors [6] in environment surveillance, security, and manufacturing systems can be noisy. Dynamic

C. Kai-Sang Leung (✉)
University of Manitoba, Winnipeg, MB, Canada
e-mail: kleung@cs.umanitoba.ca

C. C. Aggarwal, J. Han (eds.), *Frequent Pattern Mining*,
DOI 10.1007/978-3-319-07821-2_14, © Springer International Publishing Switzerland 2014

errors—such as (i) inherited measurement inaccuracies, (ii) sampling frequency of the sensors, (iii) deviation caused by a rapid change (e.g., drift, noise) of the measured property over time, (iv) wireless transmission errors, or (v) network latencies—also introduce uncertainty into the data reported by these sensors. Moreover, there is also uncertainty in survey data (e.g., number "0" vs. symbol "o" vs. letter "O" or "o"; similarly, number "1" vs. upper case letter "I" vs. lower case letter "l") and uncertainty due to data granularity (e.g., city, province) in taxonomy. Disguised missing data, which are not explicitly represented as such but instead appear as potentially valid data values, also introduce uncertainty. Furthermore, in privacy-preserving applications [10], sensitive data may be intentionally blurred via aggregation or perturbation so as to preserve data anonymity. All these sources of uncertainty lead to huge amounts of *uncertain data* in real-life applications [48, 52].

Many key models and algorithms have been developed over the past few years for various uncertain data mining tasks [3, 11]. These include (i) clustering uncertain data [2, 7, 25], (ii) classifying uncertain data [47, 53] and (iii) detecting outliers from uncertain data [5, 9]. In this chapter, we examine another data mining task—namely, *uncertain frequent pattern mining*. To mine frequent patterns from uncertain data, different methodologies (e.g., fuzzy set theory, rough set theory) can be applicable. Among them, probability theory is more popular and widely used by many researchers.

In this chapter, we focus on uncertain frequent pattern mining in a probabilistic setting. The remainder of this chapter is organized as follows. The next section describes a key model for uncertain frequent pattern mining: the probabilistic model. Then, we present those key uncertain frequent pattern mining algorithms based on (i) the candidate generate-and-test paradigm, (ii) the frequent pattern growth paradigm with hyperlinked structures, and (iii) the frequent pattern growth para-digm with tree structures in Sects. 3, 4 and 5, respectively. Sections 6, 7 and 8 describe key algorithms for uncertain frequent pattern mining (i) with constraints, (ii) from Big data, and (iii) from data streams, respectively. Section 9 examines key algorithms for mining uncertain data that are in vertical representation. We briefly discuss and compare these algorithms in Sect. 10. While Sects. 2 to 10 focus on (expected support based) frequent patterns, Sect. 11 focuses on the probabilistic frequent patterns. Finally, Sect. 12 gives conclusions.

2 The Probabilistic Model for Mining Expected Support-Based Frequent Patterns from Uncertain Data

As a building block for association rule mining [12] (which helps reveal associative relationships embedded in data), frequent pattern mining aims to discover frequently occurring sets of items, objects or events (e.g., frequently purchased merchandise items in shopper market baskets, bundles of popular books, popular courses taken by students, events that are frequently collocated). In the early days, most frequent

Table 14.1 An example of a traditional database D_1 of precise data

Transaction ID	Set of items
t_1	$\{a, b, c\}$
t_2	$\{a, b, c, d\}$
t_3	$\{a, b, d, e\}$
t_4	$\{a, b, c, e\}$

Table 14.2 An example of a probabilistic dataset D_2 of uncertain data

Transaction ID	Set of items with existential probability
t_1	$\{a{:}0.2,\ b{:}0.9,\ c{:}0.4\}$
t_2	$\{a{:}0.6,\ b{:}0.6,\ c{:}0.6,\ d{:}0.9\}$
t_3	$\{a{:}0.6,\ b{:}0.5,\ d{:}0.5,\ e{:}0.7\}$
t_4	$\{a{:}0.9,\ b{:}0.2,\ c{:}0.8,\ e{:}0.3\}$

pattern mining algorithms searched traditional databases of precise data (e.g., Table 14.1) such as shopper market basket data, in which the contents of databases are known. However, we are living in an uncertain world. Uncertain data can be found in various real-life applications, in which users may not be certain about the presence or absence of an item x in a transaction t_i in a probabilistic dataset D of uncertain data (e.g., Table 14.2). Users may suspect, but cannot guarantee, that x is present in t_i. The uncertainty of such suspicion can be expressed in terms of *existential probability* $P(x, t_i)$, which indicates the likelihood of x being present in t_i in D. The existential probability $P(x, t_i)$ ranges from a positive value close to 0 (indicating that x has an insignificantly low chance to be present in D) to a value of 1 (indicating that x is definitely present). With this notion, each item in any transaction in traditional databases of precise data (e.g., shopper market basket data) can be viewed as an item with a 100 % likelihood of being present in such a transaction.

The **probabilistic model** [1, 23] is a key model that commonly used for uncertain frequent pattern mining. When using the "possible world" interpretation of uncertain data, there are two possible worlds for an item x in a transaction t_i:

(i) a possible world W_1 where x is present in t_i (i.e., $x \in t_i$)

and

(ii) another possible world W_2 where x is absent from t_i (i.e., $x \notin t_i$).

Although it is uncertain which of these two worlds is the true world, the probability of W_1 to be the true world is $P(x, t_i)$ and the probability of W_2 to be the true world is $1 - P(x, t_i)$. To a further extent, there are multiple items in each of many transactions in a probabilistic dataset D of uncertain data. In a domain of m distinct items, when there are a total of q independent items (which include multiple occurrences of some of the m domain items, where $m \ll q$) in all transactions of D, there are $O(2^q)$ possible worlds. The **expected support** of a pattern X in D—denoted as $expSup(X, D)$—can then be computed by summing the support $sup(X, W_j)$ of X in possible world W_j (while taking into account the probability $Prob(W_j)$ of W_j to

Table 14.3 Possible worlds for the probabilistic dataset D_2 of uncertain data

Possible world W_j	$Prob(W_j)$	Transactions
W_1	6.349×10^{-5}	$\{t_1=\{a,b,c\},$ $t_2=\{a,b,c,d\},$ $t_3=\{a,b,d,e\},$ $t_4=\{a,b,c,e\}\}$
W_2	1.481×10^{-4}	$\{t_1=\{a,b,c\},$ $t_2=\{a,b,c,d\},$ $t_3=\{a,b,d,e\},$ $t_4=\{a,b,c\ \ \}\}$
W_3	1.587×10^{-5}	$\{t_1=\{a,b,c\},$ $t_2=\{a,b,c,d\},$ $t_3=\{a,b,d,e\},$ $t_4=\{a,b,\ \ e\}\}$
$\vdots$	$\vdots$	$\vdots$
W_{32767}	1.769×10^{-7}	$\{t_1=\{\}, t_2=\{\}, t_3=\{\}, t_4=\{e\}\}$
W_{32768}	4.129×10^{-7}	$\{t_1=\{\}, t_2=\{\}, t_3=\{\}, t_4=\{\}\}$
	$\sum_j Prob(W_j) = 1$	

be the true world) over all possible worlds, i.e.,

$$expSup(X,D) = \sum_j \left[sup(X, W_j) \times Prob(W_j) \right], \qquad (14.1)$$

where $sup(X, W_j)$ counts the occurrences of X (i.e., the number of transactions containing *all* the items within X) and $Prob(W_j)$ can be computed by the following equation:

$$Prob(W_j) = \prod_{i=1}^{|D|} \left[\prod_{x \in t_i \text{ in } W_j} P(x, t_i) \times \prod_{y \notin t_i \text{ in } W_j} (1 - P(y, t_i)) \right]. \qquad (14.2)$$

Table 14.3 shows all "possible worlds" of the probabilistic dataset D_2 in Table 14.2. When items within the pattern X are independent, Eq. (14.1) can be simplified [32] to become the following equation:

$$expSup(X,D) = \sum_{i=1}^{|D|} \left(\prod_{x \in X} P(x, t_i) \right). \qquad (14.3)$$

In other words, the expected support of X in D can be computed as a sum (over all $|D|$ transactions) of the product of existential probabilities of all items within X. Then, we can define the research problem of uncertain frequent pattern mining in terms of $expSup(X, D)$ as follows.

Definition 14.1 Given (i) a probabilistic dataset D of uncertain data and (ii) a user-specified support threshold *minsup*, the research problem of **uncertain frequent pattern mining** from a probabilistic dataset D of uncertain data is to find every pattern X having $expSup(X, D) \geq minsup$. Such a pattern X is called an *expected support-based frequent pattern* or just *frequent pattern* for short.

3   Candidate Generate-and-Test Based Uncertain Frequent Pattern Mining

One way to mine frequent patterns from uncertain data is to apply the candidate generate-and-test paradigm. For example, Chui et al. [19] proposed the **U-Apriori** algorithm, which mines frequent patterns from uncertain data in a levelwise breadth-first bottom-up fashion. Specifically, U-Apriori first computes the expected support of all domain items. Those items with expected supports $\geq$ *minsup* become every frequent pattern consisting of 1 item (i.e., frequent 1-itemset). Afterwards, the U-Apriori algorithm repeatedly applies the candidate generate-and-test process to generate candidate $(k+1)$-itemsets from frequent k-itemsets and test if they are frequent $(k+1)$-itemsets. Like its counterpart for mining precise data (the Apriori algorithm [8]), U-Apriori also relies on the *Apriori property* (which is also known as the *anti-monotonic property* or the *downward closure property*) that all subsets of a frequent pattern must also be frequent. Equivalently, all supersets of any infrequent pattern are also infrequent.

U-Apriori improves its efficiency by incorporating the *LGS-trimming strategy* (which includes local trimming, global pruning, and single-pass patch up) [19]. This strategy trims away every item with an existential probability below the user-specified trimming threshold (which is local to each item) from the original probabilistic dataset D of uncertain data and then mines frequent patterns from the resulting trimmed dataset D_{Trim}. On the one hand, if a pattern X is frequent in D_{Trim}, then X must be frequent in D. On the other hand, a pattern Y is infrequent in D if

$$expSup(Y, D_{Trim}) + e(Y) < minsup,$$

where $e(Y)$ is an upper bound of estimated error for $expSup(Y, D_{Trim})$. Such an infrequent pattern Y can be pruned. Moreover, a pattern Z is potentially frequent in D if

$$expSup(Z, D_{Trim}) \leq minsup \leq expSup(Z, D_{Trim}) + e(Z).$$

To patch up, U-Apriori recovers the missing frequent patterns by verifying expected supports of potentially frequent patterns with an additional single-pass scan of D. Although the LGS strategy improves the efficiency of U-Apriori, the algorithm still suffers from the following problems: (i) there is an overhead in creating D_{Trim}, (ii) only a subset of all the frequent patterns can be mined from D_{Trim} and there is overhead to patch up, (iii) the efficiency of the algorithm is sensitive to the percentage of items having low existential probabilities, and (iv) it is not easy to find an appropriate value for the user-specified trimming threshold.

Chui and Kao [18] applied the *decremental pruning technique* to further improve the efficiency of U-Apriori. The technique helps reduce the number of candidate patterns by progressively estimating the upper bounds of expected support of candidate patterns after each transaction is processed. If the estimated upper bound of a candidate pattern X falls below *minsup*, then X is immediately pruned.

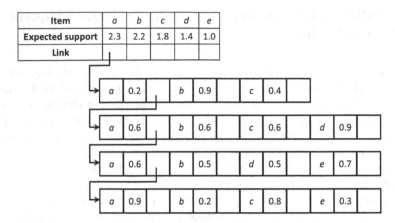

Item	a	b	c	d	e
Expected support	2.3	2.2	1.8	1.4	1.0
Link					

a	0.2		b	0.9		c	0.4			
a	0.6		b	0.6		c	0.6		d	0.9
a	0.6		b	0.5		d	0.5		e	0.7
a	0.9		b	0.2		c	0.8		e	0.3

Fig. 14.1 The UH-struct for the probabilistic dataset D_2 of uncertain data

4   Hyperlinked Structure-Based Uncertain Frequent Pattern Mining

An alternative to candidate generate-and-test based mining is *pattern-growth mining*, which avoids generating a large number of candidates. Commonly used pattern-growth mining paradigms are mostly based on (i) hyperlinked structures or (ii) tree structures. In this section, let us focus on hyperlinked structure-based uncertain frequent pattern mining. As hyperlinked structure based mining employs a pattern-growth mining paradigm, the candidate generate-and-test mining paradigm of U-Apriori is avoided. In general, hyperlinked structure based algorithms capture the contents of datasets in a hyperlinked structure, from which frequent patterns are mined in a depth-first divide-and-conquer fashion.

Aggarwal et al. [13] proposed a hyperlinked structure based algorithm called **UH-mine** to mine frequent patterns from uncertain data. This algorithm captures the contents of a probabilistic dataset D of uncertain data in a hyperlinked structure called *UH-struct*. See Fig. 14.1. Like the H-struct for mining precise data, each row in the UH-struct represents a transaction t_i in D. Unlike the H-struct, the UH-struct captures the existential probability of items. In other words, for each item $x \in t_i$, the UH-struct maintains (i) x, (ii) its existential probability $P(x, t_i)$, and (iii) its hyperlink. Once the UH-struct is built, the corresponding UH-mine algorithm mines frequent patterns by recursively extending every frequent pattern X and adjusting its hyperlinks in the UH-struct.

As a preview, when compared with tree-based uncertain frequent pattern mining (i.e., another type of mining that relies on the pattern-growth mining paradigm), the UH-structure is not as compact as the tree structure used in tree-based mining. However, on the positive side, the UH-mine algorithm keeps only one UH-struct and adjusts the hyperlinks in it. In contrast, due to their recursive nature, tree-based mining algorithms usually keep multiple tree structures. Moreover, UH-mine computes the expected support of frequent patterns on-the-fly so as to reduce the space requirement.

5 Tree-Based Uncertain Frequent Pattern Mining

Recall that (i) candidate generate-and-test based mining algorithms (e.g., the U-Apriori algorithm) use a levelwise bottom-up breadth-first mining paradigm and (ii) hyperlinked structure based algorithms (e.g., the UH-mine algorithm) recursively adjust the hyperlinks in the hyperlinked structure (e.g., UH-struct) to find frequent patterns from uncertain data in a depth-first fashion. As an alternative to Apriori-based and hyperlinked structure based mining, tree-based mining avoids generating many candidates and avoids recursively adjusting many hyperlinks. Tree-based algorithms use a depth-first divide-and-conquer approach to mine frequent patterns from a tree structure that captures the contents of the probabilistic dataset.

5.1 UF-growth

To mine frequent patterns from probabilistic datasets of uncertain data, Leung et al. [42] proposed a tree-based mining algorithm called **UF-growth**. Similar to its counterpart for mining precise data (the FP-growth algorithm [24]), UF-growth also constructs a tree structure to capture the contents of the datasets. However, it does not use the FP-tree (as in FP-growth) because each node in the FP-tree only maintains (i) an item and (ii) its occurrence count in the tree path. When mining precise data, the *actual support* of an pattern X depends solely on the occurrence counts of items within X. However, when mining uncertain data, the *expected support* of X is the sum of the product of the occurrence count and existential probability of every item within X. Hence, each node in the **UF-tree** (the tree structure for UF-growth) consists of three components: (i) an item, (ii) its existential probability, and (iii) its occurrence count in the path. See Fig. 14.2. Such a UF-tree is constructed in a similar fashion as the construction of the FP-tree, except that a new transaction is merged with a child node only if the same item *and the same existential probability* exist in both the transaction and the child node. As such, it may lead to a lower compression ratio than the original FP-tree. Fortunately, the number of nodes in a UF-tree is bounded above by the sum of the number of items in all transactions in the probabilistic dataset of uncertain data.

To reduce the memory consumption, UF-growth incorporates two improvement techniques [42]. The first technique is to discretize the existential probability of each node (e.g., round the existential probability to k decimal places such as $k = 2$ decimal places), which reduces the potentially infinite number of possible existential probability values to a maximum of 10^k possible values. The second improvement technique is to limit the construction of UF-trees to only the first two levels (i.e., only construct the global UF-tree for the original probabilistic dataset D and a UF-tree for

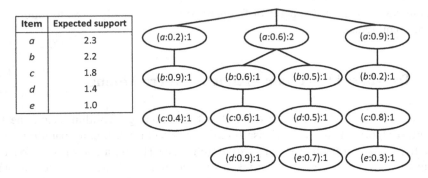

Item	Expected support
a	2.3
b	2.2
c	1.8
d	1.4
e	1.0

Fig. 14.2 The UF-tree for the probabilistic dataset D_2 of uncertain data

each frequent item—i.e., each singleton pattern) and to enumerate frequent patterns for higher levels (by traversing the tree paths and decrementing the occurrence counts) during the mining process.

On the one hand, as paths in a UF-tree are shared only if they have the same item and the same existential probability, the UF-tree accurately captures the contents (especially, the existential probabilities) of the probabilistic datasets of uncertain data so that frequent patterns can be mined without producing false positives or false negatives. On the other hand, the UF-tree may be large and may not be as compact as its counterpart for precise data (i.e., FP-tree).

5.2 UFP-growth

To make the tree more compact by reducing the tree size (via a reduction in the number of tree nodes), Aggarwal et al. [13] proposed the **UFP-growth algorithm**. Like UF-growth, the UFP-growth algorithm also scans the probabilistic dataset of uncertain data twice and builds a **UFP-tree**. As nodes for item x having similar existential probability values are clustered into a mega-node, the resulting mega-node in the UFP-tree captures (i) an item x, (ii) the *maximum* existential probability value (among all nodes within the cluster), and (iii) its occurrence count (i.e., the number of nodes within the cluster). Tree paths are shared if the nodes on these paths share the same item but *similar* existential probability values. In other words, the path sharing condition is less restrictive than that of the UF-tree. See Fig. 14.3. By extracting appropriate tree paths and constructing UFP-trees for subsequent projected databases, UFP-growth finds all truly frequent patterns at the end of the second scan of the probabilistic dataset of uncertain data. At the same time, due to the approximate nature (e.g., caused by the use of the maximum existential probability value among all the nodes clustered into a mega-node) of UFP-growth, UFP-growth also finds some infrequent patterns (i.e., some *false positives*) in addition to those truly frequent patterns (i.e., true positives). Hence, a third scan of the probabilistic dataset of uncertain data is then required to remove these false positives.

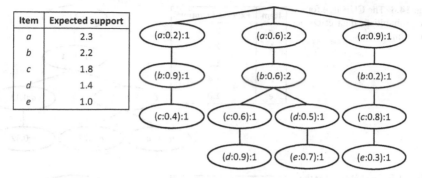

Item	Expected support
a	2.3
b	2.2
c	1.8
d	1.4
e	1.0

Fig. 14.3 The UFP-tree for the probabilistic dataset D_2 of uncertain data

Table 14.4 Transaction caps for the probabilistic dataset D_2 of uncertain data

Transaction ID	Set of items with existential probability	Transaction cap
t_1	{a:0.2, b:0.9, c:0.4}	0.36
t_2	{a:0.6, b:0.6, c:0.6, d:0.9}	0.54
t_3	{a:0.6, b:0.5, d:0.5, e:0.7}	0.42
t_4	{a:0.9, b:0.2, c:0.8, e:0.3}	0.72

5.3   CUF-growth

To further reduce the tree size (by reducing the number of tree nodes), Leung and Tanbeer [39] proposed an uncertain frequent pattern mining algorithm called **CUF-growth**, which builds a new tree structure called **CUF-tree**. Specifically, for each transaction t_i, CUF-growth computes a *transaction cap* which is defined as follows.

Definition 14.2 The **transaction cap**, denoted by $cap(t_i)$, of a transaction t_i is defined as the product of the two highest existential probability values of items within transaction t_i. Let

(i) $h = |t_i|$ represent the length of t_i,
(ii) $M_1 = \max_{q \in [1,h]} P(x_q, t_i)$, and
(iii) $M_2 = \max_{r \in [1,h], \ r \neq q} P(x_r, t_i)$.

Then,

$$cap(t_i) = \begin{cases} M_1 \times M_2 & \text{if } |t_i| > 1 \\ P(x_1, t_i) & \text{if } |t_i| = 1 \text{ (i.e., } t_i = \{x_1\}) \end{cases} \tag{14.4}$$

Table 14.4 shows the transaction cap for each transaction in a probabilistic dataset D_2 of uncertain data. The CUF-growth algorithm captures the transaction cap in the CUF-tree. Unlike the UF-tree (which captures an item, its existential probability and its occurrence count in each tree node), CUF-tree only capture (i) an item and (ii) its transaction cap. Paths in a CUF-tree are shared if the nodes on these paths share the same item. By doing so, the CUF-tree (for capturing uncertain data) can be as

Fig. 14.4 The CUF-tree for the probabilistic dataset D_2 of uncertain data

Item	Expected support
a	2.3
b	2.2
c	1.8
d	1.4
e	1.0

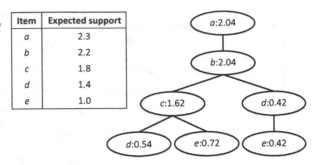

Fig. 14.5 The FP-tree for the traditional database D_1 of precise data

Item	Actual support
a	4
b	4
c	3
d	2
e	2

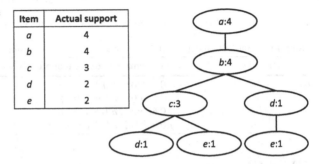

compact as the FP-tree (for capturing precise data). See Fig. 14.4, which shows the CUF-tree for probabilistic dataset D_2 of uncertain data. Note that this CUF-tree is as compact as an FP-tree (ref. Fig. 14.5) for the traditional database D_1 of precise data.

Like UFP-growth, the CUF-growth algorithm also takes three scans of the probabilistic dataset of uncertain data to mine frequent patterns. CUF-growth first scans the dataset to compute the transaction caps, and it then scans the dataset the second time to build the CUF-tree. The header table associated with the CUF-tree gives the expected support of frequent 1-itemsets (i.e., frequent singletons or frequent items). The CUF-tree stores transaction caps, which provide upper bounds to the expected support of frequent k-itemsets (for $k \geq 2$). For any k-itemset X, if the upper bound to its expected support is less than *minsup*, then X can be safely pruned.

By extracting appropriate tree paths and constructing CUF-trees for subsequent projected databases, CUF-growth finds all potentially frequent patterns at the end of the second scan of the probabilistic dataset of uncertain data. As these potentially frequent patterns include all truly frequent patterns and some infrequent patterns (i.e., some *false positives*), CUF-growth then quickly scans the dataset the third time to check each of them to verify whether or not they are truly frequent (i.e., prune false positives).

Table 14.5 Prefixed item caps for the probabilistic dataset D_2 of uncertain data

Transaction ID	Set of items w/existential probability (& prefixed item cap)
t_1	$\{a{:}0.2\ (0.2),\ b{:}0.9\ (0.18),\ c{:}0.4\ (0.36)\}$
t_2	$\{a{:}0.6\ (0.6),\ b{:}0.6\ (0.36),\ c{:}0.6\ (0.36),\ d{:}0.9\ (0.54)\}$
t_3	$\{a{:}0.6\ (0.6),\ b{:}0.5\ (0.30),\ d{:}0.5\ (0.30),\ e{:}0.7\ (0.42)\}$
t_4	$\{a{:}0.9\ (0.9),\ b{:}0.2\ (0.18),\ c{:}0.8\ (0.72),\ e{:}0.3\ (0.27)\}$

5.4 PUF-growth

Along this direction, Leung and Tanbeer [40] observed that (i) the transaction cap provides CUF-growth with an upper bound to expected support of patterns and (ii) such an upper bound can be tightened in a tree-based environment. They introduced the concept of a *prefixed item cap*, which can be defined as follows.

Definition 14.3 The **prefixed item cap**—denoted by $I^{Cap}(x_r, t_i)$—of an item x_r in a transaction $t_i = \{x_1, \ldots, x_r, \ldots, x_h\}$, where $1 \le r \le h$ (i.e., $h{=}|t_i|$ represent the length of t_i), is defined as the product of $P(x_r, t_i)$ and the highest existential probability value M of items from x_1 to x_{r-1} in t_i (i.e., in the *proper prefix* of x_r in t_i). More formally,

$$PIcap(x_r, t_i) = \begin{cases} P(x_r, t_i) \times M & \text{if } |t_i| > 1 \\ P(x_1, t_i) & \text{if } |t_i| = 1 \text{ (i.e., } t_i{=}\{x_1\}) \end{cases} \tag{14.5}$$

where $M = \max_{q \in [1, r-1]} P(x_q, t_i)$.

Assume that items are arranged in the order $\langle a, b, c, d, e \rangle$ from the root to leaves. Then, Table 14.5 shows the prefixed item cap for every item in a transaction in a probabilistic dataset D_2 of uncertain data. See Fig. 14.6 for how these prefixed item caps are captured in a new tree structure called **PUF-tree**, from which the corresponding algorithm called **PUF-growth** mines uncertain frequent patterns.

Like UFP-growth and CUF-growth, the PUF-growth algorithm also takes three scans of the probabilistic dataset of uncertain data to mine frequent patterns. With the first scan, PUF-growth computes the prefixed item caps. With the second scan, PUF-growth builds a PUF-tree to capture (i) an item and (ii) its corresponding prefixed item cap. Like those in CUF-tree, paths in the PUF-tree are shared if the nodes on these paths share the same item. Hence, the resulting PUF-tree is of the same size as the CUF-tree (also for capturing uncertain data), which can be as compact as the FP-tree (for capturing precise data). The header table associated with the PUF-tree gives the expected support of frequent 1-itemsets (i.e., singleton patterns or frequent items). The prefixed item caps in the PUF-tree provide upper bounds to the expected support of k-itemsets (for $k \ge 2$). For any k-itemset X, if the upper bound to its expected support is less than *minsup*, then X can be safely pruned.

By extracting appropriate tree paths and constructing PUF-trees for subsequent projected databases, PUF-growth finds all potentially frequent patterns at the end of the second scan of the probabilistic dataset of uncertain data. As these potentially frequent patterns include all truly frequent patterns and some infrequent patterns

Fig. 14.6 The PUF-tree for the probabilistic dataset D_2 of uncertain data

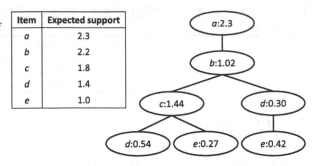

Item	Expected support
a	2.3
b	2.2
c	1.8
d	1.4
e	1.0

(i.e., some *false positives*), PUF-growth then quickly scans the dataset a third time to check each of them to verify whether or not they are truly frequent (i.e., prune false positives). As illustrated by Table 14.6, the prefixed item caps tighten the upper bound to the expected support of non-singleton patterns (when compared with the transaction caps in the CUF-tree). Consequently, the number of false positives that need to be examined by PUF-growth during the third scans of the probabilistic dataset of uncertain data is usually smaller than that by CUF-growth. Hence, PUF-growth runs faster than CUF-growth.

6   Constrained Uncertain Frequent Pattern Mining

Recall from Sect. 5 that the UF-growth, UFP-growth, CUF-growth and PUF-growth algorithms are useful in finding *all* the frequent patterns from probabilistic datasets of uncertain data in many situations. However, there are other situations in which users are interested in only *some* of the frequent patterns. In these situations, users express their interest in terms of constraints. This leads to *constrained mining* [20, 27, 30, 34, 35]. In response, Leung et al. [33, 43] extended the UF-growth algorithm to mine probabilistic datasets of uncertain data for frequent patterns that satisfy user-specified constraints. The two resulting algorithms, called **U-FPS** [33] and **U-FIC** [43], push the constraints in the mining process and exploit properties of different kinds of constraints (instead of a naive approach of first mining all frequent patterns and then pruning all uninteresting or invalid ones).

U-FPS exploits properties of (two types of) succinct constraints [31]. More specifically, by exploiting that "all patterns satisfying any *succinct and anti-monotone (SAM) constraint* C_{SAM} must comprise only items that individually satisfy C_{SAM}", U-FPS stores only these items in the UF-tree when handling C_{SAM}. Similarly, by exploiting that "all patterns satisfying any *succinct but not anti-monotone (SUC) constraint* C_{SUC} consist of at least one item that individually satisfies C_{SUC} and may contain other items", U-FPS partitions the domain items into two groups (one group contains items individually satisfying C_{SUC} and another group contains those not) and stores items belonging to each group separately in the UF-tree. See Fig. 14.7.

Table 14.6 Sample patterns mined from the probabilistic dataset D_2

Pattern X	Upper bound to $expSup(X, D_2)$ based on transaction cap	Upper bound to $expSup(X, D_2)$ based on prefixed item cap	$expSup(X, D_2)$
$\{a, c\}$	1.62	1.44	1.16
$\{a, d\}$	0.96	0.84	0.84
$\{b, d\}$	0.96	0.96	0.79
$\{c, d\}$	0.54	0.54	0.54
$\{a, c, e\}$	0.72	0.27	0.22

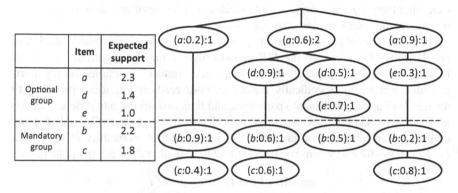

Fig. 14.7 The UF-tree for mining constrained frequent patterns from D_2

As arranging domain items in decreasing order of their support in the original FP-tree is just a heuristic, U-FIC exploits properties of (two types of) convertible constraints [29] and arranges the domain items in the UF-tree according to some monotonic order of attribute values relevant to the constraints. By doing so, U-FIC does not need to perform constraint checking against any extensions of patterns satisfying any *convertible monotone (COM) constraint* C_{COM} because all these extensions are guaranteed to satisfy C_{COM}. Similarly, U-FIC prunes all the patterns that violate any *convertible anti-monotone (CAM) constraint* C_{CAM} because these patterns and their extensions are guaranteed to violate C_{CAM}. By exploiting the user-specified constraints, computation of both U-FPS and U-FIC is proportional to the selectivity of the constraints.

7 Uncertain Frequent Pattern Mining from Big Data

As technology advances further, high volumes of valuable data—such as banking, financial, and marketing data—are generated in various real-life business applications in modern organizations and society. This leads us into the new era of Big Data [45]. Intuitively, *Big Data* are interesting high-velocity, high-value, and/or high-variety data with volumes beyond the ability of commonly-used software to capture, manage, and process within a tolerable elapsed time. Hence, new forms of processing data are needed to enable enhanced decision making, insight, knowledge discovery, and process optimization. To handle Big Data, researchers proposed the use of a high-level programming model—called *MapReduce*—to process high volumes of data by using parallel and distributed computing [54] on large clusters or grids of nodes (i.e., commodity machines), which consist of a master node and multiple worker nodes. As implied by its name, MapReduce involves two key functions: "map" and "reduce". An advantage of using the Map-Reduce model is that users only need to focus on (and specify) these "map" and "reduce" functions—without worrying about implementation details for (i) partitioning the input data, (ii) scheduling and executing the program across multiple machines, (iii) handling machine failures, or (iv) managing inter-machine communication.

To mine frequent patterns from Big probabilistic datasets of uncertain data, Leung and Hayduk [37] proposed the **MR-growth** algorithm. The algorithm uses Map-Reduce—by applying two sets of the "map" and "reduce" functions—in a pattern-growth environment. Specifically, the master node reads and divides a probabilistic dataset D of uncertain data into partitions, and then assigns them to different worker nodes. The worker node corresponding to each partition P_j (where $D = \bigcup_j P_j$) then outputs a pair consisting of an item x and its existential probability $P(x, t_i)$—i.e., $\langle x, P(x, t_i) \rangle$—for every item x in transaction t_i assigned to P_j as intermediate results:

$$\text{map:} \langle \text{ID of } t_i \text{ in } P_j, \text{contents of } t_i \rangle$$

$$\mapsto \text{list of } \langle x \in t_i, P(x, t_i) \rangle. \tag{14.6}$$

Afterwards, these $\langle x, P(x, t_i) \rangle$ pairs in the list (i.e., intermediate results) are shuffled and sorted (e.g., grouped by x). Each worker node then executes the "reduce" function, which (i) "reduces"—by summing—all the $P(x, t_i)$ values for each item x so as to compute its expected support $expSup(\{x\}, D)$ and (ii) outputs $\langle \{x\}, expSup(\{x\}, D) \rangle$ (representing a frequent 1-itemset $\{x\}$ and its expected support) if $expSup(\{x\}, D) \geq minsup$:

$$\text{reduce:} \langle x, \text{list of } P(x, t_i) \rangle$$

$$\mapsto \text{list of } \langle \text{frequent 1-itemset } \{x\}, expSup(\{x\}, D) \rangle, \tag{14.7}$$

where $expSup(\{x\}, D) = $ sum of $P(x, t_i)$ in the list for an item x.

Afterwards, MR-growth rereads the datasets to form a $\{x\}$-projected database (i.e., a collection of transactions containing x) for each item x in the list produced by the first reduce function (i.e., for each frequent 1-itemset $\{x\}$). The worker node corresponding to each projected database then (i) builds appropriate local UF-trees (based on the projected database assigned to the node) to mine frequent k-itemsets (for $k \geq 2$) and (ii) outputs $\langle X, expSup(X, D) \rangle$ (which represents a frequent k-itemset X and its expected support) if $expSup(X, D) \geq minsup$. In other words, MR-growth executes the second set of "map" and "reduce" functions as follows:

$$\text{map:} \langle \text{ID of } t_i \text{ in } P_j, \text{contents of } t_i \rangle$$

$$\mapsto \text{list of } \langle \{x\}, \{x\}\text{-proj. DB} \rangle; \tag{14.8}$$

Table 14.7 Streaming
uncertain data D_3

Batch ID	Set of items with existential probability
B_1	$\{t_1=\{f:0.7\},$
	$t_2=\{f:0.7, g:0.9, h:0.7, i:0.5\},$
	$t_3=\{f:0.7, g:0.9, h:0.8\}\}$
B_2	$\{t_4=\{f:0.7, g:0.9\},$
	$t_5=\{g:0.6\},$
	$t_6=\{g:0.6\}\}$
B_3	$\{t_7=\{f:0.7, g:0.9, h:0.7, i:0.5\},$
	$t_8=\{f:0.7, g:0.9, h:0.8\},$
	$t_9=\{f:0.7, g:0.9, i:0.5\}\}$

and

$$\text{reduce: } \langle\{x\}, \{x\}\text{-proj. DB}\rangle$$

$$\mapsto \text{list of } \langle\text{frequent } k\text{-itemset } X, expSup(X, D)\rangle. \tag{14.9}$$

To recap, by using the above two sets of "map" and "reduce" functions, the MR-growth (i) first finds all frequent 1-itemsets with their expected support and (ii) then buids appropriate local UF-trees (for projected databases) to find all frequent k-itemsets (for $k \geq 2$) with their expected support.

8   Streaming Uncertain Frequent Pattern Mining

In addition to *static* probabilistic datasets of uncertain data, *dynamic streams* of uncertain data can also be generated (e.g., by wireless sensors) in many real-life applications (e.g., environment surveillance). This leads to stream mining [22].

8.1   SUF-growth

To mine frequent patterns from streams of uncertain data, Leung and Hao [36] extended the UF-growth algorithm (ref. Sect. 5.1) to produce an exact mining algorithm called **SUF-growth**. During the mining process, the (i) sliding window model, (ii) time-fading model, or (iii) landmark model is commonly used in processing batches of transactions in the data streams. When using a sliding window model, SUF-growth captures only the contents of streaming data in batches of transactions belonging to the current window (that captures the recent w batches) in a tree structure called **SUF-tree**. When the window slides, SUF-growth removes from the SUF-tree those data belonging to older batches and adds to the SUF-tree those data belonging to newer batches. Hence, each tree node in the SUF-tree consists of three components: (i) an item, (ii) its existential probability, and (iii) a list of its w occurrence counts in the path. By doing so, when the window slides, the oldest occurrence counts (representing the oldest streaming data) are replaced by the newest occurrence

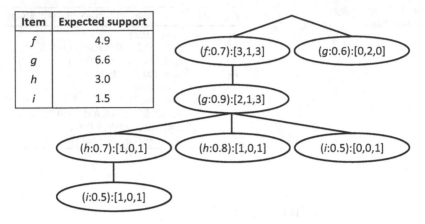

Item	Expected support
f	4.9
g	6.6
h	3.0
i	1.5

Fig. 14.8 The SUF-tree for streaming uncertain data D_3

counts (representing the newest streaming data). The SUF-tree is constructed in a similar fashion as the construction of the UF-tree, except that the occurrence count is inserted as the newest entry in the list of occurrence counts. Figure 14.8 shows an example of a SUF-tree constructed from the streaming uncertain data in Table 14.7.

Once the SUF-tree is constructed, it is always kept up-to-date when the window slides. As the window continues to slide in the dynamic streams, SUF-growth delays the mining of frequent patterns from uncertain streaming data until the user requests for the patterns. At that time, SUF-growth mines the up-to-dated SUF-tree in a fashion similar to UF-growth with the user-specified *minsup* to find all frequent patterns.

8.2 UF-streaming for the Sliding Window Model

Besides the SUF-growth algorithm, Leung and Hao [36] also extended the UF-growth algorithm to produce an approximate algorithm called **UF-streaming**. Unlike SUF-growth (which is an exact algorithm but uses a "delayed" mining mode), UF-streaming is an approximate algorithm that uses an "immediate" mining mode. Specifically, for every incoming batch of streaming uncertain data, UF-streaming applies UF-growth (ref. Sect. 5.1) to that batch with a *preMinsup* threshold (where *preMinsup* < *minsup*) to find "frequent" patterns (or more precisely, sub-frequent or potentially frequent patterns due to the use of the *preMinsup* threshold). These *"frequent" patterns* having expected support ≥ *preMinsup* are then stored in a tree structure called **UF-stream**. Each tree path represents a "frequent" pattern. Each tree node stores a list of w expected support values (one for each batch), where the i-th value indicates the expected support of that "frequent" pattern in the i-th batch. When a new batch flows in, the window slides, and the algorithm shifts the w expected support values of each node in the UF-stream structure so as to ensure that it always captures the "frequent" patterns mined from the w most recent batches of streaming uncertain data. Figure 14.9 shows an example of a UF-stream structure

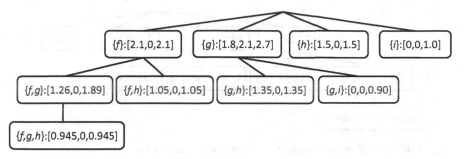

Fig. 14.9 The UF-stream structure for streaming uncertain data D_3

constructed for the streaming uncertain data in Table 14.7 with *preMinsup* = 0.9 < 1.0 = *minsup*.

As UF-streaming uses *preMinsup*, the UF-stream structure may contain some false positives (i.e., every pattern Y having *preMinsup* $\leq expSup(Y, D_w)$ < *minsup*, where D_w represents the streaming uncertain data in the current sliding window) in addition to all truly frequent patterns (i.e., every pattern X having $expSup(X, D_w) \geq$ *minsup*). At the time when the user requests for frequent patterns from uncertain streaming data, UF-streaming traverses the UF-stream structure and returns only those truly frequent patterns.

8.3   TUF-streaming for the Time-Fading Model

Besides the sliding window model (used by both SUF-growth and UF-streaming), there are also other stream processing models such as time-fading and landmark models. Leung and Jiang [38] proposed the **TUF-streaming** algorithm to mine frequent patterns from probabilistic datasets of uncertain data in a fashion similar to UF-streaming, except that TUF-streaming uses the *time-fading model* instead of the sliding window model. When using the time-fading model, TUF-streaming puts heavier weights on recent batches of streaming uncertain data than older batches. Specifically, like UF-streaming, the TUF-streaming algorithm also uses the "immediate" mining mode to apply UF-growth (ref. Sect. 5.1) to every incoming batch of streaming uncertain data with a *preMinsup* threshold to find "frequent" patterns from that batch. When using the time-fading model, the corresponding **TUF-stream** structure (i) captures all "frequent" patterns mined from all batches but weights recent batches heavier than older batches (i.e., monotonically decreasing weights from recent to older data). See Fig. 14.10, in which α (where $0 < \alpha \leq 1$) is a time-fading factor.

While TUF-streaming handles batches of streaming uncertain data that come in order, there are situations where batches may get delayed and thus arrived out of the desired order. To deal with these situations, Jiang and Leung [26] extended the TUF-streaming algorithm to mine frequent patterns from these delayed batches when using the time-fading model.

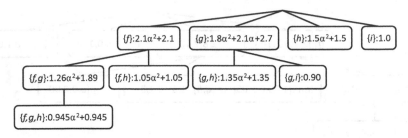

Fig. 14.10 The TUF-stream structure for streaming uncertain data D_3

Fig. 14.11 The LUF-stream
structure for streaming
uncertain data D_3

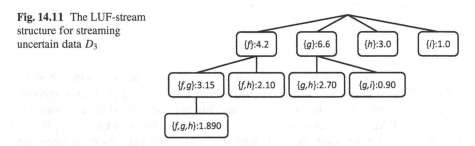

8.4 LUF-streaming for the Landmark Model

Moreover, Leung et al. [41] extended the TUF-streaming algorithm to become the
LUF-streaming algorithm, which mines frequent patterns from streaming uncertain
data in a fashion similar to the UF-streaming and TUF-streaming algorithms, except
that LUF-streaming uses the *landmark model*. When using the landmark model, the
corresponding **LUF-stream** structure (i) captures all "frequent" patterns mined from
all batches of streaming uncertain data generated from a landmark to the present time
and (ii) treats all batches with the same importance. See Fig. 14.11.

8.5 Hyperlinked Structure-Based Streaming Uncertain Frequent
Pattern Mining

So far, we have described notable exact and approximate *tree-based* streaming uncertain frequent pattern mining algorithms. Besides them, there are also hyperlinked
structure based algorithms. For instance, Nadungodage et al. [46] proposed two
false positive-oriented algorithms called **UHS-Stream** and **TFUHS-Stream**. The
UHS-Stream algorithm applies uncertain hyperlinked structure stream mining for
finding all the frequent patterns seen up to the current moment (i.e., with the landmark model). Similarly, the TFUHS-Stream algorithm applies uncertain hyperlinked
structure stream mining, but it uses the time-fading model. Hence, the TFUHS-
Stream algorithm puts heavier weights on the recent transactions than historical data
in the stream.

9   Vertical Uncertain Frequent Pattern Mining

The aforementioned candidate generate-and-test based, hyperlinked structure based, as well as tree-based mining algorithms use *horizontal mining*, for which a dataset can be viewed as a collection of transactions. Each transaction is a set of items. Alternatively, *vertical mining* can be applied, for which each dataset can be viewed as a collection of items and their associated lists (or sets or vectors) of transaction IDs (i.e., tIDs). Each list of tID of an item x represents all the transactions containing x. With this vertical representation of datasets, the support of a pattern X can be computed by intersecting the lists of tIDs of items within X.

9.1   U-Eclat: An Approximate Algorithm

To mine frequent patterns using the vertical representation of probabilistic datasets of uncertain data, Calders et al. [17] instantiated "possible worlds" of the datasets to get instantiated samples (in which data become precise) and then applied the Eclat algorithm [55] to each of these samples of instantiated databases. The resulting algorithm is called **U-Eclat**. Given a probabilistic dataset D of uncertain data, U-Eclat generates an independent random number r for each item x in a transaction t_i. If the existential probability $P(x, t_i)$ of item x in transaction t_i is no less than such a random number r (i.e., $P(x, t_i) \geq r$), then x is instantiated and included in a "precise" sampled database, which is then mined using the original Eclat algorithm. This sampling and instantiation process is repeated multiple times, and thus generates multiple sampled "precise" databases. The estimated support of any pattern X is the average support of X over the multiple sampled databases. Figure 14.12 shows three sampled databases for probabilistic dataset D_2. As a sampling-based algorithm, U-Eclat gains efficiency but loses accuracy. More instantiations (i.e., more samples) helps improve accuracy, but it comes at the cost of an increase in execution time.

9.2   UV-Eclat: An Exact Algorithm

Alternatively, to avoid instantiations and to *directly* mine frequent patterns using the vertical representation of probabilistic datasets containing uncertain data, Budhia et al. [16] proposed the **UV-Eclat** algorithm. When mining uncertain data, in addition to recording which transactions contain x in the set of tIDs (i.e., *tIDsets*), it is also important to capture additional information: If x is likely to be present in a transaction t_i, then its associated existential probability $P(x, t_i)$—which expresses the likelihood of x appearing in transaction t_i of the dataset—needs to be captured.

As it would be impractical to build a tIDset for each distinct ⟨domain item, existential probability value⟩ pair due to the huge number of such pairs, UV-Eclat builds a tIDset for each domain item x instead. In each set, UV-Eclat augments every tID (representing t_i) with its corresponding existential probability value $P(x, t_i)$. In

Probabilistic database D_2 of uncertain data:

$$D_2 = \begin{cases} t_1=\{a\text{:}0.2,\ b\text{:}0.9,\ c\text{:}0.4\}, \\ t_2=\{a\text{:}0.6,\ b\text{:}0.6,\ c\text{:}0.6,\ d\text{:}0.9\}, \\ t_3=\{a\text{:}0.6,\ b\text{:}0.5,\quad\quad\ d\text{:}0.5,\ e\text{:}0.7\}, \\ t_4=\{a\text{:}0.9,\ b\text{:}0.2,\ c\text{:}0.8,\quad\quad e\text{:}0.3\}\ \} \end{cases}$$

Samples of instantiated possible worlds for D_2 containing uncertain data:

		Sample 1					Sample 2					Sample 3			
Item	a	b	c	d	e	a	b	c	d	e	a	b	c	d	e
tID list	$\{t_1,$ $t_2,$ $t_3,$ $t_4\}$	$\{t_1,$ $t_2\}$	$\{t_2,$ $t_4\}$	$\{\}$	$\{t_3\}$	$\{t_1,$ $t_2,$ $t_4\}$	$\{t_1,$ $t_2\}$	$\{t_1,$ $t_2,$ $t_4\}$	$\{t_2,$ $t_3\}$	$\{t_3\}$	$\{t_2,$ $t_4\}$	$\{t_1,$ $t_2,$ $t_4\}$	$\{t_2,$ $t_4\}$	$\{t_2\}$	$\{t_4\}$

Observations:

- $expSup(\{e\},D_2) = 1.0$ for D_2 vs. $avg(sup(\{e\},\text{Sample } S)) = 1.0$ over the 3 samples
- $expSup(\{a,c\},D_2) = 1.16$ for D_2 vs. $avg(sup(\{a,c\},\text{Sample } S)) \approx 2.33$ over the 3 samples
- $expSup(\{b,d\},D_2) = 0.79$ for D_2 vs. $avg(sup(\{b,d\},\text{Sample } S)) \approx 0.67$ over the 3 samples

where $S = 1, 2, 3$

Fig. 14.12 Samples of instantiated "possible worlds" for D_2

Table 14.8 Augmented tIDsets for domain items in the probabilistic dataset D_2

Item	Augmented tIDset
a	$\{t_1\text{:}0.2,\ t_2\text{:}0.6,\ t_3\text{:}0.6,\ t_4\text{:}0.9\}$
b	$\{t_1\text{:}0.9,\ t_2\text{:}0.6,\ t_3\text{:}0.5,\ t_4\text{:}0.2\}$
c	$\{t_1\text{:}0.4,\ t_2\text{:}0.6,\ t_4\text{:}0.8\}$
d	$\{t_2\text{:}0.9,\ t_3\text{:}0.5\}$
e	$\{t_3\text{:}0.7,\ t_4\text{:}0.3\}$

other words, the resulting augmented tIDset for any item x is of the form $\{t_i:P(x,t_i)\}$, which is equivalent to $\{t_i:expSup(\{x\},t_i)\}$. See Table 14.8.

With the use of augmented tIDsets to vertically represent the probabilistic dataset D of uncertain data, the expected support of any 1-itemset $\{x\}$ in D can be computed by summing all $P(x,t_i)$ values in the augmented tIDset for $\{x\}$. The tIDset of any $(k+1)$-itemset $X \equiv Y \cup \{z\}$ (where Y is a k-pattern and z is an item) for $k \geq 1$ can be formed by intersecting the tIDsets of Y and $\{z\}$. Each t_i in the intersection result is associated with an expected support value $expSup(X,t_i)$, which is the product of $expSup(Y,t_i)$ and $P(z,t_i)$.

9.3 U-VIPER: An Exact Algorithm

Vertical representations for a probabilistic dataset D of uncertain data are not confined to set-based representations (e.g., augmented tIDsets used in UV-Eclat). There are

Table 14.9 Vectors for domain items in the probabilistic dataset D_2

Vector for every domain item

$$\vec{a} = \begin{pmatrix} 0.2 \\ 0.6 \\ 0.6 \\ 0.9 \end{pmatrix}, \qquad \vec{b} = \begin{pmatrix} 0.9 \\ 0.6 \\ 0.5 \\ 0.2 \end{pmatrix}, \qquad \vec{c} = \begin{pmatrix} 0.4 \\ 0.6 \\ 0 \\ 0.8 \end{pmatrix}, \qquad \vec{d} = \begin{pmatrix} 0 \\ 0.9 \\ 0.5 \\ 0 \end{pmatrix}, \text{ and } \quad \vec{e} = \begin{pmatrix} 0 \\ 0 \\ 0.7 \\ 0.3 \end{pmatrix}.$$

other representations. For instance, Leung et al. [44] proposed the **U-VIPER** algorithm, in which D is vertically represented by a collection of fixed-size vectors—one for each domain item x. The length of each vector is fixed and is equal to the number of transactions (i.e., $|D| = n$) in the probabilistic dataset of uncertain data. When mining uncertain data, in addition to using a Boolean value (say, 0 or 1) to denote whether or not transaction t_i contains x in the vector, it is also important to capture additional information: If x is likely to be present in a transaction t_i, then its associated existential probability $P(x, t_i)$—which expresses the likelihood of x appearing in transaction t_i of the dataset—needs to be captured.

As it would be a waste of space to augment $P(x, t_i)$ to the Boolean value "1" for t_i (i.e., the i-th element of the vector for x), U-VIPER replaces the Boolean value "1" by $P(x, t_i)$ as the i-th element of each vector $\vec{x}$ representing domain item x. Specifically, the i-th element of $\vec{x}$ (denoted as $\vec{x}[i]$) stores (i) "0" if x is absent from t_i and (ii) $P(x, t_i)$ if x is likely to be present in t_i:

$$i\text{-th element of } \vec{x} \text{ (i.e., } \vec{x}[i]) = expSup(\{x\}, t_i)$$
$$= \begin{cases} 0 & \text{if } x \notin t_i \\ P(x, t_i) & \text{if } x \in t_i \end{cases} \tag{14.10}$$

See Table 14.9. With this vector-based representation, the expected support of any 1-itemset $\{x\}$ can be computed by summing all non-zero $P(x, t_i)$ values in $\vec{x}$ (i.e., taking the L_1-*norm* of $\vec{x}$):

$$expSup(\{x\}, D) = \sum_{i=1}^{n} expSup(\{x\}, t_i)$$

$$= \sum_{i=1}^{n} P(x, t_i)$$

$$= \sum_{i=1}^{n} \vec{x}[i]$$

$$= ||\vec{x}||_1 \tag{14.11}$$

The i-th element of the vector of any $(k+1)$-itemset $X \equiv Y \cup \{z\}$ (where Y is a k-itemset and z is an item) for $k \geq 1$ can be formed by taking the product of $expSup(Y, t_i)$ and $P(z, t_i)$. The expected support of X is then the *dot product* of $\vec{Y}$ and $\vec{\{z\}}$.

10   Discussion on Uncertain Frequent Pattern Mining

So far, we have described various uncertain frequent pattern mining algorithms. Tong et al. [50] compared some of these algorithms.

In terms of functionality, the U-Apriori, UH-mine, UF-growth, UFP-growth, CUF-growth, PUF-growth, MR-growth, U-Eclat, UV-Eclat and U-VIPER algorithms all mine static datasets of uncertain data. Among them, the first seven mine the datasets horizontally, whereas the remaining three algorithms mine the datasets vertically. In contrast, the SUF-growth, UF-streaming, TUF-streaming, LUF-streaming, UHS-Stream and TFUHS-Stream algorithms mine dynamic streaming uncertain data. Unlike these 16 algorithms that find all frequent patterns, both U-FPS and U-FIC algorithms find only those frequent patterns satisfying the user-specified constraints.

In terms of accuracy, most algorithms return all the patterns with expected support (over all "possible worlds") meeting or exceeding the user-specified threshold *minsup*. In contrast, U-Eclat returns patterns with estimated support (over only the sampled "possible worlds") meeting or exceeding *minsup*. Hence, U-Eclat may introduce false positives (when the support is overestimated) or false negatives (when the support is underestimated). More instantiations (i.e., more samples) helps improve accuracy.

In terms of memory consumption, U-Apriori keeps a list of candidate patterns, whereas the tree-based and hyperlinked structure based algorithms construct in-memory structures (e.g., UF-tree and its variants, extended H-struct). On the one hand, a UF-tree is more compact (i.e., requires less space) than the extended H-struct. On the other hand, UH-mine keeps only one extended H-struct, whereas tree-based algorithms usually construct more than one tree. Sizes of the trees may also vary. For instance, when U-FPS handles a succinct and anti-monotone constraint C_{SAM}, the tree size depends on the selectivity of C_{SAM} because only those items that individually satisfy C_{SAM} are stored in the UF-tree. Both CUF-tree and PUF-tree (for uncertain data) can be as compact as a FP-tree (for precise data). When SUF-growth handles streams, the tree size depends on the size of sliding window (e.g., a window of w batches) because a list of w occurrence counts is captured in each node of SUF-trees (cf. only one occurrence count is captured in each node of UF-trees). Moreover, when items in probabilistic datasets take on a few distinct existential probability values, the trees contain fewer nodes (cf. the number of distinct existential probability values does not affect the size of candidate lists or the extended H-struct). Furthermore, *minsup* and density also affect memory consumption. For instance, for a sparse dataset called kosarak, different winners (requiring the least space) have been shown for different *minsup*: U-Apriori when *minsup* < 0.15 %, UH-mine when 0.15 % ≤ *minsup* < 0.5 %, and, UF-growth when 0.5 % ≤ *minsup*; for a dense dataset called connect4, UH-mine was the winner for 0.2 % ≤ *minsup* < 0.8 %.

In terms of performance, most algorithms perform well when items in probabilistic datasets take on low existential probability values because these datasets do not lead to long frequent patterns. When items in probabilistic datasets take on high existential probability values, more candidates are generated-and-tested by U-Apriori, more and bigger UF-trees are constructed by UF-growth, more hyperlinks are adjusted by UH-mine, and more estimated supports are computed by U-Eclat. Hence, longer runtimes are required. Similarly, when *minsup* decreases, more frequent patterns are returned and longer runtimes are also required. Moreover, the density of datasets also affects runtimes. For instance, when datasets are dense (e.g., connect4 augmented with existential probability), UF-trees obtain higher compression ratios and thus require less time to traverse than sparse datasets (e.g., kosarak augmented with existential probability). Some experimental results showed the following: (i) datasets with a low number of distinct existential probabilities led to smaller UF-trees and shorter runtimes for UF-growth (than U-Apriori); (ii) U-Apriori requires shorter runtimes than UH-mine when *minsup* was low (e.g., *minsup* < 0.3 % for kosarak, *minsup* < 0.6 % for connect4) but vice versa when *minsup* was high; (iii) depending on the number of samples, U-Eclat could take longer or shorter to run than U-Apriori.

11 Extension: Probabilistic Frequent Pattern Mining

The aforementioned algorithms all find (expected support-based) frequent patterns from uncertain data. These are patterns with expected support meeting or exceeding the user-specified threshold *minsup*. Note that expected support of a pattern X provides users with frequency information of X summarized over all "possible worlds", but it does not reveal the confidence on the likelihood of X being frequent (i.e., percentage of "possible worlds" in which X is frequent). However, knowing the confidence can be helpful in some applications. Hence, in recent years, there is also algorithmic development on extending the notion of frequent patterns based on expected support to useful patterns—such as probabilistic heavy hitters and probabilistic frequent patterns as described below.

11.1 Mining Probabilistic Heavy Hitters

Although the expected support of an item x (i.e., a singleton pattern $\{x\}$) provides users with an estimate of the frequency of x in many real-life applications, it is also helpful to know the confidence on the likelihood of x being frequent in the uncertain data in some other applications. Hence, Zhang et al. [56] formalized the notion of probabilistic heavy hitters (i.e., *probabilistic frequent items*, which are also known as probabilistic frequent singleton patterns) following the "possible world" semantics [21] for probabilistic datasets of uncertain data.

Definition 14.4 Given (i) a probabilistic dataset D of uncertain data, (ii) a user-specified support threshold ϕ, and (iii) a user-specified frequentness probability threshold τ, the problem of **mining probabilistic heavy hitters from uncertain data** is to find all (ϕ, τ)-probabilistic heavy hitters (PHHs). An item x is a (ϕ, τ)-probabilistic heavy hitter (PHH) if $P(sup(x, W_j) > \phi|W_j|) > \tau$ (where $sup(x, W_j)$ is the support of x in a random possible world W_j and $|W_j|$ is the number of items

Table 14.10 An example of a probabilistic dataset D_4 with mutually exclusive items

Transaction ID	Set of items with existential probability
t_1	$\{j_1:0.5, j_2:0.5\}$
t_2	$\{k_1:0.1, k_2:0.2, k_3:0.3, k_4:0.4\}$
t_3	$\{j_1:0.3, j_2:0.3, j_3:0.1\}$

Observation 1: The sum of existential probability of all items in each transaction is bounded above by 1.

Observation 2: There are 60 "possible worlds" for D_4 (cf. 512 "possible worlds" if items in each transaction were independent (i.e., not mutually exclusive)).

in W_j), which represents the probability of x being frequent exceeding the user expectation.

Equivalently, given (i) a probabilistic dataset D of uncertain data, (ii) a user-specified support threshold *minsup*, (iii) a user-specified frequentness probability threshold *minProb*, the research problem of **mining probabilistic heavy hitters (PHHs) from uncertain data** is to find every item x that is highly likely to be frequent—i.e., the probability that x occurs in at least *minsup* transactions in D is no less than *minProb*. In other words, x is a probabilistic heavy hitter if $P(sup(x, D) \geq minsup) > minProb$.

To find these PHHs from a probabilistic dataset of uncertain data where items in each transaction are mutually inclusive (e.g., D_4 in Table 14.10), Zhang et al. [56] proposed an exact algorithm and an approximate algorithm. The exact algorithm uses dynamic programming to mine offline uncertain data for PHHs. Such an algorithm runs in polynomial time when there is sufficient memory. When the memory is limited, the approximate algorithm uses sampling techniques to mine streaming uncertain data for approximate PHHs.

11.2 Mining Probabilistic Frequent Patterns

The expected support of a pattern X (that consists of one or more items) provides users with an estimate of the frequency of X, but it does not take into account the variance or the probability distribution of the support of X. In some applications, knowing the confidence on which pattern is highly likely to be frequent helps interpreting patterns mined from uncertain data. Hence, Bernecker et al. [15] extended the notion of frequent patterns and introduced the research problem of mining probabilistic frequent patterns (p-FPs). Figure 14.11 illustrates the differences between expected support and probabilistic support.

Definition 14.5 Given (i) a probabilistic dataset D of uncertain data, (ii) a user-specified support threshold *minsup*, (iii) a user-specified frequentness probability threshold *minProb*, the research problem of **mining probabilistic frequent patterns (p-FPs) from uncertain data** is to find (i) all patterns that are highly likely to be frequent and (ii) their support. Here, the support $sup(X, D)$ of any pattern X is defined by a discrete probability distribution function (pdf) or probability mass

Table 14.11 Frequent pattern vs. probabilistic frequent patterns

Let user-specified $minsup = 1.0$ and $minProb = 0.90$

$sup(\{d\})$	Set of items	$Prob(W_j)$	COUNT(W_j)
2	$(d{:}0.9) \in t_2 \wedge (d{:}0.5) \in t_3$	0.45	8192
	$(d{:}0.9) \in t_2 \wedge (d{:}0.5) \notin t_3$	0.45	8192
1	$(d{:}0.9) \notin t_2 \wedge (d{:}0.5) \in t_3$	$+0.05$	$+8192$
		$= 0.50$	$= 16384$
0	$(d{:}0.9) \notin t_2 \wedge (d{:}0.5) \notin t_3$	0.05	8192
		$\sum_j Prob(W_j) = 1$	$\sum_j$ COUNT $(W_j) = 32768$

As $expSup(\{d\}, D_2) = (2 \times 0.45) + (1 \times 0.50) + (0 \times 0.05) = 0.9 + 0.5 = 1.4 \geq minsup$, $\{d\}$ is an expected support-based **frequent pattern**.
As $Prob(sup(\{d\}, D_2) \geq minsup) = 0.45 + 0.50 = 0.95 \geq minProb$, $\{d\}$ is also a **probabilistic frequent pattern**.

$sup(\{e\})$	Set of items	$Prob(W_j)$	COUNT(W_j)
2	$(e{:}0.7) \in t_3 \wedge (e{:}0.3) \in t_4$	0.21	8192
	$(e{:}0.7) \in t_3 \wedge (e{:}0.3) \notin t_4$	0.49	8192
1	$(e{:}0.7) \notin t_3 \wedge (e{:}0.3) \in t_4$	$+0.09$	$+8192$
		$= 0.58$	$= 16384$
0	$(e{:}0.7) \notin t_3 \wedge (e{:}0.3) \notin t_4$	0.21	8192
		$\sum_j Prob(W_j) = 1$	$\sum_j$ COUNT $(W_j) = 32768$

As $expSup(\{e\}, D_2) = (2 \times 0.21) + (1 \times 0.58) + (0 \times 0.21) = 0.7 + 0.3 = 1.0 \geq minsup$, $\{e\}$ is an expected support-based **frequent pattern**.
However, as $Prob(sup(\{e\}, D_2) \geq minsup) = 0.21 + 0.58 = 0.79 < minProb$, $\{e\}$ is *not* a probabilistic frequent pattern.

function (pmf). A pattern X is highly likely to be frequent (i.e., X is a probabilistic frequent pattern) if and only if its frequentness probability is no less than $minProb$, i.e., $P(sup(X, D) \geq minsup) \geq minProb$. The frequentness probability of X is the probability that X occurs in at least $minsup$ transactions of D. (Table 14.11)

Note that frequentness probability is anti-monotonic: All subsets of a p-FP are also p-FPs. Equivalently, if X is not a p-FP, then none of its supersets is a p-FP, and thus all of them can be pruned. Moreover, when $minsup$ increases, frequentness probabilities of p-FPs decrease.

Bernecker et al. [15] used a dynamic computation technique in computing the probability function $f_X(k) = P(sup(X, D) = k)$, which returns the probability that the support of a pattern X equals to k. Summing the values of such a probability function $f_X(k)$ over all $k \geq minsup$ gives the frequentness probability of X because

$$\sum_{k \geq minsup}^{|D|} f_X(k) = \sum_{k \geq minsup}^{|D|} P(sup(X, D) \geq minsup).$$

Any pattern X having the sum no less than $minProb$ becomes a p-FP.

Sun et al. [49] proposed the top-down inheritance of support probability function (TODIS) algorithm, which runs in conjunction with a divide-and-conquer (DC) approach, to mine probabilistic frequent patterns from uncertain data by extracting patterns that are supersets of p-FPs and deriving p-FPs in a top-down manner (i.e., descending cardinality of p-FPs).

To accelerate probabilistic frequent pattern mining, Wang et al. [51] applied a model-based approach that supports both tuple uncertainty (as in the TODIS algorithm [49]) and attribute uncertainty (as in the aforementioned dynamic computation technique [15]). Specifically, they represented the support pmf of a p-FP as some existing probability models (e.g., Poisson binomial distribution model, normal distribution). By doing so, Wang et al. quickly found two types of p-FPs: (i) threshold-based p-FP (i.e., $P(sup(X) \geq minsup) \geq minProb$) and (ii) rank-based p-FP (e.g., top-k p-FP).

In addition to finding p-FPs from static datasets of uncertain data, there are also algorithms that find p-FPs from dynamic streaming uncertain data. For instance, Akbarinia and Masseglia [14] proposed an exact algorithm called FMU for fast mining of streaming uncertain data with the sliding window model.

12   Conclusions

Frequent pattern mining is an important data mining task. It helps discover implicit, previously unknown and potentially useful knowledge; it also helps reveal sets of frequently co-occurring items in numerous real-life applications (e.g., bundles of books that are frequently bought together, collections of courses that are taken in the same academic terms, events that are often co-located, groups of individuals that have common interests). Here, it has drawn the attention of many researchers over the past two decades. The research problem of frequent pattern mining was originally proposed to analyze shoppers' market basket transaction databases containing precise data, in which the contents of transactions in the databases are known. Such a research problem also plays an important role in other data mining tasks, such as the mining of interesting or unexpected patterns, sequential mining, associative classification, as well as outlier detection, in various real-life applications. As we are living in an uncertain world, data in many real-life applications are uncertain. Recently, researchers have paid more attention to the mining of frequent patterns from probabilistic datasets of uncertain data.

In this chapter, we presented some recent works on mining frequent patterns from probabilistic datasets of uncertain data. These include candidate generate-and-test based, hyperlinked structure based, tree-based, as well as vertical uncertain frequent pattern mining algorithms. Among them, the U-Apriori algorithm generates candidate patterns and tests if their expected support meets or exceeds a user-specified threshold. To avoid such a candidate generate-and-test approach, both UH-mine and UF-growth algorithms use a pattern-growth mining approach. The UH-mine algorithm keeps a UH-struct, from which frequent patterns are mined; the UF-growth algorithm constructs a UF-tree, from which frequent patterns are mined. The UFP-growth algorithm applies clustering to help reduce the number of nodes in a UFP-tree. The PUF-growth and CUF-growth algorithms respectively construct a PUF-tree and a CUF-tree, which are more compact than the corresponding UF-tree. Instead of applying horizontal mining, U-Eclat, UV-Eclat and U-VIPER use vertical mining.

Moreover, the UF-growth algorithm has also been extended for constrained mining, Big Data mining, and stream mining. The resulting U-FPS and U-FIC algorithms exploit properties of the user-specified succinct constraints and convertible constraints, respectively, to find all and only those frequent patterns satisfying the constraints from uncertain data. MR-growth uses the MapReduce model for Big Data analytics. Both SUF-growth and UF-streaming use the sliding window model for mining. SUF-growth mines all frequent patterns from a SUF-tree, which captures the contents of the current few batches of streaming uncertain data. The UF-streaming algorithm applies UF-growth to each batch and stores the mining results in the UF-stream structure, from which frequent patterns can be retrieved. The UF-streaming algorithm was extended to become the TUF-streaming and LUF-streaming algorithms, which use the time-fading and landmark models respectively for (tree-based) mining. Similarly, the TFUHS-Stream and UHS-Stream algorithms also use the time-fading and landmark models respectively, but for hyperlinked structure-based mining.

In addition to expected support-based frequent patterns, there are algorithms that mine probabilistic heavy hitters as well as probabilistic frequent patterns.

Future research directions include (i) mining frequent patterns from uncertain data in applications areas such as social network analysis [4, 28], (ii) mining frequent sequences and frequent graphs from uncertain data, as well as (iii) visual analytics of uncertain frequent patterns.

References

1. Abiteboul, S., Kanellakis, P., & Grahne, G. 1987. On the representation and querying of sets of possible worlds. In *Proceedings of the ACM SIGMOD 1987*, pages 34–48.
2. Aggarwal, C.C. 2009. On clustering algorithms for uncertain data. In C.C. Aggarwal (ed.), *Managing and Mining Uncertain Data*, pages 389–406. Springer.
3. Aggarwal, C.C. (ed.) 2009. *Managing and Mining Uncertain Data*. Springer.
4. Aggarwal, C.C. (ed.) 2011. *Social Network Data Analytics*. Springer.
5. Aggarwal, C.C. 2013. *Outlier Analysis*. Springer.
6. Aggarwal, C.C. (ed.) 2013. *Managing and Mining Sensor Data*. Springer.
7. Aggarwal, C.C. & Reddy, C.K. (eds.), *Data Clustering: Algorithms and Applications*. CRC Press.
8. Agrawal, R., & Srikant, R. 1994. Fast algorithms for mining association rules in large databases. In *Proceedings of the VLDB 1994*, pages 487–499. Morgan Kaufmann.
9. Aggarwal, C.C., & Yu, P.S. 2008. Outlier detection with uncertain data. In *Proceedings of the SIAM SDM 2008*, pages 483–493.
10. Aggarwal, C.C., & Yu, P.S. (eds.) 2008. *Privacy-Preserving Data Mining: Models and Algorithms*. Springer.
11. Aggarwal, C.C., & Yu, P.S. 2009. A survey of uncertain data algorithms and applications. *IEEE Transactions on Knowledge and Data Engineering (TKDE)*, 21(5), pages 609–623.
12. Agrawal, R., Imieliński, T., & Swami, A. Mining association rules between sets of items in large databases. In *Proceedings of the ACM SIGMOD 1993*, pages 207–216.

13. Aggarwal, C.C., Li, Y., Wang, J., & Wang, J. 2009. Frequent pattern mining with uncertain data. In *Proceedings of the ACM KDD 2009*, pages 29–38.
14. Akbarinia, R., & Masseglia, F. 2012. FMU: fast mining of probabilistic frequent itemsets in uncertain data streams. In *Proceedings of the BDA 2012*.
15. Bernecker, T., Kriegel, H.-P., Renz, M., Verhein, F., & Zuefle, A. 2009. Probabilistic frequent itemset mining in uncertain databases. In *Proceedings of the ACM KDD 2009*, pages 119–127.
16. Budhia, B.P., Cuzzocrea, A., & Leung, C.K.-S. 2012. Vertical frequent pattern mining from uncertain data. In *Proceedings of the KES 2012*, pages 1273–1282. IOS Press.
17. Calders, T., Garboni, C., & Goethals, B. 2010. Efficient pattern mining of uncertain data with sampling. In *Proceedings of the PAKDD 2010, Part I*, pages 480–487. Springer.
18. Chui, C.-K., & Kao, B. 2008. A decremental approach for mining frequent itemsets from uncertain data. In *Proceedings of the PAKDD 2008*, pages 64–75. Springer.
19. Chui, C.-K., Kao, B., & Hung, E. 2007. Mining frequent itemsets from uncertain data. In *Proceedings of the PAKDD 2007*, pages 47–58. Springer.
20. Cuzzocrea, A., Leung, C.K.-S., & MacKinnon, R.K. 2014. Mining constrained frequent itemsets from distributed uncertain data. *Future Generation Computer Systems*. Elsevier.
21. Dalvi, N., & Suciu, D. 2004. Efficient query evaluation on probabilistic databases. In *Proceedings of the VLDB 2004*, pages 864–875. Morgan Kaufmann.
22. Gaber, M.M., Zaslavsky, A.B., & Krishnaswamy, S. Mining data streams: a review. *ACM SIGMOD Record*, 34(2), pages 18–26.
23. Green, T., & Tannen, V. 2006. Models for incomplete and probabilistic information. Bulletin of the Technical Committee on Data Engineering, 29(1), pages 17–24. IEEE Computer Society.
24. Han, J., Pei, J., & Yin, Y. 2000. Mining frequent patterns without candidate generation. In *Proceedings of the ACM SIGMOD 2000*, pages 1–12.
25. Jiang, B., Pei, J., Tao, Y., & Lin, X. 2013. Clustering uncertain data based on probability distribution similarity. *IEEE Transactions on Knowledge and Data Engineering (TKDE)*, 25(4), pages 751–763.
26. Jiang, F., & Leung, C.K.-S. 2013. Stream mining of frequent patterns from delayed batches of uncertain data. In *Proceedings of the DaWaK 2013*, pages 209–221. Springer.
27. Lakshmanan, L.V.S., Leung, C.K.-S., & Ng, R.T. 2003. Efficient dynamic mining of constrained frequent sets. *ACM Transactions on Database Systems (TODS)*, 28(4), pages 337–389.
28. Lee, W., Leung, C.K.-S., Song, J.J., & Eom, C.S.-H. 2012. A network-flow based influence propagation model for social networks. In *Proceedings of the CGC/SCA 2012*, pages 601–608. IEEE Computer Society (The best paper of SCA 2012).
29. Leung, C.K.-S. 2009. Convertible constraints. In *Encyclopedia of Database Systems*, pages 494–495. Springer.
30. Leung, C.K.-S. 2009. Frequent itemset mining with constraints. In *Encyclopedia of Database Systems*, pages 1179–1183. Springer.
31. Leung, C.K.-S. 2009. Succinct constraints. In *Encyclopedia of Database Systems*, page 2876. Springer.
32. Leung, C.K.-S. 2011. Mining uncertain data. *Wiley Interdisciplinary Reviews: Data Mining and Knowledge Discovery (WIDM)*, 1(4), pages 316–329.
33. Leung, C.K.-S., & Brajczuk, D.A. 2009. Efficient algorithms for the mining of constrained frequent patterns from uncertain data. *ACM SIGKDD Explorations*, 11(2), pages 123–130.
34. Leung, C.K.-S., & Brajczuk, D.A. 2009. Mining uncertain data for constrained frequent sets. In *Proceedings of the IDEAS 2009*, pages 109–120. ACM.
35. Leung, C.K.-S., & Brajczuk, D.A. 2010. uCFS$_2$: an enhanced system that mines uncertain data for constrained frequent sets. In *Proceedings of the IDEAS 2010*, pages 32–37. ACM.
36. Leung, C.K.-S., & Hao, B. 2009. Mining of frequent itemsets from streams of uncertain data. In *Proceedings of the IEEE ICDE 2009*, pages 1663–1670.
37. Leung, C.K.-S., & Hayduk, Y. 2013. Mining frequent patterns from uncertain data with MapReduce for Big Data analytics. In *Proceedings of the DASFAA 2013, Part I*, pages 440–455. Springer.

38. Leung, C.K.-S., & Jiang, F. 2011. Frequent pattern mining from time-fading streams of uncertain data. In *Proceedings of the DaWaK 2011*, pages 252–264. Springer.
39. Leung, C.K.-S., & Tanbeer, S.K. 2012. Fast tree-based mining of frequent itemsets from uncertain data. In *Proceedings of the DASFAA 2012, Part I*, pages 272–287. Springer.
40. Leung, C.K.-S., & Tanbeer, S.K. 2013. PUF-tree: a compact tree structure for frequent pattern mining of uncertain data. In *Proceedings of the PAKDD 2013, Part I*, pages 13–25. Springer.
41. Leung, C.K.-S., Cuzzocrea, A., & Jiang, F. 2013. Discovering frequent patterns from uncertain data streams with time-fading and landmark models. *LNCS Transactions on Large-Scale Data- and Knowledge-Centered Systems (TLDKS) VIII*, pages 174–196. Springer.
42. Leung, C.K.-S., Mateo, M.A.F., & Brajczuk, D.A. 2008. A tree-based approach for frequent pattern mining from uncertain data. In *Proceedings of the PAKDD 2008*, 653–661. Springer.
43. Leung, C.K.-S., Hao, B., & Brajczuk, D.A. 2010. Mining uncertain data for frequent itemsets that satisfy aggregate constraints. In *Proceedings of the ACM SAC 2010*, pages 1034–1038.
44. Leung, C.K.-S., Tanbeer, S.K., Budhia, B.P., & Zacharias, L.C. 2012. Mining probabilistic datasets vertically. In *Proceedings of the IDEAS 2012*, pages 199–204. ACM.
45. Madden, S. 2012. From databases to big data. *IEEE Internet Computing*, 16(3), pages 4–6.
46. Nadungodage, C.H., Xia, Y., Lee, J.J., & Tu, Y. 2013. Hyper-structure mining of frequent patterns in uncertain data streams. In *Knowledge and Information Systems (KAIS)*, 37(1), pages 219–244. Springer.
47. Ren, J., Lee, S.D., Chen, X., Kao, B., Cheng, R., & Cheung, D. 2009. Naive Bayes classification of uncertain data. In *Proceedings of the IEEE ICDM 2009*, pages 944–949.
48. Suciu, D. 2009. Probabilistic databases. In *Encyclopedia of Database Systems*, pages 2150–2155. Springer.
49. Sun, L., Cheng, R., Cheung, D.W., & Cheng, J. 2010. Mining uncertain data with probabilistic guarantees. In *Proceedings of the ACM KDD 2010*, pages 273–282.
50. Tong, Y., Chen, L., Cheng, Y., & Yu, P.S. 2012. Mining frequent itemsets over uncertain databases. In *Proceedings of the VLDB Endowment (PVLDB)*, 5(11), pages 1650–1661.
51. Wang, L., Cheng, R., Lee, S.D., & Cheung, D.W. 2010. Accelerating probabilistic frequent itemset mining: a model-based approach. In *Proceedings of the ACM CIKM 2010*, pages 429–438.
52. Wasserkrug, S. 2009. Uncertainty in events. In *Encyclopedia of Database Systems*, pages 3221–3225. Springer.
53. Xu, L., & Hung, E. 2012. Improving classification accuracy on uncertain data by considering multiple subclasses. In *Proceedings of the Australasian AI 2012*, pages 743–754. Springer.
54. Zaki, M.J. 1999. Parallel and distributed association mining: a survey. *IEEE Concurrency*, 7(4), pages 14–25.
55. Zaki, M.J., Parthasarathy, S., Ogihara, M., & Li, W. 1997. New algorithms for fast discovery of association rules. In *Proceedings of the ACM KDD 1997*, pages 283–286.
56. Zhang, Q., Li, F., & Yi, K. 2008. Finding frequent items in probabilistic data. In *Proceedings of the ACM SIGMOD 2008*, pages 819–832.

Chapter 15
Privacy Issues in Association Rule Mining

Aris Gkoulalas-Divanis, Jayant Haritsa and Murat Kantarcioglu

Abstract Data mining services require accurate input data for their results to be meaningful, but privacy concerns may impel users to provide spurious information. In this chapter, we study the different aspects of privacy that arise in association rule mining, with special emphasis on *input data privacy*, *output rule privacy* and *owner privacy*. For input privacy, we examine whether users could be encouraged to provide accurate data by ensuring that the mining process cannot, with any reasonable degree of certainty, discover specific information that violates their privacy. Then, in the context of output privacy, we present a taxonomy and a survey of recent approaches that have been applied to the association rule hiding problem. Here, the objective is to minimally modify the original database in a manner that makes the sensitive association rules to disappear while retaining the non-sensitive rules. Finally, we study popular cryptographic methods for preserving the privacy of the individual sources participating in distributed association rule mining.

Keywords frequent pattern mining · privacy · randomization

1 Introduction

Privacy preserving data mining is the research area that investigates the mitigation of adverse side-effects of data mining methods whereby the privacy of individuals and organizations is compromised. In this chapter, we provide an overview of privacy issues that arise in the context of Association Rule Mining (ARM). From a general point of view, we classify privacy issues arising out of data mining into three categories: *Input Privacy*, *Output Privacy* and *Owner Privacy*, outlined below.

A. Gkoulalas-Divanis (✉)
IBM Research-Ireland, Damastown Industrial Estate, Mulhuddart, Dublin 15, Ireland
e-mail: arisdiva@ie.ibm.com

J. Haritsa
Database Systems Lab, Indian Institute of Science (IISc),
Bangalore 560012, India
e-mail: haritsa@dsl.serc.iisc.ernet.in

M. Kantarcioglu
UTD Data Security and Privacy Lab, University of Texas at Dallas,
Texas 75080–3021, USA
e-mail: muratk@utdallas.edu

C. C. Aggarwal, J. Han (eds.), *Frequent Pattern Mining*,   369
DOI 10.1007/978-3-319-07821-2_15, © Springer International Publishing Switzerland 2014

Input Privacy The first category is related to the data per se and is known as *data hiding* or *input privacy*. Specifically, data hiding tries to obfuscate the disclosed data in order to prevent the miner from reliably extracting confidential or private information. Input privacy methods aim at addressing environments where users are unwilling to provide their personal information, or deliberately provide false information, to data recipients, because they fear that their privacy may be violated. The goal of these methods is to guarantee that such personal information can be released to (potentially untrusted) data recipients in a privacy-preserving way that still allows the data recipients to build accurate data mining models from the released data. Several methods have been proposed to provide input privacy (e.g., [7, 6, 17, 45]) by employing various data transformation strategies.

Output Privacy The second category concerns the information, or the knowledge, that a data mining method may discover after having analyzed the data, and is known as *knowledge hiding* or *output privacy*. Specifically, it is concerned with the sanitization of confidential knowledge patterns derived from the data. Output privacy methods aim to eliminate the disclosure of sensitive patterns from datasets. If the datasets were shared as-is, then such patterns could easily lead to (a) the disclosure of sensitive information, such as business or trade secrets that provide competitive advantage to business competitors, or (b) discrimination, if they involve individuals in the input data who have certain unique characteristics. Several methods have been proposed to offer output privacy (e.g., [11, 15, 21, 39]) by eliminating sensitive patterns from the released data, in a way that minimizes data distortion and side-effects.

Owner Privacy Finally, a third line of research involves protocols that enable a group of data owners to collectively mine their data, in a distributed fashion, without allowing any party to reliably learn the data (or sensitive information about the data) that the other owners hold—that is, the sources of the data. For this purpose, several cryptographic methods have been recently proposed to facilitate the privacy-preserving distributed mining of data that reside in different data warehouses (e.g., [26, 51, 52]). These methods assume that the data are either horizontally or vertically partitioned among the different sites, and that any sensitive disclosures should be limited in the data mining process.

The rest of this chapter is organized as follows: Sect. 2 elaborates on input privacy methods that enable the safe discovery of association rules from large historical databases. Section 3 provides a taxonomy, along with a systematic review of related literature, on techniques for hiding sensitive association rules. Section 4 highlights important cryptographic protocols that facilitate preserving owner privacy in distributed data mining. Finally, Sect. 5 concludes the chapter.

2 Input Privacy

The knowledge models produced through data mining techniques are only as good as the accuracy of their input data. One source of data inaccuracy is when users deliberately provide false information. This is especially common with regard to customers who are asked to provide personal information on Web forms to e-commerce service providers. The compulsion for doing so may be the (perhaps well-founded) worry that the requested information may be misused by the service provider to harass the customer. As a case in point, consider a pharmaceutical company that asks clients to disclose the diseases they have suffered from in order to investigate the correlations in their occurrences—for example, "Adult females with malarial infections are also prone to contract tuberculosis". The company may be acquiring the data solely for genuine data mining purposes that would eventually reflect itself in better service to the client. But, at the same time the client might worry that if her medical records are either inadvertently or deliberately disclosed, it may adversely affect her future employment opportunities.

In this section, we study whether customers can be encouraged to provide correct information by ensuring that the mining process cannot, with any reasonable degree of certainty, violate their privacy, but at the same time produce sufficiently accurate mining results. The difficulty in achieving these goals is that privacy and accuracy are typically contradictory in nature, with the consequence that improving one usually incurs a cost in the other [3]. A related issue is the degree of trust that needs to be placed by the users in third-party intermediaries. And finally, from a practical viability perspective, the time and resource overheads that are imposed on the data mining process due to supporting the privacy requirements.

Our study is carried out in the context of extracting *association rules* from large historical databases [8], an extremely popular mining process that identifies interesting correlations between database attributes, such as the one described in the pharmaceutical example. By the end of Sect. 2, we will show that the state-of-the-art in input privacy is such that it is indeed possible to simultaneously achieve all the desirable objectives (i.e., privacy, accuracy, and efficiency) for ARM.

2.1 Problem Framework

In what follows, we describe the framework of the privacy mining problem in the context of association rules.

Database Model We assume that the original (true) database U consists of N records, with each record having M categorical attributes. Note that boolean data is a special case of this class, and further, that continuous-valued attributes can be converted into categorical attributes by partitioning the domain of the attribute into fixed length intervals.

The domain of attribute j is denoted by S_U^j, resulting in the domain S_U of a record in U being given by $S_U = \prod_{j=1}^{M} S_U^j$. We map the domain S_U to the index set $I_U = \{1, \ldots, |S_U|\}$, thereby modeling the database as a set of N values from I_U. If we denote the ith record of U as U_i, then $U = \{U_i\}_{i=1}^{N}, U_i \in I_U$.

To make this concrete, consider a database U with 3 categorical attributes *Age*, *Sex* and *Education* having the following category values:

Age	Child, Adult, Senior
Sex	Male, Female
Education	Elementary, Graduate

For this schema, $M = 3$, $S_U^1 = \{$Child, Adult, Senior$\}$, $S_U^2 = \{$Male, Female$\}$, $S_U^3 = \{$Elementary, Graduate$\}$, $S_U = S_U^1 \times S_U^2 \times S_U^3$, $|S_U| = 12$. The domain S_U is indexed by the index set $I_U = \{1, \ldots, 12\}$, and hence the set of records

<table>
<tr><td colspan="3" align="center">U</td><td></td><td align="center">U</td></tr>
<tr><td>Child</td><td>Male</td><td>Elementary</td><td></td><td>1</td></tr>
<tr><td>Child</td><td>Male</td><td>Graduate</td><td>maps</td><td>2</td></tr>
<tr><td>Child</td><td>Female</td><td>Graduate</td><td>to</td><td>4</td></tr>
<tr><td>Senior</td><td>Male</td><td>Elementary</td><td></td><td>9</td></tr>
</table>

Mining Objective The goal of the data-miner is to compute *association rules* on the above database. Denoting the set of attributes in database U by C, an association rule is a (statistical) implication of the form $C_x \implies C_y$, where $C_x, C_y \subset C$ and $C_x \cap C_y = \phi$. A rule $C_x \implies C_y$ is said to have a *support* (or frequency) factor s iff at least $s\%$ of the transactions in U satisfy $C_x \cup C_y$. A rule $C_x \implies C_y$ is satisfied in U with a *confidence* factor c iff at least $c\%$ of the transactions in U that satisfy C_x also satisfy C_y. Both support and confidence are fractions in the interval [0,1]. The support is a measure of statistical significance, whereas confidence is a measure of the strength of the rule.

A rule is said to be "interesting" if its support and confidence are greater than user-defined thresholds sup_{min} and con_{min}, respectively, and the objective of the mining process is to find all such interesting rules. It has been shown in [8] that achieving this goal is effectively equivalent to generating all subsets of C that have support greater than sup_{min} – these subsets are called *frequent* itemsets. Therefore, the mining objective is, in essence, to efficiently discover all frequent itemsets that are present in the database.

Privacy Mechanisms We now move on to considering the various mechanisms through which privacy of the user data could be provided. One approach to address this problem is for the service providers to assure the users that the databases obtained from their information would be anonymized (through the variety of techniques proposed in the statistical database literature [2, 49]), before being supplied to the data miners. For example, the swapping of attribute-values between different customer records, as proposed in [16], can be used to conceal the true value of the corresponding attribute for each customer. Such a privacy environment in which customers depend on the service provider to guarantee privacy provisioning, is referred to in the literature as a "B2B (business-to-business)" environment.

However, in today's world, most users are (perhaps justifiably) cynical about such assurances, and it is therefore imperative to demonstrably provide privacy at the point of data collection itself, that is, *at the user site*. This is referred to as the "B2C (business-to-customer)" privacy environment [57]. Note that in this environment, any technique that requires knowledge of other user records becomes infeasible, and therefore the B2B approaches cannot be applied here.

The bulk of the work in privacy-preserving data mining of association rules has addressed the B2C environment (e.g. [7, 6, 17, 45]), where the user's true data have to be anonymized at the source itself. Note that the anonymization process has to be implemented by a program which could be supplied either by the service provider or, more likely, by an independent trusted third-party vendor. Further, this program has to be verifiably secure—therefore, it must be simple in construction, eliminating the possibility of the true data being surreptitiously supplied to the service provider. In a nutshell, the goal of these techniques is to ensure the privacy of the raw local data at the source, but, at the same time, to support accurate reconstruction of the global data mining models at the destination.

Within the above framework, the general approach has been to adopt a *data perturbation* strategy, wherein each individual user's true data are altered in some manner before being forwarded to the service provider. Here, there are two possibilities: *statistical distortion*, which has been the predominant technique, and *algebraic distortion*, proposed in [57]. In the statistical approach, a common randomizing algorithm is employed at all user sites, and this algorithm is subsequently disclosed to the eventual data miner. For example, in the MASK technique [45], which is targeted towards "market-basket" type of sparse boolean databases, each bit in the true user transaction vector is independently flipped with a parametrized probability.

While there is only one-way communication from users to the service provider in the statistical approach, the algebraic scheme, in marked contrast, requires *two-way communication* between the data miner and the user. Here, the data miner supplies a user-specific perturbation vector, and the user then returns the perturbed data after applying this vector on the true data, discretizing the output and adding some noise. The vector is dependent on the *current contents* of the perturbed database available with the miner and, for large enterprises, the data collection process itself could become a bottleneck in the efficient running of the system.

Within the statistical approach, there are two further possibilities: (a) a simple *independent attribute perturbation*, wherein the value of each attribute in the user record is perturbed independently of the rest; or (b) a more generalized *dependent attribute perturbation*, where the perturbation of each attribute may be affected by the perturbations of the other attributes in the record. Most of the statistical perturbation techniques in the literature, including [18, 17, 45], fall into the independent attribute perturbation category. Notice, however, that this is in a sense antithetical to the original goal of association rule mining, which is to identify *correlations across attributes*. This limitation is addressed in [4], which employs a dependent attribute perturbation model, with each attribute in the user's data vector being perturbed based on its own value, as well as the perturbed values of the earlier attributes.

Another model of privacy-preserving data mining is the k-anonymity model [7, 46], where each record value is replaced with a corresponding generalized value. Specifically, each perturbed record cannot be distinguished from at least $k - 1$ other records in the data. However, this falls into the B2C model since the intermediate database-forming-server can learn, or recover, precise records.

Privacy Metrics Independently of the specific scheme that is used to achieve input privacy, the end result is that the data miner receives the following as input: (a) the perturbed database V, and (b) the perturbation technique T used to produce this database. From these inputs, the data miner attempts to reconstruct the original *distribution* of the true database U, and mine this reconstructed database to obtain the association rules. Given this framework, the general notion of input privacy in the ARM literature is the level of certainty with which the data miner can reconstruct the true data values of the users. The certainty can be evaluated at various levels:

(1) **Average Privacy.** This metric measures the reconstruction probability of a random value in the database.
(2) **Worst-case Privacy.** This metric measures the maximum reconstruction probability across all the values in the database.
(3) **Re-interrogated Privacy.** A common system environment is where the miner does not have access to the perturbed database after the completion of the mining process. But it is also possible to have situations wherein the miner can use the mining output (i.e., the association rules) to subsequently *re-interrogate* the perturbed database, possibly resulting in reduced privacy.
(4) **Amplification Privacy.** A particularly strong notion of privacy, called "amplification", was presented in [18]. Amplification guarantees strict limits on privacy breaches of individual user information, *independent of the distribution of the true data*. Here, the property of a data record U_i is denoted by $Q(U_i)$. For example, consider the following record from the example dataset U discussed earlier:

Age	Sex	Education
Child	Male	Elementary

Sample properties of the record include

$$Q_1(U_i) \equiv \text{``}Age = Child \text{ and } Sex = Male\text{''}, and$$
$$Q_2(U_i) \equiv \text{``}Age = Child \text{ or } Adult\text{''}.$$

In this context, the *prior probability* of a property of a customer's private information is the likelihood of the property in the absence of any knowledge about the customer's private information. On the other hand, the *posterior probability* is the likelihood of the property given the perturbed information from the customer and the knowledge of the prior probabilities through reconstruction from the perturbed database. In order to preserve the privacy of some property of a customer's private information, the posterior probability of that property should not be *unduly different* to that of the prior probability of the property for the customer. This notion of privacy is quantified in [18] through the following results, where ρ_1 and ρ_2 denote the prior and posterior probabilities, respectively:

a) **Privacy Breach:** An upward ρ_1-to-ρ_2 privacy breach exists with respect to property Q if $\exists v \in S_V$ such that

$$P[Q(U_i)] \leq \rho_1 \quad \text{and} \quad P[Q(U_i)|R(U_i) = v] \geq \rho_2.$$

Conversely, a downward ρ_2-to-ρ_1 privacy breach exists with respect to property Q if $\exists v \in S_V$ such that

$$P[Q(U_i)] \geq \rho_2 \quad \text{and} \quad P[Q(U_i)|R(U_i) = v] \leq \rho_1.$$

b) **Amplification:** Let the perturbed database be $V = \{V_1, \dots, V_N\}$, with domain S_V, and corresponding index set I_V. For example, given the sample database U discussed above, and assuming that each attribute is distorted to produce a value within its original domain, the distortion may result in

<table>
<tr><td>V</td><td></td><td colspan="3">V</td></tr>
<tr><td>5</td><td rowspan="4">which
maps
to</td><td>Adult</td><td>Male</td><td>Elementary</td></tr>
<tr><td>7</td><td>Adult</td><td>Female</td><td>Elementary</td></tr>
<tr><td>2</td><td>Child</td><td>Male</td><td>Graduate</td></tr>
<tr><td>12</td><td>Senior</td><td>Female</td><td>Graduate</td></tr>
</table>

Let the probability of an original customer record $U_i = u, u \in I_U$ being perturbed to a record $V_i = v, v \in I_V$ be $p(u \rightarrow v)$, and let A denote the matrix of these transition probabilities, with $A_{vu} = p(u \rightarrow v)$. With the above notation, a randomization operator $R(u)$

$$\forall u_1, u_2 \in S_U : \frac{p[u_1 \rightarrow v]}{p[u_2 \rightarrow v]} \leq \gamma$$

where $\gamma \geq 1$ and $\exists u : p[u \rightarrow v] > 0$. Operator $R(u)$ is at most γ-amplifying if it is at most γ-amplifying for all qualifying $v \in S_V$.

c) **Breach Prevention:** Let R be a randomization operator, $v \in S_V$ be a randomized value such that $\exists u : p[u \rightarrow v] > 0$, and ρ_1, ρ_2 ($0 < \rho_1 < \rho_2 < 1$) be two probabilities as per the above privacy breach definition. Then, if R is at most γ-amplifying for v, revealing "$R(u) = v$" will cause neither upward (ρ_1-to-ρ_2) nor downward (ρ_2-to-ρ_1) privacy breaches with respect to any property if the following condition is satisfied:

$$\frac{\rho_2(1 - \rho_1)}{\rho_1(1 - \rho_2)} > \gamma$$

If this holds, R is said to support (ρ_1, ρ_2)-privacy guarantees.

Accuracy Metrics Applying association rule mining on a perturbed database can lead to two kinds of errors. Firstly, there may be *support* errors, where a correctly-identified frequent itemset may be associated with an incorrect support value. Secondly, there may be *identity* errors, wherein either a genuine frequent itemset is mistakenly classified as rare, or the converse, where a rare itemset is claimed to be frequent.

The **Support Error** (μ) metric reflects the average relative error (in percent) of the reconstructed support values for those itemsets that are correctly identified to be frequent. Denoting the number of frequent itemsets by $|F|$, the reconstructed support by $\widehat{sup}$ and the actual support by sup, the support error is computed over all frequent itemsets as

$$\mu = \frac{1}{|F|} \Sigma_{f \in F} \frac{|\widehat{sup}_f - sup_f|}{sup_f} * 100$$

The **Identity Error** (σ) metric, on the other hand, reflects the percentage error in identifying frequent itemsets and has two components: σ^+, indicating the percentage of false positives, and σ^- indicating the percentage of false negatives. Denoting the reconstructed set of frequent itemsets with R and the correct set of frequent itemsets with F, these metrics are computed as follows

$$\sigma^+ = \frac{|R - F|}{|F|} * 100 \qquad \sigma^- = \frac{|F - R|}{|F|} * 100$$

Note that in some papers (e.g. [57]), the accuracy metrics are taken to be the worst-case, rather than average-case, versions of the above errors.

2.2 Evolution of the Literature

From the database perspective, the field of privacy-preserving data mining was catalyzed by the pioneering investigation of [6]. In that work, developing privacy-preserving *data classifiers* by adding noise to the record values was proposed and analyzed. This approach was extended in [3] and [29] to address a variety of subtle privacy loopholes.

Concurrently, the research community also began to look into extending privacy-preserving techniques to alternative mining patterns, such as association rules, clustering, etc. For association rules, two main streams of literature emerged, as mentioned earlier, one looking at providing input data privacy, and the other considering the protection of sensitive output rules (discussed in Sect. 3). An important point to note here is that unlike the privacy-preserving classifier approaches, which were based on adding a noise component to continuous-valued data, the privacy-preserving techniques in ARM are based on *probabilistic mapping* from the domain space to the range space, over categorical atttributes.

With regard to input data privacy, the early papers include [17, 45], which proposed the MASK algorithm and the Cut-and-Paste operators, respectively.

MASK In MASK [45], a simple probabilistic distortion of user data, employing random numbers generated from a pre-defined distribution function, was proposed and evaluated in the context of sparse boolean databases, such as those found in "market-baskets". The distortion technique was simply to flip each 0 or 1 bit with a parametrized probability p, or to retain as is with the complementary probability $1 - p$, and the privacy metric used was average privacy. Through a theoretical and empirical analysis, it was shown that the p parameter could be carefully tuned to simultaneously achieve acceptable average privacy and good accuracy.

However, it was also found that mining the distorted database could be orders of magnitude more time-consuming as compared to mining the original database. This issue was addressed in a followup work [9], which showed that by generalizing the distortion process to perform symbol-specific distortion (i.e., different flipping probabilities for different values), appropriately chooosing these distortion parameters, and applying a variety of set-theoretic optimizations in the reconstruction process, runtime efficiencies that are well within an order of magnitude of undistorted mining can be achieved.

Cut-and-Paste Operator The notion of a privacy breach was introduced in [17] as follows: The presence of an itemset I in the randomized transaction causes a privacy breach of level ρ if it is possible to infer, for some transaction in the true database, that the probability of some item i occuring in it exceeds ρ.

With regard to this worst-case privacy metric, a set of randomizing privacy operators were presented and analyzed in [17]. The starting point was *Uniform Randomization*, where each existing item in the true transaction is, with probability p, replaced with a new item not present in the original transaction. (Note that this means that the number of items in the randomized transaction is always equal to the number in the original transaction, and is therefore different from MASK where the number of items in the randomized transaction is usually significantly more than its source, since the flipping is done on both the 1's *and the* 0's in the transaction bit vector.) It was then pointed out that a basic deficiency of the uniform randomization approach is that while it might, with a suitable choice of p, be capable of providing acceptable average privacy, its worst case privacy could be significantly weaker.

To address this issue, an alternative *select-a-size* (**SaS**) randomization operator was proposed, which is composed of the following steps, employed on a per-transaction basis:

Step 1: For customer transaction t_i of length m, a random integer j from $[1, m]$ is first chosen with probability $p_m[j]$.

Step 2: Then, j items are uniformly and randomly selected from the true transaction and inserted into the randomized transaction.

Step 3: Finally, a uniformly and randomly chosen fraction ρ_m of the remaining items in the database that are not present in the true transaction (i.e., $C-$ items in t_i), are inserted into the randomized transaction.

In short, the final randomized transaction is composed of a subset of true items from the original transaction and additional false items from the complementary set of items in the database.

A variant of the SaS operator, studied in detail in [17], is the *cut-and-paste* (**C&P**) operator. Here, an additional parameter is a cutoff integer, K_m, with the integer j

being chosen from $[1, K_m]$, rather than from $[1, m]$. If it turns out that $j > m$, then j is set to m (which means that the entire original transaction is copied to the randomized transaction). Apart from the cutoff threshold, another difference between *C&P* and *SaS* is that the subsequent ρ_m randomized insertion (Step 3 above) is carried out on (a) the items that are not present in the true transaction (as in SaS), and (b) additionally, *on the remaining items in the true transaction* that were not selected for inclusion in Step 2.

An issue in the C&P operator is the optimal selection of the ρ_m and K_m parameters, and combinatorial formulae for determining their values are given in [17]. Through a detailed set of experiments on real-life datasets, it was shown that even with a challenging privacy requirement of not permitting any breaches with $\rho > 50 \%$, mining a C&P-randomized database was able to correctly identify around 80 to 90 % of the "short" frequent itemsets, that is frequent itemsets of lengths up to 3. The issue of how to safely randomize and mine long transactions was left as an open problem, since directly using C&P in such environments could result in unacceptably poor accuracy.

The above work was significantly extended in [18] through, as discussed in Sect.2.1.0, the formulation of strict amplification-based privacy metrics, and delineation of a methodology for limiting the associated privacy breaches.

Distributed Databases Maintaining input data privacy was also considered in [26, 52] in the context of databases that are *distributed* across a number of sites, with each site only willing to share data mining results, but not the source data. While [52] considered data that is vertically partitioned (i.e., each site hosts a disjoint subset of the matrix columns), the complementary situation where the data is horizontally partitioned (i.e., each site hosts a disjoint subset of the matrix rows) is addressed in [26]. The solution technique in [52] requires generating and computing a large set of independent linear equations—in fact, the number of equations and the number of terms in each equation is proportional to the *cardinality* of the database. It may therefore prove to be expensive for market-basket databases which typically contain millions of customer transactions. In [26], on the other hand, the problem is modeled as a secure multi-party computation [24] and an algorithm that minimizes the information shared without incurring much overhead on the mining process is presented. Note that in these formulations, a pre-existing true database at each site is assumed, i.e., a B2B model.

Algebraic Distortion Zhang et al., in [57], presented an algebraic-distortion mechanism that unlike the statistical approach of the prior literature, requires *two-way communication* between the miner and the users. If V_c is the current perturbed database, then E_k is computed by the miner, which corresponds to the eigenvectors corresponding to the largest k eigenvalues of $V_c^T V_c$, where V_c^T is the transpose of V_c. The choice of k makes a tradeoff between privacy and accuracy – large values of k give more accuracy and less privacy, while small values provide higher privacy and less accuracy. E_k is supplied to the user, who then uses it on her true transaction vector, discretizes the output, and then adds a noise component.

The privacy metric that is used in this paper is rather different, in that they evaluate the level of privacy by measuring the probability of an "unwanted" item to be included in the perturbed transaction. The definition of "unwanted" here is that it is an item that does not contribute to association rule mining in the sense that it does not appear in any frequent itemset. An implication is that privacy estimates can be *conditional* on the choices of ARM parameters (sup_{min}, con_{min}). This may encourage the miner to experiment with a variety of values in order to maximize the breach of privacy.

Frameworks A common trend in the input data privacy literature was to propose *specific* perturbation techniques, which are then analyzed for their privacy and accuracy properties. Recently, in [4], the problem was approached from a different perspective, wherein a generalized matrix-theoretic *framework*, called *FRAPP*, that facilitates a systematic approach to the *design* of random perturbation schemes for privacy-preserving mining was proposed. This framework supports amplification-based privacy, and its execution and memory overheads are comparable to that of classical mining on the true database. The distinguishing feature of FRAPP is its quantitative characterization of the *sources of error* in the random data perturbation and model reconstruction processes.

In fact, although it uses dependent attribute perturbation, it is fully decomposable into the perturbation of individual attributes, and hence has the *same run-time complexity* as any independent perturbation method. Through the framework, many of the earlier techniques are cast as special instances of the FRAPP perturbation matrix. More importantly, it was shown that through appropriate choices of matrix elements, new perturbation techniques can be constructed that provide highly accurate mining results even under strict amplification-based [18] privacy guarantees. In fact, a perturbation matrix with provably minimal condition number (in the class of symmetric positive-definite matrices), was identified, substantially improving the accuracy under the given constraints. Finally, an efficient integration of this optimal matrix with the association mining process was outlined.

3   Output Privacy

In this section, we present an overview of a specific class of methods in the knowledge hiding area, known as *frequent itemset* and *association rule hiding* (ARH), which are applied to offer output privacy. Other classes of methods, under the same area, include *classification rule hiding* [35, 36] and *sequential pattern hiding* [1, 20]. "Association rule hiding" (a term used for brevity instead of the longer title "frequent itemset and association rule hiding") has been mentioned for the first time in 1999 in a workshop paper by Atallah et al. [11]. The authors in [11] tried to apply general ideas regarding the implications of data mining in security and privacy of information—first presented by Clifton and Marks in [14]—to the association rule mining [5] framework. Clifton and Marks, following the suggestions of D.E. O'Leary [38]—who was the very first to point out the security and privacy breaches that originate from data mining algorithms—indicated the need to consider different data mining approaches under the prism of preserving information privacy.

The following scenario exemplifies the necessity of applying ARH algorithms to protect sensitive knowledge. Let us suppose that we, the purchasing directors of BigMart, a large supermarket chain, are negotiating with Dedtrees Paper Company. They offer their products with reduced prices, provided that we agree to give them access to our database of customer purchases. We accept the deal and Dedtrees starts mining our data. By using an ARM tool, they find that people who purchase skim milk also purchase Green Paper. Dedtrees now runs a coupon marketing campaign offering a 50 cents discount on skim milk with every purchase of a Dedtrees product. The campaign cuts heavily into the sales of Green Paper, which increases its prices, based on the lower sales. During our next negotiation with Dedtrees, we find out that with reduced competition they are unwilling to offer to us a low price. Finally, we start losing business to our competitors, who were able to negotiate a better deal with Green Paper. In other words, the aforementioned scenario indicates that BigMart should sanitize competitive information (and other important corporate secrets of course) before delivering their database to Dedtrees, so that Dedtrees does not monopolize the paper market.

We should emphasize here that the ARH problem can be considered as a variation of the well known *database inference control* [19] problem in statistical and multilevel databases. The primary goal, in the database inference control, is to protect access to sensitive information that can be obtained through non-sensitive data and inference rules. In ARH, it is not the data but the sensitive rules that create a breach of privacy. Given a set of sensitive association rules, which are specified by the security administrator, the task of the association rule hiding algorithms is to sanitize the data so that the ARM algorithms applied to this data will be (a) incapable of discovering the sensitive rules under certain parameter settings, and (b) able to mine all the non-sensitive rules. A recently investigated problem, known as *inverse frequent itemset mining* [33], provides a special solution to the association rule hiding problem even though it is not targeted to addressing privacy issues per se.

3.1 Terminology and Preliminaries

As stated earlier, ARM is the process involving the discovery of sets of items (*itemsets*) that frequently co-occur in a database with the goal of producing association rules that hold for the data [5, 8]. The itemset $C_x \cup C_y$ that led to the generation of an association rule $C_x \implies C_y$ is known as the *generating itemset* and consists of two parts, the *Left Hand Side* (LHS), which is the part on the left of the arrow of the rule (here C_x), and the *Right Hand Side* (RHS), which is the part on the right of the arrow of the rule (here C_y). An itemset with k items is called k–itemset. In ARH algorithms we consider that database U is given in the form of transactions, where each record (also known as *transaction*) is associated with a set of items from a domain $\mathcal{I}$. These items, for example, could refer to purchased products; thus a record of U may capture the items that were purchased together by an individual from a supermarket (e.g., $u_1 = \{$bread, milk, sugar$\}$). A similar representation that is usually adopted by ARH algorithms is that of a boolean matrix, where each column corresponds to an item from the domain of items $\mathcal{I}$ and each row is a transaction. In this representation, a transaction of U has length $|\mathcal{I}|$ and has 1's in items that are associated with it (e.g., purchased items) and 0's in the rest of the items.

Knowledge hiding, in the context of ARM, aims at sanitizing (transforming) the original dataset in a way that the following goals are accomplished to the largest possible extent:

a) **Sensitive rules are concealed.** No rule that is considered as sensitive from the data owner's perspective, can be revealed from the sanitized dataset, when the dataset is mined at pre-specified thresholds of confidence and support (or at any value higher than these thresholds).

b) **Frequent non-sensitive rules are preserved.** All the non-sensitive frequent rules can be successfully mined from the sanitized database at pre-specified thresholds of confidence and support.

c) **Ghost rules are not generated.** No rule that was not mined from the original dataset as frequent can be discovered from the sanitized database, when mining this database at pre-specified thresholds of confidence and support.

d) **Dataset distortion is minimum.** The sanitized dataset is "as similar as possible" to the original dataset, i.e., the number of data items that are affected by the hiding process is kept minimum.

The first goal requires sensitive rules to disappear. The second goal simply states that there should be no *lost rules* in the sanitized dataset. The third goal says that no *false rules* should be produced as a side-effect of the sanitization process. The fourth goal requires that the hiding process incurs minimal distortion to the original dataset. Generally speaking, in the typical case hiding scenario, the sanitization process has to be accomplished in a way that *minimally affects the original dataset, preserves the general patterns and trends*, and *successfully conceals all the sensitive knowledge*.

3.2   Taxonomy of ARH Algorithms

In this section, we present a taxonomy of frequent itemset and association rule hiding algorithms. To classify the various algorithms, we use a set of orthogonal dimensions. As a first dimension, we consider whether the hiding algorithm uses the support or the confidence of the rule to drive the hiding process. In this way we separate the hiding algorithms into *support*-based and *confidence*-based.

The second dimension in the classification is related to the modification in the raw data that is caused by the hiding algorithm. The two forms of modification comprise the *distortion* and the *blocking* of the original values. Distortion is the process of replacing 1's by 0's and 0's by 1's, while blocking refers to replacing original values by question marks (unknowns) to confuse adversaries about the actual value.

The third dimension, refers to whether a single sensitive rule or a set of sensitive rules can be hidden during an iteration of the hiding algorithm. Based on this criterion we differentiate hiding algorithms into *single rule* and *multiple rule* schemes.

The fourth dimension has to do with the nature of the hiding algorithm, which can be either *heuristic* or *exact*. Heuristic algorithms take decisions that aim at optimizing certain sub-goals in the hiding process, are computationally efficient, but do not guarantee optimality. The formulation of the ARH problem (presented in Sect. 3.1) implies that there are two specific sub-goals that need to be attained by every ARH algorithm. The first sub-goal [(**a**)], which is basically the most important, is to try to hide as many sensitive rules as possible. The second sub-goal [(**b**), (**c**), (**d**)] is to manage to hide the sensitive rules by minimizing side-effects. Different hiding algorithms give different priorities to the satisfaction of the sub-goals presented, producing in this way a list of hiding primitives.

Exact techniques, on the other hand, rely on formulating the ARH problem so that a solution satisfying all the sub-goals can be found. Of course, there is a high possibility that an exact approach fails to give a solution, and for this reason, some of the sub-goals need to be relaxed. However, this relaxation process is still part of the exact approach, which makes it different from the heuristic approaches. Although exact approaches lead to better solutions than heuristic algorithms, they are computationally demanding and can be applied only to small and medium-size datasets.

The fifth and final dimension determines whether a hiding algorithm preprocesses the user-specified sensitive rules so that a minimal set of sensitive rules are given as input to the hiding technique. The corresponding techniques make use of the *border* of the frequent itemsets [31], which provides a compact representation of the frequent itemsets mined from a database, to facilitate knowledge hiding. Specifically, given the set of frequent itemsets F mined from the transactions of database U, the *negative border* of F, denoted as $B^-(F)$, is defined as the set of all infrequent itemsets mined from U in which all proper subsets appear in F. Similarly, the *positive border* of F, denoted as $B^+(F)$, is defined as the set of all maximally frequent itemsets appearing in F. The positive and the negative border formulate collectively the *border* $B(F)$ of frequent itemsets. Formally stated, if $\mathcal{I}$ is the set of all items appearing in U, then $B^-(F) = \{C_x \subseteq \mathcal{I} : C_x \notin F \wedge \forall C_y \subset C_x : C_y \in F\}$, $B^+(F) = \{C_x \subseteq \mathcal{I} : C_x \in F \wedge \forall C_y \supset C_x : C_y \notin F\}$, and $B(F) = B^-(F) \cup B^+(F)$. Border-based hiding techniques compute the border of the frequent itemsets and modify it appropriately by recomputing it, in such a way, that a minimal set of sensitive rules joins the newly computed border. The algorithms are subsequently driven by the negative and positive border for hiding the rules [21, 31].

3.3   Heuristic and Exact ARH Algorithms

Among the different dimensions in the taxonomy of frequent itemset and association rule hiding algorithms, the prevalent dimension regards the nature of the hiding algorithms. As stated before, ARH algorithms can be divided into two broad classes, namely *heuristic* approaches and *exact* approaches. In this section, we review some of the most popular approaches that belong to each class. Furthermore, we devote a section to discuss *border-based* approaches, an important category of heuristic approaches that has also influenced the design of exact hiding algorithms. Due to their importance, in what follows we refer to *border-based* approaches as the third class of ARH algorithms.

Heuristic approaches involve efficient, fast algorithms that selectively sanitize a set of transactions from the database to hide the sensitive knowledge. Due to their efficiency and scalability, the heuristic approaches have been the focus of attention for the vast majority of researchers in the knowledge hiding field. However, there are several circumstances in which they suffer from undesirable side-effects that lead them to poor solutions.

Border-based approaches consider the task of sensitive rule hiding through modification of the original borders in the lattice of the frequent and the infrequent patterns in the dataset. In these schemes, the sensitive knowledge is hidden by enforcing the revised borders (which accommodate the hiding of the sensitive itemsets) in the sanitized database. The algorithms in this class differ both in the borders that they track and use for the hiding strategy, as well as in the methodology that they follow to enforce the revised borders in the sanitized dataset.

Finally, exact approaches contain non-heuristic algorithms which conceive the hiding process as a constraint satisfaction problem that they solve by using integer or linear programming. The main difference of these approaches, compared to the previous ones, is the fact that the sanitization process guarantees optimality in the hiding solution, provided that an optimal solution exists. On the other hand, these approaches are usually several orders of magnitude slower than the heuristic ones, particularly due to the runtime of the integer/linear programming solver. For this, they are applicable only in small and medium-size datasets.

Heuristic Approaches In this section, we review support-based and confidence-based heuristic approaches, which are based on either distortion or blocking of the original values. Between these two categories of approaches, the distortion-based are the ones commonly adopted by the overwhelming majority of researchers.

Support-based and Confidence-based Distortion Schemes Atallah et al. [11] were the first to propose an algorithm for the hiding of sensitive association rules through the reduction in the support of their generating itemsets. The authors propose the construction of a lattice-like graph in the database. Through this graph, the hiding of a large itemset, related to the existence of a sensitive rule, is achieved by a greedy iterative traversal of its immediate subsets, selection of the subset that has the maximum support among all candidates (therefore is less probable to be hidden) and setting of this itemset as the new candidate to be hidden. By iteratively following these steps, the algorithm identifies the 1-itemset ancestor of the initial sensitive itemset, having the highest support. Then, by identifying the supporting transactions for both the initial candidate and the currently identified 1-itemset, the algorithm removes the 1-itemset from the supporting transaction which affects the least number of 2-itemsets. In sequel, the algorithm propagates the results of this action to the affected itemsets in the graph. When hiding a set of sensitive rules, the algorithm first sorts the corresponding large itemsets based on their support and then proceeds to hide them in a one-by-one fashion, using the methodology presented above. One of the most significant contributions of this work is the proof regarding the *NP-hardness* of finding an optimal sanitization of a dataset. On the negative side, the proposed approach is not interested in the extent of the loss of support for a large itemset, as long as it remains frequent in the sanitized outcome.

Dasseni et al. [15] generalize the problem in the sense that they consider the hiding of both sensitive frequent itemsets and sensitive rules. The authors propose three single rule hiding approaches that are based on the reduction of either the support or the confidence of the sensitive rules, but not both. In all three approaches, the goal is to hide the sensitive rules while minimally affecting the support of the non-sensitive itemsets. The first two strategies reduce the confidence of the sensitive rule either (i) by increasing the support of the rule antecedent, through transactions that partially support it, until the rule confidence decreases below the minimum confidence threshold, or (ii) by decreasing the frequency of the rule consequent through transactions that support the rule, until the rule confidence is below the minimum threshold. The third strategy decreases the frequency of a sensitive rule, by decreasing the support of either the antecedent or the rule consequent, until either the confidence or the support lies below the minimum threshold. A basic drawback of the proposed schemes is the strong assumption that all the items appearing in a sensitive rule do not appear in any other sensitive rule. Under this assumption, hiding of the rules one at a time or altogether makes no difference. Moreover, since this work aims at hiding all the sensitive knowledge appearing in the dataset, it fails to avoid undesired side-effects, such as lost and false rules.

Verykios et al. [53] extend the work of Dasseni et al. [15] by improving and evaluating the algorithms for their performance under different sizes of input datasets and different sets of sensitive rules. Moreover, the authors propose two heuristic algorithms that incorporate the third strategy presented earlier. The first of these algorithms protects the sensitive knowledge by hiding the item having the maximum support from the minimum length transaction. The hiding of the generating itemsets of the sensitive rules is performed in a decreasing order of size and support, and in a one-by-one fashion. Similar to the first algorithm, the second algorithm first sorts the generating itemsets with respect to their size and support, and then hides them in a round-robin fashion as follows. First, for each generating itemset, a random ordering of its items and of its supporting transactions is attained. Then, the algorithm proceeds to remove the items from the corresponding transactions in a round-robin fashion, until the support of the sensitive itemset drops below the minimum support threshold. The intuition behind hiding in a round-robin fashion is fairness and the proposed algorithm (although rather naïve) serves as a baseline for conducting a series of experiments.

Oliveira and Zaïane [39] were the first to introduce multiple rule hiding approaches. The proposed algorithms are efficient and require two scans of the database, regardless of the number of sensitive itemsets to hide. During the first scan, an index file is created to speed up the process of finding the sensitive transactions and to allow for an efficient retrieval of the data. In the second scan, the algorithms sanitize the database by selectively removing the least amount of individual items that accommodate the hiding of the sensitive knowledge. An interesting novelty of this work is the fact that the proposed methodology takes into account not only the impact of the sanitization on hiding the sensitive patterns, but also the impact related to the hiding of non-sensitive knowledge. Three item restriction-based (MinFIA, MaxFIA, and IGA) algorithms are proposed that selectively remove items from sensitive transactions. The first algorithm, MinFIA, proceeds as follows. For each restrictive pattern it identifies the supporting transactions and the item having the smallest support in the pattern (called *victim item*). Then, by using a user-supplied disclosure threshold, it first sorts the identified transactions in ascending order of degree of conflict and then selects the number of transactions (among them) that need to be sanitized. Finally, from each selected transaction the algorithm removes the victim item. The MaxFIA algorithm proceeds exactly as the MinFIA with the only difference of selecting as the victim item the one that has the maximum support in the sensitive rule. Finally, IGA aims at clustering the restricted patterns into groups that share the same itemsets. By identifying overlapping clusters, the algorithm proceeds to hide the corresponding sensitive patterns at once (based on the sensitive itemsets they share) and consequently reduces the impact on the released dataset.

A more efficient approach than the one in [39] and the works of [15, 47, 48] was proposed by Oliveira and Zaïane [40]. The proposed algorithm, called SWA, is an efficient, scalable, one-scan heuristic which aims at providing a balance between the needs for privacy and knowledge discovery in ARH. It achieves to hide multiple rules in only one pass through the dataset, regardless of its size or the number of sensitive rules that need to be protected. The algorithm proceeds in five steps that are applied to every group of K transactions (thus formulating a window of size K) read from the original database. Firstly, the non-sensitive transactions are separated from the sensitive ones and copied directly to the sanitized database. For each sensitive rule, the item having the highest frequency is selected and the supporting transactions are identified. Then, a disclosure threshold, potentially different for each sensitive rule, is used to capture the severity characterizing the release of the rule. Based on this threshold, SWA computes the number of supporting transactions that need to be sanitized for each rule and then sorts them in ascending order of size. For each selected transaction, the corresponding item is removed and then the transaction is copied to the sanitized dataset. The authors present a set of computational tests to demonstrate that SWA outperforms state-of-the-art approaches in terms of concealing all the sensitive rules, while maintaining high data utility of the released dataset.

Amiri [10] proposes three effective, multiple rule hiding heuristics that outperform SWA by offering higher data utility and lower distortion, at the expense of computational cost. Although similar in the philosophy to the previous approaches, the proposed schemes do a better job in modelling the overall objective of a rule hiding algorithm. The first approach, called *Aggregate*, computes the union of the supporting transactions for all sensitive itemsets. Among them, the transaction that supports the most sensitive and the least non-sensitive itemsets is selected and expelled from the database. The same process is repeated until all the sensitive itemsets are hidden. Similarly to this approach, the *Disaggregate* approach aims at removing individual items from transactions, rather than removing the entire transaction. It achieves that by computing the union of all transactions supporting sensitive item-sets and then, for each transaction and supporting item, by calculating the number of sensitive and non-sensitive itemsets that will be affected if this item is removed from the transaction. Finally, it chooses to remove the item from the transaction that will affect the most sensitive and the least non-sensitive itemsets. The third approach, called *Hybrid*, is a combination of the previous two, since it uses *Aggregate* to iden-tify the sensitive transactions and *Disaggregate* to selectively delete items of these transactions, until the sensitive knowledge is hidden.

Wu et al. [55] propose a sophisticated methodology that removes the assumption of [15], regarding the disjoint relation among the items of the various sensitive rules. By using set theory, the authors formalize a set of constraints related to the possible side-effects of the hiding process and allow item modifications to enforce these constraints. However, the correlations among the rules can make impossible the hiding of the sensitive knowledge, without the violation of any constraints. For this reason, the user can specify which constraints she considers more significant and to relax the rest. A drawback of this approach is the simultaneous relaxation (without the user's consent) of the constraint regarding the hiding of all the sensitive itemsets. To accommodate for rule hiding, the new scheme defines a class of allowable modifications that are represented as templates and are selected in a one-by-one fashion. A *template* contains the item to be modified, the applied operation, the items to be preserved or removed from the transaction, and coverage information regarding the number of rules that are affected. Based on this, the algorithm can select and apply only the templates that are considered as beneficial since they minimize the number of side-effects.

Pontikakis et al. [43] propose two distortion-based heuristics to selectively hide the sensitive rules. On the positive side, the proposed schemes use effective data structures for the representation of the rules and effectively prioritize the selection of transactions for sanitization. However, in both algorithms the proposed hiding process may introduce a number of side-effects, either by generating ghost rules which were previously non-existent, or by eliminating existing non-sensitive rules. The first algorithm, called *Priority-based Distortion Algorithm* (PDA), reduces the confidence of a rule by reversing 1's to 0's in items belonging in its consequent. The second algorithm, called *Weight-based Sorting Distortion Algorithm* (WDA), con-centrates on the optimization of the hiding process in an attempt to achieve the least side-effects and the minimum complexity. This is achieved through the use of pri-ority values assigned to transactions based on weights. Regarding performance, the proposed schemes tend to produce hiding solutions of comparable or slightly higher quality than the algorithms in [48], by generally introducing less side-effects. How-ever, both algorithms are computationally demanding, with PDA requiring typically twice the time of the schemes in [48] to perform the hiding process.

Support-based and Confidence-based Blocking Schemes Saygin et al. [47, 48] were the first to propose the use of *unknowns* (represented as question marks in the database), instead of transforming 1's to 0's and the opposite, for the hiding of sensitive association rules. As demonstrated in [47], the use of unknowns provides a safer alternative especially in critical real life applications where the distinction between "false" and "unknown" is vital. In their work, the authors introduced three simple heuristic approaches. The first approach, relies on the reduction in the support of the generating itemsets of the rule, while the other two rely on the reduction of the rule confidence of the rule, below the minimum thresholds. The definitions of both the support and the confidence measures are extended to capture the notion of an interval instead of being crisp values, while the algorithms consider both 0 and 1 values to use for hiding (in some proportion), so that it is difficult for an adversary to conclude upon the value hidden behind a question mark. A universal *safety margin* is applied to capture how much below the minimum thresholds should the new support and confidence of a sensitive rule lie, in order to consider that the rule is safely hidden. An important contribution of this work, apart from the methodology itself, is a discussion regarding the effect of the algorithms towards hiding of the sensitive knowledge, the possibility of reconstruction of the hidden patterns by an adversary, and the importance of choosing an adequate safety margin when concealing the sensitive rules.

Wang and Jafari, in [54], propose two modification schemes that incorporate unknowns and aim at the hiding of predictive association rules, i.e., rules containing the sensitive items on their LHS. Both algorithms rely on the distortion of a portion of the database transactions to lower the confidence of the association rules. Compared to the work of Saygin et al. [47, 48], the algorithms presented in [54] require a reduced number of database scans and exhibit an efficient pruning strategy. However, by construction, they are assigned the task of hiding *all* the rules containing the sensitive items on their LHS, while the algorithms in the work of Saygin et al. can hide any specific rule. The first strategy, called ISL, decreases the confidence of a rule by increasing the support of the itemset in its LHS. The second approach, called DSR, reduces the confidence of the rule by decreasing the support of the itemset in its RHS. Both algorithms experience the *item ordering effect* under which, based on the order that the sensitive items are hidden, the produced sanitized databases are different. Moreover, the DSR algorithm seems to be more effective when the sensitive items have high support.

Pontikakis et al. [44] argue that the main disadvantage of a blocking algorithm is the fact that the dataset, apart from the blocked values (i.e., the ones replaced by unknowns), is not distorted. Thus, an adversary can disclose the hidden rules by identifying those generating itemsets that contain question marks and mine rules with a maximum confidence that lies above the minimum confidence threshold. If the number of these rules is small then the probability of identifying the sensitive ones among the discovered rules becomes high. To prohibit this threat, the authors propose a blocking algorithm that purposely creates rules that were not existent in the original dataset (a.k.a. *ghost* rules) and their generating itemsets contain unknowns. This way, the identification of the sensitive rules becomes harder, since the adversary is unable to tell which of the rules that have a maximum confidence above the minimum threshold are the sensitive, and which are the ghost ones. However, the introduction of ghost rules leads to a decrement in the data quality of the sanitized outcome. To balance the trade-off between privacy and data loss the proposed algorithm incorporates a safety margin that corresponds to the extend of sanitization that is performed in the dataset. The higher the safety margin the better the protection of the sensitive rules and the worse the data quality of the resulting dataset.

Border-based Approaches In this section, we review two border-based approaches for the hiding of sensitive rules. The work of Sun and Yu [50] was the first to introduce the process of *border revision* for the hiding of sensitive association rules. In their work, the authors propose a heuristic approach that uses the notion of the *border* (further analyzed in [31]) of the non-sensitive frequent itemsets to track the impact of altering transactions in the database. The proposed scheme, first computes the positive and the negative borders in the lattice of all itemsets and then focuses on preserving the quality of the computed borders during the hiding process. The quality of the borders directly affects the quality of the sanitized database that is produced, which can be maintained by greedily selecting those modifications that lead to minimal side-effects. In the proposed heuristic, a weight is assigned to each element of the expected positive border (which is the original positive border after it has been shaped up with the removal of the sensitive itemsets) in an attempt to quantify its vulnerability of being affected by item deletion. These weights are dynamically computed (during the sanitization process) as a function of the current support of the corresponding itemsets in the database. To reduce the support of a sensitive itemset from the negative border, the algorithm calculates the impact of the possible item deletions by computing the sum of the weights of the positive border elements that will be affected. Then, it proceeds to delete the candidate item that will have the minimal impact on the positive border.

Moustakides and Verykios [34] follow a similar approach to [50] by proposing two heuristics that use the revised positive and negative borders, produced by the removal of the sensitive itemsets and their supersets from the old frequent itemset lattice. The proposed algorithms try to remove from the database all the sensitive itemsets that belong to the revised negative border, while maintaining frequent all the itemsets of the revised positive border. For every item of a sensitive itemset, the algorithms list the set of positive border itemsets which depend on it. Then, from among all minimum border itemsets, the one with the highest support is selected as it is the one with the maximum distance from the border. This itemset, called the *max-min* itemset, determines the item through which the hiding of the sensitive itemset will incur. The proposed algorithms try to modify this item in such a way that the support of the max-min itemset is minimally affected. When hiding multiple itemsets, the algorithms perform the sanitization in a one-by-one fashion, starting from the itemsets that have lower supports. Finally, the second algorithm improves the first one and, through experimental evaluation, is shown to provide better hiding solutions than [50], in the majority of the tested settings.

Exact Approaches In this section, we review some exact approaches that have been proposed for the hiding of sensitive association rules. Exact approaches are typically capable of providing superior solutions compared to the heuristic schemes, at a high computational cost. They achieve this by formulating the sanitization process as a constraint satisfaction problem and by solving it using an integer/linear programming solver. Thus, the sanitization of the dataset is performed as an atomic operation, which avoids the local minima issues experienced by the heuristic approaches.

Menon et al., in [32], proposed a scheme that consists of an exact and a heuristic part for the hiding of sensitive frequent patterns. The exact part formulates a *Constraint Satisfaction Problem* (CSP) with the objective of identifying the minimum number of transactions that need to be sanitized for the proper hiding of all the sensitive knowledge. To avoid the *NP-hardness* issue, the authors reduce the problem size considering only the sensitive itemsets, requesting that their support remains below the minimum support threshold. The optimization process is driven by a criterion function that is inspired by the measure of accuracy [30]. Moreover, the constraints imposed in the CSP formulation capture the number of supporting transactions that need to be sanitized for the hiding of each sensitive itemset. An integer programming solver is then applied to identify the best solution of the CSP and to derive the objective. In turn, this objective is provided as input to a heuristic sanitization algorithm that is assigned the task of identifying the actual transactions within the database and performing their sanitization. An important contribution of the authors, is a discussion over the possibility of parallelization of the exact part. As demonstrated in the paper, based on the underlying properties of the dataset to be sanitized, it is possible for the produced CSP to be decomposed into parts that are solved independently. Bearing in mind the exponential complexity involving the solution of a CSP, this process can drastically reduce the required computational time for the hiding of the sensitive knowledge.

Gkoulalas-Divanis and Verykios, in [21], propose an exact approach (called *inline*) for the hiding of sensitive rules that uses the itemsets belonging to the revised positive and the revised negative borders in order to identify the candidate itemsets for sanitization. Through a set of theorems, involving existing relations among itemsets, the authors significantly reduce the set of candidates to a small fraction of its original size. The hiding process is then performed by formulating a CSP in which the status (frequent vs. infrequent) of each of the itemsets in the reduced set is controlled through a set of constraints. By using a process of *constraints degree reduction*, all the constraints in the CSP become linear and have no coefficients. Moreover, all the variables involved in the CSP are of binary nature. These facts allow solving the CSP by using binary integer programming. The provided solution is proved to lead to an exact hiding of the sensitive patterns. A heuristic approach that relaxes the initial CSP to allow for the identification of a good solution, is applied when the CSP is infeasible.

Another exact approach, called *hybrid*, was introduced by Gkoulalas-Divanis and Verykios in [22]. The goal of the hybrid approach is to allow the hiding algorithm to identify exact solutions in a wider range of problem instances than those that could be handled by the inline algorithm. The hybrid approach conceals sensitive frequent itemsets by generating a small extension of the original database, thereby introducing new transactions, and carefully controlling the items that are supported by the transactions in the extension. To control the items (i.e., 0/1s) in transactions, a similar approach to the one that is enforced by the inline algorithm is employed, while special care is taken to ensure the validity of the transactions in the extended part.

A two-phase iterative process that improves the functionality of the inline approach was proposed by Gkoulalas-Divanis and Verykios in [23]. The process consists of two phases that are executed in an iterative fashion until either (i) an exact solution of the given problem instance is found, or (ii) a pre-specified number of phase iterations (called *oscillations*) ℓ have taken place. In the first phase, the hiding algorithm uses the inline approach in an effort to conceal the sensitive knowledge without side-effects. If it succeeds, then the process terminates. Otherwise, the algorithm proceeds to the second phase, which implements the dual counterpart of the inline algorithm. In this phase, the hiding algorithm selectively removes inequalities from the infeasible CSP, until the CSP becomes feasible, and then solves the CSP to attain the sanitized dataset. This dataset is bound to suffer from side-effects (due to the removal of constraints) and the purpose of the second phase is to recover the lost itemsets by increasing their support and making them frequent again.

3.4 Metrics and Performance Analysis

In this section, we present two categories of measures related to the performance of an association rule hiding algorithm. The first category consists of measures that can either be optimized by a hiding scheme in the course of its execution, or be adopted to allow for a fair comparison among different hiding schemes under a unified framework. The measures belonging in this category are called *internal* and were proposed by Oliveira et al. [41]. They are classified as either *data sharing*-based or *pattern sharing*-based. The data sharing-based measures quantify the extent of side-effects regarding sensitive association rules that failed to be hidden, legitimate rules that were accidentally missed, and artifactual association rules that were created by the sanitization process. On the other hand, the pattern sharing-based measures quantify the extent of side-effects regarding non-sensitive association rules that were lost or sensitive rules that were improperly hidden and can be easily be recovered through the use of inference channels. Furthermore, we proceed to present another set of metrics, which measure external parameters such as the behavior of the algorithm when applied to large datasets, its computational speed, and so on and so forth. The measures of this category are called *external* and were proposed by Bertino et al. [12].

The proposed *data-sharing based measures* are the following:

(a) **Hiding Failure (HF).** This measure quantifies the percentage of the sensitive patterns that remain exposed in the sanitized dataset. It is defined as the fraction of the restrictive association rules that appear in the sanitized database divided by the ones that appeared in the original dataset. Formally,

$$\text{HF} = \frac{|R_P(U')|}{|R_P(U)|}$$

where $R_P(U')$ corresponds to the sensitive rules discovered in the sanitized dataset U', $R_P(U)$ to the sensitive rules appearing in the original dataset U and $|X|$ is the size of set X. Ideally, the hiding failure should be 0 %.

(b) **Misses Cost (MC).** This measure quantifies the percentage of the non-restrictive patterns that are hidden as a side-effect of the sanitization process. It is computed as

$$MC = \frac{|\tilde{R}_P(U)| - |\tilde{R}_P(U')|}{|\tilde{R}_P(U)|}$$

where $\tilde{R}_P(U)$ is the set of all non-sensitive rules in the original database U and $\tilde{R}_P(U')$ is the set of all non-sensitive rules in the sanitized database U'. As one can notice, there exists a compromise between the misses cost and the hiding failure, since the more sensitive association rules one needs to hide, the more legitimate association rules one is expected to miss.

(c) **Artifactual Patterns (AF).** This measure quantifies the percentage of the discovered patterns that are artifacts. AF is computed as follows:

$$AP = \frac{|P'| - |P \cap P'|}{|P'|}$$

where P is the set of association rules discovered in the original database U and P' is the set of association rules discovered in U'.

(d) **Dissimilarity (Diss).** The measure of dissimilarity quantifies the difference between the original and the sanitized datasets by comparing their histograms, where the horizontal axis contains the items in the dataset and the vertical axis corresponds to their frequencies. It is calculated as follows:

$$Diss(U, U') = \frac{1}{\sum_{i=1}^{n} f_U(i)} \times \sum_{i=1}^{n} [f_U(i) - f_{U'}(i)]$$

where $f_X(i)$ represents the frequency of the i−th item in the dataset X, and n is the number of distinct items in the original dataset D.

The proposed *pattern-sharing based metrics* are the following:

(a) **Side-Effect Factor (SEF).** Similarly to the measure of misses cost, the side-effect factor is used to quantify the amount of non-sensitive association rules that are removed as an effect of the sanitization process. It is defined as follows:

$$SEF = \frac{|P| - (|P'| + |R_P(U)|)}{|P| - |R_P|}$$

(b) **Recovery Factor (RF).** This measure expresses the possibility of an adversary to recover a sensitive rule based on the non-sensitive ones. The recovery factor of a pattern takes into account the existence of its subsets. If *all* the subsets of a sensitive rule can be recovered from the sanitized dataset, then the recovery of the rule itself is possible, thus it is assigned an RF value of 1; otherwise RF = 0. However, this measure is not certain since, for instance, an adversary may not learn an itemset despite knowing its subsets.

Bertino et al. [12] propose a set of measures that are directly related to the performance of a hiding algorithm as far as external parameters are concerned. These *process performance measures* are clustered into four categories, as follows:

(a) **Efficiency.** This category consists of measures that quantify the ability of a privacy preserving algorithm to efficiently use the available resources and execute with good performance. Efficiency is measured in terms of CPU-time, space requirements (related to the memory usage and the required storage capacity) and communication requirements.

(b) **Scalability.** This category consists of measures that evaluate how effectively the privacy preserving technique handles increasing sizes of the data from which information needs to be mined and privacy needs to be ensured. Scalability is measured based on the decrease in the performance of the algorithm or the increase of the storage requirements along with the communications cost (if in a distributed setting), when the algorithm is provided with larger datasets.

(c) **Data Quality.** The data quality of a privacy preservation algorithm depends on two parameters. There are the quality of the dataset after the sanitization process, and the quality of the data mining results when applied to this dataset, compared to the ones attained when using the original dataset. Among the various possible measures for the quantification of the data quality, the most preferable are: (i) *accuracy*, which measures the proximity of a sanitized value to the original one and is closely related to the information loss resulting from the hiding strategy, (ii) *completeness*, which is used to evaluate the degree of missed data in the sanitized database and (iii) *consistency*, which is related to the relationships that must continue to hold among the different fields of a data item or among data items in a sanitized database. Examples of data quality measures are Diss (presented earlier) and Kullback–Leibler (KL) divergence.

(d) **Privacy Level.** This category consists of measures that estimate the degree of uncertainty according to which, the protected information can still be predicted. Measures, such as the information entropy, the level of privacy and the J-measure [12], are some among the possible metrics that one can apply to quantify the privacy level attained by a hiding scheme.

4   Cryptographic Methods

Over the years, many data mining protocols have been designed to mine distributed data that reside in different data warehouses. In those protocols, data are generally assumed to be either *vertically* or *horizontally* partitioned. Table 15.1 shows a trivial example of two different data partitioning schemes for a simple transaction (binary) dataset U, consisting of four attributes.

Table 15.1 A binary dataset along with different partitioning schemes

(a) Original dataset **(b) Vertically partitioned** **(c) Horizontally partitioned**

Tr#	a	b	c	d
R_1	1	0	1	0
R_2	1	0	1	0
R_3	0	0	0	0
R_4	0	1	0	1
R_5	1	1	1	1
R_6	1	1	0	1
R_7	0	0	0	0

Tr#	a	b
R_1	1	0
R_2	1	0
R_3	0	0
R_4	0	1
R_5	1	1
R_6	1	1
R_7	0	0

Tr#	c	d
R_1	1	0
R_2	1	0
R_3	0	0
R_4	0	1
R_5	1	1
R_6	0	1
R_7	0	0

Tr#	a	b	c	d
R_1	1	0	1	0
R_2	1	0	1	0
R_3	0	0	0	0

Tr#	a	b	c	d
R_4	1	1	0	1
R_5	0	0	0	0
R_6	1	1	1	1
R_7	1	1	1	1

In the case of vertically partitioned data, shown in Table 15.1b, we assume that different sites collect information about the same set of entities, but they collect different feature sets. For example, both a university pay roll and the university's student health center may collect information about a student.

In the case of horizontally partitioned data, shown in Table 15.1c, different sites collect the same set of information about different entities. For example, different credit card companies may collect credit card transactions of different individuals.

In context of association rules mining [5], we may try to mine association rules on the horizontally partitioned data and/or vertically partitioned data. In the case of horizontally partitioned data, the traditional ARM problem can be stated as follows. Consider a set of sites S. Each site S_i ($1 \leq i \leq n$) has a private transaction database U_i where the entire database U is assumed to be of the form $U = U_1 \cup U_2 \cup \cdots \cup U_n$. The itemset C_x has *local support count* of $C_x.sup_i$ at site S_i, if and only if $C_x.sup_i$ of the transactions contain C_x. The *global support count* of C_x is given as $C_x.sup = \sum_{i=1}^{n} C_x.sup_i$. An itemset C_x is *globally supported* if $C_x.sup \geq s \times \left(\sum_{i=1}^{n} |U_i| \right)$. Similarly, the *global confidence* of a rule $C_x \Rightarrow C_y$ can be given as $\{C_x \cup C_y\}.sup / C_x.sup$.

In the case of vertically partitioned data, each U_i that resides in site S_i contains a subset of the columns that represents different items. To compute $C_x.sup$, where $C_x \subseteq \mathcal{I}$, we need to somehow combine those columns. If C_x is vertically partitioned such that sites $S_{i_1} \ldots S_{i_k}$ hold the information about the items that form C_x (i.e., $C_x = C_{x_{i_1}} \cup C_{x_{i_2}} \cup \ldots C_{x_{i_k}}$), to compute $C_x.sup$ we need to compute $\sum_{T \in U} \left(\prod_{j=1}^{k} \left(I_{C_{x_{i_j}} \subseteq T} \right) \right)$, where $I_{C_{x_{i_j}} \subseteq T}$ is the indicator function that represents whether transaction $T \in U$ contains itemset $C_{x_{i_j}}$ or not.

The main challenge arises if those databases U_i belong to different organizations and direct sharing of U_i is not feasible due to privacy concerns. For example, different credit card companies may not be able to share their data due to financial privacy regulations. Computing association rules without disclosing individual transactions is straightforward in the case of horizontally partitioned data. For example, we can compute the global support and confidence of an association rule $C_x C_y \Rightarrow C_z$

knowing only the local supports of $C_x C_y$ and $C_x C_y C_z$, and the size of each database:

$$support_{C_x C_y \Rightarrow C_z} = \frac{\sum_{i=1}^{sites} support\_count_{C_x C_y C_z}(i)}{\sum_{i=1}^{sites} database\_size(i)}$$

$$support_{C_x C_y} = \frac{\sum_{i=1}^{sites} support\_count_{C_x C_y}(i)}{\sum_{i=1}^{sites} database\_size(i)}$$

$$confidence_{C_x C_y \Rightarrow C_z} = \frac{support_{C_x C_y \Rightarrow C_z}}{support_{C_x C_y}}$$

The above approach protects individuals' data privacy, but it does require that each site discloses what rules it supports, and how much it supports each potential global rule. What if this information is sensitive? To address these challenges, cryptographic techniques have been used to develop privacy-preserving distributed ARM techniques. Below, we provide an overview of basic protocols that have been applied in different data partitioning scenarios.

4.1 Horizontally Partitioned Data

To construct a privacy-preserving ARM algorithm for horizontally partitioned data, several cryptographic sub-protocols may need to be used. Before, we summarize the algorithm proposed in [26] for three or more parties[1], in what follows we discuss a fast algorithm that has been proposed for distributed mining of association rules.

A fast algorithm for distributed ARM is given in Cheung et. al.[13]. Their procedure for Fast Distributed Mining of association rules (FDM) is summarized below.

The set of frequent itemsets $F_{(k)}$ consists of all k-itemsets that are globally supported. The set of locally frequent itemsets, $LF_{i(k)}$, consists of all k-itemsets supported locally at site S_i. $GF_{i(k)} = F_{(k)} \cap LF_{i(k)}$ is the set of globally frequent k-itemsets locally supported at site S_i. The aim of distributed ARM is to find the sets $F_{(k)}$, for all $k > 1$, and the support counts for these itemsets, and from this to compute association rules with the specified minimum support and confidence.

1 **Candidate Sets Generation**: Generate candidate sets $CG_{i(k)}$ based on $GF_{i(k-1)}$, itemsets that are supported by the S_i at the (k–1)-th iteration, using the classic a-priori candidate generation algorithm. Each site generates candidates based on the intersection of globally frequent (k-1) itemsets and locally frequent (k-1) itemsets.
2 **Local Pruning**: For each $C_x \in CG_{i(k)}$, scan the database U_i at S_i to compute $C_x.sup_i$. If C_x is locally frequent at S_i, it is included in the $LF_{i(k)}$ set. Please note that if C_x is supported globally, it will be supported in one site.

---

[1] Please see the two party case discussion given in [26].

3 **Support Count Exchange**: $LF_{i(k)}$ are broadcast, and each site computes the local support for the items in $\cup_i LF_{i(k)}$.

4 **Broadcast Mining Results**: Each site broadcasts the local support for itemsets in $\cup_i LF_{i(k)}$. From this, each site is able to compute $F_{(k)}$.

Privacy-preserving Distributed ARM In the privacy-preserving version of the FDM algorithm, we desire that information disclosure is limited. Specifically, no site should be able to learn the contents of transactions belonging to any other site, what rules are supported by any other site, or the specific value of support/confidence for any rule at any other site, unless that information is revealed by knowledge of one's own data and the final result (e.g., if a rule is supported globally but not at one's own site, we can deduce that at least one other site supports the rule.) In this basic version of the protocol, no collusion is assumed.

The method described in [26] follows the FDM algorithm given above, with special protocols for replacing the broadcasts of $LF_{i(k)}$ and the support count of items in $LF_{(k)}$. In the FDM algorithm, step 3 reveals the frequent itemsets supported by each site. To accomplish this without revealing what each site supports, we may instead exchange locally frequent itemsets in a way that obscures the source of each itemset. The main idea is that each site encrypts the locally supported itemsets, along with enough "fake" itemsets to hide the actual number supported. This is achieved by using *secure union protocols* (e.g., see [26]). Using different cryptographic tools such as *homomorphic encryption*, such protocols can compute the union of sets belonging to different parties, securely. For example, in our context, secure union protocols could be used to compute $\cup_i LF_{i(k)}$, without revealing which sites supports which itemsets and how many sites support a given itemset.

In some cases, some secure union protocols (e.g., the one given in [26]) may disclose extra information for efficiency purposes. For example, if we deem leakage of the number of commonly supported itemsets as acceptable, it can be proven that the secure union protocol described in [26] is secure under certain cryptographic definitions. Such proofs usually show that everything else seen during the protocol can be simulated based on the leaked information and the final set union. This technique can be quite powerful for generating reasonably secure and efficient protocols. A protocol that is proved not to reveal anything other than the required result and information deemed not privacy-threatening could be sufficient for many practical purposes. This approach is used to prove that the set union protocol given in [26] reveals only the union of locally frequent itemsets and a clearly bounded set of innocuous information.

Secure set union protocols give the full set of locally frequent itemsets $LF_{(k)}$. We still, however, need to determine which of these itemsets are supported globally. Step 4 of the FDM algorithm forces each site to reveal its own support count for every itemset in $LF_{(k)}$. All we need to know for each itemset $C_x \in LF_{(k)}$, is if $C_x.sup \geq s\% \times |U|$. The following observation allows us to reduce this to a comparison against a sum of local values (the *excess support* at each site):

$$C_x.sup \geq s * |U| = s * \left(\sum_{i=1}^{n} |U_i| \right)$$

$$\sum_{i=1}^{n} C_x.sup_i \geq s * \left(\sum_{i=1}^{n} |U_i| \right)$$

$$\sum_{i=1}^{n} (C_x.sup_i - s * |U_i|) \geq 0$$

Therefore, checking for support is equivalent to checking if $\sum_{i=1}^{n} (C_x.sup_i - s * |U_i|) \geq 0$. The challenge is to do this without revealing $C_x.sup_i$ or $|U_i|$. This is accomplished by first computing the sum securely, and applying a secure comparison at the end.

The first site generates a random number x_r for each itemset C_x, adds that number to its $(C_x.sup_i - s * |U_i|)$, and sends it to the next site. (All arithmetic is *mod m*, where $m \geq 2 * |U|$, for security purposes.) The random number masks the actual excess support, so the second site learns nothing about the first site's actual database size or support. The second site adds its excess support and sends the value on. The random value now hides both support counts. The last site in the chain now has $\sum_{i=1}^{n} (C_x.sup_i - s * |U_i|) + x_r (\bmod m)$.

Since the total database size is $|U| \leq m/2$, negative summation will be mapped to some number that is bigger than (or equal to) $m/2$. ($-k = m - k \bmod m$.) The last site needs to test if this sum minus $x_r \bmod m$ is less than $m/2$. This can be done securely using Yao's generic method [56]. Clearly this algorithm is secure as long as there is no collusion, as no site can distinguish what it receives from a random number. Alternatively, the first site can simply send x_r to the last site. The last site learns the actual excess support, but does not learn the support values for any single site. In addition, if we consider the excess support to be a valid part of the global result, this method is still secure.

The above basic protocol can be extended to provide privacy in the context of collusions [26]. In addition, efficiency could be improved by using fast union protocols [51].

4.2 Vertically Partitioned Data

The FDM algorithm that was described above can be modified to address the case of vertically partitioned data. First of all, each site can compute all the locally frequent itemsets that can be supported based on the items belonging to the local site. Later on, to check whether an itemset is globally frequent, vertically partitioned data need to be combined in order to compute the $\sum_{T \in U} \left(\prod_{j=1}^{k} \left(I_{C_{x_{i_j}} \subseteq T} \right) \right)$, where $I_{C_{x_{i_j}} \subseteq T}$ is the indicator function that represents whether transaction $T \in U$ contains itemset $C_{x_{i_j}}$ or not.

In the context of two parties, the above equation becomes a simple dot product. For example, suppose two parties wish to determine if an itemset C_x has the minimum support in the data set, but neither party has data on the entire set C_x. Instead, they have sets C_{x_1} and C_{x_2}, which are a disjoint cover of C_x. In order to compute this, they could each form a vector (v_1, v_2) of size n (where n is the total number of rows) of zeros and ones. For the vector v_k, the value at i is 1 if row i contains all members of C_{x_k}, and 0 otherwise. The two parties could then compute the dot product of these two vectors. This would yield the total number of rows containing the set X. The support is easily found by dividing by the total number of rows [52]. This implies that using a secure dot product protocol (e.g., [25]), we can get a two party privacy-preserving ARM algorithm for vertically partitioned data.

Secure dot product protocols could be easily obtained using additively homomorphic public key encryption. A secure public key cryptosystem is called *additive homomorphic* [42] if it satisfies the following requirements:

- Let $E_{pk}(.)$ denote the encryption function with public key pk and $D_{pr}(.)$ denote the decryption function with private key pr. Given the encryption of m_1 and m_2, $E_{pk}(m_1)$ and $E_{pk}(m_2)$, there exists an efficient algorithm to compute the public key encryption of $m_1 + m_2$, denoted by $E_{pk}(m_1 + m_2) := E_{pk}(m_1) +_h E_{pk}(m_2)$.
- Given a constant k and the encryption of m_1, $E_{pk}(m_1)$, there exists an efficient algorithm to compute the public key encryption of $k \cdot m_1$, denoted by $E_{pk}(km_1) := k \times_h E_{pk}(m_1)$.

Using such an additive homomorphic encryption scheme, we can easily compute the dot product of vectors securely. Basically, site S_1, creates a public key pk and private key pr pair. Later on, for all elements of v_1, S_1 computes $E_{pk}(v_{1j})$ and this encrypted vector is send to site S_2. Site S_2 keeps an encrypted counter C and adds $E_{pk}(v_{1j})$ using $+_h$ operation, if v_{2j} is 1. Finally, before sending the C to site S_1, S_2 generates a random value r and computes $C +_h E_{pk}(r)$ and sends this blinded value to S_1. S_1 can decrypt this value to learn the dot product result blinded with random value. Now S_1 and S_2 can use the secure comparison protocol to check whether the support threshold condition is satisfied.

Extending the above protocol to the multi-party case requires securely computing $\prod_{j=1}^{k} \left(I_{C_{x_{i_j}} \subseteq T} \right)$ for each transaction. Please note that this is equivalent to computing $\bigwedge_{j=1}^{k} \left(I_{C_{x_{i_j}} \subseteq T} \right)$, which could be easily achieved by using the secure logical $\bigwedge$ protocol [27].

One issue with the vertically partitioned data case is that all the protocols require $O(n)$ cryptographic operations, where $n = |U|$. This could be especially problematic for big data scenarios where a database contains billions of rows. Recent work has tried to address this problem by developing secure approximate dot product protocols that can provide two orders of magnitude improvement [28, 37]. These protocols (e.g., [37]) basically leverage dimensionality reduction techniques to reduce the dimension of the vectors that are provided as input to secure dot product protocols. Still more work needs to be done to provide secure protocols that can scale to billions of transactions.

5 Conclusions

Privacy preserving data mining is a new body of research focusing on the implications originating from the application of data mining algorithms to large public
databases. In this study, we focussed on several aspects of privacy, including input
privacy, output privacy and cryptographic privacy. For input privacy, we studied the
effects of privacy-preserving data publication on privacy. For output privacy, we
have surveyed a research direction that investigates how sensitive association rules
can escape the scrutiny of malevolent data miners by modifying certain values in
the database. We have also presented a thorough analysis and comparison of the
surveyed approaches, as well as a classification of association rule hiding algorithms
to facilitate the organization in our presentation. Moreover, in this chapter we studied aspects of cryptographic privacy with focus on methods for privacy-preserving
association rule mining over horizontally and vertically partitioned data.

References

1. O. Abul, F. Bonchi, and F. Giannotti. Hiding sequential and spatiotemporal patterns. *IEEE Transactions on Knowledge and Data Engineering*, 22(12):1709–1723, 2010.
2. N. Adam and J. Worthmann. Security control methods for statistical databases: A comparative study. *ACM Computing Surveys*, 21(4):515–556, Dec. 1989.
3. D. Agrawal and C. Aggarwal. On the design and quantification of privacy preserving data mining algorithms. In *Proceedings of the 20th ACM SIGMOD-SIGACT-SIGART Symposium on Principles of Database Systems*, PODS, pages 247–255, 2001.
4. S. Agrawal and J. Haritsa. A framework for high-accuracy privacy-preserving mining. In *Proceedings of the 21st IEEE International Conference on Data Engineering*, ICDE, pages 193–204, 2005.
5. R. Agrawal and R. Srikant. Fast algorithms for mining association rules in large databases. In *Proceedings of the 20th International Conference on Very Large Data Bases*, VLDB, pages 487–499, 1994.
6. R. Agrawal and R. Srikant. Privacy-preserving data mining. *ACM SIGMOD Record*, 29(2):439–450, May 2000.
7. C. Aggarwal and P. Yu. A condensation approach to privacy preserving data mining. In *Proceedings of the 9th International Conference on Advances in Database Technology*, EDBT, pages 183–199, 2004.
8. R. Agrawal, T. Imieliński, and A. Swami. Mining association rules between sets of items in large databases. In *Proceedings of the 1993 ACM SIGMOD International Conference on Management of Data*, SIGMOD, pages 207–216, 1993.
9. S. Agrawal, V. Krishnan, and J. Haritsa. On addressing efficiency concerns in privacy-preserving mining. In *Proceedings of the 2004 Database Systems for Advanced Applications*, DASFAA, pages 113–124, 2004.
10. A. Amiri. Dare to share: Protecting sensitive knowledge with data sanitization. *Elsevier Decision Support Systems*, 43(1):181–191, Feb. 2007.
11. M. Atallah, E. Bertino, A. Elmagarmid, M. Ibrahim, and V. Verykios. Disclosure limita-
tion of sensitive rules. In *Proceedings of the 1999 IEEE Workshop on Knowledge and Data Engineering Exchange*.
12. E. Bertino, I. Fovino, and L. Provenza. A framework for evaluating privacy preserving data mining algorithms. *Springer Data Mining and Knowledge Discovery*, 11(2):121–154, 2005.

13. D. Cheung, J. Han, V. Ng, A. Fu, and Y. Fu. A fast distributed algorithm for mining associa-
    tion rules. In *Proceedings of the IEEE International Conference on Parallel and Distributed
    Information Systems*, volume 1, pages 31–43, 1996.
14. C. Clifton and D. Marks. Security and privacy implications of data mining. In *ACM SIGMOD
    Workshop on Research Issues on Data Mining and Knowledge Discovery*, pages 15–19, 1996.
15. E. Dasseni, V. Verykios, A. Elmagarmid, and E. Bertino. Hiding association rules by using
    confidence and support. In *Proceedings of the 4th International Workshop on Information
    Hiding*, IHW, pages 369–383, 2001.
16. D. Denning. *Cryptography and Data Security*. Addison-Wesley, 1982.
17. A. Evfimievski, J. Gehrke, and R. Srikant. Limiting privacy breaches in privacy preserving data
    mining. In *Proceedings of the 22nd ACM SIGMOD-SIGACT-SIGART Symposium on Principles
    of Database Systems*, PODS, pages 211–222, 2003.
18. A. Evfimievski, R. Srikant, R. Agrawal, and J. Gehrke. Privacy preserving mining of association
    rules. In *Proceedings of the 2002 ACM SIGKDD Conference on Knowledge Discovery and Data
    Mining*, KDD, pages 217–228, 2002.
19. C. Farkas and S. Jajodia. The inference problem: A survey. *ACM SIGKDD Explorations*,
    4(2):6–11, Dec. 2002.
20. A. Gkoulalas-Divanis and G. Loukides. Revisiting sequential pattern hiding to enhance utility.
    In *Proceedings of the 2011 ACM SIGKDD Conference on Knowledge Discovery and Data
    Mining*, pages 1316–1324, 2011.
21. A. Gkoulalas-Divanis and V. Verykios. An integer programming approach for frequent item-
    set hiding. In *Proceedings of the 15th ACM International Conference on Information and
    Knowledge Management*, CIKM, pages 748–757, 2006.
22. A. Gkoulalas-Divanis and V. Verykios. Exact knowledge hiding through database extension.
    IEEE Transactions on Knowledge and Data Engineering, 21(5):699–713, 2009.
23. A. Gkoulalas-Divanis and V. Verykios. Hiding sensitive knowledge without side effects.
    Knowledge and Information Systems, 20(3):263–299, 2009.
24. O. Goldreich. Secure multi-party computation. www.wisdom.weizmann.ac.il/~oded/pp.html,
    1998.
25. I. Ioannidis, A. Grama, and M. Atallah. A secure protocol for computing dot-products in
    clustered and distributed environments. In *International Conference on Parallel Processing*,
    pages 379–384, 2002.
26. M. Kantarcioglu and C. Clifton. Privacy-preserving distributed mining of association rules
    on horizontally partitioned data. *IEEE Transactions on Knowledge and Data Engineering*,
    16(9):1026–1037, Sept. 2004.
27. M. Kantarcioglu, R. Nix, and J. Vaidya. An efficient approximate protocol for privacy-
    preserving association rule mining. In *PAKDD*, pages 515–524, 2009.
28. M. Kantarcioglu, W. Jiang, Y. Liu, and B. Malin. A cryptographic approach to securely share
    and query genomic sequences. *IEEE Transactions on Information Technology in Biomedicine*,
    12(5):606–617, 2008.
29. H. Kargupta, S. Datta, Q. Wang, and K. Sivakumar. On the privacy preserving properties of
    random data perturbation techniques. In *Proceedings of the 3rd IEEE International Conference
    on Data Mining*, ICDM, pages 99–106, 2003.
30. G. Lee, C.-Y. Chang, and A. Chen. Hiding sensitive patterns in association rules mining. In *Pro-
    ceedings of the 28th IEEE International Conference on Computer Software and Applications*,
    COMPSAC, pages 424–429, 2004.
31. H. Mannila and H. Toivonen. Levelwise search and borders of theories in knowledge discovery.
    Data Mining and Knowledge Discovery, 1(3):241–258, 1997.
32. S. Menon, S. Sarkar, and S. Mukherjee. Maximizing accuracy of shared databases when
    concealing sensitive patterns. *Information Systems Research*, 16(3):256–270, Sept. 2005.
33. T. Mielikainen. On inverse frequent set mining. In *Proceedings of the 2nd Workshop on Privacy
    Preserving Data Mining*, pages 18–23, 2003.
34. G. Moustakides and V. Verykios. A max-min approach for hiding frequent itemsets. In
    Proceedings of the 6th IEEE Conference on Data Mining Workshops, pages 502–506,
    Dec. 2006.

35. J. Natwichai, X. Li, and M. Orlowska. Hiding classification rules for data sharing with privacy preservation. In *Proceedings of the 7th International Conference on Data Warehousing and Knowledge Discovery*, DaWaK, pages 468–477, 2005.

36. J. Natwichai, X. Li, and M. Orlowska. A reconstruction-based algorithm for classification rules hiding. In *Proceedings of the 17th Australasian Database Conference*, ADC, pages 49–58, 2006.

37. R. Nix, M. Kantarcioglu, and K. Han. Approximate privacy-preserving data mining on vertically partitioned data. In *DBSec*, pages 129–144, 2012.

38. D. O'Leary. Knowledge discovery as a threat to database security. In *Proceedings of the 1st International Conference on Knowledge Discovery in Databases*, KDD, pages 507–516, 1991.

39. S. Oliveira and O. Zaïane. Privacy preserving frequent itemset mining. In *Proceedings of the IEEE international Conference on Privacy, Security and Data Mining*, CRPIT, pages 43–54, 2002.

40. S. Oliveira and O. Zaïane. Protecting sensitive knowledge by data sanitization. In *Proceedings of the 3rd IEEE International Conference on Data Mining*, ICDM, pages 613–616, 2003.

41. S. Oliveira and O. Zaïane. A unified framework for protecting sensitive association rules in business collaboration. *International Journal of Business Intelligence and Data Mining*, 1(3):247–287, Mar. 2006.

42. P. Paillier. Public-key cryptosystems based on composite degree residuosity classes. In *Proceedings of the 17th International Conference on Theory and Application of Cryptographic Techniques*, EUROCRYPT, pages 223–238, 1999.

43. E. Pontikakis, A. Tsitsonis, and V. Verykios. An experimental study of distortion-based techniques for association rule hiding. In *Proceedings of the 18th International Conference on Database Security*, volume 144 of *DBSEC*, pages 325–339, 2004.

44. E. Pontikakis, Y. Theodoridis, A. Tsitsonis, L. Chang, and V. Verykios. A quantitative and qualitative ANALYSIS of blocking in association rule hiding. In *Proceedings of the 2004 ACM Workshop on Privacy in the Electronic Society*, WPES, pages 29–30, 2004.

45. S. Rizvi and J. Haritsa. Maintaining data privacy in association rule mining. In *Proceedings of the 28th International Conference on Very Large Data Bases*, VLDB, pages 682–693, Aug. 2002.

46. P. Samarati and L. Sweeney. Generalizing data to provide anonymity when disclosing information. In *Proceedings of the 17th ACM SIGACT-SIGMOD-SIGART Symposium on Principles of Database Systems*, PODS, page 188, 1998.

47. Y. Saygin, V. Verykios, and C. Clifton. Using unknowns to prevent discovery of association rules. *ACM SIGMOD Record*, 30(4):45–54, Dec. 2001.

48. Y. Saygin, V. Verykios, and A. Elmagarmid. Privacy preserving association rule mining. In *Proceedings of the 12th International Workshop on Research Issues in Data Engineering*, RIDE, pages 151–158, Feb. 2002.

49. A. Shoshani. Statistical databases: Characteristics, problems, and some solutions. In *Proceedings of the 8th International Conference on Very Large Data Bases*, VLDB, pages 208–222, Sept. 1982.

50. X. Sun and P. Yu. A border-based approach for hiding sensitive frequent itemsets. In *Proceedings of the 5th IEEE International Conference on Data Mining*, ICDM, 2005.

51. T. Tassa. Secure mining of association rules in horizontally distributed databases. *CoRR*, abs/1106.5113, 2011.

52. J. Vaidya and C. Clifton. Privacy preserving association rule mining in vertically partitioned data. In *Proceedings of the 8th ACM SIGKDD International Conference on Knowledge Discovery and Data Mining*, KDD, pages 639–644, 2002.

53. V. Verykios, A. Elmagarmid, E. Bertino, Y. Saygin, and E. Dasseni. Association rule hiding. *IEEE Transactions on Knowledge and Data Engineering*, 16(4):434–447, 2004.

54. S.-L. Wang and A. Jafari. Using unknowns for hiding sensitive predictive association rules. In *Proceedings of the 2005 IEEE International Conference on Information Reuse and Integration*, IRI, pages 223–228, 2005.

55. Y.-H. Wu, C.-M. Chiang, and A. Chen. Hiding sensitive association rules with limited side effects. *IEEE Transactions on Knowledge and Data Engineering*, 19(1):29–42, 2007.
56. A. C. Yao. Protocols for secure computations. *IEEE Symposium on Foundations of Computer Science*, pages 160–164, 1982.
57. N. Zhang, S. Wang, and W. Zhao. A new scheme on privacy preserving association rule mining. In *Proceedings of the 2004 Conference on Knowledge Discovery in Databases*, PKDD, pages 484–495, 2004.

Chapter 16
Frequent Pattern Mining Algorithms for Data Clustering

Arthur Zimek, Ira Assent and Jilles Vreeken

Abstract Discovering clusters in subspaces, or subspace clustering and related clustering paradigms, is a research field where we find many frequent pattern mining related influences. In fact, as the first algorithms for subspace clustering were based on frequent pattern mining algorithms, it is fair to say that frequent pattern mining was at the cradle of subspace clustering—yet, it quickly developed into an independent research field.

In this chapter, we discuss how frequent pattern mining algorithms have been extended and generalized towards the discovery of local clusters in high-dimensional data. In particular, we discuss several example algorithms for subspace clustering or projected clustering as well as point out recent research questions and open topics in this area relevant to researchers in either clustering or pattern mining.

Keywords Subspace clustering · Monotonicity · Redundancy

1 Introduction

Data clustering is the task of discovering groups of objects in a data set that exhibit high similarity. Clustering is an unsupervised task, in that we do not have access to any additional information besides some geometry of the data, usually represented by some distance function. Useful groups should consist of objects that are more similar to each other than to objects assigned to other groups. The goal of the clustering results is that it provides information for the user regarding different categories of objects that the data set contains.

A. Zimek (✉)
Ludwig-Maximilians-Universität München, Munich, Germany
e-mail: zimek@dbs.ifi.lmu.de

I. Assent
Department of Computer Science, Aarhus University, Aarhus, Denmark
e-mail: ira@cs.au.dk

J. Vreeken
Max-Planck Institute for Informatics and Saarland University, Saarbrücken, Germany
e-mail: jilles@mpi-inf.mpg.de

C. C. Aggarwal, J. Han (eds.), *Frequent Pattern Mining*,
DOI 10.1007/978-3-319-07821-2_16, © Springer International Publishing Switzerland 2014

As there are many different intuitions on how objects can be similar, there exist many different clustering algorithms for formalizing these intuitions, and extracting such clusters from data [43–45, 51]. There are two main approaches to clustering. On the one hand we find so-called *partitional* algorithms [26, 49, 55, 56], where similarity of objects is directly expressed in a notion of spatial closeness. For example, a smaller Euclidean distance between two points than between other pairs of points in Euclidean space makes them relatively similar. On the other hand we have *density-based* approaches [8, 20, 28, 39, 40, 75, 77], where similarity is expressed in terms of density-connectivity of points. That is, points that find themselves in a densely populated area in the data space are said to be 'connected' and should be assigned to the same cluster, whereas areas of relatively low density separate different clusters.

An important point to note for unsupervised learning in general, and clustering specifically, is that the cluster structure of the data—and hence that discovered by a particular clustering algorithm—does not necessarily have to correlate with class label annotations: clusters 'simply' identify structure that exists in the data [29, 36]. This means both that clustering requires methods different from classification, as well as that for evaluating clusters we cannot rely just on class labels.

Over the last 15 years, a lot of research effort has been invested to develop clustering algorithms that can handle high-dimensional data. Compared to traditional data with only few attributes, high-dimensional data incur particular challenges, most prominently the difficulty of assessing the similarity of objects in a meaningful manner. These issues are generally known as the 'curse of dimensionality'. Important aspects of this infamous 'curse' and its consequences for clustering (and related tasks) have been discussed in various studies, surveys, and overview articles [4, 9, 17, 18, 27, 30, 41, 42, 50, 52, 76, 83, 85].

A special family of adaptations of clustering approaches to high-dimensional data is known as 'subspace clustering'. Here the idea is that clusters do not necessarily exhibit similarity over all attributes, but that their similarity may be restricted to subsets of attributes; the other attributes are not relevant to the cluster structure. In effect, there is a need for algorithms that can measure similarity of objects, and hence detect clusters, over *subspaces*. Different subspaces can be relevant for different clusters while the clusters can be obfuscated by the noise of the remaining, 'irrelevant' attributes. There exist many similarities of this problem setting to that of mining frequent patterns, and in fact algorithmic ideas originally developed for frequent pattern mining form the foundations of the paradigm of *subspace clustering* [7].

As in pattern mining, the general intuition in subspace clustering is that an object may be a member of *several* clusters, over *different* subsets of the attributes. In this manner, it is possible to group the data differently depending on the features that are considered. Figure 16.1 gives an example. As we can see, the projection to different subspaces results in different clusters, but not all dimensions contribute to the patterns. In the leftmost projection to the subspace consisting of dimensions x and y, two groups are visible that are different from the groups seen in the center projection to dimensions w and z (note that symbols are consistent across the projections shown). Interestingly, the subspace y and z does not show any clear subspace clusters. The interesting observation here is that this view of different aspects of the

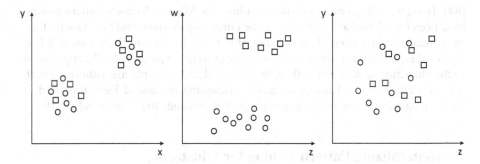

Fig. 16.1 Subspace clustering: two different groupings of the same data are seen when considering the subspace consisting of dimensions x and y (*left*) or the subspace consisting of dimensions z and w (*center*), whereas the subspace projection y and z (*right*) does not show any clear clusters

Fig. 16.2 Frequent itemset mining: transactions for the example are listed (*left*), frequent itemsets are detected when considering just the combination of item a and c, or when considering a and d, but not when considering e.g. c and d

Transactions		Example frequencies
1	a c	
2	a c e	
3	a d	a c 4 times
4	a b c	a d 4 times
5	a d	c d not found
6	a b d	
7	a d e	

data is present in frequent itemset mining as well (cf. Fig. 16.2): an item can be part of two different patterns such as $\{a, c\}$ or $\{a, d\}$, but the combination of $\{c, d\}$ does not necessarily yield frequent patterns.

There are several surveys and overview articles, discussing specifically subspace clustering [9, 50, 52, 53, 67, 74, 83], some of which also point out the connection to frequent pattern mining algorithms. The first survey to discuss the young field was presented by Parsons et al. [67], putting the research community's attention to the problem and sketching a few early algorithms. In the following years, the problem was studied in much more detail, and categories of similar approaches have been defined [50]. A short discussion of the fundamental problems and strategies has been provided by Kröger and Zimek [53]. Assent gives an overview in the context of high-dimensional data of different provenance, including time series and text documents [9]. Sim et al. [74] discuss 'enhanced' subspace clustering, i.e., they point out particular open problems in the field and discuss methods specifically addressing those problems. Kriegel et al. [52] give a concise overview and point to open questions as well. Based on this overview, an updated discussion was given by Zimek [83]. Recent textbooks by Han et al. [38], and Gan et al. [31], sketch prominent issues and example algorithms. Recent experimental evaluation studies compared some subspace clustering algorithms [60, 63].

The close relationship between the two areas subspace clustering and frequent pattern mining has been elaborated in a broader perspective by Zimek and Vreeken

[84]. Here, we will go more into detail of how the ideas of frequent pattern mining have been transferred and translated to the clustering domain, and how exactly they have found use in various clustering algorithms. To this end, in Sect. 2 we will first discuss the generalization of the reasoning about frequent patterns for the application to the clustering task. We will then, in Sect. 3, detail example algorithms for both subspace clustering and subspace search, discussing the use of ideas proposed in frequent pattern mining in these algorithms. We conclude the chapter in Sect. 4.

2   Generalizing Pattern Mining for Clustering

For a reader of a chapter in this book about frequent pattern mining, we assume familiarity with frequent pattern mining as discussed also in fundamental other chapters in this book. In particular, we assume basic knowledge of the Apriori algorithm [6]. Nevertheless, for the sake of completeness, let us briefly recapitulate the algorithmic ingredients of Apriori that are essential to our discussion.

Considering the example of market basket analysis, we are interested in finding items that are sold together (i.e., itemsets). Naïvely, the search for all frequent itemsets is exponential in the number of available items: we would simply calculate the frequency of all k-itemsets in the database over m items, resulting in $\sum_{k=1}^{m} \binom{m}{k} = 2^m - 1$ tests.

For identification of frequent patterns in a transaction database (i.e., a binary database, where each row does or does not contain a certain item), the idea of Apriori is a level-wise search for itemsets of incremental length, given a frequency threshold. Starting with all frequent itemsets of length 1 (i.e., counting all transactions containing a certain item, irrespective of other items possibly also contained in the transaction), the list of potential candidates for *frequent* itemsets of length 2 can be restricted based on the following observation: An itemset of length 2 can only be frequent if both contained items (i.e., itemsets of length 1) are frequent as well. If neither diapers nor beer is a frequent item in the transaction database, the transaction containing both diapers and beer cannot be frequent either. This holds for itemsets of all lengths n, that can only be frequent if all contained itemsets of length $n - 1$ are frequent as well. For example, an itemset may contain items A, B, C, etc. If a 1-itemset containing A is not frequent (i.e., we find such an itemset less often than a given threshold), all 2-itemsets containing A (e.g., $\{A, B\}, \{A, C\}, \{A, D\}$) cannot be frequent either (otherwise itemsets containing A would have been frequent as well) and need not be tested for exceeding the threshold. Likewise, if the itemset $\{A, B\}$ is not frequent, then all 3-itemsets containing $\{A, B\}$ (e.g., $\{A, B, C\}, \{A, B, D\}$, $\{A, B, E\}$) cannot be frequent either, etc. Theoretically, the search space remains exponential, yet practically the search is usually substantially accelerated.

This observation is a principle of monotonicity and is the most important ingredient for a heuristic speed-up of the mining for frequent patterns. More concisely, we can express this monotonicity over sets as follows:

$$\mathcal{T} \text{ is frequent} \Rightarrow \forall \mathcal{S} \subseteq \mathcal{T} : \mathcal{S} \text{ is frequent.} \qquad (16.1)$$

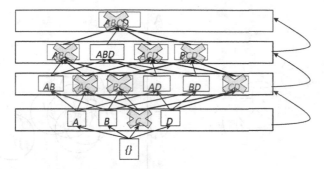

Fig. 16.3 Pruned search space during iterative database scans of Apriori (example): itemset $\{C\}$ has been found infrequent in the first scan, therefore, itemsets $\{A, C\}, \{B, C\}, \{C, D\}$ do not need to be considered in the second scan, itemsets $\{A, B, C\}, \{A, C, D\}, \{B, C, D\}$ do not need to be considered in the third scan, etc. In this example, Apriori stops scanning the database after round three, as there is no candidate of length 4 remaining

More precisely, the pruning criterion used in the Apriori algorithm is based on the equivalent anti-monotonic property, describing the opposite direction of deduction:

$$\mathcal{S} \text{ is } not \text{ frequent} \Rightarrow \forall \mathcal{T} \supseteq \mathcal{S} : \mathcal{T} \text{cannot be frequent either.} \qquad (16.2)$$

In the iterative procedure of repeated scans of the database for frequent itemsets, this anti-monotonic property allows to ignore candidates that cannot be frequent and, eventually, this pruning allows stopping at a certain size of itemsets, when no candidates of typically moderate size remain to generate larger itemsets (see Fig. 16.3).

An extension of the Apriori idea for very large itemsets has been termed 'colossal patterns' [82]. The observation is that if one is interested in finding very large frequent itemsets, then Apriori needs to generate many smaller frequent itemsets that are not relevant for the result. This effect can be used positively, in that if large patterns also have a large number of subsets, several of these subsets can be combined in order to obtain larger candidates directly. In this sense, the idea is to avoid the full search, and instead use some results at the bottom of the search space as a shortcut to particularly promising candidates higher up. This approach thus trades some of the accuracy of full search for a much more efficient frequent pattern mining algorithm. As we will see below, both the Apriori algorithm, as well as that of colossal patterns have been employed towards mining subspace clusters.

2.1   Generalized Monotonicity

In data clustering, we typically do not consider binary transaction data, or discrete data in general, but instead most often study continuous real-valued vector data, typically assuming a Euclidean vector space. In this space, attributes may be noisy, or

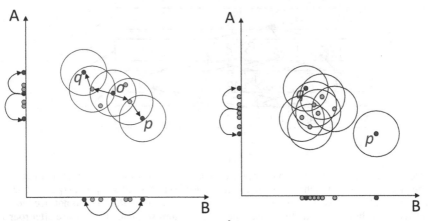

a   Objects p and q are density-connected, i.e., they can build a density-based cluster, in subspaces $\{A, B\}$, $\{A\}$, and $\{B\}$.

b   Objects p and q are not density-connected in subspace $\{B\}$. Therefore they cannot be density-connected in subspace $\{A, B\}$ either.

Fig. 16.4 Transfer of anti-monotonicity to subspace clusters

even completely irrelevant for certain clusters. If we measure similarity over the full space, i.e., over all attributes, detecting such 'subspace' clusters becomes increasingly difficult for higher numbers of irrelevant dimensions. To the end of identifying the relevant attributes, and measuring similarity only over these, the fundamental algorithmic idea of Apriori has been transferred to clustering in Euclidean spaces, giving rise to the task of 'subspace clustering', which has been defined as 'finding all clusters in all subspaces' [7].

Over time, this transfer has been done in different ways. The most important variants are to identify subspace clusters that in turn qualify some subspace as 'frequent pattern', or to identify interesting subspaces without direct clustering, but as a prerequisite for subsequent clustering in these subspaces or as an integral part of some clustering procedure.

Subspace Clusters as Patterns Let us consider the case of clusters in different subspaces with an example for density-based clusters [51], as visualized in Fig. 16.4. In the first scenario, depicted in Fig. 16.4a, we see that objects p and q are density-connected with respect to some parameters in subspaces $\{A, B\}$, $\{A\}$, and $\{B\}$. Here, the parameters capturing density are a distance threshold defining the radius of the neighborhood ball and a minimum number of points required to fall within this neighborhood ball in order to qualify as dense. That is, within these subspaces, we can reach both p and q starting at o by 'hopping' from one object with at least n neighbors within ε distance to another. This means that with these parameters, p and q belong to the same density-based cluster in each of these subspaces.

In the second scenario, depicted in Fig. 16.4b, p and q are again density-connected in subspace $\{A\}$, but not in subspace $\{B\}$. As a result from monotonicity, they therefore are also not density-connected in subspace $\{A, B\}$.

Table 16.1 Translating frequent pattern mining concepts to subspace clustering

Frequent pattern mining	Subspace clustering
Item	Dimension (attribute)
Itemset	Subspace (set of attributes)
Frequent itemset	Subspace (unit) containing cluster

Consequently, a set of points cannot form a cluster in some space $\mathcal{T}$, if it does not also form a cluster in every subspace of $\mathcal{T}$. Or, formulated as anti-monotone property that we can use to prune candidate subspaces:

$$\mathcal{S} \text{ does not contain any cluster} \Rightarrow \forall \text{ superspaces } \mathcal{T} \supseteq \mathcal{S}: \qquad (16.3)$$

$$\mathcal{T} \text{ cannot contain a cluster either.}$$

As a result, an algorithm for subspace clustering can identify all clusters in all 1-dimensional subspaces, continue to look for clusters in only those 2-dimensional subspaces that have a 1-dimensional subspace containing some cluster, and so on, following the candidate-pruning heuristic of the Apriori algorithm. Hence we see that the 'items' of Apriori translate to dimensions, 'itemsets' translate to subspaces, and 'frequent itemset' according to some frequency threshold translates to 'subspace contains some cluster' according to some clustering criterion. See Table 16.1 for a summary of this translation. This transfer of concepts requires the anti-monotonicity to hold for the clustering criterion used.

Note that the monotonicity does not hold in general for arbitrary cluster paradigms, but instead depends on the particular cluster model used. The example used here (monotonicity of density-based clusters, Fig. 16.4) has been proven for the subspace clustering approach SUBCLU [48]. However, the very idea of using a monotonicity for some cluster criterion has been used for different clustering models several times, following the seminal approach of CLIQUE [7]. We detail the specific adaptations for different clustering models in the next section.

Subspaces as Patterns In the second main variant, the setup is slightly modified, and the goal is to identify subspaces as a prerequisite for the final clustering result. These subspaces can be used in quite different ways in connection with clustering algorithms. For example, after identification of subspaces, traditional clustering algorithms are applied to find clusters within these subspaces, or distance measures can be adapted to these subspaces in the actual clustering procedure, or clusters and corresponding subspaces are refined iteratively. As such, in contrast to the setting above, here one does not identify whether subspaces are 'interesting' by the clusters they contain (which is specific to a particular clustering model), but rather defines 'interesting' more generally, for example in terms of how strongly these attributes interact.

In subspace search, just as in subspace clustering, the 'items' and 'itemsets' concepts from frequent pattern mining translate nicely to 'dimension' and 'subspace', respectively. The notion of a 'frequent itemset' according to some frequency threshold translates different here, namely to 'interesting subspace' according to some

Table 16.2 Translating frequent pattern mining concepts to subspace search	Frequent pattern mining	Subspace search
	Item	Dimension (attribute)
	Itemset	Subspace (set of attributes)
	Frequent itemset	'Interesting' subspace

measure of 'interestingness' (see Table 16.2 for a summary). How to measure this 'interestingness' in a way to satisfy anti-monotonicity is the crucial question that differs from approach to approach. Let us note that many methods follow the general idea of candidate elimination in subspace search without adhering to a criterion of *strict* anti-monotonicity, i.e., they rely on some observation that anti-monotonicity of their criterion 'usually' holds.

2.2 Count Indexes

Generalized monotonicity is a very useful property towards pruning the search space in both frequent itemset mining and subspace clustering. As part of the Apriori algorithm, however, candidate itemsets or subspaces have to be generated. For large sets of items and high-dimensional subspaces (i.e., subspaces with very many attributes), this can be a performance bottleneck [37].

Taking a different approach, the so-called FP-Growth algorithm uses a specialized index structure to maintain frequency counts of itemsets, the FP-tree [37]. As illustrated in Fig. 16.5, a node in this count index corresponds to the frequency count of a particular item, and following a path from an item to the root corresponds to the frequency count of a particular combination of items into an itemset. The index can be constructed in two data scans, where the first finds all frequent items, and the second creates nodes and updates counts for each transaction.

The FP-Growth algorithm is a depth-first approach. Starting from the most frequent item, the corresponding combinations with other items are 'grown' by recursively extracting the corresponding paths, until the index has been reduced to one path. The advantage of this method is that only frequent itemsets are generated, and that only two scans over the data are necessary in order to do so.

As we will detail in the next section, this idea of compactly representing interesting combinations in a count index and of proceeding in a depth-first traversal of the search space has also been applied to subspace clustering. This application is not straightforward due to the fact that both relevant subspace regions, as well as a notion of similarity between adjacent regions has to be defined; concepts that do not have one-to-one counterparts in frequent pattern mining.

2.3 Pattern Explosion and Redundancy

The downside to the frequency criterion and its monotonicity in frequent itemset mining is that with a threshold low enough to avoid exclusion of all but the most

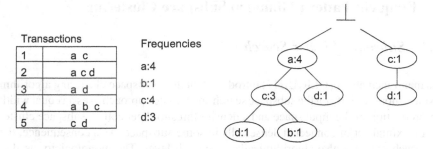

Transactions	
1	a c
2	a c d
3	a d
4	a b c
5	c d

Frequencies

a:4
b:1
c:4
d:3

Fig. 16.5 FP-tree example: the tree nodes store items and their counts, paths correspond to combinations of itemsets and their respective counts. The index is built in just two scans over the data, and the frequent itemset mining algorithm FP-Growth works exclusively on the index. Once individual item frequencies are established, the second scan updates counts for each transaction or creates new nodes where necessary

common (and therefore not really interesting) itemsets, the frequent itemsets will usually be abundant and therefore, as a result of data exploration, not be useful either. In frequent pattern mining, this phenomenon is known as the pattern explosion. By the exponential size of possible subspaces, and type of interestingness measures, subspace clustering inherited this problem with the transfer of the techniques from frequent pattern mining. For non-trivial thresholds usually huge sets of subspace clusters are discovered—which are typically quite redundant.

Different means have been studied to condense the result set of patterns or to restrict the search space further in the first place.

One approach among others is mining or keeping only those itemsets that cannot be extended further without dropping below the threshold, i.e., the *maximal* frequent itemsets [16]. An alternative approach uses borders to represent a lossless compression of the result set of frequent patterns, named *closed* frequent itemsets [68]. Another branch of summarization is that of picking or creating a number of representative results. Yan et al. [78] choose a subset of results such that the error of predicting the frequencies in the complete result set is minimized. Mampaey et al. [57] give an information theoretic approach to identifying that subset of results by which the frequencies in either the complete result set, or the data in general, can best be approximated. To this end, they define a maximum entropy model for data objects, given knowledge about itemset frequencies. The resulting models capture the general structure of the data very well, without redundancy.

Just as the basic techniques for frequent pattern mining, also these ideas for condensing the result, as well as restricting the search space, have found corresponding solutions to the problem of redundant results in subspace clustering.

3 Frequent Pattern Mining in Subspace Clustering

3.1 Subspace Cluster Search

As mentioned above, CLIQUE [7] introduced the first subspace clustering algorithm using a monotonicity on the subspace search space. The approach uses an equal width discretization of the input space and a density threshold per cell. A subspace cluster is a maximal set of connected dense cells in some subspace. As a consequence, the approach operates also algorithmically at the cell level. The monotonicity used is that a dense cell in a k-dimensional subspace is also a dense cell in all its $k - 1$ dimensional subspaces:

$$\mathcal{C} \text{ is a cluster in subspace } \mathcal{T} \Rightarrow \tag{16.4}$$
$$\mathcal{C} \text{ is part of a cluster in all subspaces } \mathcal{S} \subseteq \mathcal{T}$$

Based on the corresponding anti-monotonicity, Apriori is applied from 1-dimensional dense cells in a straightforward fashion to find all higher-dimensional dense cells. As a variation of this base scheme, an approximation is suggested that prunes subspaces from consideration if their dense cells do not cover a sufficiently large part of the data.

MAFIA [65] extends the cell-based approach by adapting the cell sizes to the data distribution. The general approach is to combine neighboring cells in one dimension if they have similar density values. The monotonicity used is the same as in CLIQUE, but additionally, a parallel algorithm is introduced that processes chunks of the data on local machines that communicate to exchange cell counts at each level of the subspace lattice. XProj [5] is an adaptation of the CLIQUE idea to clustering of graph data based on frequent sub-graphs and was applied to cluster XML data. In contrast to CLIQUE, XProj looks for a hard partitioning, rather than overlapping clusters.

CLIQUE and MAFIA may miss points or subspace clusters depending on location and resolution of the cells (see for example Fig. 16.6), so later works have proposed bottom-up algorithms that do not rely on discretization. SUBCLU [48] follows the density-based subspace clustering paradigm. As already illustrated in Fig. 16.4, subspace clusters are maximal sets of density-connected points. Any subspace cluster projection to a lower dimensional subspace is a density-connected set again (albeit not necessarily a maximal one). Anti-monotonicity is used in that if a subspace does not contain a density-based subspace cluster, then no superspace will either.

Note that this approach means that the notion of frequent patterns is also different than in CLIQUE and MAFIA: in these cell-based approaches, a (frequent) item is a (dense) cell in a particular subspace, whereas in SUBCLU (and later approaches) it is the entire subspace. In SUBCLU, the Apriori principle is used to generate candidate subspaces within which the actual subspace clusters are determined.

The DUSC [10] approach relies on a different definition of density than SUBCLU does. Based on the observation that a fixed density assessment is biased and favors

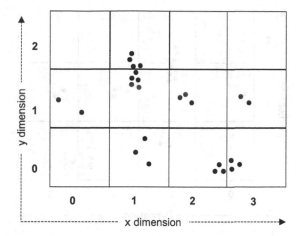

Fig. 16.6 Standard grid-based discretization as used e.g. in CLIQUE: the accuracy of subspace clustering depends on location and resolution of the grid. A minimum cell count of more than three will miss the subspace cluster at the *bottom right*, whereas a minimum cell count of three will also report cells that contain a few isolated noise points (e.g., cell at the *center right*)

low dimensional subspace clusters over high-dimensional ones, the density measure is normalized by the expected density. This means that (anti-)monotonicity is lost, and standard application of Apriori is not possible. However, as proposed in a later extension, it is possible to use the anti-monotonicity as a filtering criterion in a multistep clustering scheme (EDSC) [11]. The idea is to generate a conservative approximation of subspace clusters based on cells that are merged if potentially density-connected. Similar in spirit to the anti-monotonicity in Apriori, pruning is based on the weakest density measure as a filter step.

The idea of avoiding full lattice search in favor of more efficient runtimes (i.e., the colossal pattern idea [82] we saw above) is also found for subspace clustering [64]. Instead of analyzing all subspaces, and the entire value ranges within these subspaces, the idea is to represent subspace clusters at different levels of approximation. Using the number of objects within the current approximation as an indication, potential combinations with other subspaces are used as an indication of higher-dimensional subspace clusters. Priority queues are maintained in order to generate the most promising candidates in the lattice first. As a result, it becomes possible to avoid the generation of many relatively low-dimensional subspace clusters and to steer the search towards high-dimensional subspace clusters directly.

Another interesting connection to frequent pattern mining is discussed with the algorithm INSCY for density-based subspace clustering [12]: subspace clusters are detected based on a frequent cell count data representation, an index structure that is similar in spirit to the FP-tree from frequent itemset mining. As mentioned in the previous section, the challenge here is two-fold: first, to define an adequate representation of subspace regions (the items), and second, to identify similarities among these subspace regions. For the first part, a discretization technique as in

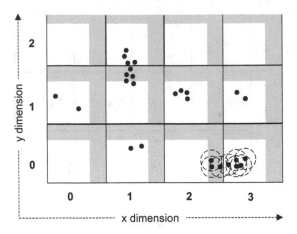

Fig. 16.7 Grid with density-preserving borders: to guarantee detection of all density-based subspace clusters, the grid is enhanced with borders (*gray shaded*) at the top of each cell in each dimension. These borders have exactly the size of the area for the density assessment (circles around points in the clusters at the *bottom right*), so that an empty border means that no cluster extends across these two cells

EDSC [11] is used, which consists of a traditional equal-width grid, plus density-preserving borders. Figure 16.7 illustrates the general idea: the density-preserving borders make it possible to determine whether points in one cell are potentially density-connected to those in a neighboring cell. They are the size of the area used for density assessment (circles around points in the figure). If a subspace cluster extends across one of these borders, this border must be non-empty. If that should be the case, these cells need to be merged during mining.

A SCY-tree is constructed, which similar to item frequency counts in FP-trees contains counts of the number of points in a particular grid cell. In addition, marker nodes are introduced to signal that the border between neighboring cells is non-empty. An example is given in Fig. 16.8. As we can see in this example, the ten points that are in the bottom '0' slice of the y-dimension (leftmost node under the root in the tree), fall into three different intervals in the x-dimension: two in cell '1', three in cell '2', and five in cell '3' (three child nodes). Additionally, a node marks the presence of one or more points in the border of cell '2' by a special node without any count information. Similar to FP-Growth, it is then possible to mine subspace clusters in a depth-first manner. Different levels of the index correspond to the dimensions in which these cells exist. As opposed to frequent itemset mining, neighboring nodes are merged if they contain cells that are potentially part of the same cluster.

3.2 Subspace Search

Subspace search based on frequent pattern mining concepts has been applied both independently of specific clustering algorithms, as well as integrated in some clustering

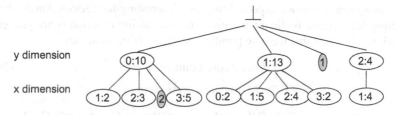

Fig. 16.8 SCY-tree index for depth-first mining of subspace clusters. Nodes contain cell counts as in frequent itemset mining. Levels correspond to different dimensions, and additional marker nodes indicate that a border is non-empty and that cells need to be merged during mining. For example, the *gray shaded node* labeled '2' at the *bottom* corresponds to the *non-empty border* of cell '2' in dimension *x* in Fig. 16.7

algorithm yet independent of the cluster model. In the first scenario, we can regard subspace search as a global identification of 'interesting' subspaces—subspaces in which we expect clusters to exist—and hence as a restriction of the search space. In the second scenario, we observe a local identification of 'interesting' subspaces. A typical use case of these 'locally interesting' subspaces is to adapt distance measures locally, that is, for different clusters, different measures of similarity are applied.

Global Subspace Search   ENCLUS [21] is based on an assessment of the subspace as a whole, i.e., a subspace search step proceeds the actual subspace clustering. In order to determine interesting subspaces, Shannon Entropy [73] is used. Entropy measures the uncertainty in a random variable, where a high value means a high level of uncertainty. A uniform distribution implies greatest uncertainty, so a low entropy value (below some threshold) is used as an indication of subspace clusters. Similar to CLIQUE, the data are discretized into equal-width cells before entropy assessment. Monotonicity is based on the fact that an additional attribute can only increase the uncertainty and thereby the Shannon Entropy:

$$\mathcal{T} \text{ has low entropy} \Rightarrow \forall \mathcal{S} \subseteq \mathcal{T} : \mathcal{S} \text{ has low entropy.} \qquad (16.5)$$

Besides this Apriori bottom-up part of the algorithm, an additional mutual information criterion is used for top-down pruning. Interesting subspaces in this sense are those with an entropy that is lower (by some threshold) than the sum of the entropy of each of its one-dimensional subspaces. Using both criteria, the most interesting subspaces for subspace clustering according to ENCLUS are located neither at the top nor at the bottom of the subspace search space, but at some medium dimensionality. This resembles the concept of borders in frequent itemset mining (Sect. 2.3). While there borders are used to derive a condensed representation of the result set, here, the result set is restricted to reduce the redundancy of too many clusters.

For RIS (Ranking Interesting Subspaces) [47], subspaces are 'interesting' if they have a large number of points in the neighborhoods of core points (i.e., points with a high local point density according to some thresholds), normalized by the expected number of points assuming uniform distribution. While this criterion adopts a density-based notion [51] of 'interesting', it is not tied to a specific clustering algorithm.

These subspaces are hence expected to prove interesting for various density-based clustering algorithms. While monotonicity of this quality criterion is not proven in general, we do know that the core point property is anti-monotonic:

$$o \text{ is not a core point in } \mathcal{S} \Rightarrow \qquad (16.6)$$
$$\forall \mathcal{T} \supseteq \mathcal{S} : o \text{ is not a core point in } \mathcal{T}.$$

A similar approach, SURFING (SUbspaces Relevant For clusterING) [15], also following a density-based notion of 'interestingness', assesses the variance of k-nearest neighbor distances. ℓ-dimensional subspaces can be rated as 'interesting', 'neutral', or 'irrelevant' in comparison to $(\ell - 1)$-dimensional subspaces, but this rating is not monotonous. Accordingly, the subspace search of SURFING follows the Apriori idea of early pruning of candidates only heuristically but does not formally implement the strictly anti-monotonic candidate elimination.

CMI (Cumulative Mutual Information) [66] is a measure to assess the correlation among the attributes of some subspace and is used to identify subspaces that are interesting w.r.t. a high contrast, that is, they are likely to contain different clusters. The authors assume monotonicity of this contrast criterion to facilitate a candidate elimination-based search starting with two dimensional subspaces. As a priority search, generating candidates from the top m subspaces only, their algorithm is more efficient than the Apriori search at the expense of completeness of the results. Finally, subspaces contained in another subspace reported as a result are dropped from the resulting set of subspaces, if the higher-dimensional subspace also has higher contrast.

Local Subspace Search Subspace search has also been incorporated locally into clustering algorithms. DiSH [1], similar to its predecessor HiCS citeclu:AchBoeKriKroetal06, follows a pattern of cluster search that is different from the Apriori-based subspace clustering idea discussed so far. Appropriate subspaces for distance computations are learned locally for each point, then the locally adapted (subspace-) distances and the dimensionality of the assigned subspace are used as a combined distance measure in a global clustering schema similar to OPTICS [8] to find hierarchies of subspaces.

For learning the most appropriate subspace for each data point both HiCS and DiSH assign a 'subspace preference vector' to each object, based on the variance of the neighborhood in each attribute. As such, the clustering procedure does not make use of an efficient frequent pattern search algorithm. However, while HiCS uses the full-dimensional neighborhood and studies the variances of the neighborhoods in attribute-wise projections, DiSH starts with attribute-wise neighborhoods and combines those neighborhoods in a bottom-up procedure. Here, an Apriori-like search strategy is one of the suggested alternatives, employing the monotonicity of neighborhoods in projections of the data. If $\mathcal{S}$ is a subspace of $\mathcal{T}$, then the cardinality of the ε-neighborhood of some object o in $\mathcal{T}$ is bound to be at most the cardinality of the ε-neighborhood of the same object o in $\mathcal{S}$:

$$\mathcal{S} \subseteq \mathcal{T} \Rightarrow \left| \mathcal{N}_\varepsilon^\mathcal{T}(o) \right| \leq \left| \mathcal{N}_\varepsilon^\mathcal{S}(o) \right| \qquad (16.7)$$

This holds, e.g., for L_P-type distances ($P \geq 1$, for example the commonly used Euclidean distance), because distances between the points can never shrink when adding more dimensions. Let us note that this is also the reason why the core point property is anti-monotone (cf. Eq. 16.6).

In a similar way, CFPC [80] ('MineClus' in an earlier version [79]) improves by a frequent pattern mining-based approach the subspace search strategy of an earlier projected clustering algorithm, DOC [71]. As projected clustering approaches, both pursue a local subspace search per (preliminary) cluster. A typical projected clustering algorithm, following the seminal approach of PROCLUS [3], starts with some initial assignment of points to clusters. Then, the optimal projection (subspace) of each cluster and the assignment of points are iteratively refined. In DOC, random-sampling was applied to find the most suitable subspace for a potential cluster. CFPC replaces this random sampling strategy by a technique related to FP-growth. A potential cluster is defined by its potential (in both approaches randomly sampled) medoid p. For all points q, an itemset includes those dimensions in which q is close to p. A large, frequent itemset would therefore correspond to a projected cluster with many points and high dimensionality. To find the best cluster and its optimal projection, FP-growth is applied over this modelling of frequent itemsets.

The projected clustering algorithm P3C [58, 59] does also incorporate an Apriori-like local subspace search, but in yet another variant. The basic idea of P3C is to find cluster cores starting with "p-signatures" that are intervals of some subset of p distinct attributes, i.e., subspace regions. Roughly, such a p-signature qualifies as a cluster core if and only if its support, i.e., the number of points falling into this subspace region, exceeds the expected support under some assumptions concerning the point distribution, and if this happens by chance (Poisson probability) less likely than specified by some (Poisson-)threshold. By these conditions, p-signatures qualifying as cluster cores can be generated using an Apriori-like candidate elimination procedure.

3.3   Redundancy in Subspace Clustering

As pointed out above, redundancy of subspace cluster results is a problem inherited from the Apriori strategy for traversing the search space of subspaces. As a consequence, for current research on subspace clustering, reducing redundancy is a major topic. As we have seen, the concept of borders found analogous use already in the early subspace search algorithm ENCLUS [21] for restricting the search space. Some approaches mine or report the most representative clusters as solutions [13, 61]. This is related to picking or creating a number of representative results in frequent pattern mining. Also the idea of restricting results of frequent pattern mining to the maximal frequent itemsets found a correspondence in subspace clustering. For example, nCluster [54], CLICKS [81], or MaPle [69] mine those subspace clusters of maximal dimensionality.

Other variants of clustering algorithms outside subspace clustering that also tackle high-dimensional data face a similar problem. For example, multiview clustering

[19, 22, 34, 46] approaches the problem from the opposite direction. It is based on the notion of semantically different subspaces, i.e., multiple representations for the same data. We cannot generally assume to know the different semantics of subspaces beforehand and, accordingly, could find results in overlapping subspaces. As a consequence, these approaches allow some redundancy between resulting clusters. A certain partial overlap between concepts is allowed in order to not exclude possibly interesting concepts.

A related though distinct way of addressing the problem of redundancy and distinctiveness of different clusters is to seek diverse clusterings by directly assessing a certain notion of distance between different partitions (so-called alternative clustering approaches [14, 23, 24, 25, 32, 33, 35, 70, 72]). Starting with one clustering solution, they search for an alternative clustering solution that provides substantially different insights. Still, alternative clustering solutions are allowed to not be absolutely orthogonal but to show some redundancy with existing clustering solutions.

Apparently, to avoid redundancy as more 'enhanced' [74] subspace clustering algorithms try to do should not be pursued as an absolute goal. Multiview clustering and alternative clustering come from the other extreme and relax the original restriction of 'no redundancy' more and more. Relationships between subspace clustering and other families of clustering approaches have been discussed by Zimek and Vreeken [84].

A question related to the redundancy issue is that of the appropriate density level. Both of these issues have decisive influence on the clusters that are selected. Determining the right density level is a general problem also in full space density-based clustering [51], but for clustering in subspaces, the problem is even more severe. Setting a fixed density threshold for an Apriori style subspace search is not appropriate for all possible subspaces. Consider for example any CLIQUE-style grid approach: the volume of a hypercube increases exponentially with the dimensionality, hence the density decreases rapidly. As a consequence, any chosen threshold introduces a bias to identify clusters of (up to) a certain dimensionality. This observation motivates research on adaptive density thresholds [10, 62]. The algorithmic challenge then comes from loss of monotonicity that would allow efficient traversal of the search space of subspaces.

When using Euclidean distance (L_2), the appropriate choice of an ε-range becomes extremely challenging as well due to the rather counter-intuitive behavior of the volume of the hypersphere with increasing dimensions. Let us note that, for outlier detection, the very same problem occurs in high-dimensional data, which has been discussed in detail by Zimek et al. [85]. Choosing the size of the neighborhood in terms of objects rather than in terms of a radius (i.e., using k nearest neighbors instead of an ε-range query) has been advocated as a workaround for this problem [2], to solve at least certain aspects such as having a well-defined (non-empty) set of objects for the density estimation or spatial properties of the neighborhood.

4   Conclusions

This chapter discusses the close relationship between frequent pattern mining and clustering, which might not be apparent at first sight. In fact, frequent pattern mining was the godfather of subspace clustering, which developed quickly into an independent and influential research area on its own. We showed how certain techniques that have been originally developed for frequent pattern mining have been transferred to clustering, how these techniques changed in their new environment, and how the drawbacks of these techniques—unfortunately transferred along—raised new research questions as well as interesting solutions in the area of data clustering.

Acknowledgments Ira Assent is partly supported by the Danish Council for Independent Research—Technology and Production Sciences (FTP), grant 10-081972. Jilles Vreeken is supported by the Cluster of Excellence 'Multimodal Computing and Interaction' within the Excellence Initiative of the German Federal Government.

References

 1. E. Achtert, C. Böhm, H.-P. Kriegel, P. Kröger, I. Müller-Gorman, and A. Zimek. Detection and visualization of subspace cluster hierarchies. In *12th International Conference on Database Systems for Advanced Applications (DASFAA), Bangkok, Thailand*, pages 152–163, 2007.
 2. E. Achtert, C. Böhm, H.-P. Kriegel, P. Kröger, and A. Zimek. Robust, complete, and efficient correlation clustering. In *7th SIAM International Conference on Data Mining (SDM), Minneapolis, MN*, pages 413–418, 2007.
 3. C. C. Aggarwal, C. M. Procopiuc, J. L. Wolf, P. S. Yu, and J. S. Park. Fast algorithms for projected clustering. In *ACM International Conference on Management of Data (SIGMOD), Philadelphia, PA*, pages 61–72, 1999.
 4. C. C. Aggarwal, A. Hinneburg, and D. Keim. On the surprising behavior of distance metrics in high dimensional space. In *8th International Conference on Database Theory (ICDT), London, UK*, pages 420–434, 2001.
 5. C. C. Aggarwal, N. Ta, J. Wang, J. Feng, and M. Zaki. Xproj: a framework for projected structural clustering of xml documents. In *13th ACM International Conference on Knowledge Discovery and Data Mining (SIGKDD), San Jose, CA*, pages 46–55, 2007.
 6. R. Agrawal and R. Srikant. Fast algorithms for mining association rules. In *20th International Conference on Very Large Data Bases (VLDB), Santiago de Chile, Chile*, pages 487–499, 1994.
 7. R. Agrawal, J. Gehrke, D. Gunopulos, and P. Raghavan. Automatic subspace clustering of high dimensional data for data mining applications. In *ACM International Conference on Management of Data (SIGMOD), Seattle, WA*, pages 94–105, 1998.
 8. M. Ankerst, M. M. Breunig, H.-P. Kriegel, and J. Sander. OPTICS: Ordering points to identify the clustering structure. In *ACM International Conference on Management of Data (SIGMOD), Philadelphia, PA*, pages 49–60, 1999.
 9. I. Assent. Clustering high dimensional data. *Wiley Interdisciplinary Reviews: Data Mining and Knowledge Discovery*, 2(4):340–350, 2012.
10. I. Assent, R. Krieger, E. Müller, and T. Seidl. DUSC: dimensionality unbiased subspace clustering. In *7th IEEE International Conference on Data Mining (ICDM), Omaha, NE*, pages 409–414, 2007.
11. I. Assent, R. Krieger, E. Müller, and T. Seidl. EDSC: efficient density-based subspace clustering. In *17th ACM Conference on Information and Knowledge Management (CIKM), Napa Valley, CA*, pages 1093–1102, 2008.

12. I. Assent, R. Krieger, E. Müller, and T. Seidl. INSCY: indexing subspace clusters with in-process-removal of redundancy. In *8th IEEE International Conference on Data Mining (ICDM), Pisa, Italy*, pages 719–724, 2008.
13. I. Assent, E. Müller, S. Günnemann, R. Krieger, and T. Seidl. Less is more: Non-redundant subspace clustering. In *MultiClust: 1st International Workshop on Discovering, Summarizing and Using Multiple Clusterings Held in Conjunction with KDD 2010, Washington, DC*, 2010.
14. E. Bae and J. Bailey. COALA: a novel approach for the extraction of an alternate clustering of high quality and high dissimilarity. In *6th IEEE International Conference on Data Mining (ICDM), Hong Kong, China*, pages 53–62, 2006.
15. C. Baumgartner, K. Kailing, H.-P. Kriegel, P. Kröger, and C. Plant. Subspace selection for clustering high-dimensional data. In *4th IEEE International Conference on Data Mining (ICDM), Brighton, UK*, pages 11–18, 2004.
16. R. Bayardo. Efficiently mining long patterns from databases. In *ACM International Conference on Management of Data (SIGMOD), Seattle, WA*, pages 85–93, 1998.
17. K. P. Bennett, U. Fayyad, and D. Geiger. Density-based indexing for approximate nearest-neighbor queries. In *5th ACM International Conference on Knowledge Discovery and Data Mining (SIGKDD), San Diego, CA*, pages 233–243, 1999.
18. K. Beyer, J. Goldstein, R. Ramakrishnan, and U. Shaft. When is "nearest neighbor" meaningful? In *7th International Conference on Database Theory (ICDT), Jerusalem, Israel*, pages 217–235, 1999.
19. S. Bickel and T. Scheffer. Multi-view clustering. In *4th IEEE International Conference on Data Mining (ICDM), Brighton, UK*, pages 19–26, 2004.
20. R. J. G. B. Campello, D. Moulavi, and J. Sander. Density-based clustering based on hierarchical density estimates. In *17th Pacific-Asia Conference on Knowledge Discovery and Data Mining (PAKDD), Gold Coast, Australia*, pages 160–172, 2013.
21. C. H. Cheng, A. W.-C. Fu, and Y. Zhang. Entropy-based subspace clustering for mining numerical data. In *5th ACM International Conference on Knowledge Discovery and Data Mining (SIGKDD), San Diego, CA*, pages 84–93, 1999.
22. Y. Cui, X. Z. Fern, and J. G. Dy. Non-redundant multi-view clustering via orthogonalization. In *7th IEEE International Conference on Data Mining (ICDM), Omaha, NE*, pages 133–142, 2007.
23. X. H. Dang and J. Bailey. Generation of alternative clusterings using the CAMI approach. In *10th SIAM International Conference on Data Mining (SDM), Columbus, OH*, pages 118–129, 2010.
24. I. Davidson and Z. Qi. Finding alternative clusterings using constraints. In *8th IEEE International Conference on Data Mining (ICDM), Pisa, Italy*, pages 773–778, 2008.
25. I. Davidson, S. S. Ravi, and L. Shamis. A SAT-based framework for efficient constrained clustering. In *10th SIAM International Conference on Data Mining (SDM), Columbus, OH*, pages 94–105, 2010.
26. A. P. Dempster, N. M. Laird, and D. B. Rubin. Maximum likelihood from incomplete data via the EM algorithm. *Journal of the Royal Statistical Society: Series B (Statistical Methodology)*, 39(1):1–31, 1977.
27. R. J. Durrant and A. Kaban. When is 'nearest neighbour' meaningful: A converse theorem and implications. *Journal of Complexity*, 25(4):385–397, 2009.
28. M. Ester, H.-P. Kriegel, J. Sander, and X. Xu. A density-based algorithm for discovering clusters in large spatial databases with noise. In *2nd ACM International Conference on Knowledge Discovery and Data Mining (KDD), Portland, OR*, pages 226–231, 1996.
29. I. Färber, S. Günnemann, H.-P. Kriegel, P. Kröger, E. Müller, E. Schubert, T. Seidl, and A. Zimek. On using class-labels in evaluation of clusterings. In *MultiClust: 1st International Workshop on Discovering, Summarizing and Using Multiple Clusterings Held in Conjunction with KDD 2010, Washington, DC*, 2010.
30. D. François, V. Wertz, and M. Verleysen. The concentration of fractional distances. *IEEE Transactions on Knowledge and Data Engineering*, 19(7):873–886, 2007.
31. G. Gan, C. Ma, and J. Wu. *Data Clustering. Theory, Algorithms, and Applications.* Society for Industrial and Applied Mathematics (SIAM), 2007.

32. D. Gondek and T. Hofmann. Non-redundant data clustering. In *4th IEEE International Conference on Data Mining (ICDM), Brighton, UK*, pages 75–82, 2004.
33. D. Gondek and T. Hofmann. Non-redundant clustering with conditional ensembles. In *11th ACM International Conference on Knowledge Discovery and Data Mining (SIGKDD), Chicago, IL*, pages 70–77, 2005.
34. S. Günnemann, E. Müller, I. Färber, and T. Seidl. Detection of orthogonal concepts in subspaces of high dimensional data. In *18th ACM Conference on Information and Knowledge Management (CIKM), Hong Kong, China*, pages 1317–1326, 2009.
35. S. Günnemann, I. Färber, E. Müller, and T. Seidl. ASCLU: alternative subspace clustering. In *MultiClust: 1st International Workshop on Discovering, Summarizing and Using Multiple Clusterings Held in Conjunction with KDD 2010, Washington, DC*, 2010.
36. S. Günnemann, I. Färber, E. Müller, I. Assent, and T. Seidl. External evaluation measures for subspace clustering. In *20th ACM Conference on Information and Knowledge Management (CIKM), Glasgow, UK*, pages 1363–1372, 2011.
37. J. Han, J. Pei, and Y. Yin. Mining frequent patterns without candidate generation. *ACM SIGMOD Record*, 29(2):1–12, 2000.
38. J. Han, M. Kamber, and J. Pei. *Data Mining: Concepts and Techniques*. Morgan Kaufmann, 3rd edition, 2011.
39. J. A. Hartigan. *Clustering Algorithms*. John Wiley & Sons, New York, London, Sydney, Toronto, 1975.
40. A. Hinneburg and D. A. Keim. An efficient approach to clustering in large multimedia databases with noise. In *4th ACM International Conference on Knowledge Discovery and Data Mining (KDD), New York City, NY*, pages 58–65, 1998.
41. A. Hinneburg, C. C. Aggarwal, and D. A. Keim. What is the nearest neighbor in high dimensional spaces? In *26th International Conference on Very Large Data Bases (VLDB), Cairo, Egypt*, pages 506–515, 2000.
42. M. E. Houle, H.-P. Kriegel, P. Kröger, E. Schubert, and A. Zimek. Can shared-neighbor distances defeat the curse of dimensionality? In *22nd International Conference on Scientific and Statistical Database Management (SSDBM), Heidelberg, Germany*, pages 482–500, 2010.
43. A. K. Jain. Data clustering: 50 years beyond k-means. *Pattern Recognition Letters*, 31(8):651–666, 2010.
44. A. K. Jain and R. C. Dubes. *Algorithms for Clustering Data*. Prentice Hall, Englewood Cliffs, 1988.
45. A. K. Jain, M. N. Murty, and P. J. Flynn. Data clustering: A review. *ACM Computing Surveys*, 31(3):264–323, 1999.
46. P. Jain, R. Meka, and I. S. Dhillon. Simultaneous unsupervised learning of disparate clusterings. *Statistical Analysis and Data Mining*, 1(3):195–210, 2008.
47. K. Kailing, H.-P. Kriegel, P. Kröger, and S. Wanka. Ranking interesting subspaces for clustering high dimensional data. In *7th European Conference on Principles and Practice of Knowledge Discovery in Databases (PKDD), Cavtat-Dubrovnik, Croatia*, pages 241–252, 2003.
48. K. Kailing, H.-P. Kriegel, and P. Kröger. Density-connected subspace clustering for high-dimensional data. In *4th SIAM International Conference on Data Mining (SDM), Lake Buena Vista, FL*, pages 246–257, 2004.
49. L. Kaufman and P. J. Rousseeuw. *Finding Groups in Data: An Introduction to Cluster Analysis*. John Wiley & Sons, 1990.
50. H.-P. Kriegel, P. Kröger, and A. Zimek. Clustering high dimensional data: A survey on subspace clustering, pattern-based clustering, and correlation clustering. *ACM Transactions on Knowledge Discovery from Data (TKDD)*, 3(1):1–58, 2009.
51. H.-P. Kriegel, P. Kröger, J. Sander, and A. Zimek. Density-based clustering. *Wiley Interdisciplinary Reviews: Data Mining and Knowledge Discovery*, 1(3):231–240, 2011.
52. H.-P. Kriegel, P. Kröger, and A. Zimek. Subspace clustering. *Wiley Interdisciplinary Reviews: Data Mining and Knowledge Discovery*, 2(4):351–364, 2012.
53. P. Kröger and A. Zimek. Subspace clustering techniques. In L. Liu and M. T. Ozsu, editors, *Encyclopedia of Database Systems*, pages 2873–2875. Springer, 2009.

54. G. Liu, J. Li, K. Sim, and L. Wong. Distance based subspace clustering with flexible dimension partitioning. In *23rd International Conference on Data Engineering (ICDE), Istanbul, Turkey*, pages 1250–1254, 2007.

55. S. P. Lloyd. Least squares quantization in PCM. *IEEE Transactions on Information Theory*, 28(2):129–136, 1982.

56. J. MacQueen. Some methods for classification and analysis of multivariate observations. In *5th Berkeley Symposium on Mathematics, Statistics, and Probabilistics*, volume 1, pages 281–297, 1967.

57. M. Mampaey, N. Tatti, and J. Vreeken. Tell me what I need to know: Succinctly summarizing data with itemsets. In *17th ACM International Conference on Knowledge Discovery and Data Mining (SIGKDD), San Diego, CA*, pages 573–581, 2011.

58. G. Moise, J. Sander, and M. Ester. P3C: A robust projected clustering algorithm. In *6th IEEE International Conference on Data Mining (ICDM), Hong Kong, China*, pages 414–425, 2006.

59. G. Moise, J. Sander, and M. Ester. Robust projected clustering. *Knowledge and Information Systems (KAIS)*, 14(3):273–298, 2008.

60. G. Moise, A. Zimek, P. Kröger, H.-P. Kriegel, and J. Sander. Subspace and projected clustering: Experimental evaluation and analysis. *Knowledge and Information Systems (KAIS)*, 21(3):299–326, 2009.

61. E. Müller, I. Assent, S. Günnemann, R. Krieger, and T. Seidl. Relevant subspace clustering: Mining the most interesting non-redundant concepts in high dimensional data. In *9th IEEE International Conference on Data Mining (ICDM), Miami, FL*, pages 377–386, 2009.

62. E. Müller, I. Assent, R. Krieger, S. Günnemann, and T. Seidl. Dens-Est:density estimation for data mining in high dimensional spaces. In *9th SIAM International Conference on Data Mining (SDM), Sparks, NV*, pages 173–184, 2009.

63. E. Müller, S. Günnemann, I. Assent, and T. Seidl. Evaluating clustering in subspace projections of high dimensional data. In *35th International Conference on Very Large Data Bases (VLDB), Lyon, France*, pages 1270–1281, 2009.

64. E. Müller, I. Assent, S. Günnemann, and T. Seidl. Scalable densitybased subspace clustering. In *20th ACM Conference on Information and Knowledge Management (CIKM), Glasgow, UK*, pages 1077–1086, 2011.

65. H. S. Nagesh, S. Goil, and A. Choudhary. Adaptive grids for clustering massive data sets. In *1st SIAM International Conference on Data Mining (SDM), Chicago, IL*, 2001.

66. H. V. Nguyen, E. Müller, J. Vreeken, F. Keller, and K. Böhm. CMI: an information-theoretic contrast measure for enhancing subspace cluster and outlier detection. In *13th SIAM International Conference on Data Mining (SDM), Austin, TX*, pages 198–206, 2013.

67. L. Parsons, E. Haque, and H. Liu. Subspace clustering for high dimensional data: A review. *ACM SIGKDD Explorations*, 6(1):90–105, 2004.

68. N. Pasquier, Y. Bastide, R. Taouil, and L. Lakhal. Discovering frequent closed itemsets for association rules. In *7th International Conference on Database Theory (ICDT), Jerusalem, Israel*, pages 398–416, 1999.

69. J. Pei, X. Zhang, M. Cho, H. Wang, and P. S. Yu. MaPle: A fast algorithm for maximal pattern-based clustering. In *3rd IEEE International Conference on Data Mining (ICDM), Melbourne, FL*, pages 259–266, 2003.

70. J. M. Phillips, P. Raman, and S. Venkatasubramanian. Generating a diverse set of high-quality clusterings. In *2nd MultiClust Workshop: Discovering, Summarizing and Using Multiple Clusterings Held in Conjunction with ECML PKDD 2011, Athens, Greece*, pages 80–91, 2011.

71. C. M. Procopiuc, M. Jones, P. K. Agarwal, and T. M. Murali. A Monte Carlo algorithm for fast projective clustering. In *ACM International Conference on Management of Data (SIGMOD), Madison, WI*, pages 418–427, 2002.

72. Z. J. Qi and I. Davidson. A principled and flexible framework for finding alternative clusterings. In *15th ACM International Conference on Knowledge Discovery and Data Mining (SIGKDD), Paris, France*, pages 717–726, 2009.

73. C. E. Shannon and W. Weaver. *The Mathematical Theory of Communication*. University of Illinois Press, 1949.
74. K. Sim, V. Gopalkrishnan, A. Zimek, and G. Cong. A survey on enhanced subspace clustering. *Data Mining and Knowledge Discovery*, 26(2):332–397, 2013.
75. P. H. A. Sneath. The application of computers to taxonomy. *Journal of General Microbiology*, 17:201–226, 1957.
76. M. Verleysen and D. François. The curse of dimensionality in data mining and time series prediction. In *8th International Work-Conference on Artificial Neural Networks (IWANN)*, *Barcelona, Spain*, pages 758–770, 2005.
77. D. Wishart. Mode analysis: A generalization of nearest neighbor which reduces chaining effects. In A. J. Cole, editor, *Numerical Taxonomy*, pages 282–311, 1969.
78. X. Yan, H. Cheng, J. Han, and D. Xin. Summarizing itemset patterns: a profile-based approach. In *11th ACM International Conference on Knowledge Discovery and Data Mining (SIGKDD)*, *Chicago, IL*, pages 314–323, 2005.
79. M. L. Yiu and N. Mamoulis. Frequent-pattern based iterative projected clustering. In *3rd IEEE International Conference on Data Mining (ICDM), Melbourne, FL*, pages 689–692, 2003.
80. M. L. Yiu and N. Mamoulis. Iterative projected clustering by subspace mining. *IEEE Transactions on Knowledge and Data Engineering*, 17(2):176–189, 2005.
81. M. J. Zaki, M. Peters, I. Assent, and T. Seidl. CLICKS: an effective algorithm for mining subspace clusters in categorical datasets. *Data & Knowledge Engineering*, 60(1):51–70, 2007.
82. F. Zhu, X. Yan, J. Han, P. S. Yu, and H. Cheng. Mining colossal frequent patterns by core pattern fusion. In *23rd International Conference on Data Engineering (ICDE), Istanbul, Turkey*, pages 706–715, 2007.
83. A. Zimek. Clustering high-dimensional data. In C. C. Aggarwal and C. K. Reddy, editors, *Data Clustering: Algorithms and Applications*, chapter 9, pages 201–230. CRC Press, 2013.
84. A. Zimek and J. Vreeken. The blind men and the elephant: On meeting the problem of multiple truths in data from clustering and pattern mining perspectives. *Machine Learning*, 2013.
85. A. Zimek, E. Schubert, and H.-P. Kriegel. A survey on unsupervised outlier detection in high-dimensional numerical data. *Statistical Analysis and Data Mining*, 5(5):363–387, 2012.

Chapter 17
Supervised Pattern Mining and Applications to Classification

Albrecht Zimmermann and Siegfried Nijssen

Abstract In this chapter we describe the use of patterns in the analysis of supervised data. We survey the different settings for finding patterns as well as sets of patterns. The pattern mining settings are categorized according to whether they include class labels as attributes in the data or whether they partition the data based on these labels. The pattern set mining settings are categorized along several dimensions, including whether they perform iterative mining or post-processing, operate globally or locally, and whether they use patterns directly or indirectly for prediction.

Keywords Rules · Classification · Subgroup discovery· Prediction · Pattern sets

1 Introduction

Although early constrained pattern mining in the form of frequent itemset mining (*FIM*) focused on an *unsupervised* setting, a natural extension is to apply these techniques in a supervised context as well. In the supervised context, one attribute (or sometimes a small set of attributes) is considered to be special, and we are only interested in finding relationships between this attribute and the other attributes. Whereas this limits the patterns that will be found, it makes the analysis more targeted and in many cases more useful. Consider for instance the context of customer defection (churn), where one wishes to find relationships between the loyalty of customers and other characteristics of the customers; or consider applications in cheminformatics, where one wishes to find relationships between molecular structures and their activity: in all these cases, a targeted analysis with respect to the indicated target attribute is likely to produce the most valuable results.

A. Zimmermann (✉)
INSA Lyon, LIRIS CNRS UMR 5205, Bâtiment Blaise Pascal,
69621 Villeurbanne CEDEX, France
e-mail: albrecht.zimmermann@insa-lyon.fr

S. Nijssen
KU Leuven, Celestijnenlaan 200 A, 3001 Leuven, Belgium

Universiteit Leiden, Niels Bohrweg 1, 2333 CA, Leiden, The Netherlands
e-mail: siegfried.nijssen@cs.kuleuven.be

C. C. Aggarwal, J. Han (eds.), *Frequent Pattern Mining*,
DOI 10.1007/978-3-319-07821-2_17, © Springer International Publishing Switzerland 2014

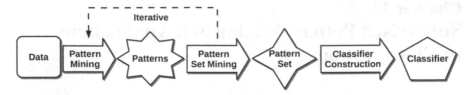

Fig. 17.1 The process of classifier construction via supervised pattern mining

In this chapter, we will provide an overview of pattern mining techniques that can be used in such a supervised context. The patterns found by these techniques can often be interpreted as *rules*: the conditions of the rule identify examples for which a certain property in the target attribute holds. The techniques are hence related to Machine Learning: many traditional Machine Learning algorithms are rule-based as well. A natural question is how to link these two fields to each other, in particular given that the focus of both areas is complementary: most traditional machine learning techniques deal with the large search space of potential rules by adopting heuristics; *pattern mining* methods, on the other hand, offer more efficient methods for traversing a search space exhaustively, promising to find better rules than those found by traditional rule learners. We will address this as well.

The earliest techniques that integrated both areas mirrored the FIM techniques closely, using support and confidence to constrain itemsets and rules, and support's anti-monotonicity to prune the search space. In addition to new challenges, supervised pattern mining also offers new opportunities, however, since the supervision allows to use additional quality measures and prune based on the properties of constraints based on these measures. By now, the field has developed far from its origins, encompassing other representations, incorporating approaches and quality measures developed in the context of Machine Learning, and paying much attention to pattern set mining.

The latter topic is not limited to supervised pattern mining but is of particular importance there: when constructing classifiers, rule lists or sets, but also decision trees, or non-symbolic classifiers, redundancy among or irrelevance of patterns is often detrimental to the classifier's performance.

We have given a unifying perspective on pattern-based classification in the past [9] in which we focused on two dimensions. The first concerned pattern set mining, specifically whether techniques performed *post-processing*, selecting some patterns out of the result set of a single pattern mining step, or whether they iterated pattern mining. The second dimension focused on whether they let the pattern mining and selection process be *guided by a particular model* or not. While these distinctions still stand, in our opinion, we have decided to structure this chapter differently, discussing each of the three steps shown in Fig. 17.1 separately: pattern mining, pattern set mining, and finally classifier construction, and surveying the different, sometimes numerous, options available.

2   Supervised Pattern Mining

The majority of texts in this book deal with different unsupervised pattern mining settings. We will quickly repeat the relevant definitions here to clarify which setting we discuss:

Definition 2.1 *Given a data language $\mathcal{L}_D$ which describes the syntax of potential transactions in the data, a transactional data set $\mathcal{D} \subseteq \mathcal{L}_D$ is of the form $\mathcal{D} = \{d_1, \ldots, d_n\}$, $d_i \in \mathcal{L}_D$. Given a pattern language $\mathcal{L}_\pi$, we define a function match : $\mathcal{L}_\pi \times \mathcal{L}_D \mapsto \{0, 1\}$, which decides whether a pattern occurs in a transaction or not. The set of transactions from a data set $\mathcal{D}$ matched by a pattern π are referred to as its cover: $cov_D(\pi) = \{d \in \mathcal{D} \mid match(\pi, d) = 1\}$, and the size of the cover is referred to as π's (absolute) support: $supp_D(\pi) = |cov_D(\pi)|$.*

The easiest instantiation of this definition is the case of itemset databases: given a set of items $\mathcal{I}$, $\mathcal{L}_\pi = \mathcal{L}_D = 2^{\mathcal{I}}$, and $match(\pi, d) = 1 \Leftrightarrow \pi \subseteq d$. For other types of data, such as for instance graph data or sequential data, alternative definitions for $\mathcal{L}_D$, $\mathcal{L}_\pi$ and *match* can be used, and most ideas presented in the rest of this paper can be applied immediately for these alternative definitions.

The biggest difference between unsupervised and supervised pattern mining is the presence of a variable of interest. This variable is often the class variable that can take on one out of several nominal *class labels*.

Definition 2.2 *Given a data language $\mathcal{L}_D$, and a set of class labels $\mathcal{C} = \{C_1, \ldots, C_k\}$, a labeled data set $\mathcal{D}_C$ is of the form $\mathcal{D}_C = \{(d_1, c_1), \ldots, (d_n, c_n)\}$, $d_i \in \mathcal{L}_D$, $c_i \in \mathcal{C}$.*

The most common setting is that of *classification*, in which the task is to learn a mechanism to predict the class label for unseen data based on rules or patterns. Alternatively, the target for prediction can also be numerical, requiring a *regression* model.

However, another popular setting is that of *subgroup discovery*, which can be generalized to *exceptional model mining* when the target attribute is not a single categorical attribute [24].

Instead of prediction, the goal in this setting is the *characterization* of subsets of the data, i.e. subgroups. The mined rules are therefore not means to the end of prediction but the end themselves, and users are expected to inspect them to gain a deeper understanding of the data. In other words, classification is concerned with *outcomes* on future data, subgroup discovery with *descriptions* of current data.

As a result of this, the quality criteria and heuristics used are sometimes different. However, many of the techniques used are also shared, and for reasons of clarity of presentation, we will mainly focus on classification in this chapter, and make differences to the other settings explicit when appropriate.

2.1   Explicit Class Labels

A first, straight-forward interpretation considers class labels just additional items in the transactional data, i.e. $\mathcal{I}' = \mathcal{I} \cup \mathcal{C}$, and imposes a *syntactical* constraint on itemsets being mined from it: each itemset has to include exactly one of those class-label items. This has the tremendous advantage that existing techniques for FIM can be used directly, e.g. Apriori [2], Eclat [40], or FPGrowth [20].

The typical FIM mining approach identifies interesting itemsets by using a *minimum support* threshold that itemsets' support has to exceed, and chooses relevant rules by using a *minimum confidence* threshold. Since specializations of patterns, e.g. extensions of an itemset with additional items, will have less than or equal support as the pattern itself, the search space can be pruned, allowing for exhaustive enumeration.

This can be adapted by using class labels explicitly as items. It allows to treat settings with more than two classes in a straightforward way:

- For all class labels C:
    1. Mine all itemsets including C that exceed the minimum support threshold
    2. Retain all association rules $r \to C$ that exceed the minimum confidence threshold

The resulting association rules are referred to as *class association rules* (*cars*) and are restricted to having only the class item as their *right-hand side*. Their quality is usually evaluated using *confidence* as in the case of general association rules:

Definition 2.3 *Given a set of items $\mathcal{I}$ and a set of class labels $\mathcal{C}$, a class association rule is of the form $r \to c, r \subseteq \mathcal{I}, c \in \mathcal{C}$. r is called its left-hand side (LHS), antecedent, or rule body, c its right-hand side, consequent, or rule head. Its confidence is defined as $conf(r \to c) = \frac{supp_{\mathcal{D}}(r \cup c)}{supp_{\mathcal{D}}(r)}$.*

Prominent examples of classification learners that build upon class association rules are the CBA [27] and CMAR [25] algorithms. The Harmony algorithm, introduced by [38], also takes this view of class labels, as does the ART technique [17]. As a direct application of FIM techniques, these methods are somewhat limited by typically using only a single minimum support and confidence threshold, which might be inappropriate in the case of skewed class distributions. They can, however, benefit from all developments in FIM research, such as better rule quality measures (replacing confidence), and the development of more efficient algorithms.

2.2   Classes as Data Subsets

A second interpretation of different *classes* in the data is to consider each class a separate data set and a whole database the union of those subsets:

Definition 2.4 *Given a labeled data set $\mathcal{D}_C$, and a set of class labels $\mathcal{C}$, the subsets $\forall C_i \in \mathcal{C} : \mathcal{D}^i = \{(d, C_i) \in \mathcal{D}_C\}$ are called classes.*

Each of these classes can be treated like a distinct data set—the ARC-BC algorithm by [3], for instance, mines *cars* from each class separately, using a single relative support threshold that is used as a constraint on each class in turn. Using this interpretation also opens up several new possibilities.

The first, and potentially most important one, is that this opens up the supervised pattern mining setting to all possible pattern languages: whether itemsets, sequences, trees, or graph-structured data and patterns, the techniques that we describe in this section are applicable to all of them.

Second, there are new ways of using significance and quality measures.

Multiple Support Thresholds There is the possibility of using support thresholds. The XRules classifier [41], for instance, uses a separate minimum support threshold for each class. It is also a first example of supervised pattern mining in a different pattern domain than itemsets, producing predictive rules the rule body of which consists of tree fragments, called *structural rules* in the work.

Instead of minimum support constraints, it is also natural to use maximum support constraints: a rule which is specific for one class should after all not cover many examples in other classes than the class it is predicting. The technique introduced by [22], for instance, exploits this observation by finding patterns that are frequent within one class, but infrequent in the other. It exploits a relationship with version space theory from machine learning.

The CCCS classifier [4] even relies only on a maximum support constraint and removes the minimum support constraint entirely. It is argued that infrequent patterns in a class can be found by enumerating small subsets of transactions in this class.

The problem that remains in each of these cases is a similar one as for single support thresholds: how to set the parameters. A pattern that occurs in 50 % of one class, and 15 % of the other, could be considered a valuable predictive pattern, as might be a pattern that occurs in 80 % of the first and 30 % of the second. Support constraints that accommodate both patterns, however, e.g. $supp_{\min} = 0.5$, $supp_{\max} = 0.3$ would allow results of questionable usefulness.

To address this, the Fitcare classifier proposed by [10] takes this idea further and uses a much larger parameter set: given k classes, each class is mined separately, parametrized by a minimum support constraint and $k - 1$ maximum support constraints on all other classes. To make this manageable, the support constraints are dynamically adjusted during mining.

Statistical Measures A popular alternative approach is the use of constraints on measures specifically designed for supervised data. These measures typically serve as a replacement for confidence in selecting relevant predictive patterns; the underlying patterns are still found using a minimum support threshold on the complete data.

As a straightforward example, consider the *accuracy* measure:

Definition 2.5 *Given two classes $\mathcal{D}^+$, $\mathcal{D}^-$, pattern r. The accuracy of r is defined as* $acc(r) = \frac{supp_{\mathcal{D}^+}(r) + (|\mathcal{D}^-| - supp_{\mathcal{D}^-}(r))}{|\mathcal{D}|}$.

In general, most measures for evaluating the predictive power of a rule can be expressed as functions from the values in the *contingency table*:

	C_+	$\neg C_+$	
r	$p = supp_{\mathcal{D}^+}(r)$	n	$supp_{\mathcal{D}}(r)$
$\neg r$	$\lvert\mathcal{D}^+\rvert - supp_{\mathcal{D}^+}(r)$	$N - n$	$supp_{\mathcal{D}}(\neg r) = \lvert\mathcal{D}\rvert - supp_{\mathcal{D}}(r)$
	$P = \lvert\mathcal{D}^+\rvert$	$N = \lvert\mathcal{D}\rvert - \lvert\mathcal{D}^+\rvert$	$\lvert\mathcal{D}\rvert$

An arrangement in a contingency table invites the use of well-established measures such as *Information Gain* or χ^2 to mine *correlating* [29], *contrast* [6], or *discriminating* patterns [11]. Similarly, the *growth rate* can be used to mine *emerging patterns* [14, 26, 37]. It divides the support in one class by the support in the other one.

A measure that is often used in *subgroup discovery* is *Weighted Relative Accuracy* [23]:

Definition 2.6 *Given a rule* $r \to C_+$, *its Weighted Relative Accuracy is defined as*
$$WRAcc(r \to C_+) = \frac{supp_{\mathcal{D}}(r)}{\lvert\mathcal{D}\rvert}\left(\frac{supp_{\mathcal{D}^+}(r)}{supp_{\mathcal{D}}(r)} - \frac{\lvert\mathcal{D}^+\rvert}{\lvert\mathcal{D}\rvert}\right).$$

It is instructive to compare *accuracy* and *WRAcc* to gain a better understanding of the conceptual differences between classification and subgroup discovery.

Since the final goal is to find rules with good predictive accuracy, accuracy treats covering one negative instance less as equal to covering one positive instance more. Consider a data set consisting of 60 instances in $\mathcal{D}^+$, and 40 in $\mathcal{D}^-$, and a rule covering 40 positive and 15 negative instances. Its accuracy is 0.65, and rules that covered 5 positive instances more, or 5 negative instances less, would both achieve a (better) accuracy of 0.7. In the case of WRAcc, the situation is different: the original rule would have a score of 0.07 and while covering 5 negative instances less improves it to 1.0, covering 5 positive instances more yields a smaller improvement (to 0.09).

Since subgroup discovery aims to characterize *differences*, this behavior makes perfect sense: the positive class is overrepresented in the entire data and coverage of this class has to increase more strongly to be interesting. Given a heavily skewed data set (e.g. $\lvert\mathcal{D}^+\rvert = 0.9\lvert\mathcal{D}\rvert$), a rule predicting all transactions to belong to the majority class might be acceptable for a classifier but would be unattractive for subgroup discovery.

WRAcc also includes a normalizing factor that weights a rule's score by its *effect size* but this is in fact *not* particular to subgroup discovery. When it comes to normalization, the difference between classification and subgroup discovery measures lies in the motivation: classification wants assurance that mined rules will work on unseen data, subgroup discovery wants rules to be representative of the data they have been mined from.

In combination with a minimum support constraint, *WRAcc* can be used in a class association rule miner instead of confidence [21]. This idea can be generalized to other subgroup discovery measures (and the measures listed above), replacing the confidence measure in class association rule miners by numerous other functions as proposed by [5]. CMAR, for instance, filters *cars* using a χ^2 minimum threshold in addition to the minimum confidence threshold.

That the differences between different types of supervised patterns mainly come down to a change in quality function has been shown in detail by [32], the authors of which coined the term "supervised descriptive rule discovery" for such approaches and has been leveraged by [44] to use one type of mining technique to address different tasks: classification, subgroup discovery, and conceptual clustering.

Eliminating Minimum Support The above settings essentially apply statistical measures in addition to minimum support. The minimum support parameter remains a parameter that needs to be set. Several approaches have successfully eliminated this parameter.

The main observation is that thresholds on quality measures can be translated into support thresholds; hence, if a support threshold is not given, it is possible to automatically determine an additional support threshold for use in a pattern mining algorithm.

Returning to the accuracy measure, we can set a minimum threshold on it: $acc(r) \geq \theta_{acc}$. This can be transformed into $p + (N - n) \geq \theta_{acc} \cdot |\mathcal{D}|$, and further into $p \geq \theta_{acc} \cdot |\mathcal{D}| - N + n \geq \theta_{acc} \cdot |\mathcal{D}| - N$. So we derive support constraints based on the threshold on the quality measure itself [31].

For measures that are convex, which includes the ones mentioned above but also many others, a similar argument is possible: convex functions take their maxima at extreme points, i.e. points with $p = 0$ or $n = 0$. Thus, based on a threshold on the minimal acceptable values for a statistical scoring function, thresholds on a pattern's p and n can be derived and enforced during mining. This makes it effective to use the quality measure to prune *during* rule mining [6, 7, 11, 12, 15, 19, 29, 31, 35, 39, 43–45].

Thus far we have discussed approaches that use thresholds and exhaustively mine all patterns that satisfy the thresholds. An even easier and often more effective approach is to perform top-k mining instead. In top-k mining, one is interested in finding only those patterns which have the k highest scores; the only parameter that needs to be specified is k. This has been leveraged by [7, 11, 12, 15, 35, 42, 45]. The nature of this mining process means that the threshold(s) increase during mining, pruning more and more candidate patterns as the search progresses. To achieve a quick increase of the threshold, it can be useful to perform a *best-first* search during which it is always the rule with the highest upper bound that is specialized.

2.3 Numerical Target Values

As opposed to the setting discussed in the preceding sections, in which each transaction is labeled with one out of a set of *discrete* labels, a numerical variable of interest can have potentially infinitely many values. As a result of this, each transaction in the data may have a different target value, and learning a predictor for particular values, or partitioning the data into subsets consisting of transactions with the same target value, are strategies that are unlikely to be successful. Nevertheless, there exist a number of techniques for *discretizing* numerical values, in which case the problem can be reduced to the classification setting.

Alternatively, one can either attempt to mine patterns that partition the data into transactions that have approximately the same numerical value, or those that can be used as elements of a regression function that outputs a numerical result based on their appearance in a transaction. An interestingness measure that can be used in the former case is *interclass variance*:

Definition 2.7 *Given a data set with numerical labels of the form* $\mathcal{D}_Y = \{(d_1, y_1),$
$\ldots, (d_n, y_n)\}$, $y_i \in R$, *pattern* π, *the average y in a subset* $\mathcal{D}_\subseteq$ *of that data set is:*

$$avg(\mathcal{D}_\subseteq) = \frac{\sum_{(d_i, y_i) \in \mathcal{D}_\subseteq} y_i}{|\mathcal{D}_\subseteq|}$$

The interclass variance of π *is defined as:*

$$var(\pi) = |cov(\pi)| \, (avg(cov(\pi)) - avg(\mathcal{D}_Y))^2$$
$$+ |\mathcal{D}_Y \setminus cov(\pi)| \, (avg(\mathcal{D}_Y \setminus cov(\pi)) - avg(\mathcal{D}_Y))^2$$

Interclass variance is convex, which means that thresholds on its value can be translated into thresholds on support values, and thresholded or top-k mining used in the same manner as for discrete target values.

In the latter case, works such as [13, 33, 34] have chosen linear regression functions that weight the contributions of individual patterns. Based on these weights, the authors define a quality function for individual patterns, and derive upper bounds that they use to perform top-k mining for component patterns of the regression model.

3 Supervised Pattern Set Mining

The result of a supervised pattern mining operation, as so often in pattern mining settings, is typically a very large set of redundant and contradictory patterns. Even when mining only the top-k patterns, many of those will cover (almost) the same instances. As we mentioned in the introduction, when constructing classifiers, redundant patterns or patterns that are irrelevant in the presence of others can be undesirable. If the classifier takes the form of an unordered rule set, for instance, which we will describe in Sect. 4, certain rules could strongly boost each other, far in excess of their actual relevance and usefulness.

Hence many techniques in the literature include a mechanism for mining or selecting a subset of the result set. Where the techniques for supervised pattern mining intended to improve on Machine Learning techniques, replacing heuristics with exhaustive search, the methods for supervised *pattern set mining* are strongly inspired by Machine Learning techniques. In particular, both sequential (covering/re-weighting) or *separate-and-conquer*, and decision tree like *divide-and-conquer* techniques can be found time and again in works on supervised pattern mining.

There are two wide-spread approaches to pattern set mining. One is *post-processing*:

1. Mine a set of supervised patterns satisfying certain constraints
2. Select some patterns out of this set following certain criteria

and *iterative pattern are derived set mining*:

1. Mine a (set of) supervised pattern(s) satisfying certain constraints
2. Modify the constraints or data
3. Return to 1.

The main argument in favor of the post-processing approach is its efficiency. It allows to run a pattern mining algorithm only once and hence avoids the possibly time consuming repeated execution of pattern mining algorithms. The main arguments in favor of iterative mining algorithms are their potentially higher accuracy and their potential to use parameter-free pattern mining algorithms; in many of these algorithms, it is not necessary to define a minimum support threshold in advance.

Both separate-and-conquer and divide-and-conquer techniques have been used within either of these categories.

Most of these techniques can be understood in terms of the partition that a set of patterns induces on the data. We therefore first need to introduce the concept of *equivalence relations* and *partitions*:

Definition 3.1 *An equivalence relation on D is a binary relation $\sim$ such that for all $d_1, d_2, d_3 \in D$, the relation is:*

1. *Reflexive: $d_1 \sim d_1$.*
2. *Symmetric: $d_1 \sim d_2 \Rightarrow d_2 \sim d_1$.*
3. *Transitive: $d_1 \sim d_2 \wedge d_2 \sim d_3 \Rightarrow d_1 \sim d_3$.*

The equivalence relation partitions D into disjunct subsets called equivalence classes or blocks. The equivalence class of an element $d \in D$ is given as $[d] = \{d' \in D \mid d \sim d'\}$. The set of blocks is called partition or quotient set, and is denoted by $D/\sim$.

Intuitively, transactions are in an equivalence class if they can not be distinguished from each other. We can use patterns to create a new database, in which each transaction is described by a list of patterns present in it. We consider two transactions equivalent in this new representation if they are described using the same lists of patterns.

More formally, an individual pattern r induces an equivalence relation $\forall d_1, d_2 \in D, d_1 \sim_r d_2 \Leftrightarrow match(r, d_1) = match(r, d_2)$, and so does a set of patterns $\mathbb{P}$: $\forall d_1, d_2 \in D, d_1 \sim_{\mathbb{P}} d_2 \Leftrightarrow (\forall r \in \mathbb{P} : match(r, d_1) = match(r, d_2))$.

In fact, the partitioning of a data set into classes that we defined in Definition 2.3 is induced by an equivalence relation based on the class labels.

In a supervised setting, it are derived is important to distinguish blocks which are *pure* and which are not pure. A block is pure if all examples in it have the same class label. Within a supervised setting it is important that the partition induced by a set of patterns contains mostly pure blocks: if two examples with different class labels contain exactly the same set of patterns, it will be impossible for a deterministic algorithm to predict both correctly.

Most pattern set mining techniques can be summarized in the following manner:

1. Mine or evaluate a (set of) pattern(s), possibly only on parts of the data
2. Based on result of 1, modify the partition, for instance by removing one block or several blocks, or by partitioning them further
3. Return to 1, unless a stopping criterion is met

The differences lie mainly in the blocks on which patterns are evaluated, and in the choice of blocks that are modified.

3.1 Local Evaluation, Local Modification

The first, and largest, class of techniques evaluates or mines patterns *locally*, i.e. only on some of the blocks of a partition, and then also modifies only some of those blocks, typically only those blocks from which the patterns have been mined. This includes in particular those techniques that draw more or less directly on machine learning forebears.

Separate-and-Conquer *Sequential* "local-local" techniques owe much to the sequential covering paradigm of early rule learners. They start from the full database and iteratively remove examples from the dataset, as follows:

1. Find the best rule on the currently remaining data
2. Remove all data covered by that rule
3. Return to 1.

This approach falls squarely into the "local-local" category. Each pattern splits the data that it has been *mined on* into two blocks (the local modification) and its successor pattern is only mined on *one* of these, the uncovered one (the local evaluation). Several early algorithms have used this approach for post-processing, for instance CBA, ARC-BC, and CMAR, whose authors refer to it as *database coverage.*

Separate-and-conquer can be applied both in the post-processing setting and in the iterative mining setting.

Post-processing can be done in two ways: (1) considering the *complete* set of previously mined patterns in each iteration of the sequential covering algorithm, or (2) fixing the order in which patterns are considered and only search for the best rule among those rules that have not been considered in the order yet. The latter means that (a) each pattern is only considered once—if it is rejected, it will never be evaluated again, and (b) the decision which patterns are "best" given certain data is effectively made before pattern set mining. In return, however, the complexity of the learning algorithm is lower.

The algorithms mentioned above (CBA, ARC-BC, CMAR) proceed by fixed order. CMAR differs from the other algorithms in removing data instances only after they have been covered by *several* rules, guided by a user-supplied parameter. CORCLASS also uses sequential covering with a fixed order as post-processing. Another variation was proposed by [1] in the context of string classification; here, the rules are processed in order of confidence, but only those instances are removed which are classified correctly by the rule under consideration.

The ART algorithm [17] learns several best *cars*, splits their respective coverage off, and re-iterates on the uncovered data. DDPMine by [12] is perhaps the algorithm that stays truest to the original sequential covering idea: it mines a highest-scoring pattern, removes all covered instances from the data, and recurs.

Divide-and-Conquer Techniques The second type of "local-local" techniques takes its cues from *decision tree induction*:

1. Find the best splitting criterion on a subset of the data
2. Split the data into two blocks corresponding to covered and uncovered instances
3. Recur on the new blocks

A potential advantage of this type of technique is that all mistakes by one pattern can be corrected by other patterns, since all data are reused in later instances to derive additional patterns. In addition, patterns that might not appear interesting on the whole data might become relevant as soon as parts of the data are removed.

This technique is most commonly used in an iterative mining setting, in which the best pattern is searched for using a branch-and-bound top-1 pattern mining algorithm. Examples are Tree^2, proposed by [7], and M^{b}T [15].

A post-processing approach can also be used. For instance, [18] developed a setting in which δ-free patterns are first mined, and then combined for use as tests in a decision tree.

3.2   Global Evaluation, Global Modification

Alternatively, patterns can be mined or evaluated on the *entire* data set, and *all* blocks in the partition are modified. While this means that mining (or selecting) patterns is done using the maximal amount of information, this usually has to be paid for by increased computational complexity, as in each iteration the complete data needs to be traversed. Additionally, the semantics of patterns' relationships are less easy to understand than in the case of "local-local" approaches.

Such techniques necessarily proceed sequentially, either post-processing or mining patterns one after another. The Picker* algorithm by [8] performs post-processing in this manner, picking the pattern that creates the most balanced partition, and splitting all blocks accordingly. It proceeds according to the first option for post-processing described above, considering all promising patterns. The FCORK [35] technique uses a measure based on *correspondences*:

Definition 3.2   *Given an equivalence relation $\sim_\mathbb{P}$ on a labeled data set $\mathcal{D}_C = \mathcal{D}^+ \cup \mathcal{D}^-$, the number of correspondences in this partition is calculated as $occ(\mathbb{P}) = \sum_{[d] \in \mathcal{D}_C/\sim_\mathbb{P}} |[d] \cap \mathcal{D}^+| \cdot |[d] \cap \mathcal{D}^-|$.*

and uses this measure both to post-process mined patterns, and to iteratively mine patterns that reduce correspondences the most. This criterion, as well as that used by the Picker* algorithm, is *sub-modular*, allowing to give a bound on the quality of greedy approximation to the optimal solution.

There are other "global-global" techniques that differ in that they do not manipulate the data explicitly: the technique introduced by [11] post-processes patterns by rewarding them for class correlation on the full data and penalizing overlap on data already covered by selected patterns. The Krimp technique, described by [36], also falls into this category since it evaluates for each pattern how much it adds to the overall, i.e. global, compression of the data, post-processing a fixed order on patterns.

Instead of removing examples, a reasonable alternative is to attach a weight to examples and modify the weights based on the current composition of a rule set, as in the following generic approach:

1. Find the best rule on the current weighted data
2. Modify the weights of the examples in the data
3. Return to 1.

A reason to give a lower weight to an example may for instance be that we already have many rules that predict this example correctly, and we would like to focus on finding rules for examples that are predicted incorrectly.

This setting performs global evaluation as each new pattern is evaluated on the complete dataset and in principle the weights of all examples can be modified.

Examples of approaches within this setting were proposed by [13, 33, 34, 44]; they can be used either in iterative mining or in post-processing. In the first work, transaction weights are adjusted directly in a subgroup discovery setting. Since subgroup discovery is more concerned with mining good descriptions of statistically different subgroups than with accurate prediction, the removal of covered instances is undesirable. The other works, comprising the GPLS and GBOOST algorithms, and a Bayesian linear regression technique, derive the transaction weights indirectly from weights for patterns involved in a linear classification or regression function and mine patterns iteratively. Since pin point prediction of a numerical value is difficult, reweighting instances based on the current performance of the function is superior to removing instances.

The upshot of these techniques is that the increased computational complexity pays off in a pattern set of smaller cardinality than for "local-local" approaches, typically of comparable or even better quality.

3.3   Local Evaluation, Global Modification

Given the faster running times yet larger pattern sets of "local-local" approaches, and the more expensive operation yet smaller, high-quality sets of "global-global" techniques, the development of "local-global" algorithms should be obvious:

1. Find a best pattern on a subset of the data

2. Based on *all* patterns, manipulate the *entire* data
3. Recur on the new blocks

Notwithstanding this statement, the REMINE algorithm proposed by [45] so far is the only one to proceed in this way to iteratively mine supervised patterns.

3.4 Data Instance-Based Selection

In addition to the partition-based techniques, there is another paradigm, which selects patterns based on individual instances. The Harmony algorithm retains for each training instance the highest-confidence rule, as does CCCS, whereas the technique described by [28], called Large Bayes (LB), selects patterns based on the instances whose labels are to be predicted. This is similar to DEEP, described by [26], and LAC, proposed by [37], which only *generate* patterns that match the instances to be predicted by projecting the data on the items contained in the unlabeled instance.

4 Classifier Construction

After supervised patterns have been mined, and suitable subsets have been selected, the remaining question is how to employ them for predictive purposes. The solutions that have been found fall into two main categories: (1) direct use of patterns as rules to predict the label of an unseen class—the techniques following this paradigm borrow heavily from rule learning approaches in machine learning, or (2) indirect use of patterns in a model; here patterns are typically treated as *features* that are used in well-established machine learning methods.

4.1 Direct Classification

There are two main methods in rule learning when it comes to making predictions. In decision lists, rules are ordered according to some criterion and the first rule that matches the unseen instance makes the prediction. For such classifiers to work requires rules with high accuracy that at the same time do not *overfit* the training data. This means that certain approaches to optimizing quality measures will work better than others: given that maximizing information gain or χ^2 trades off correlation with effect size, maximizing confidence or WRACC will be more suitable for such classifiers. CBA follows this first approach, ordering the rule list by confidence (descending), support (descending) and length (ascending), as does LAC, ordering by information gain (descending).

The second method consists of various voting mechanisms that collect all rules that match the unseen instance and has each class "gather votes" from them. This approach places less importance on the prediction of *individual* rules and is related to the *ensembles* idea from machine learning: if predictors' errors are uncorrelated, using several of them should remove many non-systematic errors.

A straightforward method consists of *majority voting*, in which the predicted class label is that predicted by the majority of rules. Alternatively, rules' votes can be weighted, by their accuracy, strength, or support in a given class, for instance, and the class with the strongest vote is predicted. Many pattern-based classifiers use this scheme: CMAR performs weighted voting, discounting rules' vote by their deviation from their potentially maximal χ^2-score, whereas FITCARE simply adds up rules relative support per class, as does ARC-BC. CAEP sums up patterns' growth rate multiplied by their relative support in a class, and DEEP takes the proportion of instances in a class that contain any of the voting patterns as the weight of the vote for that class. Harmony includes three voting options: either the *highest-confidence* rule, or *all*, or the top-k rules vote for a particular class, similar to XRULES, which also uses different rule strength measures.

CTC has used different options: the decision list, majority vote, and two weighted voting strategies, as has CORCLASS.

The analogy with machine learning is exploited most in the GBOOST algorithm [34]. In GBOOST, an analogy is observed between *weak learners* and patterns. This analogy is exploited by modifying the LPBOOST boosting algorithm, developed in the machine learning literature, to iteratively search for patterns instead of weak learners. It can be shown that under certain conditions this algorithm finds optimal linear classification and regression models, where patterns are used as features in the linear models. The boosting algorithm operates by iteratively modifying the weights of examples based on the outcome of a linear program.

A particular feature of some sets of rules is that they represent decision trees. Essentially, every path from the root of a decision tree to a leaf of a tree can be seen as a rule that predicts the label of that leaf. All the rules cover disjoint parts of the data. It is hence not surprising that patterns can also be used to represent paths in decision trees. This observation was exploited in the DL8 approach by [30], which showed that by post-processing a set of patterns found under constraints, a decision tree can be constructed that is optimal under certain conditions. The approach differs from Tree^2 (see below) in that each pattern represents a path in the tree, while in Tree^2 each pattern represents a node.

4.2   Indirect Classification

Indirect classification comes in several flavors. First, there are the techniques that partition the data, sort unseen instances into a certain block, and use the majority label of the block's instances in the training data to make the prediction, like decision trees. The Tree^2 and M^{b}T build this kind of classifier. Other machine learning formalisms can also be adopted to work with supervised patterns—the LB algorithm uses a Naïve Bayes-like formulation to derive predictions from the support of patterns in different classes—different classes have different products of probabilities and the class with the highest probability is predicted.

This is somewhat similar to the Krimp algorithm: in this technique, *coding tables* are created for each class separately, and an unseen instance's label is predicted based on the coding table that compresses it best.

These approaches are arguably still limited by what the pattern themselves can do, although the upshot is that their models are somewhat more understandable. The alternative is to mine patterns as features for use in sophisticated machine learning techniques that can add modeling and generalization capabilities that are missing from symbolic patterns themselves. This is the second big group of techniques: the technique proposed by [22] belongs to it, as does DDPMINE, the method introduced by [12], PICKER*, FCORK, and REMINE.

5 Summary

In this chapter, we have given a high level overview of supervised pattern mining and its application to prediction, specifically classification. We have abstracted from the pattern languages used and structured the chapter along the three main steps involved in building a classifier from class-labeled data: supervised pattern mining, supervised pattern set mining, and classifier construction.

Regarding the first step, we have laid out that many techniques view different classes as separate subsets of the data and evaluate patterns' co-occurrence with one of these subsets. In our opinion, this view clarifies that different quality measures will lead to similar semantical information of patterns, and that different mining approaches can be taken to find patterns that score highly with any of these measures.

Regarding the second step, we have pointed out the similarities to approaches that have been pioneered in machine learning in the context of rule learning, decision tree induction, and instance-based learning. We have interpreted the former two approaches in terms of partitions to show the similarities of existing techniques, and also identified two types of approaches that always manipulate the entire data. Although some pattern set mining techniques, in particular iterative ones, make certain demands on the pattern mining step, most of them can still be combined relatively freely with different pattern mining techniques.

Finally, when it comes to classifier building, we have made the distinction between direct and indirect classification, with the former paralleling rule-based classification in machine learning, and the latter comprising quite a few approaches that mine patterns as *features* for use in propositional learners. As a comparison of references shows, different classifiers also do not track closely with particular pattern or pattern set mining approaches.

In general, in surveying the field we find that many solutions to the three phases have been developed, most of which can be mixed-and-matched rather freely. The field is larger than the algorithms we have mentioned here yet many techniques are arguably variations of the approaches that we have contrasted.

References

1. C. C. Aggarwal. On effective classification of strings with wavelets. In *KDD*, pages 163–172. ACM, 2002.
2. R. Agrawal, H. Mannila, R. Srikant, H. Toivonen, and A. I. Verkamo. Fast discovery of association rules. In *Advances in Knowledge Discovery and Data Mining*, pages 307–328. AAAI/MIT Press, 1996. ISBN 0-262-56097-6.
3. M.-L. Antonie and O. R. Zaïane. Text document categorization by term association. In *ICDM*, pages 19–26. IEEE Computer Society, 2002.
4. B. Arunasalam and S. Chawla. CCCS: a top-down associative classifier for imbalanced class distributions. In T. Eliassi-Rad, L. H. Ungar, M. Craven, and D. Gunopulos, editors, *KDD*, pages 517–522. ACM, 2006.
5. M. Atzmüller and F. Puppe. SD-Map-a fast algorithm for exhaustive subgroup discovery. In [16], pages 6–17. ISBN 3-540-45374-1.
6. S. D. Bay and M. J. Pazzani. Detecting group differences: Mining constrast sets. *Data Mining and Knowledge Discovery*, 5 (3): 213–246, 2001.
7. B. Bringmann and A. Zimmermann. Tree^2-Decision trees for tree structured data. In A. Jorge, L. Torgo, P. Brazdil, R. Camacho, and J. Gama, editors, *9th European Conference on Principles and Practice of Knowledge Discovery in Databases*, pages 46–58. Springer, 2005.
8. B. Bringmann and A. Zimmermann. One in a million: picking the right patterns. *Knowledge and Information Systems*, 18 (1): 61–81, 2009.
9. B. Bringmann, S. Nijssen, and A. Zimmermann. Pattern-based classification: A unifying perspective. In A. Knobbe and J. Fürnkranz, editors, *From Local Patterns to Global Models: Proceedings of the ECML/PKDD-09 Workshop (LeGo-09)*, pages 36–50, 2009.
10. L. Cerf, D. Gay, N. Selmaoui-Folcher, B. Crémilleux, and J.-F. Boulicaut. Parameter-free classification in multi-class imbalanced data sets. *Data Knowl. Eng.*, 87: 109–129, 2013.
11. H. Cheng, X. Yan, J. Han, and C.-W. Hsu. Discriminative frequent pattern analysis for effective classification. In *Proceedings of the 23rd International Conference on Data Engineering*, pages 716–725. IEEE, 2007.
12. H. Cheng, X. Yan, J. Han, and P. S. Yu. Direct discriminative pattern mining for effective classification. In *Proceedings of the 24th International Conference on Data Engineering*, pages 169–178. IEEE, 2008.
13. S. Chiappa, H. Saigo, and K. Tsuda. A Bayesian approach to graphy regression with relevant subgraph selection. In *SDM*, pages 295–304. SIAM, 2009.
14. G. Dong, X. Zhang, L. Wong, and J. Li. Caep: Classification by aggregating emerging patterns. In S. Arikawa and K. Furukawa, editors, *Discovery Science*, volume 1721 of *Lecture Notes in Computer Science*, pages 30–42. Springer, 1999. ISBN 3-540-66713-X.
15. W. Fan, K. Zhang, H. Cheng, J. Gao, X. Yan, J. Han, P. S. Yu, and O. Verscheure. Direct mining of discriminative and essential frequent patterns via model-based search tree. In Y. Li, B. Liu, and S. Sarawagi, editors, *Proceedings of the 14th ACM SIGKDD International Conference on Knowledge Discovery and Data Mining*, pages 230–238. ACM, 2008. ISBN 978-1-60558-193-4.
16. J. Fürnkranz, T. Scheffer, and M. Spiliopoulou, editors. *Knowledge Discovery in Databases: PKDD 2006,10th European Conference on Principles and Practice of Knowledge Discovery in Databases, Berlin, Germany, September 18–22, 2006, Proceedings*, 2006. Springer. ISBN 3-540-45374-1.
17. F. B. Galiano, J. C. Cubero, D. Sánchez, and J.-M. Serrano. Art: A hybrid classification model. *Machine Learning*, 54 (1): 67–92, 2004.
18. D. Gay, N. Selmaoui, and J.-F. Boulicaut. Pattern-based decision tree construction. In *ICDIM*, pages 291–296. IEEE, 2007.
19. H. Grosskreutz, S. Rüping, and S. Wrobel. Tight optimistic estimates for fast subgroup discovery. In W. Daelemans, B. Goethals, and K. Morik, editors, *ECML/PKDD (1)*, volume 5211 of *Lecture Notes in Computer Science*, pages 440–456. Springer, 2008. ISBN 978-3-540-87478-2.

20. J. Han, J. Pei, and Y. Yin. Mining frequent patterns without candidate generation. In W. Chen, J. F. Naughton, and P. A. Bernstein, editors, *SIGMOD Conference*, pages 1–12. ACM, 2000. ISBN 1-58113-218-2.

21. B. Kavsek and N. Lavrac. Apriori-SD: Adapting association rule learning to subgroup discovery. *Applied Artificial Intelligence*, 20 (7): 543–583, 2006.

22. S. Kramer and L. De Raedt. Feature construction with version spaces for biochemical applications. In C. E. Brodley and A. P. Danyluk, editors, *ICML*, pages 258–265. Morgan Kaufmann, 2001. ISBN 1-55860-778-1.

23. N. Lavrač, B. Kavsek, P. A. Flach, and L. Todorovski. Subgroup discovery with CN2-SD. *Journal of Machine Learning Research*, 5: 153–188, 2004.

24. D. Leman, A. Feelders, and A. J. Knobbe. Exceptional model mining. In *ECML/PKDD (2)*, pages 1–16, 2008.

25. W. Li, J. Han, and J. Pei. CMAR: Accurate and efficient classification based on multiple class-association rules. In N. Cercone, T. Y. Lin, and X. Wu, editors, *Proceedings of the 2001 IEEE International Conference on Data Mining*, pages 369–376, San José, California, USA, Nov. 2001. IEEE Computer Society.

26. J. Li, G. Dong, K. Ramamohanarao, and L. Wong. A new instance-based lazy discovery and classification system. *Machine Learning*, 54 (2): 99–124, 2004.

27. B. Liu, W. Hsu, and Y. Ma. Integrating classification and association rule mining. In R. Agrawal, P. E. Stolorz, and G. Piatetsky-Shapiro, editors, *Proceedings of the Fourth International Conference on Knowledge Discovery and Data Mining*, pages 80–86, New York City, New York, USA, Aug. 1998. AAAI Press.

28. D. Meretakis and B. Wüthrich. Extending naïve bayes classifiers using long itemsets. In U. M. Fayyad, S. Chaudhuri, and D. Madigan, editors, *KDD*, pages 165–174. ACM, 1999. ISBN 1-58113-143-7.

29. S. Morishita and J. Sese. Traversing itemset lattices with statistical metric pruning. In *Proceedings of the Nineteenth ACM SIGACT-SIGMOD-SIGART Symposium on Principles of Database Systems*, pages 226–236, Dallas, Texas, USA, May 2000. ACM.

30. S. Nijssen and É. Fromont. Optimal constraint-based decision tree induction from itemset lattices. *Data Min. Knowl. Discov.*, 21 (1): 9–51, 2010.

31. S. Nijssen and J. N. Kok. Multi-class correlated pattern mining. In F. Bonchi and J.-F. Boulicaut, editors, *KDID*, volume 3933 of *Lecture Notes in Computer Science*, pages 165–187. Springer, 2005. ISBN 3-540-33292-8.

32. P. K. Novak, N. Lavrac, and G. I. Webb. Supervised descriptive rule discovery: A unifying survey of contrast set, emerging pattern and subgroup mining. *Journal of Machine Learning Research*, 10: 377–403, 2009.

33. H. Saigo, N. Krämer, and K. Tsuda. Partial least squares regression for graph mining. In Y. Li, B. Liu, and S. Sarawagi, editors, *Proceedings of the 14th ACM SIGKDD International Conference on Knowledge Discovery and Data Mining*, pages 230-238. ACM, 2008., pages 578–586. ISBN 978-1-60558-193-4.

34. H. Saigo, S. Nowozin, T. Kadowaki, T. Kudo, and K. Tsuda. gboost:a mathematical programming approach to graph classification and regression. *Machine Learning*, 75 (1): 69–89, 2009.

35. M. Thoma, H. Cheng, A. Gretton, J. Han, H.-P. Kriegel, A. J. Smola, L. Song, P. S. Yu, X. Yan, and K. M. Borgwardt. Discriminative frequent subgraph mining with optimality guarantees. *Statistical Analysis and Data Mining*, 3 (5): 302–318, 2010.

36. M. van Leeuwen, J. Vreeken, and A. Siebes. Compression picks item sets that matter. In [16], pages 585–592. ISBN 3-540-45374-1.

37. A. Veloso, W. M. Jr., and M. J. Zaki. Lazy associative classification. In *ICDM*, pages 645–654. IEEE Computer Society, 2006.

38. J. Wang and G. Karypis. Harmony: Efficiently mining the best rules for classification. In *SDM*, 2005.

39. G. I. Webb. Opus: An efficient admissible algorithm for unordered search. *J. Artif. Intell. Res. (JAIR)*, 3: 431–465, 1995.

40. M. J. Zaki. Scalable algorithms for association mining. *IEEE Trans. Knowl. Data Eng.*, 12 (3): 372–390, 2000.

41. M. J. Zaki and C. C. Aggarwal. XRules: an effective structural classifier for XML data. In L. Getoor, T. E. Senator, P. Domingos, and C. Faloutsos, editors, *Proceedings http://www.nakedcapitalism.com/of the Ninth ACM SIGKDD International Conference on Knowledge Discovery and Data Mining*, pages 316–325, Washington, DC, USA, Aug. 2003. ACM.

42. A. Zimmermann and B. Bringmann. Ctc-correlating tree patterns for classification. In J. Han, B. W. Wah, V. Raghavan, X. Wu, and R. Rastogi, editors, *Proceedings of the Fifth IEEE International Conference on Data Mining*, pages 833–836, Houston, Texas, USA, Nov. 2005. IEEE.

43. A. Zimmermann and L. De Raedt. Corclass: Correlated association rule mining for classification. In E. Suzuki and S. Arikawa, editors, *Proceedings of the 7th International Conference on Discovery Science*, pages 60–72, Padova, Italy, Oct. 2004. Springer.

44. A. Zimmermann and L. De Raedt. Cluster-grouping: from subgroup discovery to clustering. *Machine Learning*, 77 (1): 125–159, 2009.

45. A. Zimmermann, B. Bringmann, and U. Rückert. Fast, effective molecular feature mining by local optimization. In J. L. Balcázar, F. Bonchi, A. Gionis, and M. Sebag, editors, *ECML/PKDD (3)*, volume 6323 of *Lecture Notes in Computer Science*, pages 563–578. Springer, 2010. ISBN 978-3-642-15938-1.

Chapter 18
Applications of Frequent Pattern Mining

Charu C. Aggarwal

Abstract Frequent pattern mining has broad applications which encompass clustering, classification, software bug detection, recommendations, and a wide variety of
other problems. In fact, the greatest utility of frequent pattern mining (unlike other
major data mining problems such as outlier analysis and classification), is as an
intermediate tool to provide pattern-centered insights for a variety of problems. In
this chapter, we will study a wide variety of applications of frequent pattern mining.
The purpose of this chapter is not to provide a detailed description of every possible
application, but to provide the reader an overview of what is possible with the use of
methods such as frequent pattern mining.

Keywords Frequent pattern mining

1 Introduction

This chapter provides an overview of the key applications of frequent pattern mining.
Frequent pattern mining was first proposed by Agrawal et al in 1993 [11, 13]. Since
then, a wide variety of tree-based and pattern growth-based algorithms have been
proposed for the problem [16, 27, 59, 114, 141]. An overview of algorithms for
frequent pattern mining may be found in [60].

Frequent pattern mining is one of the unusual problems in data mining, where
the size of the output may sometimes be comparable or larger than the size of the
input. For example, in the context of a database with a few thousand transactions, it
is often possible to obtain a number of frequent patterns which are of the same or
significantly larger magnitude. Therefore, a question arises as to the utility of such
large outputs from a mining algorithm, if they do not provide a *concise* summary or
characterization of the underlying data. In practice, these outputs are often used as
intermediary steps in other data mining applications. Therefore, the greatest utility
of frequent pattern mining algorithms is as an intermediary step, rather than as a
goal in of itself. Without using some form of post-processing, it is often difficult to

C. C. Aggarwal (✉)
IBM T. J. Watson Research Center, Yorktown Heights, NY 10598, USA
e-mail: charu@us.ibm.com

C. C. Aggarwal, J. Han (eds.), *Frequent Pattern Mining,* 443
DOI 10.1007/978-3-319-07821-2_18, © Springer International Publishing Switzerland 2014

use frequent patterns directly by simply examining them manually. This is different from many other major data problems such as outlier analysis and classification in which the output of the process is concise, usually a goal in of itself, and is usually presented directly to the user for manual inspection. Therefore, this chapter will focus on applications of frequent pattern mining, which serve as the most important motivating factor for frequent pattern mining algorithms.

The applications of frequent pattern mining span a very wide variety of fields, and also incorporate several different data domains. Correspondingly, different kinds of variations of frequent pattern mining may be used to address the unique problems which are specific to each domain. For example, the kinds of patterns mined will very different in the context of temporal, spatial, multimedia or biological data. Some examples of the wide variety of problem and data domains are as follows:

- **Customer Analysis:** Customer analysis is the original and motivating application for frequent pattern mining. The idea is that frequent correlations between customer buying behavior can be used in order to make useful business decisions.
- **Facilitator for other major data mining problems:** Frequent pattern mining has close connections with other major data mining problems such as clustering and classification. This is because frequent pattern mining is closely related to the problem of subspace clustering. Furthermore, discriminative frequent patterns can often be used to construct classifiers. Since the clustering problem is closely related to outlier analysis, frequent patterns are often used in order to determine outliers from the underlying data.
- **Indexing and Retrieval:** Frequent pattern mining algorithms can be used in order to design signature-based techniques for indexing and retrieval of market basket data. Since indexes often depend upon a concise representation of the underlying data, frequent pattern mining methods serve as an important intermediate step in the process.
- **Web Mining Tasks:** Sequential pattern mining algorithms are frequently used to determine important traversal patterns from Web logs. Such traversal patterns can be used in order to design and organize Web sites.
- **Software Bug Detection:** Frequent patterns can be used to determine bugs in software programs by using frequent pattern mining in order to determine the most relevant patterns in the underlying data.
- **Event Detection and Other Temporal Applications:** A variety of temporal applications such as event detection use frequent pattern mining methods. Many techniques have been designed for periodic pattern mining, event detection, and other related applications which use variants of frequent pattern mining methods as subroutines.
- **Spatial and Spatiotemporal Analysis:** Spatial data is one in which both spatial and non-spatial attributes are attached to objects (e.g. temperature readings on the sea surface). In such cases, association rules can characterize useful relationships between the spatial and non-spatial properties of the attributes. Spatio-temporal data such as trajectories can often be analyzed with the use of frequent pattern

mining methods. In particular, frequent patterns can be used to determine the key segments in the trajectories which are used frequently over time.

- **Image and Multimedia Data Mining:** In this case, features of images can be treated as attributes in transactions, and frequent patterns can be determined from these transactions in order to determine the important characteristics of images. Such characteristics can be used for a variety of mining tasks. Image data are closely related to spatial data, since the pixels in an image have spatial attributes as well as non-spatial attributes.
- **Chemical and Biological Applications:** Frequent patterns can be used to determine important motifs in a variety of chemical and biological applications. In many cases, these correspond to frequent patterns in graphs and structured data. Examples include toxicological analysis, chemical compound prediction, phylogenetic and RNA analysis.

This chapter will provide an overview of the afore-mentioned applications of frequent pattern mining. The number of possible applications of frequent pattern mining are varied, and arise in many domains. For example, different kinds of applications are possible within the context of set-based data (e.g. market baskets), graph-based data, or graphs represented as trees. While this chapter provides an idea of the landscape, the main goal is to cover the key scenarios in which frequent pattern mining can be applied. This will provide the reader the machinery for understanding how these techniques can be useful in different contexts.

This chapter is organized as follows. Customer analysis applications are discussed in Sect. 2. In Sect. 3, we discuss the problem of using frequent patterns for clustering. The problem of using frequent pattern mining for classification is discussed in Sect. 4. Applications of frequent pattern mining to outlier analysis are discussed in Sect. 5. Methods for using frequent pattern mining methods in indexing are discussed in Sect. 6. The use of frequent pattern mining methods in Web-related mining tasks is discussed in Sect. 7. Text applications of frequent pattern mining are discussed in Sect. 8. Applications for temporal data are discussed in Sect. 9. Methods for using frequent pattern mining for analyzing spatial and spatio-temporal data are discussed in Sect. 10. Methods for software bug detection are discussed in Sect. 11. Methods for mining biological and chemical data are discussed in Sect. 12. Section 13 discusses resources for the practitioner, which includes the key commercial and open-source software available for frequent pattern mining. The conclusions and summary are discussed in Sect. 14.

2 Frequent Patterns for Customer Analysis

The motivating application for frequent pattern mining was proposed in the context of supermarket and customer analysis [13]. In this case customer behavior is captured either by baskets of items bought together or by sequences of items which are bought in succession. Frequent patterns can be used in order to determine the common patterns of buying behavior. A rather old, but much used example of a frequent pattern is the 2-tuple $\{Beer, Diapers\}$, which suggests that the items $Beer$ and $Diapers$ are often bought together. This suggests that it may be useful to stock these items in shelves which are located close to each other. Furthermore, such information is also useful to making promotion decisions based on previous customer buying behavior.

Sequential pattern mining [12] is used in the context of very similar scenarios, except that a temporal component may exist in the transactions. In some cases, the temporal aspect of the data may be significant from the perspective of analysis. For example, a customer is more likely to buy a particular kind of printer ink, only *after* she has already bought the relevant printer. Therefore, the temporal aspect of the buying behavior provides more refined information for targeting purposes, when information about earlier periods is available.

In general however, frequent pattern mining is more useful as subroutine even in these applications. For example, in a customer targeting application, a rule-based classifier can be constructed from the discovered frequent patterns. In some cases, constraints may be used in order to further refine the discovered patterns [107, 109], whereas in other cases the sequential patterns may be used in order to make recommendations. This distinction is important because the vanilla problem of frequent pattern mining is almost never used in applications on a stand-alone basis. Some of these applications are discussed in detail in the following subsections.

3   Frequent Patterns for Clustering

The problem of frequent pattern mining is closely related to other data mining problems such as clustering. The simplest relationship between clustering and frequent patterns is discussed in [124], where large items are used in order to enable the clustering process. The idea is that clusters of transactions will have a large overlaps between their frequent items. Much more sophisticated methods for clustering are possible if correlations among the items are used directly in the clustering process.

In particular, the original definition of subspace clustering [14] is closely related to the problem of association rule mining. The CLIQUE algorithm discretizes the original data into intervals, and uses these intervals as pseudo-items in order to determine relevant patterns. A density measure is used as a surrogate for the support in the order to determine the frequent patterns. Specifically, the density measure requires that each cell should contain a particular minimum number of data points in order to be considered a relevant candidate. The subsequent k-dimensional grid structures are then re-constructed together in order to create the broader contours of the subspace clusters in the data. A related method known as ENCLUS [32] was proposed, in which the subspace clusters are quantified with the use of an entropy measure, rather than a density-based measure. Such an entropy-based measure sometimes has some advantages because of better normalization. Since then, a significant amount of work has been done in the area of subspace clustering. These techniques have been used both in the context of biclustering [93, 126] of discrete data, and in the context of projected clustering [137]. Such methods have also been used for a variety of applications such as grouping Web transactions [135]. Subsequently, a very large number of methods have been designed for clustering high-dimensional data with the use of pattern-based methods. A detailed discussion of the connections between such high dimensional clustering algorithms and the frequent pattern mining problem may be found in the survey article [106] and in chapter on high dimensional data in [4].

A second problem is on using pattern mining methods for clustering discrete attributes such as the case of biological data. Clusters can be considered as an orthogonal representation of the *localized* associations, as is the case for all subspace clustering methods. Such a technique for finding localized associations and clusters simultaneously is discussed in [9]. In this work, it is shown that localized associations can be enhanced, when local regions of the data are explored simultaneously with the association analysis process. At the same time, the clustering process is enhanced as well. This is also the general principle in many clustering methods such as matrix factorization and co-clustering [4]. Biological data is often represented as a sequence of discrete values corresponding to the amino-acids or the DNA/RNA bases. The sequences are usually too long to be clustered purely by similarity computations alone. Therefore, the use of pattern or motif-mining can be very useful in these cases. An example of a sequence-based clustering approach is the CLUSEQ method [136]. A common class of algorithms in this context is those of biclustering, in which clusters are constructed from frequent patterns in biological data [93, 99]. An excellent survey on biclustering methods may be found in [93]. The problem of motif discovery is very closely related to that of clustering in such domains. A discussion of different methods which connect the frequent pattern mining problem to the clustering problem in the context of biological data may be found in [4].

4   Frequent Patterns for Classification

The problem of data classification is closely related to that of frequent pattern mining, particularly in the context of *rule-based methods*. A classification rule is a condition of the form:

$$A_1 = a_1, A_2 = a_2 \Rightarrow C = c$$

In the case, the left hand side of the rule implies that attributes A_1 and A_2 should take on values a_1 and a_2 respectively, and the right hand side implies that the class value should be c. The training phase creates a set of rules from the labeled data, whereas the testing phase determines the relevant (or *fired*) rules, for which the left-hand side of the rule matches the test instance. The final class label for the test instance is determined as a carefully designed combination of the class labels on the right-hand side of the fired rules. In addition, a default (or catch-all) label may be defined, if no rules are fired by a test instance, in order to ensure full coverage.

Since classification rules are of a very similar form as association rules, it is possible to determine relevant patterns from the data with the use of association rule mining techniques. The main goal is to ensure that the patterns are sufficiently discriminative for classification, and the support criterion does not become too dominant in the rule selection process. The earliest work on the connections between classification and association rule mining was provided in [18]. Subsequently, one of the most popular methods for classification based on associations was the CBA (or *Classification Based on Associations*) method proposed in [87]. This method is also available as a practical software package [147]. Subsequently, another technique for classification on the basis of the FP-Growth method for association rule mining was the CMAR method [77]. Some techniques focus more directly on finding *discriminative* patterns, with a special focus on the discriminative power of the patterns with respect to the class labels. Discriminative frequent pattern mining methods, which are particularly tailored to classification are discussed in [33]. Such methods have also been used for software bug detection [90]. Methods for using discriminative frequent patterns in order to create decision trees are discussed in [49].

Such techniques have also been extended to other data domains. For examples methods for classification of structural data and graphs with the use of rule-based patterns are discussed in [140]. In these methods, discriminative subtrees and subgraphs are discovered from the underlying structured data, and are used for the purposes of classification. Some methods have also been designed for constructing classification rules from spatio-temporal data, in order to determine anomalies in the form of rare classes [82]. Rule-based methods have also been used in order to classify strings with the use of the wavelet representation [1]. The idea is that the wavelets provide a multi-granularity representation of the data on which the rules are constructed. Test sequences are classified by first converting them to the wavelet representation, and then using the relevant rules for classification purposes. The relevant rules are determined by matching the test instance with the predicates on the left hand side of the rules. Association rules have also been used for medical image classification in the context of spatial data [20].

The typical approach in all of these methods is quite similar. The first step is to mine all frequent patterns above a given support, as in standard classification mining algorithms. Such patterns may either be mined on either the entire database or on each class-specific database. The latter is preferred when there is a significant imbalance between the classes in order to ensure that the patterns relevant to the rare class are not lost in the pattern mining process. Subsequently, the confidence of each of these frequent patterns with respect to the class variable is determined. The patterns which have high confidence with respect to the class variable are then determined and reported. Since the number of possible rules which satisfy the support and confidence constraints may be very high, it is usually desirable to pick a small subset of rules which reflect the behavior in the training data effectively. In some methods such as in [122], the best rules for classification are mined directly, rather than as a post-processing phase in order to ensure better efficiency. This set of rules defines the training model for the classification process. For a given test instance, the set of rules for which the pattern on the left hand side match with the test instance are identified. These rules are prioritized with one or more criteria such as the confidence and support. This priority is used to determine which class is most relevant to the test instance by combining the votes from the different rules in a prioritized or weighted way. The precise method for prioritizing and weighting the rules may vary quite significantly in different applications.

5 Frequent Patterns for Outlier Analysis

Frequent pattern mining techniques are frequently used for outlier analysis in binary and transaction data. Since transaction data is inherently high-dimensional, it is natural to utilize subspace methods in order to identify the relevant outliers. The challenge in subspace methods is that it is no longer computationally practical or statistically feasible to define subspaces (or sets of items), which are sparse for outlier detection. For example, in a sparse transaction database containing hundreds of thousands of items, sparse itemsets are the norm rather than the rule. Therefore, a subspace exploration for sparse itemsets is likely to report the vast majority of patterns. The work in [62] addresses this challenge by working in terms of the relationship of transactions to dense subspaces, rather than sparse subspaces. In other words, this is a reverse approach of determining transactions, which are *not included* in most of the relevant dense subspace clusters of the data. In the context of transaction data, subspace clusters are essentially frequent patterns.

The idea in such methods is that frequent patterns are less likely to occur in outlier transactions, as compared to normal transactions. Therefore, a measure has been proposed in [63], which sums up the support of all frequent patterns occurring in a given transaction in order to provide the outlier score of that transaction. The total sum is normalized by dividing with the number of frequent patterns. However, this term can be omitted from the final score, since it is the same across all transactions.

Let $\mathcal{D}$ be a transaction database containing the transactions denoted by $T_1 \ldots T_N$. Let $s(X, \mathcal{D})$ represent the support of itemset X in $\mathcal{D}$. Therefore, if $FPS(\mathcal{D}, s_m)$ represents the set of frequent patterns in the database $\mathcal{D}$ at minimum the support level s_m, then, the frequent pattern outlier factor $FPOF(T_i)$ of a transaction $T_i \in \mathcal{D}$ at minimum support s_m is defined as follows:

$$FPOF(T_i) = \frac{\sum_{X \in FPS(\mathcal{D}, s_m), X \subseteq T_i} s(T_i, \mathcal{D})}{|FPS(\mathcal{D}, s_m)|}$$

Intuitively, a transaction containing a large number of frequent patterns with high support will have high value of $FPOF(T_i)$. Such a transaction is unlikely to be an outlier, because it reflects the major patterns in the data.

As in other subspace methods, such an approach can also be used in order to describe, why a data point may not be considered an outlier. Intuitively, the frequent patterns with the largest support, which are also not included in the transaction T_i are considered *contradictory patterns* to T_i. Let S be a frequent pattern not contained in T_i. Therefore, $S - T_i$ is non-empty, and the *contradictiveness* of frequent pattern S to the transaction T_i is defined by $s(S, \mathcal{D}) * |S - T_i|$. Therefore, a transaction which does not have many items in common with a very frequent itemset is likely to be one of the explanatory patterns for the T_i being an outlier. The patterns with the top-k values of contradictiveness are reported as the corresponding explanatory patterns.

At an intuitive level, such an approach is analogous to non-membership of data points in clusters in order to define outliers, rather than directly trying to determine the deviation or sparsity level of the transactions. As was discussed in the chapter on clustering-based methods, such an approach may sometimes not be able to distinguish between noise and anomalies in the data. However, the approach in [63] indirectly uses the weight and number of clusters in the outlier score. Furthermore, it uses *multiple* patterns in order to provide an ensemble score. This is at least partially able to alleviate the noise effects. In the context of very sparse transactional data, in which direct exploration of rare subspaces is infeasible, such an approach would seem to be a reasonable adaptation of subspace methods.

Frequent pattern mining methods are closely related to information theoretic measures for anomaly detection. This is because frequent patterns can be viewed as a code-book in terms of which to represent the data in a compressed form. It has been shown in [115], how frequent patterns can be used in order to create a compressed representation of the data set. Therefore, a natural extension is to use the description length [116] of the compressed data in order to compute the outlier scores. This approach was further improved in [17]. Pattern mining methods have also been used recently in the context of temporal data for outlier analysis [54].

6 Frequent Patterns for Indexing

Indexing algorithms typically require a concise representation of the data for mining purposes. Typically, clustering methods are used in order to create concise representations of the data for indexing purposes [8, 103]. The idea here is that the database of transactions are partitioned into groups on the basis of the broad patterns in them. This grouping is then used in order to perform branch-and-bound search during similarity-based query processing. Frequent pattern mining methods can also be an effective method to create such representations, since it is closely connected to the clustering problem. However, the methods in [8, 103] directly use clustering methods. Nevertheless, the use of frequent pattern mining would be a natural approach in the context of market basket data.

Such methods have however been used quite successfully in the context of graph indexing methods. A particular indexing structure which uses discriminative frequent patterns is the *gIndex* method [132]. The key idea here is that structures which are very similar will contain similar kinds of discriminative patterns. Therefore, such an approach defines similarity in the context of discriminative patterns in the underlying data. This work is also able to handle the fact that infrequent patterns may sometimes be relevant to similarity by using a size increasing support function, in which the support level depends upon the size of the pattern. Such an approach has low support for small patterns, but higher support for longer patterns. This broader approach can also be applied to sequences in the form of methods such as *SeqIndex*. A variety of methods such as *Grafil* [133] and *PIS* [134] have been developed in this context.

7   Web Mining Applications

In these cases, Web logs, linkage patterns and content are processed in order to determine important frequent and sequential patterns [46, 69]. A discussion of frequent pattern mining algorithms for Web log data may be found in [65]. A variety of different patterns can be mined from Web data. The key types of Web log mining correspond to *Web log mining*, and *linkage structure mining*. These are described in the subsections below.

7.1   Web Log Mining

Web logs contain data about user accesses in a standard format. Each log typically contains the IP address of the accessing host, the time stamp, the Web page accessed, the referrer, and a few other pieces of meta-information about the data. In such cases, it is useful to determine frequent access patterns in the logs. Such information can be very useful for designing the site in order to maximize accesses. Furthermore, Web log mining can also be used in the context of problems such as anomaly detection, in which unusual sequences of patterns which do not conform to the normal patterns in the logs, are determined for outlier analysis. Web log mining has also been used by educators in order to evaluate and discriminate between learner's access behaviors, especially in the context of scenarios such as distance-learning. The earliest work on Web log mining was performed in [30], in which frequent and sequential pattern analysis was used for determining important Web log patterns. The algorithm in this paper distinguishes between forward references and backward references during traversal by the user over the Web graph. Forward references may correspond to a user clicking on a Web page to traverse forward, whereas backward references correspond to the user revisiting the same object. Correspondingly, it defines the concept of *maximal forward references*, which correspond to the maximal sequence of forward traversals. The first step is to use the Web logs in order to create a database of maximal forward sequences in a pre-processing stage. Subsequently, frequent pattern mining algorithms are applied to this database in order to determine the most relevant patterns.

A different method for finding path traversals has been proposed in [102]. In this method, the major assumption is that irrelevant patterns may be interleaved with other more relevant patterns. This work defines a relevant pattern on the basis of the notion of *subpath* containment. The algorithm takes into account the underlying graph structure in order to determine the most relevant patterns. One major difference from the work in [30] is that the the candidate patterns need to be paths in the underlying

Web graph, and not any arbitrary sequence of vertices. This ensures that irrelevant vertices are less likely to be considered in the process of mining relevant patterns. An Apriori-like algorithm is used in [102] for this purpose, except that it is modified to ensure that the candidates also correspond to paths in the underlying Web graph.

Methods for data preparation of Web traversal patterns are proposed in [37]. Data preparation is a key issue in the process of finding the correct traversal patterns, because Web logs are inherently noisy. The ability to find the correct patterns therefore depends upon the ability to process these logs properly. The work in [37] provides an excellent overview of methods for processing these logs. Other methods for Web log usage mining are discussed in [36, 108, 110, 119, 129, 138].

A useful application of association rule mining is that of *personalization*. Personalization is a very natural application of association rule mining, because correlations between user behavior can be used in order to group their interests and perform recommendations. Methods for using associations in order to perform recommendations are discussed in [100, 101]. Recommendations can also be viewed as supervised learning problems, which can be effectively solved with the use of rule-based methods.

7.2   Web Linkage Mining

In Web linkage mining methods, the structure of the Web graph is mined for patterns, rather than the user traversal patterns. Mining the Web graph for patterns is closely related to the problem of community detection on the Web graph. In fact, such an approach can also be used for other kinds of large scale graphs such as social networks. Frequent patterns can be used to compress very large scale graphs, and then use the compressed representation for clustering. Such an approach has been proposed in [26] for mining communities with the use of compressed patterns. The method known as *VirtualNodeMiner* achieves graphs compression by generating virtual nodes from frequent itemsets in vertex adjacency lists. Another algorithm, which is focussed on mining frequent patterns from massive networks in the *gApprox* algorithm [31]. The key in this approach is that the approximation process allows the creation of an anti-monotonicity constraint, which can be pushed into the mining process. Another method has also been proposed in [10] for mining communities from multiple graphs (rather than a single large graph) with the use of frequent patterns, though this method is designed for smaller graphs, and not particularly focussed on the scenario of the World Wide Web.

8   Frequent Patterns for Text Mining

Frequent patterns have significant applications to text mining, both in terms of positional and non-positional co-occurrence. Positional co-occurrence corresponds to scenarios in which certain words co-occur together from a perspective of adjacency of occurrence. Such patterns can typically be found either by adapting methods from sequence pattern mining, or by using constrained frequent pattern mining methods. The latter case is related to the problem of finding frequent bigrams, trigrams, or phrases in the underlying data. Such frequent patterns can be used in order to enrich the underlying text representation for a variety of indexing and mining problems. For example, clustering and classification algorithms can typically benefit from a richer feature representation, which contains the frequent phrases in the collection. A specific example of improved text classification with the use of n-gram representations is discussed in [29]. It has been shown [19] that the expansions of query terms with relevant phrases can significantly enrich a variety of search applications. Therefore, frequent patterns can be used in order to expand the search phrases and enhance the quality of the search. It has been shown in [85] that such rules can be applied not only to individual words, but also to the paths in the dependency trees of a parsed corpus.

Frequent patterns have also been used in order to explore interesting patterns in text collections in terms of temporal and sequential co-occurrence, especially when the text arrives in the form of a stream. An example of such an approach is discussed in [42], where frequently occurring trends in text phrases are discovered in conjunction with visualization methods. Phrases whose frequency increases or decreases over time provide valuable hints about the key trends in the underlying text streams. Since many forms of social network content and news wire services provide text streams, such methods can provide useful tools in terms of exploring the changes in the behavior of the underlying collection. In addition association rules have been shown to be very useful in providing visual representations of the underlying text collection [42, 91, 128].

A significant number of applications also exist for mining of frequent patterns without adjacency constraints. Such frequent patterns can be used for co-clustering of text documents [4], or for indexing text documents with the use of conceptual phrases [6]. The idea in these methods is that simultaneous discovery of relevant word patterns and clusters is generally more effective than the discovery of each of them individually on a global basis. In the context of clustering, numerous methods have been proposed, which use the frequent itemsets in the text collection [21, 51, 83] in order to measure the similarities between the documents for the clustering process. A detailed discussion of many of these applications of frequent pattern mining to text collections may be found in [7].

9 Temporal Applications

Temporal applications correspond to scenarios in which the data is presented either in the form of continuous time series or discrete sequences. The two cases are quite similar, since continuous time series can be discretized into discrete sequences with the use of a variety of methods such as SAX [86]. The SAX method discretizes the average values in small time windows into a set of discrete values. Subsequently, all analysis is performed only on the discrete representations of the underlying data, since it is directly suited to the frequent pattern mining framework. It should be pointed out that the work done on mining patterns from biological sequences can also be applied directly to temporal sequences with a few modifications [34, 35, 89, 104, 105, 111, 125]. Another interesting similarity between biological and temporal data is that (unlike other frequent pattern mining scenarios) biological data often contains a small number of very long rows. As a result, row-enumeration techniques are often used in these methods.

One of the most common applications of pattern analysis techniques in the temporal context is that of event detection [23, 64, 66, 80, 121]. It should be noted that event detection can be considered a temporal version of the classification problem, in which labels are associated with time-stamps rather than records. The connection between pattern mining and classification has already been discussed in a previous section. Therefore, it is natural to utilize sequential pattern mining methods in the context of rule-based methods. In these cases, the data consists of a set of sequences defined on *base feature events*, and a *class event* which needs to be predicted from the patterns in the sequences defined by the feature events. The idea in most such techniques is that events can be predicted by particular sequences in the underlying data. This is used to construct temporal classification rules, which correspond to events. Such temporal classification rules will typically contain a sequence of feature events on the left hand side and a class event on the right hand side. In addition, the rule may contain a numerical lag value associated with it, which indicates the time lag with which the event will occur after a particular sequence of feature events. Once such rules have been mined, they are used for the prediction process, as in the case of all pattern-based classifiers. Event detection with the use of frequent pattern mining methods has been used frequently in the context of intrusion detection [73–75]. The goal in these methods is to relate the temporal patterns of the features in the network packets to the intrusion events. This model is then used in order to predict events. It should be pointed out that these methods can be used more generally for a variety of event detection problems beyond the intrusion scenario. An overview of classification methods for sequential data with the use of rule-based methods is provided in [131]. A closely related problem is that of mining process models from workflow logs [15]. Sequence mining is also used in order to predict customer behavior in telecommunications [44].

Time series data and sequence data are often mined for characteristic motifs. Such motifs may often describe the important trends in the underlying data, and can even be used for classification. An example of such an approach is discussed in [1], where pattern-based rule mining is used to determine the class labels of the underlying sequences. In this case, wavelet decomposition is applied to the sequences in order to create a multi-granularity representation, in terms of which the rules can be represented. The multi-granularity representation allows the construction of rules which span different lengths of the time-series, as long as they are relevant to the classification process. Note that this application is somewhat different from the event detection problem, since labels are associated with individual time series, rather than with specific instants in the time series.

A closely related problem is that of mining frequent episodes in sequences [55, 56, 95, 94, 144]. While this problem is not the same[1] as frequent or sequential pattern mining, it is very closely related, and often uses similar frameworks which use support in order to quantify the significance of the underlying patterns. Other related problems include that of periodic pattern mining in which patterns are constructed on the basis of the seasonality in time-series sequences [43, 58, 92]. Such patterns are often useful for clinic diagnosis in time-series patterns such as ECG measurements. Sequence data provides a much richer domain than non-sequential data for mining purposes. A method known as MARBLES proposes methods for finding association rules between episodes [38]. The problem of mining train delays with the use of sequence mining is discussed in [39]. Such methods are useful for finding how different episodes are related to one another.

10 Spatial and Spatiotemporal Applications

With advances in mobile sensing technology, an important emerging scenario is of social sensing [3]. In this case, the data is collected from mobile phones continuously over time, and much of this data is in the form of GPS-based location data. GPS-based location data can also be used in order to construct trajectories. In many cases, it is desirable to determine clusters and frequent patterns from the underlying trajectories.

Frequent pattern mining methods have frequently been used for clustering spatiotemporal data. An example of such a technique is the *Swarm* method proposed in [84]. In this approach, the data is first pre-processed into different snapshots. In each snapshot, a discrete value is used to indicate the cluster membership of an object. For example, clustering could be applied to each snapshot in order to obtain a discrete value for the cluster membership. Objects which have the same discrete identifier over multiple snapshots clearly correspond to a SWARM which moves together. Therefore, the approach in [84] defines a pattern-based model in which frequently occurring sequences of discrete values are reported together with the objects, which correspond to such sequences. Other related models for pattern mining in spatiotemporal data are discussed in [22, 52, 53]. Many of these models can benefit from the use of frequent or sequential pattern mining methods.

Methods have also been designed for performing classification from trajectory data with the use of frequent-pattern mining methods. In particular, the method in [82] determines important patterns which are related to the rare classes. These patterns are then used in order to predict the rare class. Thus, this approach can be used for supervised anomaly detection in spatiotemporal data. Another method proposed in [81] finds movement fragment patterns by spatial overlay. These are then used in order to identify outliers with the use of pattern-based classification.

---

[1] Many other kinds of methods such as Markov Models [55] are used in order to solve this problem.

Another interesting application in such contexts is the discovery of interesting *spatial association rules*. The integration of such methods with commercial database systems in order to determine spatial association rules has been a key research goal [47, 48]. Spatial association rules determine the implicit correlations between objects which contain both spatial and non-spatial attributes. Note that a spatial object may contain both spatial attributes and other non-spatial ones. For example sea-surface temperatures may correspond to spatial locations and temperature values. In such cases, consider the rule:

The temperature in the northern regions is usually low.

This is a spatial association rule, which contains both spatial and non-spatial attributes. In some applications, temporal components may also be associated with such rules. It is important to note that it is often quite complex to find the appropriate resolution at which the spatial association rules my be found. This may also result in challenges in terms of computational efficiency. An important methodology in order to substantially reduce the computational cost is that of progressive refinement. Therefore, the spatial association rules are first determined at a coarser resolution, and only promising candidates are explored for further mining [67]. The idea is to use rough spatial approximations such as minimum bounding rectangles in order to determine the frequent pattern candidates. These candidates are then further explored at a finer spatial resolution. A system prototype for this class of spatial data mining methods is provided in [57]. A similar methodology has been used in order to mine co-location patterns [131, 143]. In these methods, the idea of spatial continuity is used in order to refine the pattern mining process. The idea is that spatially close objects are often more likely to exhibit interesting correlations that objects, which are spatially further apart. An overview of methods for spatial data mining is provided in [97, 68]. A further temporal component to this analysis in the form of spatiotemporal patterns is provided in [28].

It should be pointed out that many forms of image and multimedia data may be considered spatial data, since spatial locations are often associated with pixels. Similarly, other non-spatial attributes such as color may also be associated with the different locations. Many of the techniques, which have been discussed above for the case of spatial data, can also be applied to such kinds of multimedia data. For example, methods for classification of images with the use of association rules is discussed in [20]. In general, spatial methods for classification may be extended to images by using appropriate features to represent the pixels in the images.

11 Software Bug Detection

Software programs can be considered structured entities which can be represented as graphs. Such graphs often have a "typical" structure in the case of normal software programs. These can be represented in the form of normal patterns in the underlying graph structured data. These are referred to as *software behavior graphs*. Different kinds of bugs may be present in a software program corresponding to either core dumps, memory violations and logical errors. The last of these kinds of errors is often the most difficult, because they may be caused by subtle changes in the structure of the program, and are typically *non-crashing* bugs, which do not provide an immediate warning of the kinds of errors in the program.

Typically, the methods used to detect such logical errors design classifiers which can distinguish structured traces of program executions (using software behavior graphs) with logical errors from the correct ones. The work in [88] uses a combination of frequent graph pattern mining and SVM classification in order to determine such logical errors. The idea is that different regions of the program execution trace (or parts of software behavior graph) may show different levels of classification accuracy, corresponding to whether or not bugs are present in that region. Another method known as *CP-miner* [78] finds copy pasted code. This is then used for determining the location of bugs in the data. Another natural direction of exploration is use rule-based methods which can extract application-specific rules from the data. A big may correspond to scenarios in which these rules are not satisfied. An example of such an approach *PR-Miner* [76]. A procedure for mining edge-weighted call graphs for localizing non-crashing bugs in software programs is provided in [45]. A detailed discussion on graph mining methods for software bug localization is provided in the chapter on software bug detection in [5]. A closely related problem is that of finding block access correlations in storage systems. A method called *C-Miner* was proposed by in [79], which finds frequent sequential patterns of correlated blocks from block access traces. Pattern mining techniques are also useful for determining bugs in the execution of software in sensor networks. A detailed discussion of such methods may be found in the chapter on using data mining methods for sensor bug diagnosis in [2].

12 Chemical and Biological Applications

Chemical and biological data can often be represented as graphs. For example, chemical compounds can be represented as graphs, in which nodes correspond to the atoms and the bonds between the nodes correspond to the edges between nodes. Similarly, biological data can be represented either as sequences or as graphs in many different ways such as complex biological molecules, microarrays or protein interaction networks. The variety of structural representations is typically more diverse in the case of biological data, as compared to chemical data. In all these cases, frequent pattern mining can play an important role in identifying useful (and common) properties of the underlying compounds or networks. These properties can be used to explore the data in either an unsupervised or supervised way, depending upon the kinds of patterns which are found. Such patterns are also referred to as *motifs* in biological and chemical compound analysis. For biological data, both sequence mining and structural mining may be relevant, depending upon the underlying scenario. Some of the earliest work on frequent subgraph discovery was performed in [70], and this approach is applicable to both chemical and biological data. An example of an application which is relevant to both domains is that of finding relevant substructures of molecules [25, 40]. In the following, we will discuss some useful applications in both domains.

12.1   Chemical Applications

Since frequent pattern mining is closely related to that of classification, as discussed earlier, many methods have been developed for predictive tasks with the use of frequent pattern mining. Examples of such tasks include carcinogenesis prediction [117] and predictive toxicology evaluation [118]. Key characteristics of compound representations can often be characterized by descriptor-based representations [24, 72]. The properties which are tracked are generally structure-driven, and may correspond to activity, toxicity, absorption, distribution, metabolism and excretion [24]. A natural way of mining these descriptors is with the use of algorithms such as frequent subgraph mining. Frequent subgraphs of a chemical graph database are defined as all subgraphs that are present in at least a certain minimum number of compounds in the database. This is essentially the minimum support requirement, and define the descriptors for the compounds in the database. The main challenge here is that the optimum value of the minimum support to be used may not be known a-priori for a given database. Nevertheless, since different data sets may contain different number of descriptors, with different supports, sizes, and shapes, such an approach provides some flexibility with the sue of the minimum support parameter, as long as an effective approach for tuning is available. Such descriptors are quite useful for chemical compound classification, since they encode important properties of the chemical compound, which may be very relevant to classification. An example of such an approach is discussed in [41], which uses the descriptors defined by frequent subgraphs for chemical compound classification.

12.2   Biological Applications

Biological data is available either in the form of sequence data or graph-structured data. In both cases, frequent pattern mining methods can be very helpful in discovering different kinds of insights. Much of biological and microarray data can be expressed as sequences in its most simplified form. In these cases, many algorithms have been developed in order to determine useful frequent patterns from these sequences [34, 35, 89, 104, 105, 111, 125, 123]. One special characteristic of biological data is that the number of rows may not be too large, but each individual row may be very long. As a result, row-enumeration techniques are often used in such scenarios. Such patterns provide an idea of the characteristics of the underlying data, and may also be used for other data mining tasks such as classification. The issue of scalability is particularly important in the context of pattern mining of biological sequences, because such sequences are typically very long [127]. As discussed earlier, frequent pattern mining algorithms are used for biclustering of biological data [93]. It should be pointed out that many of the sequential pattern mining algorithms, which were originally designed for temporal data, can also be applied directly to biological sequences.

In the context of *graph structured* biological data, a key problem is that of clustering protein-protein interaction networks. Such networks can be rather large, and the problem shares a number of similarities with that of community detection in social networks. Since frequent pattern mining algorithms are closely related to that of clustering, such methods can also be used for community detection in interaction networks. Such a method for using frequent subgraph mining algorithms for community detection in interaction networks has been discussed in [26].

Trees are often used to represent many biological structures such as glycans, RNA, and phylogenies. Frequent subgraph mining is often used on all of these biological structures in the context of different kinds of applications. In many cases, when phylogenies are inferred with the use of different techniques, many different trees are produced for a given set input genes. As a result, it becomes hard to assimilate and understand the relationships between such trees. Typically, while the goal is to understand evolutionary relationships between entities, the large number of possible trees makes this very difficult. Therefore, it is often desirable to find the broader patterns in these trees, a problem which is closely related to that of frequent subtree mining. Such trees are also referred to as *consensus trees* or *supertrees* [96, 120]. The common relations between the different trees provides an idea of the commonly occurring patterns in the underlying data. For example, pairs (or groups)of nodes which share the same ancestral node are useful in discovering common patterns in multiple phylogenies. Methods for finding such frequent patterns are discussed in [113]. Frequent subtree mining algorithms are useful for extending such methods to more complex data [139, 142], which are not necessarily represented as trees.

Frequent pattern mining is also used for mining different kinds of RNA data. Multiple species often have common substructures due to common evolutionary origins [98]. These similarities are often expressed in the form of functional similarities among RNAs. Therefore, it is useful to apply frequent pattern mining algorithms for predictive mining. In particular, the discovery of common RNA substructures has been used for prediction of RNA folding and processing mechanisms [71, 112]. Note that such predictive learning methods are closely related to the classification problem, which is commonly solved by frequent pattern mining in the context of rule-based methods.

Frequent subtree mining methods are also very useful for mining glycan databases [61]. These methods can be used to develop a classification method for glycan databases by using pattern-based classification methods. As in all pattern based classification methods, rules can be constructed in order to determine whether or not a particular glycan belongs to a given class. In this case, the left hand sides of the rules correspond to the subtrees in the glycan database.

13   Resources for the Practitioner

Since frequent pattern mining methods are used frequently for different applications, it is helpful to use off-the-shelf software for frequent pattern mining in many of these applications. In addition, software for some of the applications discussed in this chapter are also available.

A general Web site containing pointers to different resources on frequent pattern mining is the *KDD Nuggets* site [146]. This can be considered an excellent meta-repository containing pointers to frequent pattern mining software. The *Weka* repository [154] contains many implementations of different data mining algorithms including frequent pattern mining. In terms of specific implementations of different algorithms, an implementation by Bart Goethals of some of the more well known frequent pattern mining algorithms such as *Apriori, Eclat, DIC* and *FP-Growth* may be found at [155]. A fast implementation of *Apriori* which uses prefix trees may be found in [149]. This site also contains a significant amount of software for other algorithms such as *Eclat, FP-Growth*, closed pattern mining, and maximal pattern mining. Methods for fault tolerant and sequence mining are also covered by this software collection. A good set of implementation of the *FP-Growth* family of algorithms may also be found in [157]. The *ARtool* [156] is an open-source software which is available under the GNU public license and is a collection of software and tools for performing association analysis in market basket data sets. A well known repository for different implementations of frequent pattern mining algorithms is the *FIMI* repository [145]. This is an open source repository containing many efficient implementations of frequent pattern mining algorithms. In addition, a free R-software package *arules*, which can perform frequent pattern mining of different kinds is available in [150]. A significant amount of software is also available for rule-based classification. The CBA system for classification with frequent pattern mining [87] is available as an implementation at [147]. The rule-based system known as RIPPER is available at [148].

In terms of commercial software, the *IBM SPSS system* contains software for discovering frequent patterns and association rules from transaction and market basket data. While this software is a general data mining platform, it also contains significant parts which are tailored to market basket analysis. *Oracle Data Mining* is a general purpose data mining tool, which also provides association mining capabilities [158]. SAS provides an *Enterprise Miner*, which provides the capability to mine both associations and sequential patterns [159]. The *SmartBundle* commercial software [151] provides different ways of exploring associations and frequent patterns in market basket data. This software is particularly tailored towards transaction data. The *WizRule* software [152] by *WizSoft* performs data mining with the use of association rules. Specifically, this software can also perform classification with the use of the discovered rules. Thus, this software explores the power of association rules for data mining in different ways. The *XAffinity (TM)* software [153] is suited to click-stream and Web log data. This software can be used for effective click-stream and Web log analysis.

A number of data benchmarks are also available in order to test the efficiency of different mining algorithms. Along these, the *QUEST synthetic data generator* [160] is one of the earliest data generators for pattern analysis. This data generator uses an intuitive model to create transactions as a combination of smaller baskets. Among real data sets, numerous data sets from the UCI machine learning repository [50] have been frequently used for efficiency analysis.

14   Conclusions and Summary

This chapter provides an overview of the key applications of frequent pattern mining. Frequent pattern mining has a variety of applications to many data mining problems such as clustering and classification. It also has applications to database problems such as indexing. Many specific domains such as Web mining and recommendation analysis, spatiotemporal analysis, multimedia analysis, software bug analysis and biological analysis can be addressed with frequent pattern mining algorithms. The main challenge in applying frequent pattern mining to the different domains is that the constraints and data representations are very different in these domains. Correspondingly, the vanilla frequent pattern mining problem needs to be appropriately adapted to these domains. It is expected that numerous other applications of frequent pattern mining algorithms may be found, as new forms of hardware and software technology create different kinds of data.

References

1. C. C. Aggarwal. On Effective Classification of Strings with Wavelets, *ACM KDD Conference*, 2002.
2. C. C. Aggarwal. Managing and Mining Sensor Data, *Springer*, 2013.
3. C. C. Aggarwal, T. Abdelzaher. Social Sensing, *Managing and Mining Sensor Data, Springer*, 2013.
4. C. C. Aggarwal, C. K. Reddy. Data Clustering: Algorithms and Applications, *CRC Press*, 2013.
5. C. C. Aggarwal, and H. Wang. Managing and Mining Graph Data, *Springer*, 2010.
6. C. Aggarwal and P. Yu. On Effective Conceptual Indexing and Similarity Search in Text Data, *ICDM Conference*, 2001.
7. C. Aggarwal and C. Zhai. Mining Text Data, *Springer*, 2012.
8. C. C. Aggarwal, J. Wolf, P. Yu. A New Method for Similarity Indexing of Market Basket Data, *ACM SIGMOD Conference*, 1999.
9. C. Aggarwal, C. Procopiuc, and P. Yu. Finding Localized Associations in Market Basket Data, *IEEE Transactions on Knowledge and Data Engineering*, 14(1), pp. 51–62, 2002.
10. C. C. Aggarwal, N. Ta, J. Wang, J. Feng, M. Zaki. Xproj: A framework for projected structural clustering of XML documents, *ACM KDD Conference*, 2007.
11. R. Agrawal, and R. Srikant. Fast Algorithms for Mining Association Rules in Large Databases, *VLDB Conference*, pp. 487–499, 1994.
12. R. Agrawal, and R. Srikant. Mining Sequential Patterns, *ICDE Conference*, 1995.

13. R. Agrawal, T. Imielinski, and A. Swami. Mining association rules between sets of items in large databases. *ACM SIGMOD Conference*, 1993.
14. R. Agrawal, J. Gehrke, D. Gunopulos, P. Raghavan. Automatic Subspace Clustering of High Dimensional Data for Data Mining Applications, *ACM SIGMOD Conference*, 1998.
15. R. Agrawal, D. Gunopulos, and F. Leymann. Mining Process Models from Workflow Logs, *Springer*, 1998.
16. R. Agarwal, C. C. Aggarwal, and V. V. V. Prasad. A Tree Projection Algorithm for Generation of Frequent Itemsets, *JPDC Journal*, 2001.
17. L. Akoglu, H. Tong, J. Vreeken, and C. Faloutsos. Fast and Reliable Anomaly Detection in Categorical Data, *CIKM Conference*, 2012.
18. K. Ali, K. Manganaris, R. Srikant. Partial Classification using Association Rules, *KDD Conference*, 1997.
19. P. Anick, and S. Tipirneni. The Paraphrase Search Assistant: Terminological Feedback for Iterative Information Seekings, *ACM SIGIR*, 1999.
20. M.-L. Antonie, O. Zaiane, and A. Coman. Application of Data Mining Techniques for Medical Image Classification, *Second International Workshop on Multimedia Data Mining at KDD*, 2001.
21. F. Beil, M. Ester, and X. Xu. Frequent Term-based Text Clustering, *ACM KDD Conference*, 2002.
22. M. Benkert, J. Gudmundsson, F. Hubner, and T. Wolle. Reporting flock patterns, *COMGEO*, 2008.
23. C. Bettini, X. S. Wang, S. Jajodia, and J. L. Lin. Discovering Frequent Event Patterns with Multiple Granularities in Time Sequences, *IEEE Transactions on Knowledge and Data Engineering*, 10(2), pp. 222–237, 1998.
24. H. Bohm and G. Schneider. *Virtual Screening for Bioactive Molecules*. Wiley-VCH, 2000.
25. C. Borgelt, M. Berthold. Mining molecular fragments: finding relevant substructures of molecules. *ICDM Conference*, 2002.
26. G. Buehrer, and K. Chellapilla. A Scalable Pattern Mining Approach to Web Graph Compression with Communities. *WSDM Conference*, 2009.
27. T. Calders, and B. Goethals. Mining all non-derivable frequent itemsets *Principles of Data Mining and Knowledge Discovery*, pp. 1–42, 2002.
28. H. Cao, N. Mamoulis, D. W. Cheung. Mining Frequent Spatiotemporal Sequential Patterns, *ICDM Conference*, 2005.
29. W. Cavnar, and J. Trenkle. N-Gram based Text Categorization, *Proceedings of SDAIR*, pp. 161–174, 1994.
30. M. S. Chen, J. S. Park, and P. S. Yu. Efficient data mining for path traversal patterns, *IEEE Transactions on Knowledge and Data Engineering*, 10(2), pp. 209–221, 1998.
31. C. Chen, X. Yan, F. Zhu, and J. Han. gapprox: Mining Frequent Approximate Patterns from a Massive Network, *ICDM Conference*, 2007.
32. C. Cheng, A. Fu, Y. Zhang. Entropy-based Subspace Clustering for Mining Numerical Data, *ACM KDD Conference*, 1999.
33. H. Cheng, X. Yan, J. Han, and C.-W. Hsu. Discriminative Frequent Pattern Analysis for Effective Classification, *ICDE Conference*, 2007.
34. G. Cong, A. Tung, X. Xu, F. Pan, and J. Yang. FARMER: Finding Interesting Rule Groups in Microarray Data Sets, *ACM SIGMOD Conference*, 2004.
35. G. Cong, K.-L. Tan, A. K. H. Tung, X. Xu. Mining Top-*k* covering Rule Groups for Gene Expression Data. *ACM SIGMOD Conference*, 2005.
36. R. Cooley, B. Mobasher, and J. Srivasatava. Web mining: Information and pattern discovery on the world wide web. *Ninth International Conference on Tools with Artificial Intelligence*, 1997.
37. R. Cooley, B. Mobaser, and J. Srivastava. Data preparation for mining world wide web browsing patterns. *Knowledge and information systems*, 1(1), pp. 5–32, 1999.
38. B, Cule, N. Tatti, and B. Goethals. MARBLES: Mining Association Rules Buried in Long Event Sequences. *SDM Conference*, 2002.

39. B. Cule, B. Goethals, S. Tassenoy and S. Verboven. Mining Train Delays. *Proc. 10th International Symposium on Intelligent Data Analysis (IDA 2011)*, 2011.
40. L. Dehaspe, H. Toivonen, and R. King. Finding Frequent Substructures in Chemical Compounds. *ACM KDD Conference*, 1998.
41. M. Deshpande, M. Kuramochi, N. Wale, and G. Karypis. Frequent substructure-based approaches for classifying chemical compounds. *IEEE TKDE.*, 17(8), pp. 1036–1050, 2005.
42. A. Don, E. Zheleva, M. Gregory, S. Tarkan, L. Auvil, T. Clement, B. Schneiderman, C. Plaisant. Discovering Interesting Usage Patterns in Text Collections: Integrating Text Mining with Visualization, *CIKM Conference*, 2007.
43. G. Dong, and J. Li. Efficient Mining of Emerging Patterns: Discovering Trends and Differences, *ACM KDD Conference*, 1999.
44. F. Eichinger, D. Nauck, and F. Klawonn. Sequence Mining for Customer Behaviour Predictions in Telecommunications, *Workshop on Practical Data Mining: Applications, Experiences and Challenges*, 2006.
45. F. Eichinger, K. Bohm and M. Huber. Mining Edge-Weighted Call Graphs to Localize Software Bugs, *Machine Learning and Knowledge Discovery in Databases*, Springer, 2008.
46. M. Eirinaki, M. Vazirgiannis. Web mining for web personalization. *ACM Transactions on Internet Technology*, 3: pp. 1–27, 2003.
47. M. Ester, H.-P. Kriegel, and J. Sander. Spatial Data Mining: A Database Approach, *Advances in Spatial Databases*, pp. 47–66, Springer, 1997.
48. V. Estivill-Castrol, and A. T. Murray. Discovering Associations in Spatial Data—An Efficient Medoid-Based Approach, *DMKD Workshop*, 1998.
49. W. Fan, K. Zhang, H. Cheng, J. Gao, X. Yan, J. Han, P. Yu, and P. Verscheure. Direct Mining of Discriminative and Essential Frequent Patterns via Model-based Search Tree, *ACM KDD Conference*, 2008.
50. A. Frank, and A. Asuncion. UCI Machine Learning Repository, Irvine, CA: University of California, School of Information and Computer Science, 2010. http://archive.ics.uci.edu/ml.
51. B. Fung, K. Wang, and M. Ester. Hierarchical Document Clustering using Frequent Itemsets, *SDM Conference*, 2003.
52. J. Gudmindsson, M. van Krevald, B. Speckmann. Efficient detection of motion patterns in spatiotemporal data sets, *GIS*, 2004.
53. J. Gudmundsson, M. van Krewald. Computing Longest Duration Flocks in Trajectory Data, *GIS*, 2006.
54. M. Gupta, J. Gao, Y. Sun, and J. Han. Community Trend Outlier Detection Using Soft Temporal Pattern Mining. *ECML/PKDD Conference*, 2012.
55. R. Gwadera, M. J. Atallah, and W. Szpankowski. Markov Models for Identification of Significant Episodes, *SDM Conference*, 2005.
56. R. Gwadera, M. J. Atallah, and W. Szpankowski. Reliable detection of episodes in event sequences. *Knowledge and Information Systems*, 7(4), pp. 415–437, 2005.
57. J. Han, K. Koperski, and N. Stefanovic. GeoMiner: a system prototype for spatial data mining. *ACM SIGMOD Record* 26(2), pp. 553–556, 1997.
58. J. Han, G. Dong, and Y. Yin. Efficient Mining of Partial Periodic Patterns in Time Series Database, *ICDE Conference*, 1999.
59. J. Han, J. Pei, and Y. Yin. Mining Frequent Patterns without Candidate Generation, *ACM SIGMOD Conference*, 2000.
60. J. Han, H. Cheng, D. Xin, and X. Yan. Frequent Pattern Mining: Current Status and Future Directions, *Data Mining and Knowledge Discovery*, 15(1), pp. 55–86, 2007.
61. K. Hashimoto, I. Takigawa, M. Shiga, M. Kanehisa, and H. Mamitsuka. Mining significant tree patterns in carbohydrate sugar chains. *Bioinformatics*, 24(16), pp. 1167, 2008.
62. Z. He, S. Deng, and X. Xu. Outlier Detection Integrating Semantic Knowledge. *Web Age Information Management (WAIM)*, 2002.
63. Z. He, X. Xu, J. Huang, and S. Deng. FP-Outlier: Frequent Pattern-based Outlier Detection, *COMSIS*, 2(1), 2005.

64. J. Hellerstein, S. Ma, and C.-S. Perng. Discovering Actionable Patterns in Event Data, *IBM Systems Journal*, 41(3), pp. 475–493, 2002.
65. R. Ivancy, and I. Vajk. Frequent Pattern Mining in Web Log Data, *Acta Polytechnica Hungarica*, 3(1), pp. 77–90, 2006.
66. P.-S. Kam, A. W.-C. Fu. Discovering Temporal Patterns for Interval-based Events, *Springer*, Berlin, 2000.
67. K. Koperski, and J. Han. Discovery of Spatial Association Riles in Geographic Information Databases, *Advances in Spatial Databases*, 1995.
68. K. Koperski, J. Adhikary, and J. Han. Spatial Data Mining: Progress and Challenges Survey Paper, *ACM SIGMOD Workshop on Research Issues in Data Mining and Knowledge Discovery*, 1996.
69. R. Kosala, H. Blockeel. Web Mining Research: A Survey. *ACM SIGKDD Explorations*, 2000.
70. G. Kuramuchi and G. Karypis. Frequent Subgraph Discovery, *ICDM Conference*, 2001.
71. S. Le, J. Owens, R. Nussinov, J. Chen, B. Shapiro, and J. Maizel. RNA secondary structures: comparison and determination of frequently recurring substructures by consensus. *Bioinformatics*, 5(3), pp. 205–210, 1989.
72. A. R. Leach and V. J. Gillet. *An Introduction to Chemoinformatics*. Springer, 2003.
73. W. Lee, and S. Stolfo. Data Mining Approaches for Intrusion Detection. *Proceedings of the 7th USENIX Security Symposium*, 1998.
74. W. Lee, S. Stolfo, and P. Chan. Learning Patterns from Unix Execution Traces for Intrusion Detection, *AAAI workshop on AI methods in Fraud and Risk Management*, 1997.
75. W. Lee, S. Stolfo, and K. Mok. A Data Mining Framework for Building Intrusion Detection Models, *IEEE Symposium on Security and Privacy*, 1999.
76. Z. Li, Y. Zhou. PR-Miner: Automatically extracting implicit programming rules and detecting violations in large software code, *ACM SIGSOFT Symposium on Foundations of Software Engineering*, 2005.
77. W. Li, J. Han, and J. Pei. CMAR: Accurate and Efficient Classification based on Multiple Association Rules, *ICDM Conference*, 2001.
78. Z. Li, S. Lu, S. Myagmar, and Y. Zhou. CP-Miner: A Tool for Finding Copy-Paste and Related Bugs in Operating System Code, *Symposium on Operating Systems Design and Implementation*, 2004.
79. Z. Li, Z. Chen, S. M. Srinivasan, Y. Zhou. C-Miner: Mining Block Correlations in Storage Systems, *USENIX Conference on File and Storage System Technologies*, 2004.
80. T. Li, F. Liang, S. Ma, and W. Peng. An Integrated Framework on Mining Log Files for Computing System Management. *ACM KDD Conference*, 2005.
81. X. Li, J. Han, and S. Kim. Motion-alert: Automatic Anomaly Detection in Massive Moving Objects, *IEEE Conference in Intelligence and Security Informatics*, 2006.
82. X. Li, J. Han, S. Kim, and H. Gonzalez. ROAM: Rule- and Motif-based Anomaly Detection in Massive Moving Object Data Sets, *SDM Conference*, 2007.
83. Y. Li, S. Chung, and J. Holt. Text document clustering based on frequent word meaning sequences. *Data and Knowledge Engineering*, 64(1), pp. 381–404, 2008.
84. Z. Li, B. Ding, J. Han, R. Kays. Swarm: Mining Relaxed Temporal Object Moving Clusters, *VLDB Conference*, 2010.
85. D. Lin, and P. Pantel. DIRT@ SBT@ Discovery of Inference Rules from Text, *ACM KDD Conference*, 2001.
86. J. Lin, E. Keogh, S. Lonardi, and B. Y.-C. Chiu. A Symbolic Representation of Time Series, with Implications for Streaming Algorithms. *DMKD Workshop*, 2003.
87. B. Liu, W. Hsu, Y. Ma. Integrating Classification and Association Rule Mining, *ACM KDD Conference*, 1998.
88. C. Liu, X. Yan, H. Lu, J. Han, and P. S. Yu. Mining Behavior Graphs for "backtrace" of non-crashing bugs, *SDM Conference*, 2005.
89. H. Liu, J. Han, D. Xin, and Z. Shao. Mining frequent patterns on very high dimensional data: a top-down row enumeration approach. *SDM Conference*, 2006.

90. D. Lo, H. Cheng, J. Han, S.-C. Khoo, and C. Sun. Classification of Software Behaviors for Failure Detection: A Discriminative Pattern Mining Approach, *ACM KDD Conference*, 2009.
91. A. Lopes, R. Pinho, F. Paulovich, and R. Minghim. Visual Text Mining using Association Rules, *Computers and Graphics*, 31(3), pp. 316–326, 2007.
92. S. Ma, and J. Hellerstein. Mining Partially Periodic Event Patterns with Unknown Periods, *IEEE International Conference on Data Engineering*, 2001.
93. S. Madeira, and A. Oliveira. Biclustering Algorithms for Biological Data Analysis; A Survey, *IEEE/ACM Transactions on Computational Biology and Bioinformatics*, 1(1), pp. 24–45, 2004.
94. H. Mannila, and H. Toivonen. Discovering Generalized Episodes using Minimal Occurrences, *KDD Conference*, 1996.
95. H. Mannila, H. Toivonen, and A. I. Verkamo. Discovering Frequent Episodes in Sequences, *ACM KDD Conference*, 1995.
96. T. Margush, F. McMorris. Consensus-trees. *Bulletin of Mathematical Biology*, 43(2), pp. 239–244, 1981.
97. H. J. Miller, and J. Han. Geographic Data Mining and Knowledge Discovery. *CRC Press*, 2003.
98. J. Mitchell, J. Cheng, and K. Collins. A box H/ACA small nucleolar RNA-like domain at the human telomerase RNA end. *Molecular and cellular biology*, 19(1), pp. 567–576, 1999.
99. S. Mitra, and H. Banka. Multi-objective Evolutionary Biclustering of Gene Expression Data, *Pattern Recognition*, 39(12), pp. 2464–2477, 2006.
100. B. Mobasher, R. Cooley, and J. Srivastava. Automatic personalization based on Web usage mining *Communications of the ACM*, 43(8), pp. 142–151, 2000.
101. B. Mobasher, H. Dai, T. Luo, and M. Nakagawa. Effective personalization based on association rule discovery from web usage data, *Proceedings of the 3rd international workshop on Web information and data*, 2001.
102. A. Nanopoulos, and Y. Manolopoulos. Finding generalized path patterns for web log data mining, *Lecture notes in computer science*, pp. 215–228, 2000.
103. A. Nanopoulos, and Y. Manolopoulos. Efficient similarity search for market basket data, *VLDB Journal*, 11(2), 2002.
104. F. Pan, G. Cong, A. Tung, J. Yang, and M. Zaki. CARPENTER: Finding closed patterns in long biological datasets. *ACM KDD Conference*, 2003.
105. F Pan, A. K. H. Tung, G. Cong, X. Xu. COBBLER: Combining column and Row Enumeration for Closed Pattern Discovery. *SSDBM*, 2004.
106. L. Parsons, E. Haque, and H. Liu. Subspace Clustering for High Dimensional Data: A Review, *ACM SIGKDD Explorations*, 6(1), pp. 90–105, 2004.
107. J. Pei, and J. Han. Can we push more constraints into frequent pattern mining? *ACM KDD Conference*, 2000.
108. J. Pei, J. Han, B. Mortazavi-Asl and H. Zhu. Mining access patterns efficiently from web logs. *PAKDD*, 2000.
109. J. Pei, J. Han, and L. V. S. Lakshmanan. Mining Frequent Patterns with Convertible Constraints in Large Databases, *ICDE Conference*, 2001.
110. J. Punin, M. Krishnamoorthy, M. Zaki. Web usage mining: languages and algorithms. *Springer-Verlag*, 2001.
111. I. Rigoutsos, and A. Floratos. Combinatorial Pattern Discovery in Biological Sequences: The TEIRESIAS algorithm, *Bioinformatica*, 14(1), pp. 55–67, 1998.
112. B. Shapiro, and K. Zhang. Comparing multiple RNA secondary structures using tree comparisons. *Bioinformatics*, 6(4), pp. 309–318, 1990.
113. D. Shasha, J. Wang, and S. Zhang. Unordered tree mining with applications to phylogeny. *ICDE Conference*, pp. 708–719, 2004.
114. P. Shenoy, J. Haritsa, S. Sudarshan, G. Bhalotia, M. Bawa, D. Shah. Turbo-charging Vertical Mining of Large Databases. *ACM SIGMOD Conference*, pp. 22–33, 2000.
115. A. Siebes, J. Vreeken, and M. van Leeuwen. Itemsets than Compress, *SIAM Conference on Data Mining*, 2006.

116. K. Smets and J. Vreeken. The Odd One Out: Identifying an Characterising Anomalies, *SIAM Conference on Data Mining*, 2011.
117. A. Srinivasan, R. King, S. Muggleton, and M. J. E. Sternberg. Carcinogenesis predictions using ILP. *Workshop on Inductive Logic Programming*, Vol. 1297, pp. 273–287, 1997.
118. A. Srinivasan, R. King, S. Muggleton, and M. J. E. Sternberg. The predictive toxicology evaluation challenge, *IJCAI*, 1997.
119. J. Srivastava, R. Cooley, M. Deshpande, and P. N. Tan. Web usage mining: Discovery and applications of usage patterns from web data. *ACM SIGKDD Explorations Newsletter*, 1(2), pp. 12–23, 2000.
120. C. Stockham, L. Wang, T. Warnow. Statistically based postprocessing of phylogenetic analysis by clustering. *Bioinformatics*, 18(3), pp. 465–469, 2002.
121. R. Vilalta, and S. Ma. Predicting Rare Events in Temporal Domains, *ICDM Conference*, 2002.
122. J. Wang, G. Karypis. HARMONY: Efficiently Mining the Best Rules for Classification. *SDM Conference*, 2005.
123. J. T.-L. Wang, G.-W. Chirn, T. G. Marr, B. Shapiro, D. Shasha, and K. Zhang. Combinatorial Pattern Discovery for Scientific Data: Some Preliminary Results, *ACM SIGMOD Record*, 23(2), pp. 115–125, 1994.
124. K. Wang, C. Xu, and B. Liu. Clustering Transactions using Large Items, *CIKM Conference*, 1999.
125. J. Wang, D. Shasha, and B. Shapiro. Pattern Discovery in Biomolecular Data: Tools, Techniques, and Applications. *Oxford University Press*, 1999.
126. H. Wang, W. Wang, J. Yang, and P. S. Yu. Clustering by pattern similarity in large data sets, *ACM SIGMOD Conference*, 2002.
127. K. Wang, Y. Xu, and J. X. Yu. Scalable Sequential Pattern Mining for Biological Sequences, *ACM KDD Conference*, 2004.
128. P. C. Wong, P. Whitney, and J. Thomas. Visualizing Association Rules for Text Mining, *InfoVis*, 1999.
129. Y. Xiao, M. Dunham. Efficient mining of traversal patterns, *Data and Knowledge Engineering*, 39(2), pp. 191–214, 2001.
130. Z. Xing, J. Pei, and E. Keogh. A Brief Survey on Sequence Classification, *ACM SIGKDD Explorations*, 12(1), 2010.
131. H. Xiong, S. Shekhar, Y. Huang, V. Kumar, X. Ma, J. Yoo. A framework for discovering co-location patterns in data sets with extended spatial objects, *SDM Conference*, pp. 78–89, 2004.
132. X. Yan, P. S. Yu, and J. Han. Graph indexing: A frequent structure-based approach. *ACM SIGMOD Conference*, 2004.
133. X. Yan, P. S. Yu, and J. Han. Substructure similarity search in graph databases. *ACM SIGMOD Conference*, 2005.
134. X. Yan, F. Zhu, J. Han, and P. S. Yu. Searching substructures with superimposed distance, *ICDE Conference*, 2006.
135. Y. Yang, and B. Padmanabhan. GHIC: A Hierarchical Pattern-based Clustering for Grouping Web Transactions, *IEEE TKDE*, 17(9), pp. 1300–1304, 2005.
136. J. Yang, and W. Wang. CLUSEQ: Efficient and Effective Sequence Clustering. *ICDE Conference*, 2003.
137. M. Yiu, and N. Mamoulis. Frequent-pattern based iterated projected clustering, *ICDM Conference*, 2003.
138. O. Zaiane, M. Xin, and J. Han. Discovering Web Access Patterns and Trends by applying OLAP and Data Mining Technology on Web Logs. *Research and Technology Advances in Digital Libraries*, pp. 19–29. 1998.
139. M. Zaki. Efficiently mining frequent trees in a forest: Algorithms and applications. *IEEE Transactions on Knowledge and Data Engineering*, 17(8), pp. 1021–1035, 2005.
140. M. Zaki, C. Aggarwal. XRules: An Effective Classifier for XML Data, *ACM KDD Conference*, 2003.

141. M. Zaki, S. Parthasarathy, M. Ogihara, and W. Li. New Algorithms for Fast Discovery of Association Rules. *KDD Conference*, pp. 283–286, 1997.
142. S. Zhang, T. Wang. Discovering Frequent Agreement Subtrees from Phylogenetic Data. *IEEE Transactions on Knowledge and Data Engineering*, 20(1), pp. 68–82, 2008.
143. X. Zhang, N. Mamoulis, D. W. Cheung, Y. Shou. Fast mining of spatial collocations. *ACM KDD Conference*, pp. 384–393, 2004.
144. W. Zhou, H. Liu, and H. Cheng. Mining closed episodes from event sequences efficiently, *PAKDD Conference*, 2010.
145. http://fimi.ua.ac.be/.
146. http://www.kdnuggets.com/software/associations.html.
147. http://www.comp.nus.edu.sg/~dm2/.
148. http://www.cs.cmu.edu/~wcohen/#sw.
149. http://www.borgelt.net//software.html
150. http://cran.at.r-project.org/web/packages/arules/index.html
151. http://www.decidyn.com/SmartBundle.php?id=91005.
152. http://www.wizsoft.com/.
153. http://www.xore.com/.
154. http://www.cs.waikato.ac.nz/ml/weka/.
155. http://adrem.ua.ac.be/~goethals/software/.
156. http://www.cs.umb.edu/~laur/ARtool/.
157. http://cgi.csc.liv.ac.uk/~frans/KDD/Software/FPgrowth/fpGrowth.html.
158. http://www.oracle.com/technetwork/database/options/advanced-analytics/odm/index.html.
159. http://www2.sas.com/proceedings/forum2007/132-2007.pdf.
160. http://sourceforge.net/projects/ibmquestdatagen/.

Index

Printed in the United States
By Bookmasters